CONTENTS

PREFACE

While this second edition of *Modern Elementary Algebra* retains all the strong features of the first edition, it includes various modifications resulting from a wide source of classroom usage.

The discussion and explanations have been made more intuitive and informal, and proofs, considered too advanced for a first course in algebra, have been omitted. The stress on mechanics has been increased by the inclusion of many more worked-out examples and many more drill exercises.

Topics on polynomials and rational expressions have been rearranged to appear before linear systems in order to afford the student more development in algebraic techniques before learning a new major concept. The formal presentation of functions and relations has been deleted because these topics are considered more appropriate for the intermediate course in algebra. Set concepts have been introduced within the text at the time they are pertinent to the discussion. Scientific notation has been included in the section on exponents.

The verbal problems have been expanded to include more relevant and more realistic problems insofar as this is possible.

The abundant and continuous presentation of verbal problems remains an important feature of the text. The problems have been designed to stimulate learning and to develop the ability to translate words into symbols. This translation starts with the first chapter and is expanded in each successive chapter.

Whenever it has been possible, the material has been given an added dimension by the inclusion of geometric models, diagrams, summary charts, and historical notes.

Each chapter has a chapter summary and a chapter review. Two exercise sets are included in each chapter. The answers to the A exercise sets and the chapter reviews are in the back of the book; a solutions manual is available for the B exercise sets.

The authors would like to express their appreciation to the following people for their very valuable suggestions on improving the first edition:

Mary Carter Smith, Laney College, California

Mary Louise Adams, Chabot College, California, and to her algebra students, who suggested the name "First and Ten Method" for one of the rapid calculations

P. B. Sampson, East Los Angeles College, California

Lloyd J. Rochon, Chabot College

Henry Harmeling, Jr., North Shore Community College, Massachusetts

Michael J. MacCallum, Long Beach, California

V. S. G. and S. S.

CHAPTER **1**

BASIC CONCEPTS

HISTORICAL NOTE

The word "algebra" first appeared around A.D. 825 in the work *Hisab al-jabr wal-muqabala*, written by the Arabic scholar al-Khowarizmi. Literally, it meant "reduction" or "restoration." A more common meaning of al-jabr was introduced into Spain by the Moors: An algebrista was known as a person who reset or "restored" broken bones. The name "algebra" became established by means of the book *Vollstaendige Anleitung zur Algebra* (*Complete Introduction to Algebra*), written in 1770 by the Swiss mathematician Leonhard Euler. His book served as the model of algebra texts for many years.

Elementary algebra is a study of **numbers** and their properties.

Number problems have played an important role in man's existence for thousands of years. At first these problems were written in words. Gradually the words were replaced by abbreviations of the words, and finally the words and abbreviations were replaced by **symbols**. The use of symbols reduced the amount of writing that was necessary and enabled the problem to be seen and understood more rapidly.

For example, contrast the verbal account of the financial transaction stated below with its symbolic representation, which follows.

Verbal Account

A dealer sold a car for two thousand three hundred ninety-five dollars. The car cost him one thousand nine hundred eighty dollars. His profit was four hundred fifteen dollars.

Symbolic Account

$$\begin{array}{lc} \$2395.00 & S \\ \underline{\$1980.00} & \underline{C} \\ \$\ 415.00 & P \end{array}$$

The letter S is written at the right of \$2395.00 to indicate that this amount is the selling price of the car. Similarly, the letter C is used to indicate the cost, and the letter P is used to indicate the profit.

It is easy to see that, in general, the profit is found by subtracting the cost from the selling price. In symbols, one writes

$$P = S - C$$

In summary, symbols are used in algebra for the following reasons:

1. To economize the amount of writing that is necessary.
2. To promote clarity and understanding of the problem.
3. To facilitate the discovery and generalization of number properties.

In this chapter the student will be introduced to the symbols used for numbers, for the operations of addition, subtraction, multiplication, division, squaring, cubing, finding square roots and cube roots, and for the equal, less than, and greater than relations.

Then the student will learn the meaning of signed numbers and how to find sums, products, differences, and quotients of signed numbers.

Finally, the concept of substitution and how to use this concept in the evaluation of algebraic expressions and formulas will be presented.

By the end of the chapter the student should be able to read, understand, and use the basic language of algebra.

1.1 NUMBERS, NUMERALS, SETS

1.1.1 Numbers and Numerals

Man developed the abstract concept of number to answer the question "How many?" The first numbers to be invented were the **counting numbers**: one, two, three, four, five, and so on.

To keep records of numbers, man invented symbols or numerals to serve as the names of numbers.

Numerals are symbols that name numbers according to some specified system, such as the Hindu-Arabic system of numeration. The Hindu-Arabic system of numeration is the modern decimal system in use today. The symbols 1, 2, 3, 4, 5, 6, 7, 8, and 9, which name the first nine counting numbers, and the zero symbol, 0, are called the **digits**. The numerals that name the other counting numbers are formed from these ten digits by using the principles of a positional numeral system with base ten.

For example, 346 is the *numeral name* of a number whose *word name* is three hundred forty-six.

Letters of the alphabet are also used as the names of numbers. A letter is used to designate a number whose numeral name is unspecified.

For example, in the formula $P = S - C$, the letters P, S, and C are used to designate numbers that represent the profit, the selling price, and the cost, respectively.

As another example, $y = x + 5$ means that y represents a number that is five more than the number represented by x.

1.1.2 The Set Concept

A **set** is a well-defined collection of objects called "elements" or "members" of the set.

A collection is **well-defined** if it is always possible to determine whether or not a particular object or element belongs to the set.

Set membership is indicated by the notation $a \in S$, which means that the object whose name is "a" is an element or member of the set whose name is "S."

For example, if S is the set of counting numbers and if $a = 5$, then $5 \in S$ means that 5 is a counting number.

A set for which there is a counting number that indicates how many elements are in the set is called a **finite set**. For example, the set of digits is a finite set consisting of ten numbers. (The word "digits" is used for the numbers as well as for the numerals.)

An **infinite set** is a set that is not finite. An infinite set has a never-ending list of elements. For example, the set of counting numbers is infinite.

Particular sets may be defined by two general methods, the *listing* method and the *description* method.

In the **listing method**, the set is defined by listing or stating the names of its members enclosed by braces and separated by commas.

For example, the set of **digits** may be defined by the listing method as follows:

$$\{0, 1, 2, 3, 4, 5, 6, 7, 8, 9\}$$

As another example, the set of **natural numbers** (another name for the set of counting numbers) may be defined as follows:

$$\{1, 2, 3, 4, 5, 6, 7, 8, 9, 10, \ldots\}$$

where the three dots indicate that this list continues in this manner without ending.

Note that 0 is *not* a natural number.

The set of natural numbers is an infinite set, whereas the set of digits is a finite set.

In the **description method,** the set is defined by "describing" the set or stating a property possessed by each member of the set. For example,

S is the set of digits exactly divisible by 3

Using the listing method,

$$S = \{0, 3, 6, 9\}$$

A formal variation of the description method is often referred to as the **rule** or **set-builder method.** The braces are used to enclose the description, which consists of one or more letters used to designate a member of the set, a vertical bar, |, read "such that," and then a statement describing the property in terms of the letter or letters.

For example, letting $N =$ the set of natural numbers,

	$\{$	x	$\vert$	$x \in N$	and x is exactly divisible by 3$\}$
is read	the set of	all elements x	such that	x is a natural number	and x is exactly divisible by 3.

Using the listing method, this set may be expressed as

$$\{3, 6, 9, 12, 15, \ldots\}$$

The **empty set**, also called the **null set**, is the set that has no members. It is denoted symbolically by $\varnothing$ or by $\{\ \}$.

For example, the set of digits that are greater than 17 is the empty set.

The empty set is considered a finite set.

A **universal set**, U, is a set to which the elements of all other sets in a particular discussion must belong.

For example, let $U = \{1, 2, 3, 4, 5, 6, 7, 8, 9\}$. Now if A is the set of numbers that are exactly divisible by 3, then $A = \{3, 6, 9\}$. The number 12 cannot be a member of A because 12 does not belong to the universal set, U, that was selected.

In a particular discussion, the universal set that is selected is always stated or understood.

Use the listing method to define each described set in Examples 1.1.1–1.1.8. State whether each set is finite or infinite.

EXAMPLE 1.1.1 The set of counting numbers between 3 and 7

Solution {4, 5, 6}, finite

EXAMPLE 1.1.2 The set of natural numbers exactly divisible by 5

Solution {5, 10, 15, 20, 25, . . .}, infinite

EXAMPLE 1.1.3 The set of digits exactly divisible by 13

Solution {0}, finite

EXAMPLE 1.1.4 The set of natural numbers greater than 12 and less than 20

Solution {13, 14, 15, 16, 17, 18, 19}, finite

EXAMPLE 1.1.5 The set of natural numbers greater than 10

Solution {11, 12, 13, 14, 15, . . .}, infinite

EXAMPLE 1.1.6 The set of digits exactly divisible by 8

Solution {0, 8}, finite

EXAMPLE 1.1.7 The set of natural numbers that are greater than 7 and also less than 2

Solution $\varnothing$, finite

EXAMPLE 1.1.8 The set of natural numbers that are exactly divisible by both 6 and 9

Solution {18, 36, 54, 72, 90, . . .}, infinite

EXAMPLE 1.1.9 Describe the following set by the listing method, where D = the set of digits:

$$\{y \mid y \in D \text{ and } y \text{ is less than } 6\}$$

Solution {0, 1, 2, 3, 4, 5}

EXAMPLE 1.1.10 Describe the following set by the listing method, where N = the set of natural numbers:

$$\{x \mid x \in N \text{ and } x \text{ is between 10 and 20}\}$$

Solution {11, 12, 13, 14, 15, 16, 17, 18, 19}

EXERCISES I.1 A

In Exercises 1–20, use the listing method to describe the indicated set. State whether the set is finite or infinite.

$$D = \textit{the set of digits} = \{0, 1, 2, 3, 4, 5, 6, 7, 8, 9\}$$
$$N = \textit{the set of natural numbers} = \{1, 2, 3, 4, 5, \ldots\}$$

1. The set of natural numbers exactly divisible by 4.
2. The set of natural numbers between 2 and 8
3. The set of natural numbers less than 6
4. The set of natural numbers greater than 10
5. The set of digits greater than 12
6. The set of digits exactly divisible by 2 (even digits)
7. The set of natural numbers exactly divisible by 2 (even numbers)
8. The set of natural numbers not divisible by 2 (odd numbers)
9. The set of digits exactly divisible by 23
10. The set of natural numbers greater than 18 and less than 25
11. $\{x \mid x \in D$ and x is exactly divisible by $4\}$
12. $\{x \mid x \in N$ and x is exactly divisible by $4\}$
13. $\{x \mid x \in N$ and x is between 9 and $16\}$
14. $\{x \mid x \in D$ and x is less than $5\}$
15. $\{x \mid x \in D$ and x is greater than $20\}$
16. $\{y \mid y \in N$ and y is greater than 25 and less than $36\}$
17. $\{y \mid y \in N$ and y is greater than 25 and less than $20\}$
18. $\{z \mid z \in N$ and z is exactly divisible by 4 and by $6\}$
19. $\{n \mid n \in N$ and n divides 12 exactly$\}$
20. $\{n \mid n \in N$ and n divides both 12 and 18 exactly$\}$

EXERCISES I.1 B

In Exercises 1–20, use the listing method to describe the indicated set. State whether the set is finite or infinite.

$$D = \textit{the set of digits} = \{0, 1, 2, 3, 4, 5, 6, 7, 8, 9\}$$
$$N = \textit{the set of natural numbers} = \{1, 2, 3, 4, 5, \ldots\}$$

1. The set of natural numbers less than 9
2. The set of digits less than 9
3. The set of digits greater than 7

4. The set of natural numbers greater than 7
5. The set of natural numbers exactly divisible by 6
6. The set of digits exactly divisible by 6
7. The set of natural numbers between 15 and 20
8. The set of digits between 15 and 20
9. The set of natural numbers less than 1
10. The set of digits that divide 60 exactly
11. $\{x \mid x \in N$ and x is exactly divisible by $12\}$
12. $\{x \mid x \in N$ and x is between 50 and $60\}$
13. $\{x \mid x \in D$ and x is greater than $5\}$
14. $\{x \mid x \in D$ and x is greater than 2 and less than $15\}$
15. $\{y \mid y \in N$ and y is greater than 2 and less than $15\}$
16. $\{y \mid y \in N$ and y is greater than 15 and less than $2\}$
17. $\{n \mid n \in N$ and n divides 30 exactly$\}$
18. $\{n \mid n \in N$ and n divides both 30 and 40 exactly$\}$
19. $\{d \mid d \in D$ and d is exactly divisible by $19\}$
20. $\{d \mid d \in D$ and d divides 28 exactly$\}$

1.2 OPERATIONS

Elementary algebra is concerned with six operations that are performed on numbers:

1. Addition
2. Subtraction
3. Multiplication
4. Division
5. Raising to a power
6. Root extraction

All these operations are indicated by symbols. The set of numbers on which the operations are performed is always designated. In this section, all numbers will be restricted to belong to the set of **whole numbers**—that is, the number 0 and the natural numbers (counting numbers).

Furthermore, the operations discussed in this section will be restricted to addition, subtraction, multiplication, division, two special cases of raising to a power—namely, squaring and cubing; and two special cases of root extraction—namely, extracting square roots and cube roots.

1.2.1 Addition $x + y$

The result of adding two numbers is called the **sum** of the two numbers. The numbers that are added are called **terms**.

In symbols, $x + y$ indicates the sum of the terms x and y and is read "x plus y" or "the sum of x and y."

For example, the sum of x and 5 is $x + 5$.

1.2.2 Subtraction: $x - y$

The result of subtracting one number from another is called the **difference** or **remainder**.

In arithmetic, the name **subtrahend** is given to the number that is subtracted from the other number, called the **minuend**.

In algebra, both the subtrahend and the minuend are called **terms**.

In symbols, $x - y$ indicates the difference between the terms x and y, when y is subtracted from x. Thus $x - y$ is read "x minus y" or "y subtracted from x."

For example, the difference when 5 is subtracted from x is $x - 5$.

1.2.3 Multiplication: xy, $3x$, $3 \cdot 4$, $3(4)$, $(3)(4)$

The result of multiplying two numbers is called the **product**. The numbers that are being multiplied are called **factors**.

In arithmetic, multiplication is indicated by the "cross" symbol, $\times$, such as 3×4. However, because this symbol can be confused with the letter x used to name an unspecified number, the $\times$ symbol is not used in algebra to indicate multiplication.

In algebra, multiplication is indicated in three different ways. If two numerals are used to name the factors, then the dot symbol, $\cdot$, is used, as in $3 \cdot 4$, which means the product of 3 and 4, or 12. The dot is written at half the vertical height of the numerals so that it will not be confused with the decimal point.

If two letters or a numeral and a letter are used to name the factors, then multiplication is indicated by juxtaposition—that is, the symbols for the numbers are written side by side. For example, $3x$ means the product of 3 and x, and xy means the product of x and y. Twice x is written $2x$.

Multiplication is also indicated by writing the name of one factor next to the name of the other factor with either factor or both enclosed in parentheses. For example, $3(4)$ and $(3)(4)$ mean the product of 3 and 4.

1.2.4 Division: $\frac{x}{y}$

The result of dividing one number by another is called the **quotient**. The name **dividend** is given to the number that is being divided by the other number, called the **divisor**.

In arithmetic, division is indicated by the symbol $\div$. In algebra, division is indicated by using the fractional notation, $\frac{x}{y}$. The expression $\frac{x}{y}$ is read "x divided by y" or, more informally, "x over y."

For example, the quotient when x is divided by 5 is written $\frac{x}{5}$. This can also be read "x over five."

1.2.5 Squaring: x^2

The result of multiplying a number by itself is called the **square** of the number. The counting numeral 2, written to the upper right of the number symbol, indicates that the number is to be used twice as a factor.

For example, 3^2 (read "three square") means $3 \cdot 3$, or 9. Similarly, 5^2 means $5 \cdot 5$, or 25. In general, x^2 means $x \cdot x$.

1.2.6 Cubing: x^3

The product obtained by using a number as a factor three times is called the **cube** of the number. The counting numeral 3 written to the upper right of the number symbol indicates that the number is to be used three times as a factor.

For example, 5^3 (read "five cube") means $5 \cdot 5 \cdot 5$, or 125. In general, x^3 means $x \cdot x \cdot x$.

1.2.7 Root Extraction: Square Root and Cube Root

To extract the square root of a number means to find a number whose square is the given number. The operation of square-root extraction is indicated by the symbol $\sqrt{}$, which has evolved from the letter "r," the first letter of the word "root." For example, $\sqrt{9}$, read "the square root of nine," designates the number 3, since 3^2 is 9.

Similarly, $\sqrt[3]{x}$, read "the cube root of x," designates the number whose cube is x. For example, $\sqrt[3]{125}$ names the number 5, since 5^3 is 125.

Tables are useful for finding squares, cubes, square roots, and cube roots of some natural numbers. An extensive table is found on the inside cover of this book.

TABLE 1.1 SUMMARY OF THE OPERATION SYMBOLS

Operation \ Number Symbol	Numerals	Numeral and Letter	Two Letters
Addition	$3+4$	$x+4$	$x+y$
Subtraction	$7-3$	$x-4$	$x-y$
Multiplication	$3 \cdot 4$ or $3(4)$	$3x$	xy
Division	$\frac{12}{3}$	$\frac{x}{3}$	$\frac{x}{y}$
Squaring	3^2	x^2	(Not applicable)
Cubing	5^3	x^3	
Square root	$\sqrt{9}$	$\sqrt{x}$	(Not applicable)
Cube root	$\sqrt[3]{27}$	$\sqrt[3]{x}$	

HISTORICAL NOTE

During the Middle Ages abbreviations began to be used to indicate the operations. The symbols gradually developed after the appearance of printed works. (The printing press was invented in 1438.)

The plus sign, +, is a contraction of the Latin word *et*, meaning "and." When *et* is written rapidly, it looks like [illegible].

The minus sign, −, was used as an equivalent form for "m̄" and "m," the abbreviations of the word "minus."

The marks + and − were used in medieval warehouses on sacks, crates, or barrels to indicate whether the contents were more or less than what they were supposed to be. The signs + and − made their first appearance in print in Widman's *Commercial Arithmetic*, published in 1489.

The dot for multiplication was introduced in 1583 by Christopher Clavius of Germany (1537–1612). It became established with its use by the German mathematician Gottfried Wilhelm Leibniz (1646–1716).

The division notation of algebra can be traced to the Hindus through the works of Bhaskara, who in 1150 wrote $\begin{matrix}1\\3\end{matrix}$ for $\frac{1}{3}$. The bar appears in the works of Viète, who also indicated multiplication by juxtaposition.

The symbol $\sqrt{}$ for root extraction was introduced in 1525 by Christoff Rudolff because it resembles a small "r," the initial letter of *radix*, the Latin word for "root."

Viète wrote *A*, *Aq*, and *Acu* for our modern x, x^2, and x^3. Pierre Herigone, in his *Cursus Mathematicus* (1634–1637), used $a2$, $a3$, and $a4$ for a^2, a^3, and a^4. Our modern notation is first found in Descartes' *La Géométrie* (1637), in which he used x^3, x^4, x^5, and so on. However, Descartes used both xx and x^2 to indicate the square of x. Until the latter part of the eighteenth century it was common to find xx for x^2 and xxx for x^3.

EXAMPLE 1.2.1 Perform each of the following operations:

$$6\cdot 4,\ 20 - 8,\ \frac{15}{3},\ 9 + 7,\ 6^2,\ 2^3,\ \sqrt{81},\ \sqrt[3]{64}$$

Solution

$6\cdot 4 = 24$	(Multiplication)
$20 - 8 = 12$	(Subtraction)
$\frac{15}{3} = 5$	(Division)
$9 + 7 = 16$	(Addition)
$6^2 = 6\cdot 6 = 36$	(Squaring)
$2^3 = 2\cdot 2\cdot 2 = 8$	(Cubing)
$\sqrt{81} = 9$	(Square root)
$\sqrt[3]{64} = 4$	(Cube root)

EXAMPLE 1.2.2 Translate each of the following into algebraic expressions:

	Solution
The product of 3 and y	$3y$
The quotient when t is divided by 3	$\frac{t}{3}$
The cube of a	a^3
The sum of 4 and z	$4 + z$
The square of b	b^2
The difference when c is subtracted from 8	$8 - c$
The square root of the sum of x and 3	$\sqrt{x + 3}$
The cube root of the product of m and 2	$\sqrt[3]{2m}$

EXERCISES 1.2 A

Perform each of the indicated operations in Exercises 1–20.

1. $3\cdot 5$

2. $\frac{20}{2}$

3. $7+8$

4. $12-4$

5. 5^2

6. 4^3

7. $9\cdot 7$

8. 7^2

9. $\frac{36}{12}$

10. $23 - 19$

11. 9^3

12. $(10)(20)$

13. $\frac{6}{1}$

14. $8(125)$

15. $\sqrt{121}$

16. $(25)(4)$

17. $\frac{200}{10}$

18. $100 - 36$

19. $\sqrt[3]{125}$

20. 6^3

Translate Exercises 21–40 into algebraic (symbolic) expressions.

21. The sum of x and 5

22. The product of 5 and x

23. The square of n

24. The cube of t

25. The difference when 4 is subtracted from y

26. The quotient when x is divided by 6

27. n minus m

28. The product of s and t

29. c square

30. z cube

31. x plus y

32. x times y

33. The difference when x is subtracted from y

34. The quotient when x is divided by y

35. The product of 25 and 45

36. The square root of the sum of 15 and y

37. The cube root of the difference when p is subtracted from 7

38. The product of m and the square root of t

39. The product of the cube of t and the cube root of s

40. The quotient of x over the square of y

In Exercises 41–50, list the members of each set and state if the set is finite or infinite. The universal set is N, the set of natural numbers. This means $x \in N$ for each problem.

41. $\{x \mid 10 - x \in N\}$

42. $\{x \mid \frac{x}{3} \in N\}$

43. $\{x \mid x = n^2 \text{ where } n \in N\}$

44. $\{x \mid x = 5n \text{ where } n \in N\}$

45. $\{x \mid \frac{12}{x} \in N\}$

46. $\{x \mid x = n^3 \text{ where } n \in N\}$

47. $\{x \mid x + 12 \text{ is less than } 20\}$

48. $\{x \mid x + 12 \text{ is greater than } 20\}$

49. $\{x \mid x = \sqrt{n}, n = 4, 9, 16, \text{ or } 25\}$

50. $\{x \mid x = \sqrt[3]{n}, n = 8, 64, 216, \text{ or } 1000\}$

EXERCISES 1.2 B

In Exercises 1–20, perform each of the indicated operations.

1. $5 \cdot 9$

2. $\frac{15}{3}$

3. $6 + 7$

4. $14 - 3$

5. 3^2

6. 2^3

7. $\sqrt[3]{27}$

8. $\frac{24}{1}$

9. $\sqrt{64}$

10. $\sqrt[3]{64}$

11. $(4)(25)$

12. $35 + 65$

13. $4(15)$

14. 6^2

15. $\frac{200}{5}$

16. $100 - 14$

17. 3^3

18. $5 \cdot 4$

19. $(\sqrt{4})(\sqrt{9})$

20. $\sqrt{(4)(9)}$

Translate Exercises 21–40 into algebraic (symbolic) expressions.

21. The sum of r and s

22. The product of r and s

23. The square root of m

24. The square of m

25. The difference when x is subtracted from 7

26. The quotient when 5 is divided by y

27. The product of a and the square of b

28. The sum of the square of x and the cube of x

29. p square times q

30. 5 minus p

31. The product of 15 and 5

32. The difference when x is subtracted from 10

33. The difference when 10 is subtracted from x

34. The cube of x

35. The cube root of x

36. 8 more than x

37. 8 less than x

38. 8 times x

39. The quotient when 8 is divided by x

40. The sum of the square root of y and the cube root of y

In Exercises 41–50, list the members of each set and state if the set is finite or infinite. The universal set is N, the set of natural numbers. This means $x \in N$ for each problem.

41. $\{x \mid x + 5 \text{ is between 5 and 10}\}$

42. $\{x \mid 12 - x \in N\}$

43. $\{x \mid x + 100 \in N\}$

44. $\left\{x \,\middle|\, \frac{x}{4} \in N\right\}$

45. $\{x \mid \sqrt{x} = n \text{ and } n \in N\}$

46. $\{x \mid \sqrt[3]{x} = n \text{ and } n \in N\}$

47. $\{x \mid x + 5 \text{ is less than 20}\}$

48. $\{x \mid x + 12 \text{ is greater than 100}\}$

49. $\{x \mid x = 2n \text{ and } n \in N\}$

50. $\{x \mid 3x \in N\}$

1.3 CONSTANTS, VARIABLES, SPECIAL SETS

1.3.1 Constants and Variables

A **constant** is a letter or a numeral that names a number belonging to a set consisting of exactly one member. Letters at the beginning of the alphabet, such as a, b, or c, are usually used to name a constant. The Greek letter π (read "pi") is used to designate the constant ratio of the circumference C of any circle to its diameter d—that is, $\pi = \frac{C}{d}$.

A **variable** (also called an unknown) is a letter that names a number belonging to a set consisting of more than one member.

Any letter can be used to name a variable, but usually the letters at the end of the alphabet, such as x, y, and z, are used, following the example of the French mathematician René Descartes, who perfected much of the symbolism of algebra.

1.3.2 Set Relations

DEFINITION

The set A is a **subset** of the set B if every element of A is an element of B. In symbols, $A \subset B$.

(Some authors prefer to write $A \subseteq B$, reserving the notation $A \subset B$ to mean A is a **proper** subset of B—that is, A is not identical to B. The set B is called an **improper** subset of itself.)

EXAMPLE 1.3.1 List all of the subsets of $\{2, 4, 6\}$.

Solution $\{2\}, \{4\}, \{6\}, \{2, 4\}, \{2, 6\}, \{4, 6\}, \{\ \}, \{2, 4, 6\}$

The set $\{2, 4, 6\}$ is an improper subset of $\{2, 4, 6\}$. All of the other sets listed are proper subsets of $\{2, 4, 6\}$. Note that the empty set is also a subset.

The empty set is a subset of every set.

EXAMPLE 1.3.2 Express in symbols that the set $A = \{1, 2, 3, 4, 5\}$ is a subset of the set of natural numbers, N.

Solution $\{1, 2, 3, 4, 5\} \subset N$ or $A \subset N$

DEFINITION

Two sets A and B are said to be **equal, or identical, if and only if**

$$A \subset B \quad \text{and} \quad B \subset A$$

In symbols, $A = B$.

Thus $\{3, 6, 9\} = \{9, 6, 3\}$. Note that the order in which the elements are written is not important; it is only necessary that the two sets have exactly the same objects in them.

HISTORICAL NOTE

The French lawyer François Viète (1540–1603) was one of the first persons to use letters to represent numbers. In 1591 he used capital vowels, such as A, E, I, for the variables. Viète is often called the "father of algebra" because of the improvements he made in the symbolism of algebra.

In 1637 the French mathematician René Descartes (1596–1650) used a notation similar to our modern one, in which he used the letters x, y, and z for the variables. Descartes' influence was so great that his notation became the generally accepted one.

1.3.3 Special Sets

DEFINITIONS

If r and s are natural numbers and if there exists a natural number n such that

$$r = sn$$

then r is a **multiple** of s

and s is a **factor** of r

and r is **divisible** by s

and s is a **divisor** of r

EXAMPLE 1.3.3 Use the set-builder method to describe the natural numbers that are multiples of 5.

Solution $\{x \mid x = 5n \text{ where } n \in N\}$

EXAMPLE 1.3.4 List the natural numbers in the set

$$S = \{x \mid x \text{ is a factor of } 12\}$$

Solution $S = \{1, 2, 3, 4, 6, 12\}$

DEFINITION

A natural number n is **even** if and only if there is a natural number k so that $n = 2k$.

The set of even numbers $= \{2, 4, 6, 8, \ldots\} = \{x \mid x = 2k \text{ and } k \in N\}$.

DEFINITION

A natural number is **odd** if and only if it is not even.

The set of odd numbers $= \{1, 3, 5, 7, \ldots\} = \{x \mid x = 2k - 1 \text{ and } k \in N\}$.

There are special numbers called *prime* numbers that can be thought of as the basic building units from which all the natural numbers are made. It is thought that a sufficient knowledge of the properties of prime numbers could provide a key for understanding the properties of all numbers.

DEFINITION

A natural number p is a prime if and only if $p \neq 1$ and p has no divisors different from 1 and itself.

The set of the first ten primes $= \{2, 3, 5, 7, 11, 13, 17, 19, 23, 29\}$.

DEFINITION

A natural number n is a composite if and only if $n \neq 1$ and n is not a prime.

The set of the first ten composite numbers $= \{4, 6, 8, 9, 10, 12, 14, 15, 16, 18\}$.

1.3.4 Set Operations

Addition, subtraction, multiplication, division, raising to a power, and root extraction are *operations* which are defined on *numbers*. There are also operations defined on *sets*, the most important of which are *union* and *intersection*.

DEFINITION

The union of two sets A and B is the set of all elements that are in A, *or* in B, *or* in both A and B. In symbols, $A \cup B$ designates the union of A and B.

For example, if $A = \{1, 2, 3\}$ and $B = \{2, 4, 6\}$, $A \cup B = \{1, 2, 3, 4, 6\}$. Elements which appear in both sets are listed only once in the union.

EXAMPLE 1.3.5 If $A = \{x \mid x = 2k$ and k is a natural number less than 8$\}$ and $B = \{3, 6, 9, 12, 15\}$, list $A \cup B$.

Solution First list the elements of A:

$$A = \{2, 4, 6, 8, 10, 12, 14\}$$

To explain how this was done, note that k is a natural number less than 8. This makes 1, 2, 3, 4, 5, 6, 7 the eligible numbers for k. Since

$$x = 2k, x = 2, 4, 6, 8, 10, 12, \text{ or } 14$$

Now

$$A \cup B = \{2, 4, 6, 8, 10, 12, 14, 3, 6, 9, 12, 15\}$$

Rearranging the elements in order from smallest to largest and eliminating duplication of elements,

$$A \cup B = \{2, 3, 4, 6, 8, 9, 10, 12, 14, 15\}$$

DEFINITION

The intersection of two sets A and B is the set of all elements that are in both A *and* B. In symbols, $A \cap B$ designates the intersection of A and B.

For example, if $A = \{1, 2, 3, 4\}$ and $B = \{3, 4, 5\}$, then $A \cap B = \{3, 4\}$.

EXAMPLE 1.3.6 For the sets A and B described in Example 1.3.5, list $A \cap B$.

Solution $A \cap B = \{6, 12\}$

EXAMPLE 1.3.7 If $S = \{2, 4, 6, 8\}$ and $T = \{1, 3, 5\}$, list $S \cup T$ and $S \cap T$.

Solution

$$S \cup T = \{2, 4, 6, 8, 1, 3, 5\} = \{1, 2, 3, 4, 5, 6, 8\}$$

Since the sets have no elements in common,

$$S \cap T = \varnothing$$

DEFINITION

If A and B are sets such that $A \cap B = \varnothing$, then A and B are said to be disjoint sets.

HISTORICAL NOTE

The study of the properties of the natural numbers is an important branch of mathematics known as number theory. It originated in the school of the Greek Pythagoras around 540 B.C. The distinction between odd and even numbers was known to the Pythagoreans. In Book VII of *Elements* (300 B.C.), Euclid defines the concepts "unit," "number," "odd," "even," "prime," and "composite," as well as many theorems concerned with these concepts.

EXERCISES 1.3 A

In Exercises 1–16, list the elements of the following subsets of the set of natural numbers, N. (Note that $x \in N$.)

1. $\{x \mid 8 - x \in N\}$

2. $\{x \mid \frac{30}{x} \in N\}$

3. $\{x \mid \frac{x}{30} \in N\}$

4. $\{x \mid x - 15 \in N\}$

5. $\{x \mid \sqrt{x} \in N\}$

6. $\{x \mid \sqrt[3]{x} \in N\}$

7. $\{x \mid \frac{60}{x} \text{ is a prime}\}$

8. $\{x \mid x \text{ is prime and } \frac{60}{x} \in N\}$

9. The set of primes between 30 and 50

10. The set of odd numbers between 30 and 40

11. The set of even numbers between 17 and 25

12. The set of composites between 19 and 27

13. $\{x \mid x \text{ is a factor of } 28\}$

14. $\{x \mid x \text{ is a multiple of } 13\}$

15. $\{x \mid x \text{ is a divisor of } 18\}$

16. $\{x \mid x \text{ is divisible by } 25\}$

For Exercises 17–20, list (a) $A \cup B$ and (b) $A \cap B$. The universal set is N, the set of natural numbers.

17. $A = \{1, 5, 12, 20\}$
$B = \{5, 15, 20, 25\}$

18. $A = \{x \mid x = 2k, k \in N\}$
$B = \{x \mid x = 2k - 1, k \in N\}$

19. A is the set of composites between 1 and 10
B is the set of primes between 1 and 10

20. A is the set of odd numbers between 30 and 40
B is the set of multiples of 3 between 30 and 40

21. $A = \{x \mid x \text{ is a factor of } 30\}$
$B = \{x \mid x \text{ is a prime less than } 30\}$

22. List all subsets of $\{2, 3, 5\}$

23. List all subsets of $\{1, 3, 5, 7\}$

In Exercises 24–33, let N = *set of natural numbers*

P = *set of primes*

C = *set of composites*

E = *set of even numbers*

O = *set of odd numbers*

Determine whether each statement is true or false.

24. $P \subset N$

25. $O \subset P$

26. $C \subset O$

27. $43 \in P$

28. $37 \in C$

29. $P \cup C = N$

30. $E \cup O = N$

31. $P \cap E = \{2\}$

32. $E \cap N = E$

33. $E \cup N = N$

EXERCISES 1.3 B

In Exercises 1–16, list the elements of the following subsets of the set of natural numbers, N. (Note that $x \in N$.)

1. $\{x \mid x - 20 \in N\}$

2. $\{x \mid 20 - x \in N\}$

3. $\left\{x \,\middle|\, \frac{28}{x} \in N\right\}$

4. $\left\{x \,\middle|\, \frac{x}{28} \in N\right\}$

5. $\{x \mid \sqrt{2x} \in N\}$

6. $\{x \mid \sqrt[3]{2x} \in N\}$

7. $\left\{x \,\middle|\, \frac{210}{x} \text{ is a prime}\right\}$

8. $\left\{x \,\middle|\, x \text{ is prime and } \frac{210}{x} \in N\right\}$

9. The set of primes between 50 and 70

10. The set of even numbers between 25 and 35

11. The set of odd numbers between 60 and 70

12. The set of composites between 71 and 83

13. $\{x \mid x \text{ is a multiple of } 15\}$

14. $\{x \mid x \text{ is a factor of } 30\}$

15. $\{x \mid x \text{ is divisible by } 17\}$

16. $\{x \mid x \text{ is a divisor of } 36\}$

For Exercises 17–21, list (a) $A \cup B$ and (b) $A \cap B$. The universal set is N, the set of natural numbers.

17. $A = \{3, 6, 9, 12, 15\}$
$B = \{2, 4, 6, 8\}$

18. $A = \{x \mid x = y^2, y \in N\}$
$B = \{x \mid x^2 = y, y \in N\}$

19. A is the set of primes
B is the set of all composites

20. $A = \{x \mid x \text{ is a factor of } 12\}$
$B = \{x \mid x = 2n, n \in N \text{ and } n \text{ is less than } 10\}$

21. $A = \{1, 4, 9, 16, 25\}$
B is the set of primes less than 25

22. List all subsets of $\{2, 4, 6, 8\}$.

23. List all subsets of $\{x \mid x \in N \text{ and } x \text{ is less than } 6\}$.

In Exercises 24–33, let N = *set of natural numbers*

P = *set of primes*

C = *set of composites*

E = *set of even numbers*

O = *set of odd numbers*

Determine whether each statement is true or false.

24. $E \subset C$

25. $P \subset E$

26. $E \subset N$

27. $15 \in O$

28. $51 \in P$

29. $E \cap O = \varnothing$

30. $P \cap C = N$

31. $N \cup C = N$

32. $C \cap N = C$

33. $C \cup N = N$

1.4 GROUPING SYMBOLS

After an operation has been performed on two numbers, an operation can be performed again on the resulting number and a third number. This process can be continued as often as desired.

In order to indicate which operation was performed first, which second, which third, and so on, grouping symbols are used. The grouping symbols that are most frequently used are parentheses, ( ), braces, { }, brackets, [ ], and the bar (also called vinculum), ——.

For example, to indicate that 3 is to be subtracted from the difference $12 - 5$, any one of the following symbolic expressions can be used:

$$(12 - 5) - 3$$
$$\{12 - 5\} - 3$$
$$[12 - 5] - 3$$
$$\overline{12 - 5} - 3$$

If the operations are performed as indicated, then first one obtains that $12 - 5$ is 7, and then that $7 - 3$ is 4. Thus $(12 - 5) - 3$ results in the number 4.

If the parentheses or other grouping symbols were not used, the order in which the operations are to be performed would not be clear, and in some cases it would be possible to obtain as a result a different number than the one intended.

For example, consider $12 - 5 - 3$. If 3 is first subtracted from 5, then the difference is 2. Now if 2 is subtracted from 12, then the result is 10.

Thus the expression $12 - 5 - 3$ is ambiguous—that is, more than one number can be obtained as a final result, depending on how the expression is interpreted.

It is concluded that grouping symbols must be used to avoid ambiguity and to make the intended meaning clear.

Thus $(12 - 5) - 3$ results in 4 only

and $12 - (5 - 3)$ results in 10 only

since it is understood that the operation within the parentheses is to be performed first.

Several kinds of grouping symbols are used to promote clarity in reading an algebraic expression that involves more than two operations. For example, one way to indicate that the sum of 3 and the difference $8 - 2$ is to be subtracted from 12 is as follows:

$$12 - \{3 + (8 - 2)\}$$

This means that the operation within the innermost grouping symbols, the subtraction of 2 from 8, is to be performed first. Next, the result 6 is to be added to 3. Finally, the sum of 6 and 3, or 9, is to be subtracted from 12, giving a final result of 3.

This combination of operations could also have been written, using parentheses only, as follows:

$$12 - (3 + (8 - 2))$$

Note that the use of two kinds of grouping symbols causes the algebraic expression to be read more clearly and more rapidly.

The use of several kinds of grouping symbols becomes more advantageous as the number of operations involved increases. For example, compare

$$25 - \{3 + [10 - (2 + \overline{4 - 1})]\}$$

with

$$25 - (3 + (10 - (2 + (4 - 1))))$$

In reading an algebraic expression involving two or more sets of grouping symbols, it is understood that the innermost set indicates the operation that is to be performed first. After this operation has been performed, the innermost set remaining indicates the next operation that is to be performed, and so on.

EXAMPLE 1.4.1 Perform the indicated operations:

$$25 - \{3 + [10 - (2 + \overline{4 - 1})]\}$$

Solution

1. Perform the operation under the bar and obtain

$$25 - \{3 + [10 - (2 + 3)]\}$$

2. Perform the operation within the parentheses and obtain

$$25 - \{3 + [10 - 5]\}$$

3. Perform the operation within the brackets and obtain

$$25 - \{3 + 5\}$$

4. Perform the operation within the braces and obtain

$$25 - 8$$

5. Subtract 8 from 25 and obtain 17 as the final result.

Conventions

By adopting certain conventions, some grouping symbols can be omitted from certain expressions. This yields an expression which is simpler to read, and the intended meaning is still clear. The following convention is adopted:

CONVENTION

Unless the grouping symbols indicate otherwise, the operations are to be performed in the following order:

1. **Taking square roots and/or cube roots as read from left to right**
2. **Squaring and/or cubing as read from left to right**
3. **Multiplication and/or division as read from left to right**
4. **Addition and/or subtraction as read from left to right**

Thus $5 + (2^3)$ can be written more simply as $5 + 2^3$, since the cubing must be done first, in agreement with the above convention. Thus $5 + 2^3$ means $5 + 8$, or 13, whereas $(5 + 2)^3$ means 7^3, or 343.

Similarly, $5 \cdot (2^3)$ can be written as $5 \cdot 2^3$, with $5 \cdot 8$, or 40, as the intended final result, since the cubing must be performed before the multiplication.

If one wants to indicate that 5 is to be multiplied by 2 and then the resulting product is to be cubed, grouping symbols must be used as follows: $(5 \cdot 2)^3$, which means 10^3, or 1000.

The number indicated by $5 \cdot 2 + 7$ is the sum of 10 and 7, or 17, since the convention requires that the multiplication be performed before the addition. Thus $5 \cdot 2 + 7$ is another name for 17.

To indicate that 2 is to be added to 7 and then the resulting sum is to be multiplied by 5, one writes $5(2 + 7)$, which names the number $5 \cdot 9$, or 45.

The product $5(2 + 7)$ could also have been written as $5 \cdot (2 + 7)$. However, since the numerals 5 and $(2 + 7)$ are written in juxtaposition (side by side), the operation of multiplication is clearly indicated and the dot is not necessary. Greater simplicity and clarity are achieved by writing only the symbols that are necessary.

EXAMPLE 1.4.2 Perform the indicated operation: $13 - (5 - 2)^2$.

Solution

1. Simplify the expression inside parentheses: $5 - 2 = 3$. Thus

$$13 - 3^2$$

2. Raise to power: $3^2 = 9$. Thus

$$13 - 9$$

3. Perform subtraction: $13 - 9 = 4$. Therefore

$$\begin{aligned} &13 - (5 - 2)^2 \\ &= 13 - 3^2 \\ &= 13 - 9 \\ &= 4 \end{aligned}$$

EXAMPLE 1.4.3 Simplify $5 + 2(3)^2 - (12 - 9)$.

Solution

$$\begin{aligned} &5 + 2(3)^2 - (12 - 9) \\ &= 5 + 2(3)^2 - 3 && \text{(Simplifying parentheses)} \\ &= 5 + 2 \cdot 9 - 3 && \text{(Raise to power)} \\ &= 5 + 18 - 3 && \text{(Multiply)} \\ &= 20 && \text{(Add and subtract from left to right)} \end{aligned}$$

EXAMPLE 1.4.4 Simplify

$$\frac{8^2 + 8 \cdot 6}{2 \cdot 8 + 2 \cdot 6}$$

Solution The bar used in division is a grouping symbol, so the operations in the numerator and in the denominator must be done before the division:

$$\frac{8^2 + 8 \cdot 6}{2 \cdot 8 + 2 \cdot 6} = \frac{64 + 8 \cdot 6}{2 \cdot 8 + 2 \cdot 6} \qquad \text{(Squaring first)}$$

$$= \frac{64 + 48}{16 + 12} \qquad \text{(Multiply)}$$

$$= \frac{112}{28} \qquad \text{(Add)}$$

$$= 4 \qquad \text{(Divide)}$$

EXAMPLE 1.4.5 Simplify $2\sqrt{5^2 - 3^2}$.

Solution The bar used in root extraction is a grouping symbol, so the operations under the bar must be done first:

$$2\sqrt{5^2 - 3^2} = 2\sqrt{25 - 9}$$

$$= 2\sqrt{16}$$

$$= 2(4)$$

$$= 8$$

EXAMPLE 1.4.6 Translate into symbols: The product of 4 and the sum of x and 6.

Solution Since the entire expression is a product, multiplication is the last operation performed. Therefore, parentheses are needed to indicate that addition is done first:

The product of 4 and the (sum of x and 6)
The product of 4 and $(x + 6)$

$$4(x + 6)$$

EXAMPLE 1.4.7 Translate into symbols: The difference obtained by subtracting 8 from the quotient resulting when x is divided by 8.

Solution Since the entire expression is a difference, subtraction is the last operation to be done. First, abbreviating the words:

Difference subtracting 8 from x over 8

Difference subtracting 8 from $\frac{x}{8}$

$$\frac{x}{8} - 8$$

EXERCISES 1.4 A

In Exercises 1–40, perform the indicated operations.

1. $9 + (8 + 7)$

2. $(9 + 8) + 7$

3. $15 - (10 - 3)$

4. $(15 - 10) - 3$

5. $3\cdot(4\cdot5)$

6. $(3\cdot4)\cdot5$

7. $3(8 + 2)$

8. $3(8) + 2$

9. $(9 - 3)(6 - 4)$

10. $9 - 3(6 - 4)$

11. $\dfrac{(\frac{24}{6})}{2}$

12. $\dfrac{24}{(\frac{6}{2})}$

13. $(2^3)^2$

14. $2^{(3^2)}$

15. $(15 - 3)^2$

16. $15 - 3^2$

17. $2\cdot5^2$

18. $(2\cdot5)^2$

19. $2 + 5^2$

20. $(2 + 5)^2$

21. $15 - (10 + 3)$

22. $(15 - 10) + 3$

23. $5(10 - 2^3)$

24. $5(10 - 2)^3$

25. $\dfrac{7^2 - 4^2}{7 - 4}$

26. $\dfrac{7^2}{7} - \dfrac{4^2}{4}$

27. $\dfrac{2^3 + 3^3}{2 + 3}$

28. $\dfrac{2^3}{2} + \dfrac{3^3}{3}$

29. $2\sqrt{5^2 - 4^2}$

30. $2\sqrt{5^2} - \sqrt{4^2}$

31. $2(\sqrt{5^2} - \sqrt{4^2})$

32. $9 - [7 + (5 - 3)]$

33. $8 - 2(5 - [2 + 1])$

34. $(8 - 2)(5 - [2 + 1])$

35. $2(7^2 - 3^2) - (4^2 - 3^2)$

36. $10 - \{8 - [6 - (4 - 2)]\}$

37. $2([5^2 - 3\{4 + 2\}] + 1)$

38. $5\{(9 + 4)^2 - (9 - 4)^2\}$

39. $1 + \{1 - [1 - (1 - \sqrt{1 - 1})]\}$

40. $7 - [6 - \{5^2 - 4(2 + 3)\}]$

Translate the verbal expressions in Exercises 41–50 into symbols.

41. The sum of x and twice y

42. Twice the sum of x and y

43. Twice the product of x and y

44. The product of 5 and the square of x

45. The difference obtained by subtracting the sum of x and 3 from y

46. The quotient obtained by dividing x by the sum of x and 3

47. 7 less than the square of the sum of y and 2

48. One half the product of 3 and the sum of a and b

★ **49.** 6 times the difference obtained by subtracting the sum of x and 2 from 9

★ **50.** The sum obtained by adding 6 to 3 times the difference obtained by subtracting 2 from the sum of x and 9

EXERCISES 1.4 B

In Exercises 1–40, perform the indicated operations.

1. $\frac{18}{6} + \frac{6}{18}$

2. $(10 - 5) - 5$

3. $10 - (5 - 5)$

4. $\frac{1 + (2 + 3)}{(1 + 2) + 3}$

5. $\frac{2(8 - 3)}{2 \cdot 8 - 2 \cdot 3}$

6. $\frac{2(8 - 3)}{2 \cdot 8 - 3}$

7. $\frac{6^2 + 4^2}{2}$

8. $\frac{6^2}{2} + \frac{4^2}{2}$

9. $\frac{20}{2^2 + 1^2}$

10. $\frac{20}{2^2} + \frac{20}{1^2}$

11. $\frac{5^2 - 4^2}{5 + 4}$

12. $\frac{5^3 - 4^3}{5 - 4}$

13. $3(7^2)$

14. $(3 \cdot 7)^2$

15. $3 + 7^2$

16. $(3 + 7)^2$

17. $(3^2)^3$

18. $(3^3)^2$

19. $3^{(2^3)}$

20. $8(3^2 + 2^2)$

21. $8(3 + 2)^2$

22. $24 - (10 + 6)$

23. $(24 - 10) + 6$

24. $4 + 2\{5 - 2[3 - (2 - 1)]\}$

25. $3\{(4 + 5)^2 - 41\}$

26. $(15 - 2)(7 - 2)$

27. $15 - 2(7 - 2)$

28. $2(3 \cdot 4)$

29. $(2 \cdot 3)(2 \cdot 4)$

30. $30 - 2(5 - 2)$

31. $30 - (2 \cdot 5 - 2)$

32. $\frac{3\sqrt{6^2 + 8^2}}{5^2}$

33. $12\left[\frac{1}{2} - \left(\frac{1}{3} - \frac{1}{4}\right)\right]$

34. $\sqrt[3]{\frac{225 - 100}{8}}$

35. $\frac{\sqrt[3]{225 - 100}}{8}$

36. $5 - [4 - \{3 - (2 - 1)\}]$

37. $(5 - 4) - ([3 - 2] - 1)$

38. $7^2 - (6^2 - [5^2 - \{4^2 - (3^2 - 2^2)\}])$

39. $\frac{\frac{1}{2} + \frac{1}{3}}{1 - (\frac{1}{2})(\frac{1}{3})}$

40. $\frac{20 - [10 - 2(7 - 3)]}{20 - [10 + 2(7 - 3)]}$

Translate the verbal expressions in Exercises 41–50 into symbols.

41. The result of adding y to 3 times x

42. 3 times the sum of x and y

43. The quotient obtained by dividing the sum of 7 and a by the product of 7 and a

44. The difference obtained when one half the sum of a and b is subtracted from the square of b

45. The square root of the difference obtained by subtracting the square of 5 from the square of x

46. The cube of the product of 4 and y

47. 1 less than the remainder obtained by subtracting the square of 2 from the difference between the squares of 4 and 3

48. A number N is multiplied by 3. Then 15 is added to the product. The resulting sum is divided by 3, and then 5 is subtracted from this quotient.

★ **49.** The product of 6 and the square root of the difference obtained by subtracting the square of 1 from the square of x

★ **50.** The quotient obtained by dividing 8 times the square of t by the sum of 8 and the square of t

1.5 EQUALITY, SUBSTITUTION, FORMULAS

1.5.1 The Equality Relation

An equation is a statement that has the form $A = B$, read "A equals B." The symbol $=$ indicates that A and B are names of the same number.

As examples,

$12 + 3 = 15$ means that $12 + 3$ (the sum of 12 and 3) names the same number as the numeral 15

$12 \cdot 3 = 36$ means that $12 \cdot 3$ (the product of 12 and 3) names the same number as the numeral 36.

It is often convenient to use the symbol $\neq$, which means "does not equal." For example, $3 \neq 5$ is read "three does not equal five" and means that the numerals 3 and 5 do not name the same number.

1.5.2 Substitution

Since the relation $r = s$ is understood to mean that the variables r and s are names of the same number, it seems reasonable that one name can replace the other in a given statement without changing the truth or falsity of the statement. Furthermore, if one name replaces the other in an algebraic expression, the number that is named by the algebraic expression remains the same. This property is called the **substitution axiom** for the equal relation. An axiom is a statement that is accepted without proof.

THE SUBSTITUTION AXIOM

If $r = s$, then r may replace s or s may replace r in an algebraic expression without changing the number that is being named, or in an algebraic statement without changing the truth or falsity of the statement.

For example, consider the statement $y = x + 5$.

Now if $x = 3$, then 3 can replace x in this statement, and one obtains $y = 3 + 5$. This substitution process may be repeated by using the fact that $3 + 5 = 8$. Replacing $3 + 5$ by its equal, 8, in the statement $y = 3 + 5$, one obtains $y = 8$.

The substitution axiom may be used to *evaluate* algebraic expressions. To evaluate an algebraic expression means to replace each letter by the numeral to which it is equal and then to perform the operations that are indicated, replacing the names of the numbers by simpler names that do not involve the operation symbols.

EXAMPLE 1.5.1 Evaluate

$$3(x + 4) - \frac{x}{2} \text{ for } x = 6$$

Solution

1. Rewrite the given expression:

$$3(x + 4) - \frac{x}{2}$$

2. Remove the letter and hold its place with open parentheses:

$$3((\ ) + 4) - \frac{(\ )}{2}$$

3. Insert the numeral to which the letter is equal within the parentheses:

$$3((6) + 4) - \frac{(6)}{2}$$

4. Do the indicated operations:

$$\begin{aligned} 3(10) &- 3 \\ 30 &- 3 \\ &= 27 \end{aligned}$$

EXAMPLE 1.5.2 Evaluate

$$\frac{x^2 + 2xy - 3y^2}{x + 3y} \text{ for } x = 8 \text{ and } y = 4$$

Solution

1. Rewrite the given expression:
$$\frac{x^2 + 2xy - 3y^2}{x + 3y}$$
2. Replace one letter, say x, by open parentheses:
$$\frac{(\ )^2 + 2(\ )y - 3y^2}{(\ ) + 3y}$$
3. Insert the value for x:
$$\frac{(8)^2 + 2(8)y - 3y^2}{(8) + 3y}$$
4. Replace the other letter, y, by open parentheses:
$$\frac{(8)^2 + 2(8)(\ ) - 3(\ )^2}{(8) + 3(\ )}$$
5. Insert the value for y:
$$\frac{(8)^2 + 2(8)(4) - 3(4)^2}{(8) + 3(4)}$$
6. Evaluate (do the operations):
$$\begin{aligned} &\frac{64 + 2(32) - 3(16)}{8 + 12} \\ &= \frac{64 + 64 - 48}{20} \\ &= \frac{128 - 48}{20} \\ &= \frac{80}{20} \\ &= 4 \end{aligned}$$

1.5.3 Formulas

Many problems in mathematics, science, business, and other areas are most easily solved by using formulas.

A **formula** is a symbolic statement indicating what operations are to be performed on certain numbers with specialized meanings determined by the particular problem.

EXAMPLE 1.5.3 (Geometry: Perimeter of a Square) Express the following statement as a formula: The length of the perimeter of a square is obtained by multiplying the length of a side by 4.

Solution Let P represent the length of the perimeter. Let s represent the length of a side.
Formula: $P = 4s$.

EXAMPLE 1.5.4 Using the formula $P = 4s$, find P if $s = 3$ inches.

Solution $P = 4(3) = 12$ inches

EXAMPLE 1.5.5 (Geometry: Area of a Square) Express as a formula: The area A of a square is the square of the length of a side s.

Solution $A = s^2$

EXAMPLE 1.5.6 Evaluate $A = s^2$ for $s = 5$ centimeters.

Solution $A = (5)^2 = 25$ square centimeters

EXAMPLE 1.5.7 (Uniform Motion) Express as a formula: The distance d that an object travels is the product of the rate of speed r and the time t that is traveled.

Solution $d = rt$

EXAMPLE 1.5.8 Given $d = rt$, find d if $r = 30$ mph and $t = 2$ hours.

Solution $d = 30 \cdot 2 = 60$ miles

EXAMPLE 1.5.9 (Resistance of an Electrical Circuit) Write a formula to express the following relation: The resistance R, measured in ohms, of an electrical circuit is the quotient obtained by dividing the electromotive force E, measured in volts, by the intensity I of the current, measured in amperes.

Solution

$$R = \frac{E}{I}$$

EXAMPLE 1.5.10 Using $R = \frac{E}{I}$, find R if $E = 60$ volts and $I = 15$ amperes.

Solution

$$R = \frac{60}{15} = 4 \text{ ohms}$$

EXAMPLE 1.5.11 (Business Profit) Express as a formula: The profit P of a business transaction is obtained by subtracting the cost C from the selling price S.

Solution $P = S - C$

EXAMPLE 1.5.12 Find the profit made on selling an end table for \$55 if the cost of the table was \$35.

Solution

Using $P = S - C$ with $S = 55$ and $C = 35$, $P = 55 - 35 = 20$ dollars.

EXERCISES 1.5 A

Evaluate the algebraic expressions in Exercises 1–30.

1. $4x^3 + 5x^2 + 7x + 6$ for $x = 10$

2. $3(x - 4)^2$ for $x = 8$

3. $\dfrac{x^2 - 25}{x - 5}$ for $x = 7$

4. $\dfrac{6x}{2x + 4}$ for $x = 4$

5. $x(x^2 - x)$ for $x = 6$

6. $y^2 - \dfrac{2y}{3}$ for $y = 9$

7. $\dfrac{y(y - 1)}{2}$ for $y = 10$

8. $\dfrac{y^3 + 8}{y + 2}$ for $y = 3$

9. $(n + 1)(n - 1)$ for $n = 11$

10. $(n - 3)(n^2 + 3n + 9)$ for $n = 5$

11. $x^2 + 2xy + y^2$ for $x = 5, y = 3$

12. $(x + y)^2$ for $x = 5, y = 3$

13. $x^2 + y^2$ for $x = 5, y = 3$

14. $\dfrac{xy}{x + y}$ for $x = 10, y = 15$

15. $c^2 - a^2$ for $c = 13, a = 12$

16. $\dfrac{a + b}{a - b}$ for $a = 6, b = 4$

17. $\dfrac{a^3 - b^3}{a - b}$ for $a = 10, b = 6$

18. $3r^2 - rs - 2s^2$ for $r = 7, s = 3$

19. $(x + y)^2 + (x + y)(x - y) + (x - y)^2$ for $x = 4, y = 1$

20. $\dfrac{x}{y}(x^2 - y^2)$ for $x = 10, y = 5$

21. $\dfrac{x + y + z}{3}$ for $x = 10, y = 12, z = 20$

22. $x^2 + y^2 + z^2$ for $x = 3, y = 4, z = 12$

23. $\dfrac{a}{2}(b + c)$ for $a = 8, b = 6, c = 9$

24. $a(b - c)^2$ for $a = 5, b = 12, c = 3$

25. $a + (n - 1)d$ for $a = 20, n = 10, d = 2$

26. $(r - s)[(r - t)(s - t)]$ for $r = 12, s = 8, t = 4$

★ **27.** $\dfrac{(r^2 - rt) + (rs - st)}{r - t}$ for $r = 5, s = 4, t = 3$

★ **28.** $n\{a + [a + (n - 1)d]\}$ for $a = 1, d = 1, n = 100$

★ **29.** $[x(y - z)^2]^3$ for $x = 5, y = 8, z = 6$

★ **30.** $2^3a + 2^2b + 2c + d$ for $a = 1, b = 0, c = 0, d = 1$

Write a formula to express each of the relations in Exercises 31–40.

31. The area A of a triangle is one half the product of its base b and its height h.

32. The area A of a trapezoid is one half the product of its height h and the sum of its bases a and b.

33. The volume V of a cube is the cube of its side s.

34. The volume V of a cone is one third the product of the number π, the height h, and the square of the radius r.

35. The illumination E at a point on a surface in lumens is equal to the quotient obtained by dividing the intensity I of the source in candle-power by the square of the distance r between the point on the surface and the source of light.

36. The centrigrade temperature C in degrees is equal to five ninths the difference obtained by subtracting 32 from the Fahrenheit temperature F in degrees.

37. The period T of a pendulum is twice the product of π and the square root of the quotient obtained by dividing the length L by the gravitational force g.

38. The total resistance R in an electrical circuit with two resistances, having values S and T, connected in parallel is obtained by dividing the product of the values of the two resistances by their sum.

39. Using the straight-line depreciation method, the annual contribution R to a sinking fund is obtained by dividing the difference between the original cost C and the probable scrap value S by the number n of years representing the probable useful life of the article.

40. In anthropology, the cephalic index C is obtained by dividing the product of 100 and the width of a head W by the length of the head L.

Evaluate the formulas in Exercises 41–50 by using the values indicated.

41. $A = \dfrac{bh}{2}$ for $b = 18, h = 20$ (see Exercise 31)

42. $A = \frac{h}{2}(a + b)$ for $h = 7$, $a = 13$, $b = 5$ (see Exercise 32)

43. $V = s^3$ for $s = 12$ (see Exercise 33)

44. $V = \frac{\pi h r^2}{3}$ for $\pi = \frac{22}{7}$, $h = 28$, $r = 5$ (see Exercise 34)

45. $E = \frac{I}{r^2}$ for $I = 360$ candlepower, $r = 6$ feet (see Exercise 35)

46. $C = \frac{5}{9}(F - 32)$ for $F = 212$ degrees (see Exercise 36)

★ 47. $T = 2\pi\sqrt{\frac{L}{g}}$ for $\pi = \frac{22}{7}$, $L = 98$, $g = 32$ (see Exercise 37)

48. $R = \frac{ST}{S + T}$ for $S = 20$ ohms, $T = 30$ ohms (see Exercise 38)

49. $R = \frac{C - S}{n}$ for $C = \$3000$, $S = \$600$, $n = 12$ years (see Exercise 39)

50. $C = \frac{100W}{L}$ for $W = 18$ centimeters, $L = 24$ centimeters (see Exercise 40)

EXERCISES 1.5 B

Evaluate the algebraic expressions in Exercises 1–30.

1. $3x^3 + 2x^2 - 10x + 4$ for $x = 3$
2. $4(x - 2)^2$ for $x = 5$
3. $\frac{x^3 + 4}{x + 4}$ for $x = 2$
4. $\frac{3x}{5(x - 7)}$ for $x = 10$
5. $x^2(x - 3)$ for $x = 4$
6. $a^3 - \frac{a}{2}$ for $a = 6$
7. $\frac{a(a + 1)}{3}$ for $a = 5$
8. $\frac{a^3 - 8}{a - 2}$ for $a = 3$
9. $(y + 3)(y - 3)$ for $y = 7$
10. $(y - 2)(y^2 + 2y + 4)$ for $y = 3$
11. $x^2 + 3xy + 5y^2$ for $x = 5, y = 2$
12. $(x + y)^3$ for $x = 2$, $y = 3$
13. $x^3 + y^3$ for $x = 2$, $y = 3$
14. $\frac{xy + x^2}{xy + y^2}$ for $x = 8$, $y = 2$
15. $\sqrt{x^2 - y^2}$ for $x = 25$, $y = 24$
16. $\frac{x^3 + y^3}{x + y}$ for $x = 5$, $y = 2$
17. $\frac{a^2 - b^2}{a + b}$ for $a = 3$, $b = 2$
18. $5m^2 - ms + 6s^2$ for $m = 4, s = 3$
19. $ab(a + b)^2$ for $a = 3$, $b = 5$
20. $5ab^2 - 3a^2b$ for $a = 3$, $b = 5$

21. $r^2 - (s^2 + t^2)$ for $r = 3, s = 2, t = 1$

22. $(r - s)^2 + t^2$ for $r = 3, s = 2, t = 1$

23. $5(m - n)^2$ for $m = 10, n = 7$

24. $\dfrac{m^3 - n^3}{2m - n}$ for $m = 6, n = 5$

25. $30 - [a^2 - (3b - a)]$ for $a = 3, b = 2$

26. $\dfrac{m^2n - mn^2}{m - n}$ for $m = 7, n = 3$

★ **27.** $\dfrac{(x^2 + yx) + (xy - yz)}{2x - y}$ for $x = 5, y = 3, z = 2$

★ **28.** $a\{b - [c - (d - b)]\}$ for $a = 2, b = 1, c = 3, d = 4$

★ **29.** $[a(b - c)^2]^2$ for $a = 3, b = 4, c = 2$

★ **30.** $\dfrac{x^2 - x(y - z)}{(x + z) - y}$ for $x = 7, y = 6, z = 2$

Write a formula to express each of the relations in Exercises 31–40.

31. The perimeter P of a rectangle is twice the sum of the base b and the height h.

32. The circumference C of a circle is the product of a constant, π, and the length of the diameter d.

33. The area A of a circle is the product of a constant, π, and the square of the radius r.

34. The volume V of a pyramid with a square base is one third the product of the height h and the square of a side b of the base.

35. The focal length f of a camera lens is the quotient obtained by dividing the product of the distance a from an object to the lens and the distance b from the lens to the image by the sum of these two distances.

36. Neglecting air resistance, the distance d in feet that a dropped object falls in t seconds is equal to the product of 16 and the square of the time t.

37. The efficiency E of an engine is equal to the quotient when the difference of heat input I, reduced by the heat output O, is divided by the heat input.

38. In psychology, the intelligence quotient Q is obtained by dividing 100 times the mental age M by the chronological age C.

39. The amount of medication C for a child over 1 year of age can be obtained by dividing the product of the age of the child y in years and the adult dosage A by the sum of y and 12.

40. The amount of money A that results from investing a sum P at a simple interest rate r for n interest periods is the sum of P and the product of P, r, and n.

Evaluate the formulas in Exercises 41–50 by using the values indicated.

41. $P = 2(b + h)$ for $b = 10$, $h = 14$ (see Exercise 31)

42. $C = \pi d$ for $\pi = \frac{22}{7}$, $d = 21$ (see Exercise 32)

43. $A = \pi r^2$ for $\pi = \frac{22}{7}$, $r = 14$ (see Exercise 33)

44. $V = \frac{1}{3} hb^2$ for $h = 5$, $b = 6$ (see Exercise 34)

45. $f = \frac{ab}{a + b}$ for $a = 4$, $b = 0.05$ (see Exercise 35)

46. $d = 16t^2$ for $t = 9$ (see Exercise 36)

47. $E = \frac{I - O}{I}$ for $I = 45$, $O = 30$ (see Exercise 37)

48. $Q = \frac{100M}{C}$ for $M = 18$, $C = 15$ (see Exercise 38)

49. $C = \frac{yA}{y + 12}$ for $y = 8$, $A = 10$ (see Exercise 39)

50. $A = P + Prn$ for $P = 500$, $r = 0.06$, $n = 8$ (see Exercise 40)

1.6 REAL NUMBERS, NUMBER LINES, ABSOLUTE VALUE

1.6.1 Special Sets of Numbers

The answers to most number problems are found in a set of numbers called the set of real numbers. The set of real numbers and some of its important subsets will be introduced in this section.

Man developed the counting numbers, $\{1, 2, 3, 4, 5, \ldots\}$, at a very early stage in his existence. By 3000 B.C. he had invented several numeral systems for these numbers. The early Egyptians had a simple grouping system something like our Roman numerals, and the early Babylonians had a positional system something like our modern Hindu-Arabic numerals.

DEFINITION

The set of counting numbers, $\{1, 2, 3, 4, 5, \ldots\}$, is also called the set of **natural numbers,** which may be designated by the letter N.

$$N = \{1, 2, 3, 4, 5, \ldots\}$$

As civilization progressed, man's requirements rose above the necessity of simply recording numbers of objects. In particular, he needed a technique for measuring quantities that were opposite in nature.

For example, in order to measure temperature, man invented a thermometer on which he established a scale. He called his starting point 0 (zero) and then selected a unit of measurement, such as 1 degree Fahrenheit. Then he marked off equal units of measurement on one side of the zero point to obtain his scale, as in Figure 1.6.1.

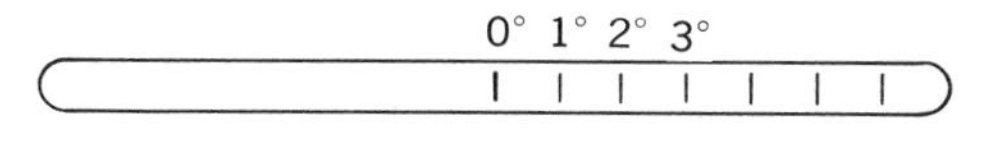

FIGURE 1.6.1

To record temperatures below 0 degrees—that is, when the mercury fell lower than the zero point—he marked off units in the opposite direction from the one he started with.

To distinguish between a temperature above zero and one below zero, the numbers above zero were tagged with the symbol + (plus), and the ones below zero were tagged with the symbol − (minus) (Figure 1.6.2).

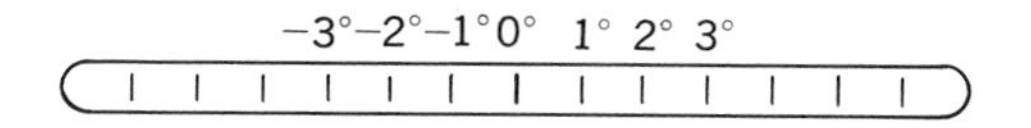

FIGURE 1.6.2

Thus negative numbers, such as -1, -2, -3, and so on, came into existence as measures of quantities opposite in nature to other quantities.

Some examples of opposite quantities that can be measured by using negative numbers are shown in the table below.

+30	*−30*
30 degrees above zero	30 degrees below zero
30 feet above sea level	30 feet below sea level
30 miles east	30 miles west
30 dollars bank deposit	30 dollars withdrawal

DEFINITION

The union of the set of natural numbers and their negatives and the number zero is called the set of integers, which may be designated by the letter I.

$$I = \{\ldots, -3, -2, -1, 0, +1, +2, +3, \ldots\}$$

The natural numbers $\{1, 2, 3, \ldots\}$ are also called the **positive integers.** The set $\{1, 2, 3, \ldots\}$ is considered to be the same set of numbers as the set $\{+1, +2, +3, \ldots\}$, in which the names of the natural numbers are tagged with the $+$ symbol. In other words, $+n$ and n are to be considered as different names for the same number.

The numbers in the set $\{-1, -2, -3, \ldots\}$ are called the **negative integers.**

The numbers in the set $\{0, 1, 2, 3, \ldots\}$ are called the **whole numbers**, or the **nonnegative integers.**

The set of natural numbers, N, is a subset of the set of integers, I. In symbols, $N \subset I$.

When one natural number is divided by another natural number, the resulting quotient is not always a natural number. In arithmetic, these nonintegral quotients are called **fractions**, and each fraction is thought of as the number of subunits of a basic unit.

HISTORICAL NOTE

Historically, fractions came into existence as a necessity for obtaining more accurate measurements. The ancient Egyptians had a well-developed system of fractions called *unit fractions*. The numerators were restricted to be the number 1 with one exception, the fraction $\frac{2}{3}$. The ancient Babylonians had a sexagesimal (based on 60) system of fractions similar to our decimal fractions. Instead of using denominators which are powers of 10 (10, 100, 1000, and so on), their denominators were restricted to be powers of 60 (60; 3600; 216,000; and so on). The Hindus originated the common fractions of our present-day arithmetic, in which any natural number may be a numerator and any natural number may be a denominator.

Many visual representations are possible, such as those in Figure 1.6.3.

In algebra, this meaning of a fraction is retained. In addition to the fractions of arithmetic, called the *positive fractions*, their negatives, called the *negative fractions*, are also included in the number system of algebra.

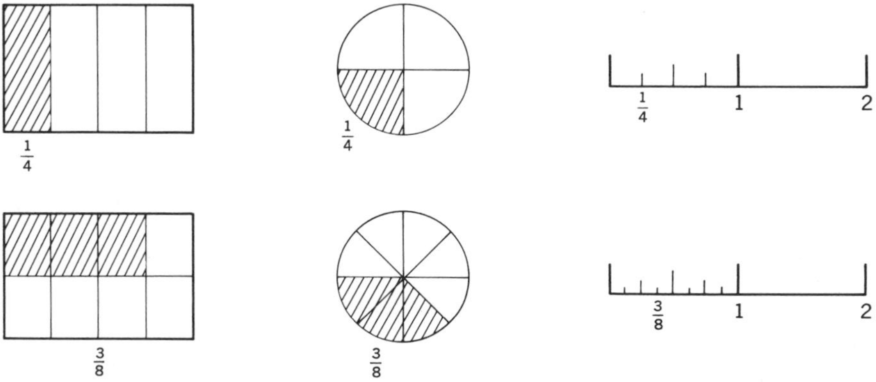

FIGURE 1.6.3

DEFINITION

The union of the set of integers and the positive and negative fractions is called the set of rational numbers, which may be designated by the letter Q.

The set Q may also be described as the set of numbers that can be expressed as the quotient of two integers, $\frac{p}{q}$, where p and q are integers and $q \neq 0$.

$$Q = \left\{\frac{p}{q}\,\middle|\, p \in I \text{ and } q \in I \text{ and } q \neq 0\right\}$$

Examples of rational numbers are $\frac{2}{3}, \frac{5}{9}, \frac{-2}{3}, \frac{5}{-9}, -\frac{16}{17}$. In each of these numbers the numerator and denominator is an integer, and no denominator is zero. In other words, each number fits the description

$$\left\{\frac{p}{q}\,\middle|\, p \in I, q \in I, q \neq 0\right\}$$

The set of integers can be identified as a subset of the rationals—that is, those quotients $\frac{p}{q}$ with $q = 1$.

$$I = \left\{\frac{p}{q}\,\middle|\, p \in I \text{ and } q = 1\right\}$$

This means, for example, that $\frac{5}{1}$ and 5 are names of the same number. In symbols, $I \subset Q$.

1.6.2 Number Lines

A visual representation of numbers may be obtained by using the names of numbers as the names of points on a straight line.

A point is selected as the starting point. This point is called the **origin** and is given the name 0 (zero).

A unit of measurement and a direction, called the positive direction, are selected. Using this unit of length, points are marked off in succession in the positive direction, which is to the right in Figure 1.6.4.

The arrow at the right indicates that the line continues indefinitely in this direction.

The negative integers $\{-1, -2, -3, \ldots\}$ are assigned to the points in the opposite direction in a similar manner. This direction is to the left in Figure 1.6.4.

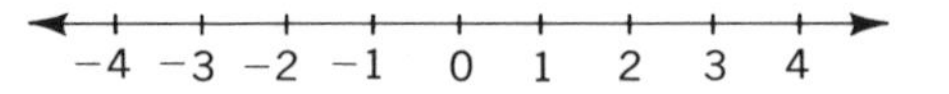

FIGURE 1.6.4 A horizontal number line.

The arrow at the left indicates that the line continues indefinitely in this direction.

A line whose points are named by using numbers is called a **number line.**

The number that names a point is called the **coordinate** of the point.

The point that is given a number name is called the **graph** of this number.

A number line does not have to be horizontal. A vertical arrangement is often used, as shown in Figure 1.6.5. In this arrangement, the positive direction is usually upward.

All the rational numbers can be identified as coordinates on a number line by considering subdivisions of the basic unit and by using the concept of opposites (Figure 1.6.6).

Thus it is observed that each rational number can be made to correspond to exactly one point on a number line.

One might be tempted to believe that the rational numbers exhaust the number line. However, this is not the case. For example, $\sqrt{2}$ is *not* a rational number, but it can still be located on the number line by the following process.

The length of the diagonal of a unit square is known to be $\sqrt{2}$. Using the origin as the center of a circle with radius equal to the diagonal of a unit square, the point on the number line with coordinate $\sqrt{2}$ can be located as

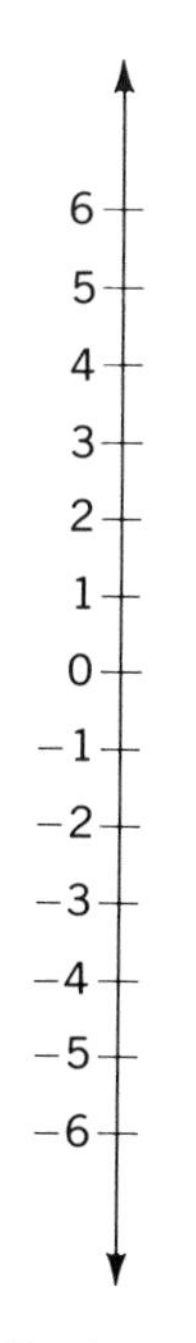

FIGURE 1.6.5 A vertical number line.

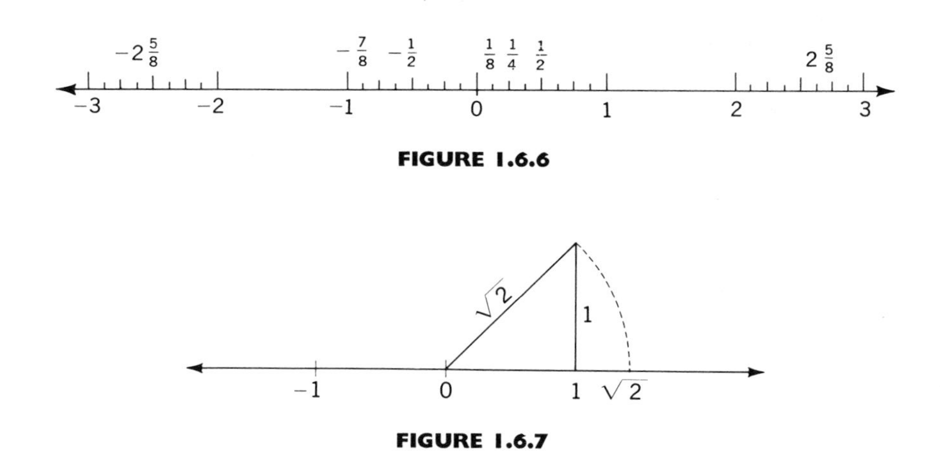

FIGURE 1.6.6

FIGURE 1.6.7

the intersection of this circle with the positive portion of the number line (Figure 1.6.7).

The real number $\sqrt{2}$ is called an *irrational* number. Some other examples of irrational numbers are $\sqrt{3}$, $\sqrt{5}$, $\sqrt[3]{2}$, $-\sqrt{2}$, $-\sqrt{3}$, and π.

DEFINITION

An irrational number is a real number that is not rational—that is, it cannot be expressed as the quotient of two integers.

The totality of all the numbers that can be associated with points on a number line is called the set of **real numbers**. Thus a number line is called a **real number line.**

Each point on a real number line can be made to correspond to exactly one real number, and each real number can be made to correspond to exactly one point on a real number line. Thus there is a **one-to-one correspondence between the set of real numbers and the set of points on a real number line.** This important property of the set of real numbers is called the **axiom of completeness.** The set of real numbers is "complete" in the sense that all the real numbers are "used up" in naming the points on a number line, and every point has exactly one real number name.

The numbers associated with points to the right of 0 are called the **positive real numbers**, and those to the left of 0 are called the **negative real numbers** (Figure 1.6.8).

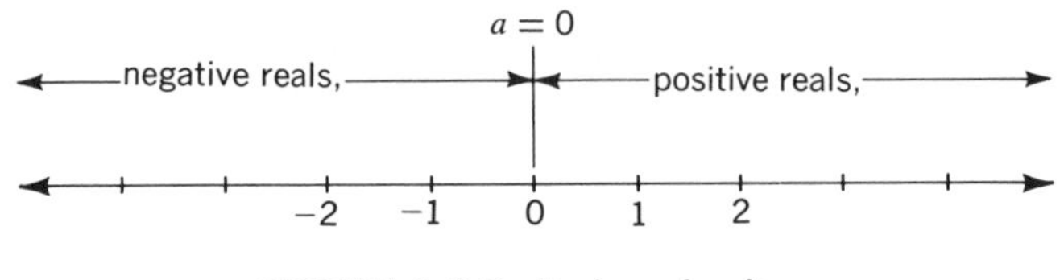

FIGURE 1.6.8 Real number line.

Numbers that are tagged with a + or − sign are often called **signed numbers**. The numbers x and $-x$ are called **opposites** of each other. For example, the positive number 5 and the negative number -5 are opposites of each other. They are also called **additive inverses.**

1.6.3 Absolute Value

The number line provides a geometric model for the set of real numbers where points on the number line correspond to real numbers. Any point r and its opposite, or additive inverse, $-r$ are the same *distance* from the origin, but they are on opposite sides of the origin. The algebraic sign of a real number indicates on which side of the origin the corresponding point is located. The *distance* of the point from the origin, regardless of the side on which it is located, is called the *absolute value* of the number, denoted symbolically by $|r|$.

DEFINITION OF ABSOLUTE VALUE

If p is a positive real number, then $|p| = p$ and $|-p| = p$ and $|0| = 0$.

EXAMPLE 1.6.1 Simplify $|-6| + |+3|$.

Solution $|-6| + |+3| = 6 + 3 = 9$

EXAMPLE 1.6.2 Simplify $8\,|-5|$.

Solution $8\,|-5| = 8(5) = 40$

EXAMPLE 1.6.3 Simplify $2\,|7 - 7|$.

Solution $2\,|7 - 7| = 2\,|0| = 2(0) = 0$

HISTORICAL NOTE

The Chinese were familiar with negative numbers by at least 200 B.C. They were known to write positive numbers in red and negative numbers in black.

The Greek Diophantus (ca. 275), in his *Arithmetica*, calls the equation $4x + 20 = 4$ "absurd," since its solution would be -4. Diophantus did not seem to comprehend the abstract concept of a negative number.

The negative number is first mentioned as such in the works of the Hindu Brahmagupta (ca. 628). The Hindus indicated a negative number be placing a dot over a numeral or placing a circle above or around the numeral, such as $\dot{5}$, $\mathring{5}$, or ⑤.

Around 1225 the Italian Leonardo Fibonacci interpreted the negative solution of a financial problem to mean a loss.

In his *Ars Magna* (1545), Cardan presented the elementary properties of negative numbers, although he was unable to give a clear interpretation of these numbers. He referred to the positive numbers as "true" numbers and to the negative numbers as "fictitious" numbers.

In 1544, Stifel stated that numbers such as $0 - 3$ were "absurd."

Finally, because of the works of Viète, Fermat, Descartes, and other mathematicians, the concept of the negative number became thoroughly understood.

EXERCISES 1.6 A

In Exercises 1–10, determine which of the following would best be measured by signed numbers and which by positive numbers alone.

1. Heights of people

2. Entries on a bank statement

3. Latitudes of cities

4. Historical dates from 4000 B.C. to now

5. Direction and speed of cars on a highway

6. Temperatures of cities

7. Number of calories in various foods

8. Areas of plots of land

9. Daily changes in barometric pressure

10. Scores in a card game in which points can be gained or lost

If each phrase in Exercises 11–20 is represented by a positive number, state what is represented by the corresponding negative number.

11. 80 degrees east longitude

12. 50 dollars profit

13. 70 feet above sea level

14. A gain of 5 yards by a football team

15. A 15-mph south wind

16. A velocity of 2800 mph upward, of a rocket

17. 6 steps upward

18. 8 hours later

19. 35 pounds overweight

20. An increase in volume of 20 cubic centimeters

For Exercises 21–30, name the coordinate of the point on the number line indicated by the dot.

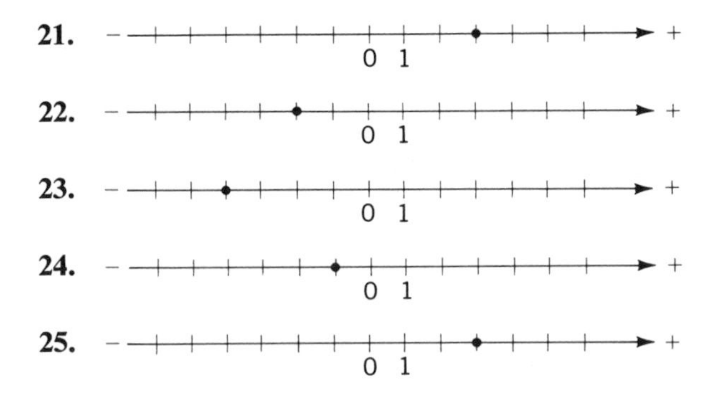

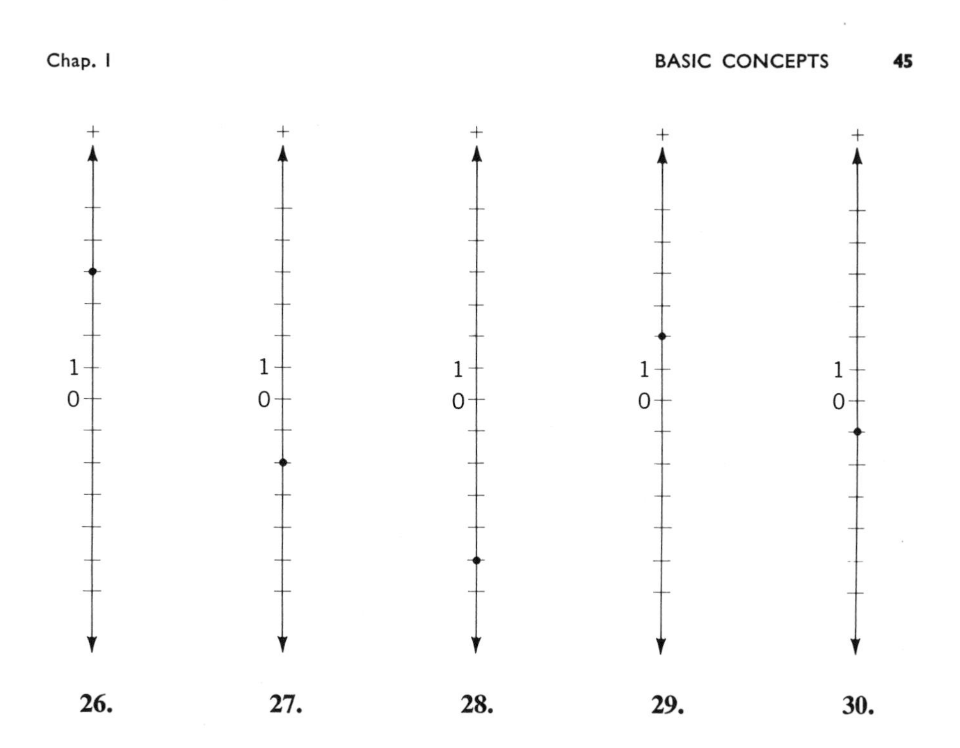

For Exercises 31–35, draw the graph of each of the following numbers on a horizontal number line whose positive direction is to the right.

31. $+8$ **32.** -8 **33.** $+5$ **34.** 0 **35.** -2

For Exercises 36–40, draw the graph of the following numbers on a vertical number line whose positive direction is upward (see Exercises 26–30).

36. $+4\frac{1}{2}$ **37.** $-2\frac{1}{3}$ **38.** $-\frac{3}{4}$ **39.** 0 **40.** $+2\frac{3}{8}$

Simplify the expressions in Exercises 41–60.

41. $|5|$ **42.** $|-4|$ **43.** $|0|$ **44.** $|-1|$

45. $\left|-1\frac{3}{4}\right|$ **46.** $\left|7\frac{1}{2}\right|$ **47.** $|-6| + |+6|$

48. $|-7| + |-2|$ **49.** $|+8| + |-3|$ **50.** $3\,|-5|$

51. $|-6| \cdot |-4|$ **52.** $|12| - |-5|$ **53.** $|-15| - |+8|$

54. $2(|-9| - |-1|)$ **55.** $4(|-8| - |+8|)$ **56.** $|10 - 6|$

57. $|4 - 4|$ **58.** $-|-6|$ **59.** $|-5| \cdot |-8|$

60. $-(|-12| - |-7|)$

In Exercises 61–65, list the elements of each set where

$$U = \left\{-15, -4\tfrac{1}{2}, -\sqrt{2}, -1, -\tfrac{2}{3}, 0, \tfrac{3}{4}, 1, \sqrt{2}, 5\tfrac{7}{8}, 14\right\}.$$

61. $\{x \mid x \text{ is an integer and } x \in U\}$

62. $\{x \mid x \text{ is a rational number and } x \in U\}$

63. $\{x \mid x \text{ is a rational number that is } \textit{not} \text{ an integer and } x \in U\}$

64. $\{x \mid x \text{ is a real number and } x \in U\}$

65. $\{x \mid x \text{ is a real number that is not rational and } x \in U\}$

EXERCISES 1.6 B

In Exercises 1–10, determine which would best be measured by signed numbers and which by positive numbers alone.

1. Daily minimum temperatures at Glacier National Park

2. Volumes of containers

3. Populations of cities **4.** Daily changes in stock prices

5. Weekly changes in the weight of a person on a diet

6. Longitudes of cities

7. Ages of people **8.** Electric charges on ions

9. Freezing points of chemical compounds, in degrees centigrade

10. Barometric pressures

If each phrase in Exercises 11–20 is represented by a positive number, state what is represented by the corresponding negative number.

11. 35 degrees north latitude **12.** 250 A.D.

13. 20 degrees above 0 Fahrenheit

14. A score of 35 in a pinochle game in which points may be won or lost

15. A force of 20 pounds toward an object

16. An acceleration of 32 ft/sec^2

17. A gain in heat of 150 Btu **18.** An increase in price of $5

19. A counterclockwise rotation of 40 degrees

20. An image distance of 30 centimeters in front of a mirror

For Exercises 21–30, name the coordinate of the point on the number line indicated by the dot.

21.

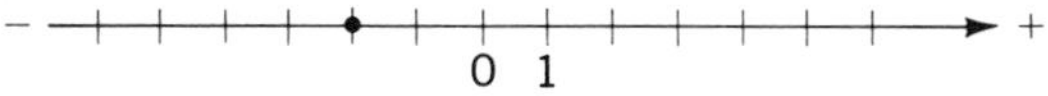

22.

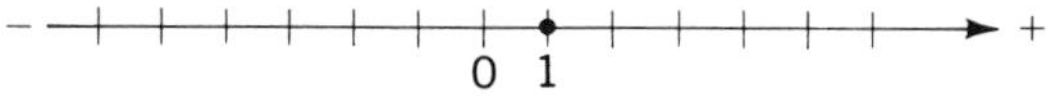

23.

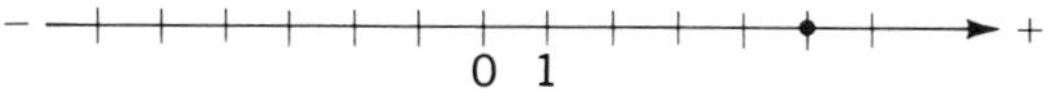

24.

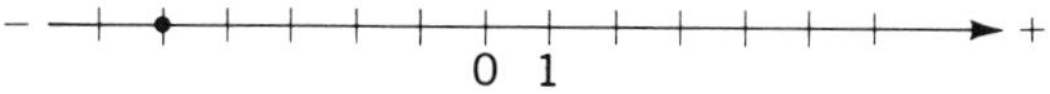

25.

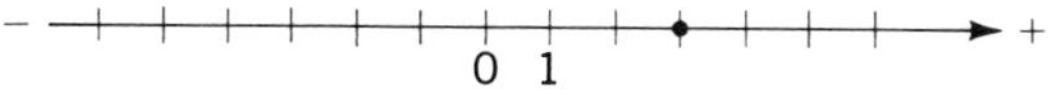

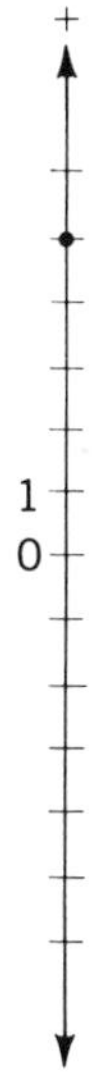

26. 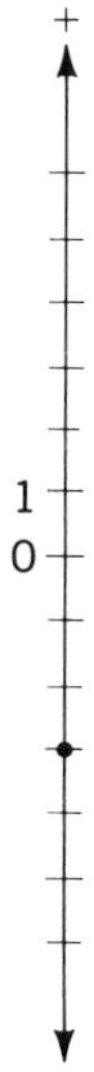 **27.** 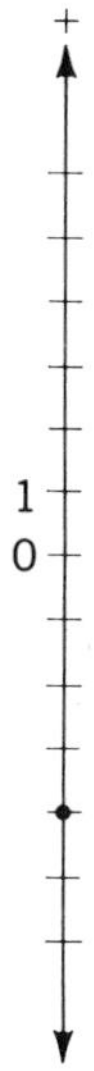 **28.** 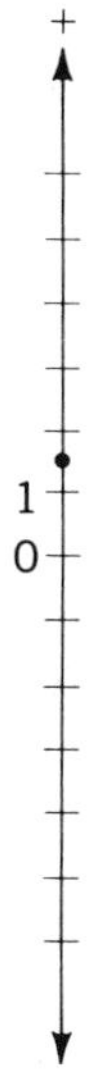 **29.** 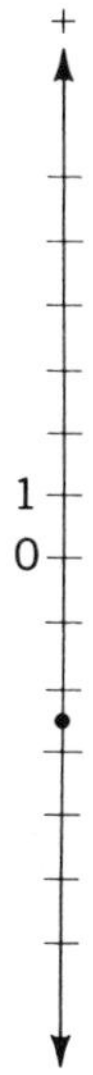 **30.**

For Exercises 31–35, draw the graph of each of the following numbers on a horizontal number line whose positive direction is to the right.

31. -4 **32.** 0 **33.** $+4$ **34.** -6 **35.** -10

For Exercises 36–40, draw the graph of each of the following on a vertical number line whose positive direction is upward (*see Exercises 26–30*).

36. -3 **37.** $+2\frac{1}{2}$ **38.** $-2\frac{1}{2}$ **39.** $+5\frac{5}{8}$ **40.** $-1\frac{3}{4}$

Simplify the phrases in Exercises 41–60.

41. $|3|$ **42.** $|-3|$ **43.** $\left|-2\frac{1}{2}\right|$

44. $-|-3|$ **45.** $\left|-\left|\frac{1}{3}\right|\right|$ **46.** $|-3| + |2|$

47. $|-3| - |2|$ **48.** $|3 - 3|$ **49.** $2(|3| + |-2|)$

50. $|-7| - |4|$ **51.** $|-4|\cdot|-2|$ **52.** $3\,|-2|$

53. $4 + |-2|$ **54.** $6 - |-2|$ **55.** $2(|-4| - |+4|)$

56. $|8 - 2|$ **57.** $-\,|-|-5|\,|$ **58.** $-\,(|6|\cdot|-2|)$

59. $|14 - 2|$ **60.** $14 - 2\,|-3|$

In Exercises 61–65, list the elements of each set where

$$U = \left\{-10, -5, -\sqrt{3}, -1, -\frac{1}{5}, 0, \frac{2}{3}, \sqrt{3}, 4\frac{1}{2}, 100\right\}.$$

61. $\{x \mid x \in I \text{ and } x \in U\}$ **62.** $\{x \mid x \in Q \text{ and } x \in U\}$

63. $\{x \mid x \in Q$ and x is *not* an integer and $x \in U\}$

64. $\{x \mid x$ is a real number and $x \in U\}$

65. $\{x \mid x$ is a real number that is not rational and $x \in U\}$

1.7 ORDER RELATIONS

The symbols $<$ and $>$ are used to indicate that one number is less than (smaller than) another number and that one number is greater than (larger than) another number.

For example, $3 < 5$ means "3 is less than 5," and $5 > 3$ means "5 is greater than 3."

The expression $r < s$ means "the number r is less than the number s."

The expression $r > s$ means "the number r is greater than the number s."

The set of real numbers is an ordered set, and a number line provides a means to visualize this order. If points r and s correspond to two real numbers, r and s, then there are exactly three ways in which the placement of r and s can occur on a horizontal number line:

1. r is to the left of s
2. r is the same point as s
3. r is to the right of s

Symbolically, these three possibilities are expressed as follows:

1. $r < s$
2. $r = s$
3. $r > s$

It should be obvious that no two of these three situations can occur simultaneously. This property is called the **trichotomy axiom.**

THE TRICHOTOMY AXIOM

If r and s are any real numbers, then exactly one of the following is true: $r < s$ or $r = s$ or $r > s$.

Symbolic Statement	*Algebraic Meaning*	*Geometric Meaning (for horizontal number line)*
1. $r < s$	r is less than s	r is to the left of s
2. $r = s$	r is equal to s	r is the same point as s
3. $r > s$	r is greater than s	r is to the right of s

For the special case when $s = 0$, the trichotomy axiom states that the set of real numbers can be divided into three **mutually disjoint** subsets (no element in common) (Figure 1.7.1).

−4 −3 −2 −1 0 1 2 3 4

$r < 0$ $r = 0$ $r > 0$

FIGURE 1.7.1

1. $r > 0$ The set of positive real numbers
2. $r = 0$ The set consisting of the number 0 only
3. $r < 0$ The set of negative real numbers

Sometimes it is desirable to indicate that two numbers are related in one way or another. The combination of symbols $\leq$ is used to indicate that the number named on the left may be equal to or smaller than the one on the right. For example, $x + y \leq 25$ means that either the sum of x and y is equal to 25 or the sum is smaller than 25.

In general, $x \leq y$ means $x = y$ or $x < y$.

Similarly, $x \geq y$ means $x = y$ or $x > y$.

A slash drawn through an order symbol has the effect of the word "not." See Table 1.2.

TABLE 1.2 SUMMARY OF RELATION SYMBOLS

Symbol	*Verbal Translation*
$=$	Equals, is equal to
$\neq$	Does not equal
$<$	Is less than
$>$	Is greater than
$\nless$	Is not less than
$\ngtr$	Is not greater than
$\leq$	Is less than or is equal to
$\geq$	Is greater than or is equal to

EXAMPLE 1.7.1 State the algebraic meaning and the geometric meaning of each of the following statements:

(a) $2 > -5$ (b) $-3 < -1$

Solution

(a) 2 is greater than -5; 2 is to the right of -5 on a horizontal number line.
(b) -3 is less than -1; -3 is to the left of -1 on a horizontal number line.

EXAMPLE 1.7.2 Insert either the symbol $<$ or the symbol $>$ between each pair of numbers so that the resulting statement is true:

a. $-7, 3$ b. $-7, -8$

Solution a. $-7 < 3$ b. $-7 > -8$

EXAMPLE 1.7.3 Translate into an algebraic statement: 5 times the sum of x and 4 is less than or equal to 6.

Solution $5(x + 4) \leq 6$

EXAMPLE 1.7.4 Translate into a verbal statement:

$$\frac{x - 2}{5} \geq 9$$

Solution The quotient obtained by dividing the difference of 2 subtracted from x by 5 is greater than or equal to 9.

HISTORICAL NOTE

The symbol =, which indicates equality, was introduced in 1557 by Robert Recorde in his work *The Whetstone of Witte* because, as he put it, "Noe 2 thynges can be moare equalle."

Descartes preferred $\propto$ or ∞, probably a modification of "ae," the first two letters of the Latin word *aequalis*, meaning "equal."

Thomas Harriot was the first to use the symbols < and >. These symbols appeared in his work *Artis Analyticae Praxis*, published ten years after his death in 1631. These symbols were not immediately accepted. Other writers preferred and , the symbols suggested by Oughtred, also in 1631.

In the preceding section, it was shown that the number line can be used to graph points such as $x = 2$, $x = -\frac{1}{2}$, or $x = \sqrt{2}$.

Statements such as $x < 5$ or $x > 2$ can also be illustrated on a number line, but more than one point is involved. Consider the first statement: $x < 5$. Many numbers make this statement true: $3 < 5$, $0 < 5$, $-15 < 5$, $\frac{1}{2} < 5$; as a matter of fact, *all* numbers whose representation is to the left of 5 on a horizontal number line are less than 5. In the same manner it can be shown that *all* numbers to the right of 2 on a horizontal number line make $x > 2$ a true statement for x.

The following examples show how a set of real numbers described by an order relation can be graphed on a number line.

EXAMPLE 1.7.5 Graph $\{x \mid x > 3\}$ on a number line.

Solution The set $\{x \mid x > 3\}$ is graphed as shown in Figure 1.7.2. The circle above the numeral 3 indicates that 3 is excluded from the solution set; the solution set is indicated by the half-line starting at 3 (but not including 3) and including all values greater than 3, as shown by the direction of the line.

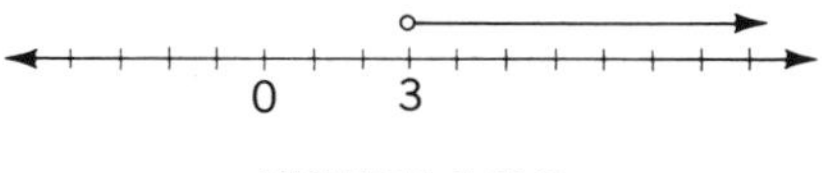

FIGURE 1.7.2

EXAMPLE 1.7.6 List the integers in the set $\{x \mid x > 3\}$.

Solution $\{4, 5, 6, 7, 8, \ldots\}$

EXAMPLE 1.7.7 Graph $\{x \mid x \leq -2\}$ on a number line.

Solution The set $\{x \mid x \leq -2\}$ is graphed as shown in Figure 1.7.3. This time a solid dot over the -2 coordinate indicates that -2 is included in the solution set as well as all points to the left of -2, since x is less than or equal to -2.

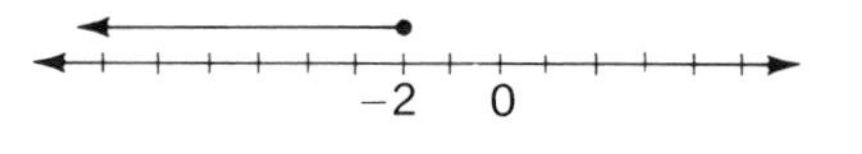

FIGURE 1.7.3

EXAMPLE 1.7.8 List the integers in the set $\{x \mid x \leq -2\}$.

Solution $\{\ldots, -5, -4, -3, -2\}$

Often it is desirable to indicate that x can be any number *between* two given numbers. For example, "x is any number between 1 and 4" is expressed symbolically as

$$1 < x < 4$$

where 1 and 4 are called the *end points* of the interval containing x. If either end point or both end points are to be included, the $\leq$ symbol is used to indicate this fact. For example,

$$1 \leq x < 4$$

means "x is between 1 and 4 and x may equal 1."

EXAMPLE 1.7.9 Graph $\{x \mid 1 \leq x < 4\}$ on a number line.

Solution The set $\{x \mid 1 \leq x < 4\}$ is shown in Figure 1.7.4. A solid dot over the 1 indicates that 1 is included, whereas the circle over the 4 indicates that 4 is excluded.

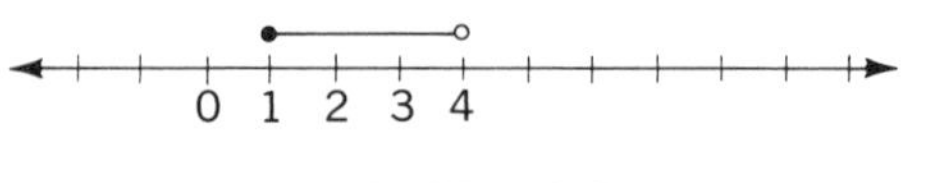

FIGURE 1.7.4

EXAMPLE 1.7.10 List the integers in the set $\{x \mid 1 \leq x < 4\}$.

Solution $\{1, 2, 3\}$

EXERCISES 1.7 A

For each of the symbolic statements in Exercises 1–10, state the algebraic meaning and the geometric meaning with respect to a horizontal number line.

1. $3 < 7$ **2.** $8 > 5$ **3.** $-7 < 0$

4. $4 > -7$ **5.** $7 > 0$ **6.** $-5 < 1$

7. $-6 < -2$ **8.** $-9 > -15$ **9.** $-1 > -4$

10. $-8 < -6$

State the algebraic meaning in Exercises 11–15.

11. $|-8| > |-6|$ **12.** $|-5| > 0$ **13.** $|-4| < |-7|$

14. $0 < |-9|$ **15.** $4 < 7 < 10$

In Exercises 16–30, insert the symbol $<$*, the symbol* $>$*, or the symbol* $=$ *between each pair of numbers so that the resulting statement is true.*

16. $12, 5$ **17.** $4, 9$ **18.** $-4, 2$

19. $5, -6$ **20.** $0, -3$ **21.** $0, 8$

22. $-5, -7$ **23.** $-9, -7$ **24.** $-1, -2$

25. $1, 2$ **26.** $|-3|, |-6|$ **27.** $|-2|, |+2|$

28. $|-2|, |+1|$ **29.** $|-7|, |-4|$ **30.** $|-5|, 0$

In Exercises 31–40, translate the verbal statements into algebraic (symbolic) statements.

31. The sum of x and 8 is less than 20.

32. The product of 5 and y is greater than 6.

33. One half of x is less than or equal to 7.

34. The result obtained by subtracting 4 from x is greater than or equal to 12.

35. 3 times x is positive.

36. The sum of x and y is negative.

37. The sum of x and 2 is between 1 and 8.

38. Twice the sum of n and 3 is greater than or equal to 25.

39. The difference obtained by subtracting 9 from the square of x is positive.

40. The absolute value of a real number x is positive or zero.

In Exercises 41–50, translate the symbolic statements into verbal statements.

41. $x < 7$ **42.** $x \geq 4$ **43.** $x \leq -3$
44. $x > -5$ **45.** $x + y > 10$ **46.** $x + y \leq 15$
47. $-3 < x < 4$ **48.** $1 \leq x < 6$ **49.** $0 < 2x \leq 5$
50. $|x + 2| \leq 7$

For Exercises 51–65:

a. Graph each of the following sets of real numbers on a horizontal number line.
b. List the integers in each given set.

51. $\{x \mid x < 4\}$ **52.** $\{x \mid x \geq 4\}$
53. $\{x \mid x \leq -5\}$ **54.** $\{x \mid x > -5\}$
55. $\{x \mid x > -2\}$ **56.** $\{x \mid x \leq 6\}$
57. $\{x \mid x < -3\}$ **58.** $\{x \mid x \geq -1\}$
59. $\{x \mid x \leq 0\}$ **60.** $\{x \mid x \geq 0\}$
61. $\{x \mid 3 < x < 5\}$ **62.** $\{x \mid -1 < x \leq 3\}$
63. $\{x \mid -4 \leq x < -1\}$ **64.** $\{x \mid 0 \leq x \leq 2\}$
65. $\{x \mid -1 < x \leq 1\}$

EXERCISES 1.7 B

For each of the symbolic statements in Exercises 1–10, state the algebraic meaning and the geometric meaning with respect to a vertical number line.

1. $5 < 9$ **2.** $3 > 1$ **3.** $-4 < 0$
4. $5 > -1$ **5.** $6 > 0$ **6.** $-1 < 1$
7. $-8 < -6$ **8.** $-3 > -4$ **9.** $0 > -2$
10. $-3 < -2$

State the algebraic meaning in Exercises 11–15.

11. $|-3| > |-2|$ **12.** $|-5| > 0$ **13.** $|-6| < |-8|$
14. $0 < |-2|$ **15.** $-3 < 0 < 3$

In Exercises 16–30, insert the symbol $<$, the symbol $>$, or the symbol $=$ between each pair of numbers so that the resulting statement is true.

16. 15, 2 **17.** 3, 10 **18.** $-2, -4$

19. $-2, 4$ **20.** $2, -4$ **21.** $|2|, |-4|$
22. $0, -3$ **23.** $-5, -7$ **24.** $-10, -1$
25. $|-6|, 0$ **26.** $-6, 0$ **27.** $|-3|, |+3|$
28. $|-4|, |-1|$ **29.** $7, 0$ **30.** $|-3|, |+1|$

In Exercises 31–40, translate the verbal statements into algebraic (symbolic) statements.

31. The sum of x and 7 is greater than 4.
32. The product of 2 and y is less than negative 6.
33. x is positive.
34. The product of 3 and n is negative.
35. The sum of 3 times x and 4 times y is less than or equal to 13.
36. When 9 is subtracted from n, the difference is greater than 14.
37. Twice y is between 0 and 7 or equals 0.
38. The absolute value of negative 4 is between 0 and 6.
39. The absolute value of the sum of x and 4 is greater than or equal to 43.
40. The absolute value of a real number x cannot be negative.

In Exercises 41–50, translate the symbolic statements into verbal statements.

41. $x < 5$ **42.** $x \geq 9$ **43.** $x \leq -2$
44. $x > -\frac{1}{2}$ **45.** $x + y < 7$ **46.** $0 < 2x < 3$
47. $-1 \leq x + y < 1$ **48.** $0 < \frac{1}{2}x \leq 1$ **49.** $|x| \geq 0$
50. $0 < |x + 1| < 3$

For Exercises 51–65:

a. Graph each of the following sets of real numbers on a horizontal number line.
b. List the integers in each given set.

51. $\{x \mid x < 2\}$ **52.** $\{x \mid x \geq 2\}$
53. $\{x \mid x \leq -1\}$ **54.** $\{x \mid x > -3\}$
55. $\{x \mid x \geq -1\}$ **56.** $\{x \mid x > 0\}$
57. $\{x \mid x < 0\}$ **58.** $\{x \mid x = 0\}$
59. $\{x \mid 0 < x < 2\}$ **60.** $\{x \mid x \ngtr 2\}$
61. $\{x \mid -1 \leq x < 4\}$ **62.** $\{x \mid x \nless -1\}$
63. $\{x \mid -3 \leq x \leq 0\}$ **64.** $\{x \mid 2 < x \leq 5\}$
65. $\{x \mid -5 < x < -3\}$

1.8 SUMS OF SIGNED NUMBERS

The meaning of the sum of two real numbers is motivated by the interpretation of a negative number as measuring a quantity opposite in nature to that measured by the corresponding positive number.

Since a $5 gain followed by a $3 gain results in a net gain of $8, it is desirable to assign the meaning

$$(+5) + (+3) = +8$$

Since a $5 loss followed by a $3 loss results in a net loss of $8, it is desirable to assign the meaning

$$(-5) + (-3) = -8$$

Noting also that a $3 gain followed by a $5 gain results in a net gain of $8 and that a $3 loss followed by a $5 loss results in a net loss of $8, it follows that

$$(+5) + (+3) = (+3) + (+5) = +8$$

and

$$(-5) + (-3) = (-3) + (-5) = -8$$

In other words, the order in which these numbers are added does not change the sum.

In general, the sum of two real numbers having the same sign is obtained by adding the absolute values of the numbers and prefixing their common sign.

In symbols, where r and s are positive real numbers,

$$(+r) + (+s) = +(|r| + |s|) = +(r + s)$$

$$(-r) + (-s) = -(|-r| + |-s|) = -(r + s)$$

Since a $5 gain followed by a $3 loss results in a net gain of $2,

$$(+5) + (-3) = +2$$

Similarly, a $3 loss followed by a $5 gain results in a net gain of $2,

$$(-3) + (+5) = +2$$

In other words,

$$(+5) + (-3) = (-3) + (+5) = +(5 - 3) = 2$$

On the other hand, since a $5 loss followed by a $3 gain results in a net loss of $2,

$$(-5) + (+3) = -2$$

Also, since a $3 gain followed by a $5 loss results in a net loss of $2,

$$(+3) + (-5) = -2$$

In other words,

$$(-5) + (+3) = (+3) + (-5) = -(5 - 3) = -2$$

In general, the sum of two real numbers having opposite signs and unequal absolute values is obtained by subtracting their absolute values, the smaller from the larger, and prefixing the sign of the number having the larger absolute value.

In symbols, where r and s are positive real numbers,

$$(+r) + (-s) = +(r - s) \text{ if } r > s$$

and

$$(+r) + (-s) = -(s - r) \text{ if } s > r$$

If $r = s$, where r and s are positive, then

$$(+r) + (-s) = (+r) + (-r) = 0$$

and

$$(-r) + (+s) = (-r) + (+r) = 0$$

Since the addition of 0 can be interpreted as no change in measurement, for any real number x, x + 0 = x and 0 + x = x.

Formalizing the above considerations, some properties of the set of real numbers with respect to addition can now be stated.

THE ADDITION AXIOMS FOR REAL NUMBERS

Let r, s, and t be any real numbers.

1. **Closure axiom** $r + s$ is a real number
2. **Commutative axiom** $r + s = s + r$
   (The order in which the numbers are added does not change the sum.)
3. **Associative axiom** $(r + s) + t = r + (s + t)$
   (The way in which the numbers are grouped does not change the sum.)
4. **Identity axiom** There is a special number, 0, called the additive identity, so that $r + 0 = r$ and $0 + r = r$
5. **Inverse axiom** For each real number r there exists exactly one number, $-r$, called the opposite of r (or the additive inverse of r), so that $r + (-r) = 0$ and $(-r) + r = 0$

By using these addition axioms, the rules for the addition of real numbers can be derived as the following theorem.

THE ADDITION OF REAL NUMBERS THEOREM

If r and s are any positive real numbers, then

1. $(+r) + (+s) = +(r + s)$
2. $(-r) + (-s) = -(r + s)$
3. $(+r) + (-s) = +(r - s)$ if $r > s$
4. $(+r) + (-s) = -(s - r)$ if $s > r$

EXAMPLE 1.8.1 Perform the indicated additions:

a. $(+8) + (+2)$ b. $(-8) + (-2)$ c. $(+8) + (-2)$
d. $(-8) + (+2)$

Solution

a. $(+8) + (+2) = 8 + 2 = 10$
b. $(-8) + (-2) = -(8 + 2) = -10$
c. $(+8) + (-2) = +(8 - 2) = 6$
d. $(-8) + (+2) = -(8 - 2) = -6$

EXAMPLE 1.8.2 Find the indicated sum: $(+5) + [(-8) + (+3)]$.

Solution

$$\begin{aligned}(+5) + [(-8) + (+3)] &= (+5) + [-(8 - 3)] \\ &= (+5) + (-5) \\ &= 0\end{aligned}$$

EXAMPLE 1.8.3 Find the indicated sum: $[(-6) + (+6)] + (-7)$.

Solution

$$\begin{aligned}[(-6) + (+6)] + (-7) &= (0) + (-7) \\ &= -7\end{aligned}$$

EXAMPLE 1.8.4 Find the indicated sum: $(+20) + (-30) + (-40)$.

Solution

$$\begin{aligned}(+20) + (-30) + (-40) &= [(+20) + (-30)] + (-40) \\ &= [-(30 - 20)] + (-40) \\ &= (-10) + (-40) \\ &= -(10 + 40) = -50\end{aligned}$$

Since the associative axiom for the addition of real numbers states that the way in which the numbers are grouped does not affect the sum, the sum may be obtained by grouping one way, and then this result may be checked by grouping the other way.

Check (for Example 1.8.4):

$$\begin{aligned}(+20) + (-30) + (-40) &= (+20) + [(-30) + (-40)] \\ &= (+20) + [-(30 + 40)] \\ &= (+20) + (-70) \\ &= -(70 - 20) \\ &= -50\end{aligned}$$

EXERCISES 1.8 A

In Exercises 1–5, determine the daily result for the bank transactions.

1. Monday: $400 deposit, $200 deposit
2. Tuesday: $400 deposit, $50 withdrawal
3. Wednesday: $100 deposit, $275 withdrawal
4. Thursday: $50 withdrawal, $75 withdrawal
5. Friday: $25 deposit, $80 withdrawal

Five men went on a diet. Their original weight was recorded, and their weekly changes in weight for two weeks were recorded, a positive number indicating a gain and a negative number indicating a loss. For Exercises 6–10, find the weight at the end of the two-week period for each of the five men.

6. Al. 210 pounds, $+5$, $+2$
7. Bill. 190 pounds, -6, -4
8. Carl. 235 pounds, -10, $+2$
9. Doug. 185 pounds, $+2$, -8
10. Ed. 220 pounds, -15, $+3$

In Exercises 11–40, express each as a single integer.

11. $(+7) + (+3)$
12. $(-7) + (-3)$
13. $(+7) + (-3)$
14. $(-7) + (+3)$
15. $(-5) + 0$
16. $0 + (-8)$
17. $(-12) + (-18)$
18. $(-12) + (+18)$
19. $(+14) + (-9)$
20. $(-25) + (+15)$

21. $(-3) + [(-5) + (-2)]$

22. $[(-4) + (-5)] + (-6)$

23. $[(+10) + (-20)] + (-30)$

24. $(+15) + [(-25) + (-40)]$

25. $[(-8) + (+8)] + (-7)$

26. $(+9) + [(-12) + (+3)]$

27. $-(|-7| + |+5|)$

28. $(|-6| + |-8|)$

29. $-(|+12| + |-3|)$

30. $|-9| + |(+7) + (-10)|$

31. $(+5) + (-10) + (-15)$

32. $(-7) + (+8) + (-9)$

33. $(+12) + (+4) + (-20)$

34. $(-60) + (-70) + (+90)$

35. $(-11) + (-9) + (-7)$

36. $(+18) + (-15) + (+17)$

37. $(-1) + (+1) + (-1)$

38. $5 + [(-3) + (-2)]$

39. $[5 + (-3)] + (-2)$

40. $|-3| + |-2| + |-5|$

41. A certain bank statement contains the following entries—a positive number means a deposit and a negative number means a withdrawal:

$$+220, \ +70, \ -95, \ -22, \ +15, \ -84$$

If the account originally contained $300, how much was in the account after the above deposits and withdrawals were made?

42. In the card game of pinochle, it is possible to gain points and to lose points. Find the final score of a player whose scores on successive hands are listed as follows:

$$+10, \ -25, \ +32, \ +8, \ -34$$

43. The weekly changes in weight for six weeks of a person on a diet are recorded below:

$$+2, \ -4, \ -2, \ +3, \ -5, \ -1$$

If the person originally weighed 156 pounds, find his weight at the end of the six-week period.

★ 44. At the end of a week, the closing price of a certain stock was quoted in the newspaper as $57\frac{1}{4}$ (meaning $57\frac{1}{4}$ dollars, or $57.25). The changes in price for the next week were quoted as listed below, in dollars and fractions of dollars:

$$\text{Monday } -1\tfrac{1}{4}, \text{ Tuesday } +\tfrac{3}{4}, \text{ Wednesday } +\tfrac{3}{8}, \text{ Thursday } -1\tfrac{1}{8}, \text{ Friday } -\tfrac{3}{4}$$

Find the closing price of the stock at the end of this week.

★ 45. In a certain mechanical system, a counterclockwise torque is measured by a positive number and a clockwise torque by a negative number. The following torques were recorded:

$$+20 \text{ lb-ft}, \ +35 \text{ lb-ft}, \ -25 \text{ lb-ft}, \ -40 \text{ lb-ft}$$

If the sum of the torques is 0, the system is said to be balanced (in equilibrium). What additional torque is needed to balance the system? Should it be clockwise or counterclockwise?

EXERCISES 1.8 B

In Exercises 1–5, determine the net yardage for the two football plays described in each exercise.

1. Gain of 5 yards, loss of 3 yards
2. Loss of 7 yards, loss of 15 yards
3. Gain of 12 yards, gain of 9 yards
4. Loss of 10 yards, gain of 4 yards
5. Gain of 2 yards, loss of 10 yards

In chemistry, the valence of an atom (or a set of atoms) is an integer that determines how the atom (or set) will unite chemically as compared to the hydrogen atom, assigned a valence of $+1$. *Find the valence (the sum of the component valences) for the ions in Exercises 6–10, whose component valences are given.*

6. Sulfate ion, SO_4, where the valence of $S = +6$, of $O_4 = -8$
7. Hydroxyl ion, OH, where the valence of $O = -2$, of $H = +1$
8. Ammonium ion, NH_4, where the valence of $N = -3$, of $H_4 = +4$
9. Hydronium ion, H_3O, where the valence of $H_3 = +3$, of $O = -2$
10. Nitrate ion, NO_3, where the valence of $N = +5$, of $O_3 = -6$

In Exercises 11–40, express each as a single integer.

11. $(+8) + (+7)$
12. $(-8) + (-7)$
13. $(-8) + (+7)$
14. $(+8) + (-7)$
15. $(-4) + (+4)$
16. $0 + (-5)$
17. $(-50) + (+35)$
18. $(+35) + (-50)$
19. $(-75) + (-25)$
20. $(-25) + (+75)$
21. $(-6) + (-9) + (-5)$
22. $(-12) + (+4) + (-2)$
23. $(+50) + (-70) + (+90)$
24. $(+25) + (-35) + (-45)$
25. $(-9) + [(-6) + (+6)]$
26. $[(+7) + (-7)] + (+8)$
27. $-(|-9| + |-8|)$
28. $-(|-15| - |+6|)$
29. $[(+20) + (-14)] + (-10)$
30. $(+20) + [(-14) + (-10)]$

31. $[(-3) + (-2)] + (-5)$

32. $(-3) + [(-5) + (-2)]$

33. $(-2) + (-15) + (7) + (9)$

34. $[(-2) + (9)] + [(-15) + (7)]$

35. $0 + (-8) + (+8)$

36. $(-1.52) + (-3.25)$

37. $(4.213) + (-1.561)$

38. $(0.123) + (-5.146)$

39. $(-14) + (-14)$

40. $(-14) + (+14)$

41. Determine the net yardage for the following sequence of football plays, where a positive number (+) means a gain in yardage and a negative number (−) means a loss in yardage:

$$+5,\ -3,\ -6,\ +15,\ +2,\ -10,\ +4$$

42. Find the resulting speed of an object if its original speed was 40 mph and it was subject to the following successive accelerations (+) and decelerations (−):

$$+15,\ -30,\ -10,\ +15,\ +20$$

43. The following is a treasurer's report for a certain organization (a positive number indicates dollars received, and a negative number indicates dollars paid out):

Dues, +150; Stationery, −8; Stamps, −15; Food sale, +60;
Gifts, −80; Party expenses, −75

Find the amount of money in the treasury.

44. A piston is moving upward and downward from a central position C. A positive number indicates the number of centimeters above C, and a negative number indicates the number of centimeters below C. Find the position of the piston with respect to the central position C after the following motions have taken place (the piston started at C):

$$-5,\ -3,\ +2,\ +6,\ -4,\ -1$$

45. A chemical compound is such that the sum of the valences of its components is 0. Find the valence of the component whose valence is not given.

a. Baking soda, $NaHCO_3$; $Na = +1$, $H = +1$, $O_3 = -6$

b. Ammonium sulfate (fertilizer), NH_4HSO_4; $H_4 = +4$, $H = +1$, $S = +6$, $O_4 = -8$

c. Jade (jadeite), $NaAlSi_2O_6$; $Na = +1$, $Si_2 = +8$, $O_6 = -12$

1.9 DIFFERENCES OF SIGNED NUMBERS

In arithmetic the difference $12 - 3$ was defined as 9 because $12 = 9 + 3$. The difference between two real numbers is defined in such a way that this inverse relationship is preserved. The definition of subtraction involves the concept of the additive inverse. The inverse axiom for addition states that for any real number r there is exactly one real number, $-r$, called the additive inverse of r (or opposite of r), so that $r + (-r) = 0$.

For example, if $r = 5$, then $5 + (-r) = 0$ implies that $-r = -5$ because $5 + (-5) = 0$ and because each real number has only one additive inverse.

If $r = 0$, then $0 + (-r) = 0$ implies that $-r = 0$ since $0 + 0 = 0$—that is, 0 is the additive inverse of itself.

If $r = -5$, then $(-5) + (-r) = 0$ implies that $-r = +5$ since $(-5) + (+5) = 0$. Since $r = -5$, $-r = -(-5)$. This means

$$-(-5) = (+5)$$

In general, $-(-r) = r$—that is, the additive inverse of the additive inverse of a real number is the original real number. This statement can be proved and is thus classified as a theorem.

THEOREM

For any real number r, $-(-r) = r$.

DEFINITION OF SUBTRACTION

If r and s are any real numbers, then $r - s = r + (-s)$.

The definition of subtraction states that the difference between two real numbers is obtained by the addition of the additive inverse of the subtrahend to the minuend.

Sometimes this definition is expressed by saying, "Change the sign of the subtrahend and add."

EXAMPLE 1.9.1 Perform the indicated subtractions:

a. $(+4) - (+9)$ b. $(+4) - (-9)$ c. $(-4) - (+9)$
d. $(-4) - (-9)$

Solution

a. $(+4) - (+9) = (+4) + (-9) = -(9 - 4) = -5$
b. $(+4) - (-9) = (+4) + (+9) = 13$
c. $(-4) - (+9) = (-4) + (-9) = -13$
d. $(-4) - (-9) = (-4) + (+9) = 9 - 4 = 5$

In arithmetic, it is stated that $r - s = t$ if and only if $r = t + s$. For example, $10 - 3 = 7$ because $10 = 7 + 3$.

The next theorem indicates that the definition of subtraction preserves the arithmetical meaning of $r - s$ as the number that must be added to s to obtain r.

THEOREM

If r and s are any real numbers, then $(r - s) + s = r$.

Proof:

$(r - s) + s = [r + (-s)] + s$	(Definition of subtraction)
$= r + [(-s) + s]$	(Associative axiom, addition)
$= r + 0$	(Inverse axiom, addition)
$= r$	(Identity axiom, addition)

The preceding theorem is useful for checking a subtraction problem.

Since $+r$ and r are names of the same number, r is usually written instead of $+r$ in order to use fewer symbols.

For example, $5 - 8$ has the same meaning as $(+5) - (+8)$—that is, $5 - 8 = (+5) - (+8)$.

In general, $r - s = (+r) - (+s) = (+r) + (-s)$.

EXAMPLE 1.9.2 Subtract and check:

a. $5 - 8$ b. $5 - (-8)$ c. $(-5) - (+8)$
d. $(-5) - (-8)$ e. $0 - 5$

Solution	*Check*
a. $5 - 8 = 5 + (-8) = -3$	$(-3) + 8 = 5$
b. $5 - (-8) = 5 + [-(-8)]$ $= 5 + 8$ $= 13$	$13 + (-8) = 5$
c. $(-5) - (+8) = (-5) + (-8)$ $= -13$	$(-13) + (+8) = -5$
d. $(-5) - (-8) = (-5) + [-(-8)]$ $= (-5) + (+8)$ $= +3$	$(+3) + (-8) = -5$
e. $0 - 5 = 0 - (+5)$ $= 0 + (-5)$ $= -5$	$(-5) + (+5) = 0$

EXAMPLE 1.9.3 Write as a single integer: $-(2-3)$.

Solution

$$\begin{aligned} -(2-3) &= -(2+[-3]) \\ &= -(-1) \\ &= 1 \end{aligned}$$

EXERCISES 1.9 A

In Exercises 1–20, subtract and check.

1. $(+7)-(+3)$
2. $(+3)-(+7)$
3. $(+7)-(-3)$
4. $(-3)-(+7)$
5. $(-6)-(+4)$
6. $(-6)-(-4)$
7. $(+6)-(+6)$
8. $(-4)-(-4)$
9. $(-7)-(+7)$
10. $(+8)-(-8)$
11. $(-20)-(+8)$
12. $(-15)-(-10)$
13. $(+22)-(+12)$
14. $(+17)-(+19)$
15. $0-(+2)$
16. $0-(-5)$
17. $(-8)-(-15)$
18. $(+8)-(+15)$
19. $(-14)-(+9)$
20. $(-12)-(+7)$

Perform the indicated operations in Exercises 21–30.

21. $(-5+5)-9$
22. $(-1)-[(-2)-(-3)]$
23. $2+(8-20)$
24. $0-(6-12)$
25. $27+[59+(-27)]$
26. $(-38)+(38-26)$
27. $(-65)+[46-(-65)]$
28. $(-6)+(7-10)$
29. $50-(20-20)$
30. $[-(-8)-(+2)]-(+3)$

On a certain day, a newspaper recorded the high and low temperatures for different cities, as shown in Exercises 31–35. For each day, find the difference between the high temperature and the low temperature.

	City	*High*	*Low*
31.	San Francisco	59	48
32.	New York	25	0
33.	Chicago	10	−6

34. Fairbanks	-15	-40
35. Denver	-4	-12

On a certain day, a newspaper listed the closing prices in dollars for the day and the net changes in dollars for the day of different stocks, as shown in Exercises 36–40. For each stock, find in dollars and cents the closing price for the previous day.

Stock	*Close*	*Net*
36. Gulf Oil	$22\frac{3}{4}$	$+\frac{1}{8}$
37. Coleco Ind	$29\frac{5}{8}$	$-1\frac{1}{8}$
38. Lubrizol	$42\frac{1}{4}$	$-\frac{3}{4}$
39. Extend Car	20	$+1\frac{1}{4}$
40. Am Tel and Tel	$41\frac{5}{8}$	$-\frac{1}{8}$

Determine the difference in latitudes between each pair of cities in Exercises 41–45 (a positive number means north of the equator, and a negative number means south of the equator).

41. Sacramento, California	$+39$ degrees
Lima, Peru	-12 degrees
42. Los Angeles, California	$+34$ degrees
Hanoi, North Vietnam	$+21$ degrees
43. Johannesburg, S. Africa	-26 degrees
Perth, Australia	-32 degrees
44. Moscow, U.S.S.R.	$+56$ degrees
Seattle, Washington	$+48$ degrees
45. Rio de Janeiro, Brazil	-23 degrees
New York, N.Y.	$+43$ degrees

Determine whether the statements in Exercises 46–55 are true or false by computing each side and comparing.

46. $20 - 8 = 8 - 20$

47. $6 - 12 = -12 + 6$

48. $-(6 - 12) = 12 - 6$
49. $20 - (8 - 2) = (20 - 8) - 2$
50. $2 + (8 - 20) = (2 + 8) - 20$
51. $0 - (-9) = 9 + 0$
52. $-17 - 5 = -5 + (-17)$
53. $4 - (6 - 8) = (4 - 6) - 8$
54. $58 + (76 - 58) = 76 + (58 - 58)$
55. $(-7 + 15) + 4 + (-5) = (15 + 4) - (7 + 5)$

In Exercises 56–65, simplify.

56. $n + 0$
57. $n - 0$
58. $0 - n$
59. $n - n$
60. $-n - (-n)$
61. $-(-n)$
62. $0 - (-n)$
***63.** $n - (n - m)$
64. $(m - n) + (n - m)$
***65.** $0 - (m - n)$
***66.** Does $s - r = -(r - s)$ for all real numbers r and s? Why?

EXERCISES 1.9 B

In Exercises 1–20, subtract and check.

1. $(+8) - (+2)$
2. $(-9) - (+4)$
3. $(+10) - (-4)$
4. $(-15) - (-19)$
5. $(-17) - (-23)$
6. $0 - (-7)$
7. $2 - 9$
8. $-3 - 8$
9. $0 - 6$
10. $-2 - (-7)$
11. $10 - 9$
12. $6 - 10$
13. $(+50) - (+15)$
14. $(+40) - (-25)$
15. $(-3) - (-3)$
16. $5 - (-7)$
17. $7 - (-7)$
18. $(-7) - (-7)$
19. $0 - (-3)$
20. $0 - 3$

Perform the indicated operations in Exercises 21–30.

21. $20 - (10 - 5)$

22. $(20 - 10) - 5$

23. $-12 - (14 - 17)$

24. $-8 - (-4 + 4)$

25. $35 - [(-20) - (-20)]$

26. $[-65 - (-65)] - 40$

27. $-3 - (4 + [-2])$

28. $-3 - (4 - [-2])$

29. $(-3 - 4) + (-2)$

30. $(-3 - 4) - (-2)$

Find the difference in elevation between each pair of geographic locations whose elevations are given in Exercises 31–35. A positive number indicates an elevation above sea level, and a negative number indicates an elevation below sea level.

31. Mount Kilimanjaro (highest point in Africa) — +19,321 feet
Qattara Depression, Egypt (lowest point in Africa) — − 436 feet

32. Mount Whitney — +14,495 feet
Death Valley — − 282 feet
(highest and lowest points in California)

33. Qattara Depression — − 436 feet
Death Valley — − 282 feet

34. Brawley, California — − 119 feet
El Centro, California — − 45 feet

35. Dead Sea (surface level) — − 1,292 feet
Mount Everest (highest point in the world) — +29,028 feet

For Exercises 36–40, determine the number of years that elapsed between each pair of dates. A positive number means time A.D. *and a negative number means time* B.C. (*Note: There was no year* 0 *since the number* 0 *had not been invented when the calendar was reformed. In other words, the year* 1 A.D. *followed the year* 1 B.C. *For Exercises 38–40, calculate the difference in the usual way and then subtract* 1 *from the result to obtain the answer.*)

36. −2900, the building of the Great Gizeh Pyramid, to
−1650, date of the Rhind papyrus (mathematical work)

37. −550, during lifetime of Pythagoras, to
−300, during lifetime of Euclid

38. -212, fall of Syracuse to Rome, to $+400$, fall of Rome (western part)

39. -425, Golden Age of Athens, to $+500$, Golden Age of India

40. -2400, first historical records, to $+1980$

In Exercises 41–45, determine the difference in longitudes between each given pair of cities. A positive number means east longitude (east of Greenwich, England) and a negative number means west longitude.

41. San Francisco, California	-122 degrees
Tokyo, Japan	$+140$ degrees
42. Denver, Colorado	-105 degrees
London, England	0 degrees
43. Mexico City, Mexico	-99 degrees
Nairobi, Kenya	$+37$ degrees
44. Hong Kong, Asia	$+114$ degrees
Manila, Philippines	$+121$ degrees
45. San Francisco, California	-122 degrees
Boston, Massachusetts	-71 degrees

Determine whether the statements in Exercises 46–55 are true or false by computing each side and comparing.

46. $0 - 4 = -4 + 0$

47. $-8 - (-5) = -8 + 5$

48. $8 + (-5) = -5 + 8$

49. $9 - (3 - 8) = (9 - 3) - 8$

50. $9 - (3 - 8) = (9 - 3) + 8$

51. $9 - (3 - 8) = 9 + (8 - 3)$

52. $8 - 3 = -(3 - 8)$

53. $(82 - 27) - 82 = -27 + (82 - 82)$

54. $13 - (7 + 11) = (13 - 7) + 11$

55. $(3 - 7) + (-4 + 6) = (3 + 6) - (7 + 4)$

In Exercises 56–65, simplify.

56. $-(-t) + (-t)$

57. $0 - (-t)$

58. $-t - (-t)$

59. $-(s - t) - t$

60. $(s - t) + (t - s)$

61. $s - (s - t)$

62. $(s - t) - s$

63. $0 - (s - t)$

64. $(a - b) - (a - c)$

65. $r - (r + s)$

1.10 PRODUCTS OF SIGNED NUMBERS

Although there are many practical applications involving products of signed numbers, it is not easy to provide a simple geometric illustration of such a product.

The rules for obtaining products of signed numbers can be derived from the basic properties of the set of real numbers.

THE MULTIPLICATION AXIOMS FOR REAL NUMBERS

Let r, s, and t be any real numbers.

1. **Closure axiom** rs is a real number
2. **Commutative axiom** $rs = sr$
   (The order in which the numbers are multiplied does not change the product.)
3. **Associative axiom** $(rs)t = r(st)$
   (The way in which the numbers are grouped does not change the product.)
4. **Identity axiom** There is a special number, 1, called the multiplicative identity, so that $r \cdot 1 = 1 \cdot r = r$
5. **Inverse axiom** For each real number r different from 0, there is exactly one real number, $\frac{1}{r}$, called the reciprocal of r (or the multiplicative inverse of r), so that

$$r\left(\frac{1}{r}\right) = 1 \quad \text{and} \quad \left(\frac{1}{r}\right)r = 1$$

THE ZERO-FACTOR THEOREM

If r is any real number, then $r \cdot 0 = 0$ and $0 \cdot r = 0$.

THE MULTIPLICATION OF REAL NUMBERS THEOREM

If r and s are any real numbers, then

1. $(+r)(+s) = +rs = rs$
2. $(-r)(-s) = +rs = rs$
3. $(+r)(-s) = -rs$
4. $(-r)(+s) = -rs$

Informally, the following may be stated:

1. **The product of two nonzero real numbers having the same sign is positive.**
2. **The product of two nonzero real numbers having different signs is negative.**

EXAMPLE 1.10.1 Find the indicated products:

a. $-5 \cdot 0$ b. $(+5)(+3)$ c. $(-5)(+3)$
d. $(+5)(-3)$ e. $(-5)(-3)$

Solution

a. $-5 \cdot 0 = 0$
b. $(+5)(+3) = +15$
c. $(-5)(+3) = -15$
d. $(+5)(-3) = -15$
e. $(-5)(-3) = +15$

EXAMPLE 1.10.2 Find the product in two different ways:

$$(+4)(-6)(-2)$$

Solution

1. $$\begin{aligned}(+4)(-6)(-2) &= [(+4)(-6)](-2)\\ &= (-24)(-2)\\ &= +48\end{aligned}$$
2. $$\begin{aligned}(+4)(-6)(-2) &= (+4)[(-6)(-2)]\\ &= (+4)(+12)\\ &= +48\end{aligned}$$

EXAMPLE 1.10.3 Perform the indicated operation: $(-9)^3$.

Solution

$$\begin{aligned}(-9)^3 &= (-9)(-9)(-9)\\ &= [(-9)(-9)](-9)\\ &= (+81)(-9)\\ &= -729\end{aligned}$$

EXAMPLE 1.10.4 Perform the indicated operations:

a. -3^2 b. $(-3)^2$

Solution

a. -3^2 means $-(3)^2$, or the negative of 3^2, which is $-(3 \cdot 3) = -9$; therefore, $-3^2 = -9$.
b. $(-3)^2$ means $(-3)(-3) = +9$; therefore, $(-3)^2 = 9$.

EXERCISES 1.10 A

Perform the indicated operations in Exercises 1–30.

1. $(+5)(+7)$
2. $(+5)(-7)$
3. $(-5)(-7)$
4. $(-8)(+2)$
5. $(-8)(-2)$
6. $(+4)(-25)$
7. $(-8)(+125)$
8. $(-25)(-4)$
9. $(-125)(+8)$
10. $(+3)(-5)(-7)$
11. $(-2)(-4)(-6)$
12. $(-5)(+46)(-2)$
13. $(+9)(-2)(+3)$
14. $(-4)(-4)(-4)$
15. $(-5)^3$
16. $(-6)^2$
17. $(-2)^3(-2)^2$
18. $4(-3)^2$
19. $2(-7)^3$
20. $(-2)(5 - 8)$
21. $(+3)(4 - 9)$
22. $(7 - 8)(6 - 10)$
23. $(-4 + 8)(-7 + 7)$
24. $5[-3 - (-5)]$
25. $(3 - 9)(9 - 3)$
26. $[-2 - (-2)]\cdot(3 - 9)$
27. $(-3)(7 - 15)$
28. $(-3)(7) - (-3)(15)$
29. $8 - 4(3 - 9)$
30. $(8 - 4)(3 - 9)$

For Exercises 31–35, find the position of each airplane at the given time and velocity if each airplane is at the airport at noon. Use the formula

$$d = vt$$
$$\textit{(directed distance)} = \textit{(velocity)} \times \textit{(directed time)}$$

The directed distance of a plane is positive (+) if it is east of the airport and negative (−) if it is west of the airport, the velocity is positive (+) if the plane is moving toward the east and negative (−) if the plane is moving toward the west, and the directed time is counted from noon, with time p.m. indicated by a positive number and time a.m. indicated by a negative number. For example, 3 p.m. would be +3, 9 a.m. would be −3, and noon would be 0.

★ **31.** 2 p.m. flying 550 mph east
★ **32.** 10 a.m. flying 550 mph east
★ **33.** 9 a.m. flying 620 mph west
★ **34.** 3 p.m. flying 620 mph west
★ **35.** 10:30 a.m. flying 500 mph west

In a television set, the change c in the capacitance of a condenser is given by

$$c = kCT$$

where k *is the temperature coefficient*

C *is the capacitance of the condenser*

T *is the change in the temperature*

Find c for each set of values in Exercises 36–40.

★ **36.** $k = \frac{-22}{100{,}000}$, $C = 0.002$, $T = +20$ degrees centigrade

★ **37.** $k = \frac{-75}{100{,}000}$, $C = 0.004$, $T = -10$ degrees centigrade

★ **38.** $k = \frac{3}{100{,}000}$, $C = \frac{1}{10{,}000}$, $T = -4$ degrees centigrade

★ **39.** $k = \frac{-47}{100{,}000}$, $C = 0.0006$, $T = -2$ degrees centigrade

★ **40.** $k = \frac{-33}{100{,}000}$, $C = 0.0005$, $T = +6$ degrees centigrade

EXERCISES 1.10 B

Perform the indicated operations in Exercises 1–30.

1. $(+8)(+125)$

2. $(+11)(-8)$

3. $(-8)(+11)$

4. $(-17)(-13)$

5. $(-35)^2$

6. $(-20)^3$

7. $(+10)(-25)$

8. $(-25)(-100)$

9. $(0)(-19)$

10. $(-3)(6 - 6)$

11. $(-3)(6) + (-3)(-6)$

12. $(+5)(+2)(+7)$

13. $(-2)(-5)(+9)$

14. $(-5)(-8)(-2)$

15. $(+4)(-25)(+7)$

16. $(-4)(+9)(-25)$

17. $4(-5)^2$

18. $4(-5)^3$

19. $(10 - 4)(4 - 10)$

20. $(5 - 8)(6 - 6)$

21. $8(4 - 9)$

22. $8(4) - 8(9)$

23. $(-7)(10 - 15)$

24. $(-7)(10) - (-7)(15)$

25. $(-2)(-2)(-2)(-2)$

26. $(-2)(-5)(-2)(-5)(-2)(-5)$

27. $[(-2)(-5)]^3$

28. $(-7 + 3)^2 - (-7 - 3)^2$

29. $-(6 - 5)(6 - 4)(6 - 3)$

30. $(10 - 1)(10 - 2)(10 - 3)(10 - 4)$

In physics, the moment M of a force F is given by $M = Fd$, where d is the directed distance of the force from a fixed position. Find M in Exercises 31–35.

31. $F = +200$ pounds, $d = +15$ feet
32. $F = +200$ pounds, $d = -10$ feet
33. $F = -70$ pounds, $d = +8$ feet
34. $F = -120$ pounds, $d = -6$ feet
35. $F = -500$ grams, $d = -15$ centimeters

In chemistry, a subscript at the lower right of a symbol for an atom indicates the number of atoms. For example, sulfuric acid is expressed in symbols as H_2SO_4, *meaning* 2 *atoms of hydrogen* (H), 1 *atom of sulfur* (S), *and* 4 *atoms of oxygen* (O). *The valence of an ion, a positive or negative integer indicating how the ion combines chemically, is the sum of the valences of the atoms composing the ion. For example, the valence of the sulphate ion,* SO_4, *equals* (*valence of* S) + 4(*valence of* O) $= (+6) + 4(-2) = 6 - 8 = -2$. *Find the valence of the ions in Exercises 36–40, given the valence of the component atoms.*

36. Nitrate ion, NO_3; valence of N $= +5$, valence of O $= -2$
37. Phosphate ion, PO_4; valence of P $= +5$, valence of O $= -2$
38. Bichromate ion, Cr_2O_7; valence of Cr $= +5$, valence of O $= -2$
39. Ammonium ion, NH_4; valence of N $= -3$, valence of H $= +1$
40. Permanganate ion, MnO_4; valence of Mn $= +7$, valence of O $= -2$

1.11 QUOTIENTS OF SIGNED NUMBERS

Division may be defined for real numbers just as it was defined in arithmetic as the inverse operation to multiplication. For example, the quotient of 12 divided by 3 is defined to be 4, $\frac{12}{3} = 4$, because 4 is the only number such that $12 = 3 \cdot 4$.

DEFINITION OF DIVISION

If r and s are any real numbers such that $s \neq 0$, then $\frac{r}{s} = t$ if and only if there is exactly one real number t so that $r = st$.

EXAMPLE I.11.1 Find the indicated quotients:

a. $\frac{+12}{+3}$ b. $\frac{-12}{-3}$ c. $\frac{+12}{-3}$ d. $\frac{-12}{+3}$

Solution

a. $\frac{+12}{+3} = +4$ because $+12 = (+3)(+4)$

b. $\frac{-12}{-3} = +4$ because $-12 = (-3)(+4)$

c. $\frac{+12}{-3} = -4$ because $+12 = (-3)(-4)$

d. $\frac{-12}{+3} = -4$ because $-12 = (+3)(-4)$

EXAMPLE I.11.2 Find the indicated quotients, if they exist:

a. $\frac{0}{+5}$ b. $\frac{0}{-5}$ c. $\frac{5}{0}$ d. $\frac{0}{0}$

Solution

a. $\frac{0}{+5} = 0$ because $0 = (+5)(0)$

b. $\frac{0}{-5} = 0$ because $0 = (-5)(0)$

c. $\frac{5}{0}$ is undefined. There is no number t such that $5 = 0 \cdot t$ since $0 \cdot t = 0$ for all real numbers t.

d. $\frac{0}{0}$ is undefined. By definition, if $\frac{0}{0} = t$, then there must be exactly one real number t such that $0 = 0 \cdot t$. In this case, there are many possibilities since $0 = 0 \cdot t$ for all real numbers t.

The special cases shown in the preceding examples can be generalized as the following theorem.

THE QUOTIENTS OF REAL NUMBERS THEOREM

Let s and t be any nonzero real numbers.

1. $\frac{+s}{+t} = +\frac{s}{t}$ 2. $\frac{-s}{-t} = +\frac{s}{t}$

3. $\frac{+s}{-t} = -\frac{s}{t}$ 4. $\frac{-s}{+t} = -\frac{s}{t}$

5. $\frac{0}{s} = 0$

6. $\frac{s}{0}$ is undefined

7. $\frac{0}{0}$ is undefined

Informally, it may be stated that the quotient of two nonzero real numbers having the same sign is positive, and the quotient of two nonzero real numbers having different signs is negative. It may be noted that this mechanical rule is similar to the one for the product of two signed numbers.

EXAMPLE I.11.3 Perform the indicated operations, if possible:

a. $\frac{7 - 22}{-5}$ b. $\frac{6 - 6}{3}$ c. $\frac{4 - (-4)}{4 - (+4)}$ d. $\frac{8 - 2}{2 - 8}$

Solution

a. $\frac{7 - 22}{-5} = \frac{-15}{-5} = +3$

b. $\frac{6 - 6}{3} = \frac{0}{3} = 0$

c. $\frac{4 - (-4)}{4 - (+4)} = \frac{4 + (+4)}{4 + (-4)} = \frac{8}{0}$, undefined

d. $\frac{8 - 2}{2 - 8} = \frac{6}{-6} = -1$

EXERCISES I.11 A

In Exercises 1–40, perform the indicated operations, if possible. If the expression is undefined, write "undefined."

1. $\frac{+21}{-7}$ **2.** $\frac{-21}{+7}$ **3.** $\frac{-21}{-7}$ **4.** $\frac{+21}{+7}$

5. $\frac{0}{-6}$ **6.** $\frac{-6}{0}$ **7.** $\frac{-56}{+8}$ **8.** $\frac{+63}{-7}$

9. $\frac{-55}{-11}$ **10.** $\frac{+42}{-7}$ **11.** $\frac{-28}{+4}$ **12.** $\frac{+36}{+4}$

13. $\frac{-132}{+12}$ **14.** $\frac{+169}{-13}$ **15.** $\frac{-100}{-4}$ **16.** $\frac{-1000}{+8}$

17. $\frac{-125}{-25}$ 18. $\frac{+15}{0}$ 19. $\frac{0}{25}$ 20. $\frac{0}{0}$

21. $\frac{7-2}{2-7}$ 22. $\frac{4-12}{-4}$ 23. $\frac{-120}{(-4)+(+6)}$ 24. $\frac{8-8}{8-(-8)}$

25. $\frac{120}{5-17}$ 26. $\frac{10-30}{40-45}$ 27. $\frac{25-15}{25-35}$ 28. $\frac{0-9}{0+(-9)}$

29. $\frac{+12}{\left(\frac{-6}{+2}\right)}$ 30. $\frac{\left(\frac{+12}{-6}\right)}{+2}$ 31. $\frac{-24}{\left(\frac{-6}{-3}\right)}$ 32. $\frac{\left(\frac{-24}{-6}\right)}{-3}$

33. $\frac{(-10)^2}{-5}$ 34. $\frac{56-89}{89-56}$ 35. $\frac{95-48}{48-95}$ 36. $\frac{(-10)^3}{125}$

37. $\frac{7-(-7)}{-7}$ 38. $\frac{-6}{6-(-6)}$ 39. $\frac{(-7)^2-(5)^2}{-7-5}$ 40. $\frac{(-3)^3-(-2)^3}{-3-(-2)}$

EXERCISES I.II B

In Exercises 1–40, perform the indicated operations, if possible. If the expression is undefined, write "undefined."

1. $\frac{+63}{+9}$ 2. $\frac{+63}{-9}$ 3. $\frac{-63}{+9}$ 4. $\frac{-63}{-9}$

5. $\frac{-1000}{125}$ 6. $\frac{-1000}{-8}$ 7. $\frac{100}{-5}$ 8. $\frac{0}{-60}$

9. $\frac{-50}{0}$ 10. $\frac{+56}{+7}$ 11. $\frac{-56}{+7}$ 12. $\frac{+56}{-4}$

13. $\frac{-132}{-22}$ 14. $\frac{8-8}{8}$ 15. $\frac{8-(-8)}{-8}$ 16. $\frac{-3}{-3-(-3)}$

17. $\frac{18-25}{25-18}$ 18. $\frac{96-39}{39-96}$ 19. $\frac{(-5)^2}{-5}$ 20. $\frac{(-5)^3}{-5}$

21. $\frac{40-55}{55-58}$ 22. $\frac{60}{\left(\frac{-6}{2}\right)}$ 23. $\frac{\left(\frac{60}{-6}\right)}{2}$ 24. $\frac{-60}{\left(\frac{-10}{-2}\right)}$

25. $\frac{\left(\frac{-60}{-10}\right)}{-2}$ 26. $\frac{(-4)-(-4)}{(-4)+(-4)}$

27. $\frac{(-1)+(-1)}{(-1)-(-1)}$ 28. $\frac{(-7)^2-(-3)^2}{(-7)-(-3)}$

29. $\dfrac{(-10)^3 + (-5)^3}{(-10) + (-5)}$

30. $\dfrac{(5 - 1)(5 - 2)(5 - 3)}{(-1)(-2)(-3)}$

31. $\dfrac{(1 - 2)(1 - 4)(1 - 6)}{(-2)(-4)(-6)}$

32. $\dfrac{(-5)^2 + 2(-5)(-3) + (-3)^2}{(-5) + (-3)}$

33. $-\dfrac{-72}{-8}$

34. $-\dfrac{60}{-15}$

35. $-\dfrac{-154}{11}$

36. $\dfrac{(2 - 8)^3}{(8 - 2)^3}$

37. $(26 - 19)\dfrac{(2 - 6)}{(26 - 19)}$

38. $\dfrac{(3 - 9)(4 - 7)}{(4 - 7)(3 - 9)}$

39. $\dfrac{(5 - 1)(6 - 1)}{(1 - 6)(1 - 5)}$

40. $\dfrac{(-6)^2 - 2(-6)(-4) + (-4)^2}{(-6) - (-4)}$

1.12 COMBINED OPERATIONS: EVALUATION

Since the associative axiom for addition states that the way in which the numbers are grouped does not affect the sum, it is not necessary to use parentheses to indicate a sum of three or more terms. In other words,

$$x + y + z = (x + y) + z = x + (y + z)$$

for all real numbers x, y, and z.

On the other hand, subtraction is *not* associative. For example,

$$(10 - 5) - 2 \neq 10 - (5 - 2)$$

However, since by definition $x - y = x + (-y)$, it is convenient to use this idea and adopt certain conventions that reduce the number of symbols that must be written.

CONVENTIONS

For all real numbers r, s, and t

$$r + s - t = r + s + (-t)$$

$$r - s + t = r + (-s) + t$$

$$r - s - t = r + (-s) + (-t)$$

EXAMPLE 1.12.1 Use the preceding convention to calculate each of the following expressions in two ways, thus showing that the result is independent of the way in which the terms are grouped:

a. $7 + 6 - 5$ b. $12 - 8 + 3$ c. $10 - 5 - 2$

Solution

a. $$\begin{aligned} 7 + 6 - 5 &= 7 + 6 + (-5) \\ &= (7 + 6) + (-5) \\ &= 13 + (-5) = 8 \end{aligned}$$

$$\begin{aligned} 7 + 6 - 5 &= 7 + 6 + (-5) \\ &= 7 + [6 + (-5)] \\ &= 7 + 1 = 8 \end{aligned}$$

b. $$\begin{aligned} 12 - 8 + 3 &= 12 + (-8) + 3 \\ &= [12 + (-8)] + 3 \\ &= 4 + 3 = 7 \end{aligned}$$

$$\begin{aligned} 12 - 8 + 3 &= 12 + (-8) + 3 \\ &= 12 + [(-8) + 3] \\ &= 12 + (-5) = 7 \end{aligned}$$

c. $$\begin{aligned} 10 - 5 - 2 &= 10 + (-5) + (-2) \\ &= [10 + (-5)] + (-2) \\ &= 5 + (-2) = 3 \end{aligned}$$

$$\begin{aligned} 10 - 5 - 2 &= 10 + (-5) + (-2) \\ &= 10 + [(-5) + (-2)] \\ &= 10 + (-7) = 3 \end{aligned}$$

Certain expressions and formulas require the substitution of one or more negative values. This is illustrated in the following examples.

EXAMPLE 1.12.2 Evaluate

$$F = \frac{9C}{5} + 32$$

for $C = -20$ degrees.

Solution

$$F = \frac{9C}{5} + 32$$

$$F = \frac{9(\ )}{5} + 32$$

$$F = \frac{9(-20)}{5} + 32$$

$$F = 9(-4) + 32 = -36 + 32 = -4 \text{ degrees}$$

EXAMPLE 1.12.3 Evaluate

$$\frac{100 + 2x - y}{5}$$

for $x = -35$, $y = -15$.

Solution

$$\frac{100 + 2x - y}{5} = \frac{100 + (+2x) + (-y)}{5}$$

$$= \frac{100 + (+2[-35]) + [-(-15)]}{5}$$

$$= \frac{100 + (-70) + (+15)}{5}$$

$$= \frac{30 + 15}{5}$$

$$= \frac{45}{5} = 9$$

EXERCISES 1.12 A

Perform the indicated operations in Exercises 1–10.

1. $12 - 3 - 15$
2. $1 - 6 + 3$
3. $20 + 40 - 30$
4. $-15 - 25 + 5$
5. $4 - 7 + 3$
6. $6 - 5 - 4 - 3$
7. $15 - 17 + 19 - 21$
8. $8 - 2 + 2 - 8$
9. $50 - 40 - 30 - 20$
10. $75 - 64 + 25 - 36$

Evaluate in Exercises 11–20.

11. $x + y + z$ for $x = 7, y = -5, z = -9$
12. $x - y - z$ for $x = 12, y = 3, z = -8$
13. $x - y + z$ for $x = -6, y = -7, z = 8$
14. $x + y - z$ for $x = -15, y = 19, z = 21$
15. $a^2 + b^2 - c^2$ for $a = 2, b = -2, c = -1$
16. $a^2 - ab + b^2$ for $a = 7, b = -6$
17. $\frac{rs}{r - s - 1}$ for $r = 10, s = 7$

18. $\dfrac{r^2 + s^2 + 1}{r - s + 1}$ for $r = 4$, $s = -2$

19. $(n + 1)^2 - (n + 1)(n - 1) + (n - 1)^2$ for $n = -5$

20. $b^2 - 4ac$ for $a = -3$, $b = -5$, $c = -4$

21. (Temperature scales: Fahrenheit to centigrade)

$$C = \frac{5(F - 32)}{9}$$

Find C for $F = -4$ degrees.

22. (Temperature scales: centigrade to Fahrenheit)

$$F = \frac{9C}{5} + 32$$

Find F for $C = -15$ degrees.

23. (Center of mass)

$$x = \frac{MD + md}{M + m}$$

Find x for $M = 150$ grams, $D = 40$ centimeters, $m = 50$ grams, $d = -20$ centimeters.

24. Find x in Exercise 23 for $M = 25$ pounds, $D = -6$ feet, $m = 15$ pounds, $d = -2$ feet.

In Exercises 25–26 (Mirrors),

$$f = \frac{pq}{p + q}$$

where f = *focal length of mirror (positive for concave mirrors and negative for convex mirrors)*

p = *object distance from mirror (positive when object is in front of the mirror)*

q = *image distance from mirror (positive when image is in front of the mirror and negative when image is behind the mirror)*

25. Find f for $p = 45$ centimeters, $q = -180$ centimeters.

26. Find f for $p = 6$ feet, $q = -3$ feet.

27. (Acceleration A of a body whose velocity is increasing or decreasing uniformly. A negative value for an acceleration means a deceleration.)

$A = \dfrac{V - v}{t}$ where V = final velocity

v = initial velocity

t = elapsed time

Find A for $V = 15$ feet per second, $v = 75$ feet per second, $t = 12$ seconds.

28. (Acoustics)

$$F = \frac{fV}{V + v}$$

where $F =$ observed frequency (pitch) of a moving source of sound waves

$V =$ speed of sound in air

$v =$ speed of the moving source (positive if moving away from the observer and negative if moving toward the observer)

Find F for $f = 460$ cycles per second, $V = 1100$ feet per second, $v = -88$ feet per second.

In Exercises 29–30 (Chemistry: capillary tubes),

$$h = \frac{10SC}{49dr}$$

where $h =$ *height in millimeters that the liquid rises* $(+)$ *or is depressed* $(-)$

$S =$ *surface tension in dynes per centimeter*

$C =$ *constant depending on the liquid and the material of the tube*

$d =$ *density of the liquid in grams per cubic centimeter*

$r =$ *radius of the tube in millimeters*

29. Find h for $S = 490$, $C = -0.68$, $d = 13.6$, $r = 2$ and state whether the liquid rises or is depressed (mercury in a glass tube).

30. Find h for $S = 72.8$, $C = 1$, $d = 1$, $r = 4$ (water in a glass tube).

EXERCISES 1.12 B

Perform the indicated operations in Exercises 1–10.

1. $8 - 4 + 2$

2. $15 - 8 - 3$

3. $9 + 6 - 5$

4. $2 - 5 - 7$

5. $1 - 2 + 3$

6. $4 + 8 - 2 - 6$

7. $27 - 3 - 2 + 24$

8. $15 - 22 - 6 - 2$

9. $5 - 2 + 3 - 6$

10. $-8 - 10 + 5 + 7$

Evaluate Exercises 11–20, using the given values.

11. $\frac{x + y + z}{3}$ for $x = 20, y = -17, z = -15$

12. $2x + 2y - 1$ for $x = 5, y = -4$

13. $\frac{c}{a} + \frac{d}{b} + \frac{a}{b}$ for $a = -6, b = -3, c = +18, d = -15$

14. $c^2 - a^2 - b^2$ for $c = 11, a = 2, b = 6$

15. $\frac{ac + bd}{a + b}$ for $a = 8, c = 3, b = 10, d = -6$

16. $a^2 - ab + b^2$ for $a = -4, b = -5$

17. $\frac{p + q - w}{2}$ for $p = 20, q = 0, w = -25$

18. $\frac{xyz}{x - y - z}$ for $x = 50, y = 60, z = -20$

19. $xy + xz + yz$ for $x = 5, y = -4, z = -2$

20. $a^2 + b^2 + c^2$ for $a = 2, b = -3, c = -4$

21. Use $F = \frac{9C}{5} + 32$ to find F for $C = -95$ degrees (melting point of acetone)

22. Use $C = \frac{5(F - 32)}{9}$ to find C for $F = -109$ degrees (temperature of Dry Ice)

23. (Thermodynamics) $E = Q - W$

where E = change in internal energy

Q = change in heat (positive when added to the system and negative when removed)

W = work done (positive when done by the system and negative when done on the system)

Find E for each of the following cases:

a. $Q = 0$, $W = -6$ ft-lb

b. $Q = 0$, $W = 50$ joules

c. $Q = 400$ joules, $W = -175$ joules

d. $Q = -20$ Btu, $W = 0$

e. $Q = 810$ Btu, $W = 470$ Btu

24. (Corrective lenses)

$$P = \frac{100(p + q)}{pq}$$

where P = power in diopters (positive for farsighted persons and negative for nearsighted)

p = object distance from lens (positive) in centimeters

q = image distance from lens (positive when on the opposite side of the lens as the object and negative when on the same side) in centimeters

Find P for each of the following cases:

a. $p = 20, q = -50$

b. $p = 300, q = -75$

25. (Flow of liquids)

$$P = \frac{d}{2}(V^2 - v^2)$$

where P = change in pressure (positive for increase and negative for decrease)

V = original speed of the liquid

v = new speed of the liquid

d = density of the liquid

Find P for each of the following cases:

a. $d = 62.5$, $V = 3$ feet per second, $v = 6$ feet per second

b. $d = 50$, $V = 4$ feet per second, $v = 2$ feet per second

SUMMARY

- ☐ **Numerals** are symbols that name numbers according to a specified system.
- ☐ A **variable** is a letter that may designate the name of any number in a given set of numbers having more than one member.
- ☐ A **constant** is a letter or numeral that designates the name of exactly one number during a particular discussion.
- ☐ The **sum** of two numbers is the result obtained by adding the two numbers.
- ☐ The **difference**, or **remainder**, is the result obtained by subtracting one number from another.
- ☐ The **terms** are the numbers that are being added or subtracted.
- ☐ The **product** is the result obtained by multiplying two numbers.

☐ The **factors** are the numbers that are being multiplied.
☐ The **quotient** is the result obtained when one number, the **dividend**, is divided by another number, the **divisor**.
☐ The **square** of a number is the product obtained by using the number as a factor two times.
☐ The **cube** of a number is the product obtained by using the number as a factor three times.

TABLE 1.1 SUMMARY OF THE OPERATION SYMBOLS

Number Symbol / *Operation*	*Numerals*	*Numeral and Letter*	*Two Letters*
Addition	$3+4$	$x+4$	$x+y$
Subtraction	$7-3$	$x-4$	$x-y$
Multiplication	$3\cdot 4$ or $3(4)$	$3x$	xy
Division	$\frac{12}{3}$	$\frac{x}{3}$	$\frac{x}{y}$
Squaring	3^2	x^2	(Not applicable)
Cubing	5^3	x^3	
Square root	$\sqrt{25}$	$\sqrt{x}$	(Not applicable)
Cube root	$\sqrt[3]{8}$	$\sqrt[3]{x}$	

☐ **Grouping symbols** are used to indicate the order in which operations are to be performed. The operation within the innermost set is to be done first, then the innermost set remaining, and so on.
☐ Grouping symbols commonly used are the **parentheses** ( ), **brackets** [ ], **braces** { }, and the **bar** ____.
☐ Unless grouping symbols indicate otherwise, operations are to be performed in the following order:
1. Taking square roots and/or cube roots as read from left to right.
2. Squaring and/or cubing as read from left to right.
3. Multiplication and/or division as read from left to right.
4. Addition and/or subtraction as read from left to right.

☐ A **set** is a well-defined collection of objects.
☐ In the **listing method**, a set is defined by listing its members enclosed by braces and separated by commas.
☐ In the **rule** or **set-builder method**, a set is defined as follows:

$$\{x \mid x \text{ has a certain property}\}$$

- ☐ A **universal set** is a set to which the elements of all other sets in a particular discussion must belong.
- ☐ The **empty set**, denoted by $\varnothing$ or { }, is the set that has no members.
- ☐ $a \in S$ means a is a member of set S.
- ☐ $A \subset B$ means set A is a subset of set B.
- ☐ $A \cup B$ means "A union B," the set of all elements in A or B or in both sets.
- ☐ $A \cap B$ means "A intersection B," the set of all elements sets A and B have in common.
- ☐ The set of **digits**, D, $= \{0, 1, 2, 3, 4, 5, 6, 7, 8, 9\}$.
- ☐ The set of **natural numbers,** N, $= \{1, 2, 3, 4, 5, \ldots\}$.
- ☐ The set of **integers,** I, $= \{\ldots, -3, -2, -1, 0, +1, +2, +3, \ldots\}$.
- ☐ The natural numbers, $\{1, 2, 3, \ldots\}$, are called the **positive integers**.
- ☐ The numbers in the set $\{\ldots, -3, -2, -1\}$ are called the **negative integers.**
- ☐ The numbers in the set $\{0, 1, 2, 3, \ldots\}$ are called the **whole numbers** or the **nonnegative integers.**
- ☐ A line whose points are named by using numbers is called a **number line.**
- ☐ The number that names a point on the number line is called the **coordinate** of the point.
- ☐ The set of **rational numbers,** Q, $= \left\{\frac{p}{q} \;\middle|\; p \in I \text{ and } q \in I \text{ and } q \neq 0\right\}$.
- ☐ The set of **real numbers,** R, is the totality of all the numbers that can be associated with points on a number line.
- ☐ The natural numbers, N, are a subset of the integers, I.
- ☐ The integers, I, are a subset of the rational numbers, Q.
- ☐ The rational numbers, Q, are a subset of the real numbers, R.

$$N \subset I \subset Q \subset R$$

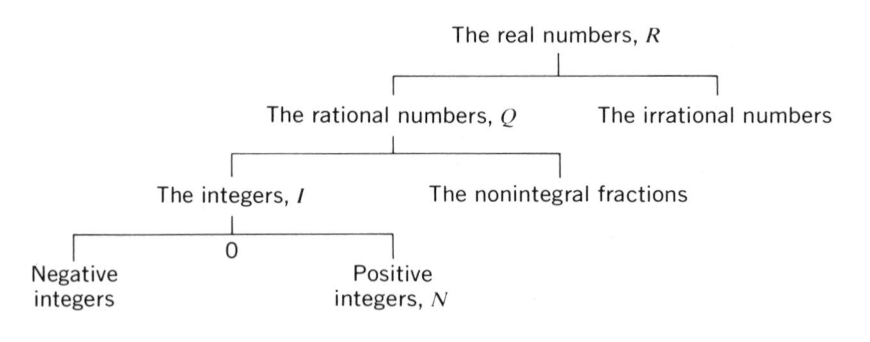

Diagram of the real number system.

The Substitution Axiom

☐ If $r = s$, then r may replace s or s may replace r in an algebraic expression without changing the number being named or in an algebraic statement without changing the truth or falsity of the statement.

Order Relations

☐ $r < s$ means "r is less than s." On a horizontal number line, point r is to the left of point s.

☐ $r > s$ means "r is greater than s." On a horizontal number line, point r is to the right of point s.

Absolute Value, $|p|$

☐ If p is a positive real number, then $|+p| = p$, $|-p| = p$, and $|0| = 0$.

Addition of Real Numbers

☐ If r and s are any positive real numbers, then:

1. $r + 0 = r, 0 + r = r, -r + 0 = -r, 0 + (-r) = -r$
2. $(+r) + (+s) = +(r + s)$
3. $(-r) + (-s) = -(r + s)$
4. $(+r) + (-s) = +(r - s)$ if $r > s$
5. $(+r) + (-s) = -(s - r)$ if $r < s$

Subtraction of Real Numbers

☐ If r and s are any real numbers:

1. $-(-r) = r$
2. $r - s = r + (-s)$

Multiplication of Real Numbers

☐ If r and s are any real numbers:

1. $r \cdot 0 = 0$ and $0 \cdot r = 0$
2. $(+r)(+s) = +rs = rs$
3. $(-r)(-s) = +rs = rs$
4. $(+r)(-s) = -rs$
5. $(-r)(+s) = -rs$

Division of Real Numbers

☐ If r and s are any nonzero real numbers:

1. $\dfrac{+r}{+s} = \dfrac{r}{s}$
2. $\dfrac{-r}{-s} = \dfrac{r}{s}$
3. $\dfrac{+r}{-s} = -\dfrac{r}{s}$
4. $\dfrac{-r}{+s} = -\dfrac{r}{s}$
5. $\dfrac{0}{r} = 0$
6. $\dfrac{r}{0}$ is undefined
7. $\dfrac{0}{0}$ is undefined

Conventions

1. $r + s + t = (r + s) + t = r + (s + t)$
2. $rst = (rs)t = r(st)$
3. $r + s - t = r + s + (-t)$
4. $r - s + t = r + (-s) + t$
5. $r - s - t = r + (-s) + (-t)$

REVIEW EXERCISES

If 5 and N represent two numbers, express in symbols each verbal expression in Exercises 1–8.

1. The sum of 5 and N
2. The product of 5 and N
3. The difference when 5 is subtracted from N
4. The square of N
5. The quotient when 5 is divided by N
6. The cube of N
7. The square root of 5
8. The cube root of N

Express the verbal expressions in Exercises 9–17 in symbols.

9. 5 times the sum of x and 8
10. The difference when 4 is subtracted from the product of 6 and x
11. The square of the quotient of y divided by 5

12. One half the difference when the square root of y is subtracted from the square of x

13. The sum of the cube of n and the square of the difference when 1 is subtracted from n

14. The quotient when twice the product of x and y is divided by the sum of x and y

15. The difference when the product of 3 and x is subtracted from 10 and is equal to 7

16. The square of the sum of x and 4 is less than the quotient when x is divided by 4

17. The sum of 9 and the product of 6 and x is greater than or equal to 8

18. Write an algebraic expression to indicate that if one begins with a number x, then multiplies it by 3, then adds 6, then multiplies by 4, then subtracts 12 times the original number x, the result is always equal to 24

19. Write in symbols: If five is subtracted from four fifths of a certain number x, then the difference is 3 times the sum of the number x and 2

In Exercises 20–32, evaluate each.

20. $2(x - 3)$ for a. $x = 5$, b. $x = 0$, c. $x = -2$

21. $(x + 2)(x - 3)$ for a. $x = 5$, b. $x = 0$, c. $x = -5$

22. $\dfrac{4x}{6 - x}$ for a. $x = 8$, b. $x = 0$, c. $x = -2$

23. $\dfrac{x^2 - 16}{x - 4}$ for a. $x = 2$, b. $x = 0$, c. $x = -5$

24. $3x - (7 - x)$ for a. $x = 4$, b. $x = 0$, c. $x = -2$

25. $x - [x - (x - 2)]$ for a. $x = 7$, b. $x = 0$, c. $x = -1$

26. $10 - 2[x - (5 - 3x)]$ for a. $x = 1$, b. $x = 3$, c. $x = -2$

27. $\dfrac{(x - 1)(x - 2)(x - 3)}{(-1)(-2)(-3)}$ for a. $x = 9$, b. $x = 1$, c. $x = -3$

28. $\dfrac{x^2 + 3xy - 10y^2}{x - 2y}$ for a. $x = 3, y = 2$
b. $x = -4, y = -5$

29. $x - y + z$ for a. $x = 7, y = 9, z = 11$
b. $x = 5, y = -2, z = -6$

30. $x - y - z$ for a. $x = 25, y = 15, z = 35$
b. $x = -5, y = -10, z = +15$

31. xyz for a. $x = 7, y = -4, z = -25$
b. $x = -4, y = -3, z = -2$

32. $\frac{xy}{xz}$ for a. $x = -5, y = -8, z = +2$
b. $x = -6, y = +12, z = -9$

In Exercises 33–42, answer true or false.

33. $4 < 10$
34. $-4 < -10$
35. $8 > -3$
36. $0 > -6$
37. $-12 > -15$
38. $-7 > -6$
39. $|-5| < 0$
40. $|-5| > 0$
41. $|-8| = 8$
42. $|-5| < |-9|$

For Exercises 43–52, graph each on the real number line.

43. $\{x \mid x < 4\}$
44. $\{x \mid x \leq -2\}$
45. $\{x \mid x > 0\}$
46. $\{x \mid x > -3\}$
47. $\{x \mid x \geq 5\}$
48. $\{x \mid x \leq 0\}$
49. $\{x \mid 3 < x < 6\}$
50. $\{x \mid -2 \leq x < 0\}$
51. $\{x \mid 1 < x \leq 5\}$
52. $\{x \mid -2 \leq x \leq 3\}$

53. (Slope of a line)

$$m = \frac{Y - y}{X - x}$$

Find m for $Y = 7, y = 10, X = -1, x = -7$.

54. (Flow of liquids)

$$P = p + \frac{d}{2}(v^2 - V^2)$$

Find P for $p = 50, d = 1, v = 20, V = 40$.

55. (Medication for child)

$$C = \frac{WA}{150}$$

Find C for $W = 60$ pounds, $A = 200$ milligrams.

56. (Radio and television)

$$r = \frac{R(E - G)}{E}$$

Find r for $R = 2$ megohms, $E = -12$ volts, $G = -9$ volts.

CHAPTER 2

LINEAR EQUATIONS

There are two main reasons for studying algebra:

1. To acquire a tool for the study of other disciplines, such as physics, biology, statistics, or economics.
2. To pursue algebra for its own sake as a branch of pure mathematics.

One of the most important skills needed for both purposes is the solution of equations. It is the aim of this chapter to introduce the student to the solution of the simplest type of equation: the linear equation in one variable. In preparation for this, some preliminary axioms and theorems must be stated. Some of these were introduced in the previous chapter. In this chapter they will be treated in more detail.

There are three major categories of general statements in mathematics: definitions, axioms, and theorems. A **definition** is a precise explanation of the meaning of a term. An **axiom** is a statement that is assumed to be true. A **theorem** is a statement that is proved.

2.1 SOME AXIOMS

2.1.1 Equality Axioms

Three important axioms of the equal relation are stated on the next page.

THE EQUALITY AXIOMS

For all real numbers A, B, C,

1. **Reflexive axiom** $A = A$
2. **Symmetric axiom** If $A = B$, then $B = A$
3. **Transitive axiom** If $A = B$ and $B = C$, then $A = C$

The reflexive axiom states that the number whose name is A is the same as the number whose name is A. This is a general way of stating that any algebraic equation having this form is true.

For example, $3 = 3$ is a true statement

$3 + 5 = 3 + 5$ is a true statement

$2(x + 3) = 2(x + 3)$ is a true statement

One particular use of the reflexive axiom is to establish a check of a solution of an open equation.

EXAMPLE 2.1.1 Check that $x = 4$ is a solution of $x(x + 3) = 28$.

Solution If $x = 4$, then

$$x(x + 3) = 4(4 + 3) = 4 \cdot 7 = 28$$

Thus if $x = 4$, then $x(x + 3) = 28$. Since $28 = 28$ is a true statement, 4 is a solution.

The symmetric axiom states that an equation is reversible. In other words, this axiom permits the left and right sides of an equation to be exchanged.

For example, $7 = 2x - 3$ may be replaced by $2x - 3 = 7$. This is often convenient in solving equations.

The transitive axiom means that if two quantities are each equal to a third quantity, then these two quantities are equal to each other.

It is often desirable to replace names of numbers by simpler names that involve fewer symbols. For example, $2(3 + 5) = 2 \cdot 8$ and $2 \cdot 8 = 16$. Thus $2(3 + 5) = 16$.

The transitive axiom is also useful in checking the solution of an equation. For example, check that $x = 3$ is a solution of

$$5x - 7 = 2(x + 1)$$

$$5x - 7 = 5 \cdot 3 - 7 = 15 - 7 = 8$$

and $$2(x + 1) = 2(3 + 1) = 2 \cdot 4 = 8$$

Thus when $x = 3$, $5x - 7 = 2(x + 1)$, since $8 = 8$.

The reflexive and symmetric axioms distinguish the equal relation from the less than and greater than relations. For example, $3 < 3$ and $3 > 3$ are both false. Also, if $3 < 5$ is true, then $5 < 3$ is false. Thus the order relations, $<$ and $>$, are neither reflexive nor symmetric. However, these relations are transitive. For example, if $3 < 5$ and $5 < 8$, then $3 < 8$.

EXAMPLE 2.1.2 For each statement, if the statement is true, write "reflexive," "symmetric," or "transitive" to indicate the axiom that is illustrated; if the statement is false, write "false."

Given Statement	*Solution*
a. If $-5 + 0 = -5$, then $-5 = -5 + 0$.	Symmetric
b. $x + 4 = x + 4$.	Reflexive
c. If $4 + (-9) = -9 + 4$, then $-9 + 4 = 4 + (-9)$.	Symmetric
d. $4 - 9 = 9 - 4$.	False
e. If $\frac{2x}{2} = \frac{1}{2}(2x)$ and $\frac{1}{2}(2x) = x$, then $\frac{2x}{2} = x$.	Transitive

2.1.2 The Commutative Axioms

The equality axioms are axioms concerning the equal relation. There are also very important axioms which concern operations. The commutative axiom and the associative and distributive axioms discussed in subsequent sections are operational axioms dealing with the operations "addition" and "multiplication" on real numbers.

THE COMMUTATIVE AXIOM FOR ADDITION

For all real numbers r and s,

$$r + s = s + r$$

THE COMMUTATIVE AXIOM FOR MULTIPLICATION

For all real numbers r and s,

$$rs = sr$$

The commutative axiom for addition states that the order in which two real numbers are added does not change the value of the sum.

Thus $3 + 5 = 8$ and $5 + 3 = 8$. Therefore, $3 + 5 = 5 + 3$. Also, if p stands for some number, then $p + 2 = 2 + p$, $p + 9 = 9 + p$, and so forth.

Similarly, the commutative axiom for multiplication states that the order in which two numbers are multiplied does not change the value of the product.

Thus $3 \cdot 5 = 15$ and $5 \cdot 3 = 15$. Therefore, $3 \cdot 5 = 5 \cdot 3$. Also, if x stands for some number, then $2 \cdot x = x \cdot 2$, $9 \cdot x = x \cdot 9$, and so forth.

The student may wonder why the commutative axioms are stated only for addition and multiplication and not for subtraction and division. To see if they apply to subtraction, ask the question: Does $r - s = s - r$ for all values of r and s? Let $r = 3$ and $s = 5$. Then $3 - 5 = -2$ and $5 - 3 = 2$. Since $-2 \neq 2$, the statement is not true, and subtraction is not a commutative operation. Similarly, it may be shown that since $\frac{r}{s} \neq \frac{s}{r}$ for all nonzero real numbers r and s, division is not commutative. For example, $\frac{12}{3} \neq \frac{3}{12}$ since $4 \neq \frac{1}{4}$.

2.1.3 The Associative Axioms

THE ASSOCIATIVE AXIOM FOR ADDITION

For all real numbers r, s, and t,

$$(r + s) + t = r + (s + t)$$

THE ASSOCIATIVE AXIOM FOR MULTIPLICATION

For all real numbers r, s, and t,

$$(rs)t = r(st)$$

The associative axiom for addition states that terms of a sum may be regrouped without changing the value of the sum.

The associative axiom for multiplication states that the factors of a product may be regrouped without changing the value of the product.

Since the value of a sum does not depend on the way the terms are grouped, and since the value of a product does not depend on the way the factors are grouped, it is not necessary to use the grouping symbols shown in the associative axioms to indicate a sum of three or more terms or a product of three or more factors.

In other words,

$$x + y + z \text{ means } (x + y) + z \text{ or } x + (y + z)$$

and $\quad xyz \quad$ means $(xy)z \quad$ or $x(yz)$

Similarly, $a + b + c + d$ can be calculated in any one of the following ways:

$$(a + b) + (c + d),\quad a + (b + c + d),\quad a + (b + c) + d,$$
$$a + b + (c + d),\quad (a + b + c) + d$$

Likewise, $abcd$ can be calculated in any one of the following ways:

$$(ab)(cd),\quad a(bcd),\quad a(bc)d,\quad ab(cd),\quad (abc)d$$

EXAMPLE 2.1.3 Find the value of each indicated sum or product and check by applying an associative axiom.

Given	*Solution*	*Check*
a. $6 + 7 + 8$	$(6 + 7) + 8$ $= 13 + 8 = 21$	$6 + (7 + 8)$ $= 6 + 15 = 21$
b. $5 + (-4) + (-9)$	$[5 + (-4)] + (-9)$ $= 1 + (-9) = -8$	$5 + [(-4) + (-9)]$ $= 5 + (-13) = -8$
c. $3\cdot 4\cdot 5$	$(3\cdot 4)\cdot 5$ $= 12\cdot 5 = 60$	$3\cdot(4\cdot 5)$ $= 3\cdot 20 = 60$
d. $(-2)(6)(-3)$	$[(-2)(6)](-3)$ $= (-12)(-3) = 36$	$(-2)[(6)(-3)]$ $= -2(-18) = 36$
e. $7 - 2 + 4 - 9$	$= 7 + (-2) + 4 + (-9)$ $= [7 + (-2)] + [4 + (-9)]$ $= 5 + (-5) = 0$	 $= [7 + (-2 + 4)] + (-9)$ $= 9 + (-9) = 0$
f. $(-2)(-3)(-4)(-5)$	$[(-2)(-3)]\cdot[(-4)(-5)]$ $= 6\cdot 20$ $= 120$	$(-2)[(-3)(-4)](-5)$ $= (-2)(12)(-5)$ $= (-24)(-5) = 120$

Two important consequences result from combining a commutative axiom with the associative axiom for the same operation:

CONSEQUENCE 1. The terms of a sum can be rearranged in any order without changing the value of the sum.

CONSEQUENCE 2. The factors of a product can be rearranged in any order without changing the value of the product.

There are certain conventions for writing the terms of a sum or the factors of a product. These conventions, stated below, are especially useful for comparing the results obtained by one person with the results obtained by another person (such as checking the answers in the back of the book).

CONVENTIONS

The terms of a sum are written so that:

1. **A numerical term is to the right of all literal terms.**
2. **The letters involved are written in alphabetical order whenever possible.**

The factors of a product are written so that:

1. **A numerical factor is to the left of all literal factors.**
2. **The letters involved are written in alphabetical order whenever possible.**

EXAMPLE 2.1.4 Rearrange the terms of each sum and the factors of each product according to the sum and product conventions. Simplify each result if possible.

Given Expression	*Solution Rearrangement*	*Simplification*
a. $y + 5 + x$	$x + y + 5$	None possible
b. $-6 + n + 6$	$n + 6 + (-6)$	n
c. $a - 4 - b$	$a - b - 4$	None possible
d. $6x(-2)$	$(-2)(6)x$	$-12x$
e. $(-4xy)(-7x)$	$(-4)(-7)xxy$	$28x^2y$

EXAMPLE 2.1.5 Using the commutative and associative axioms, rearrange the numerals in each of the following so that the computation can be performed mentally in the easiest way possible:

a. $(79 + 56) + 21$ b. $4(38 \cdot 25)$

c. $1 + 2 + 3 + 4 + 5 + 6 + 7 + 8 + 9$

d. $625(24)$

Solution The idea is to use certain facts of base 10. For example,
Addition: $1 + 9 = 10$, $2 + 8 = 10$, $3 + 7 = 10$, $4 + 6 = 10$, $5 + 5 = 10$
Multiplication: $2 \cdot 5 = 10$, $4 \cdot 25 = 100$, $8 \cdot 125 = 1000$

a. $(79 + 56) + 21 = (79 + 21) + 56$
$= 100 + 56 = 156$

b. $4(38 \cdot 25) = (4 \cdot 25)38$
$= 100(38) = 3800$

c. $1 + 2 + 3 + 4 + 5 + 6 + 7 + 8 + 9 = (1 + 9) + (2 + 8) + (3 + 7) + (4 + 6) + 5$
$= 10 + 10 + 10 + 10 + 5$
$= 45$

d. $625(24) = (5 \cdot 5 \cdot 5 \cdot 5)(2 \cdot 2 \cdot 2 \cdot 3)$
$= (5 \cdot 2)(5 \cdot 2)(5 \cdot 2)(5 \cdot 3)$
$= 10 \cdot 10 \cdot 10 \cdot 15$
$= 15{,}000$

EXERCISES 2.1 A

For each statement in Exercises 1–10, if the statement is true, write "reflexive," "symmetric," or "transitive" to indicate the axiom which is illustrated; if the statement is false, write "false."

1. $5 = 5$

2. If $x + 2 = 5$, then $5 = x + 2$.

3. If $10 - 2 = 8$, then $8 = 10 - 2$.

4. If $x = \frac{y}{3}$ and $\frac{y}{3} = -9$, then $x = -9$.

5. $(-3)(-3)(-3) = (3)(3)(3)$

6. $4 - 10 = 10 - 4$

7. $4 - 10 = 4 - 10$

8. If $1 \cdot x = x \cdot 1$ and $x \cdot 1 = x$, then $1 \cdot x = x$.

9. If $2 + (x + 5) = x + (2 + 5)$ and $x + (2 + 5) = x + 7$, then $2 + (x + 5) = x + 7$.

10. $(-1)^2 = (-1)^3$

State the operational axiom or axioms which are used in the equations in Exercises 11–20.

11. $3 + 5 = 5 + 3$

12. $a(bc) = (ab)c$

13. $a(b + c) = (b + c)a$

14. $[3 + (-2)] + 1 = 3 + [(-2) + 1]$

15. $8(5 + x) = 8(x + 5)$

16. $(x + 2) + 2x = (x + 2x) + 2$

17. $3(2 + 5) = (5 + 2)3$

18. $3x + 5y = 5y + 3x$

19. $3yx + (-2)(3x) = 3xy + (-6x)$

20. $2x(3) = 6x$

Rearrange the terms of each sum and the factors of each product in Exercises 21–30 according to the sum and product conventions stated in this section. Simplify each result if possible.

21. $y + 4 + x$

22. $(3x)(-5x)$

23. $(-8 - x) + 8$

24. $(y - 6) + (x - 3)$

25. $(-4xy)(-9xy)$

26. $(x - 5) + (8 - x)$

27. $8 + b(2a) - 12$

28. $(4c)(-3b)(-2a)$

29. $n - 7 + m$

30. $(-na)(-nb)(-nc)$

Using the commutative and associative axioms, rearrange the numbers in Exercises 31–35 so that each computation can be performed mentally in the easiest way possible.

31. $(35 + 87) + 65$

32. $(25 \cdot 69) \cdot 4$

33. $7\frac{3}{8} + \left(9 + 2\frac{5}{8}\right)$

34. $20(45 \cdot 0.5)$

35. $(5 + 6) + (195 + 194)$

EXERCISES 2.1 B

For each statement in Exercises 1–10, if the statement is true, write "reflexive," "symmetric," or "transitive" to indicate the axiom which is illustrated; if the statement is false, write "false."

1. $-3 = 3$
2. If $x = y$ and $y = 7$, then $x = 7$.
3. If $5 - x = x + 2$, then $x + 2 = 5 - x$.
4. $3(x + 5) = 3(x + 5)$
5. If $0 = x + 2$, then $x + 2 = 0$.
6. If $n + 0 = 0 + n$ and $0 + n = 0$, then $n + 0 = 0$.
7. $\frac{-2}{-1} = \frac{-1}{-2}$

8. $\frac{-2}{-1} = \frac{-2}{-1}$

9. If $5 + 7 = 7 + 5$, then $5 - 7 = 7 - 5$.

10. If $6x(5) = (5 \cdot 6)x$ and $(5 \cdot 6)x = 30x$, then $6x(5) = 30x$.

State the operational axiom or axioms which are used in the equations in Exercises 11–20.

11. $x(a + 2) = (a + 2)x$

12. $5 + x = x + 5$

13. $b + (-5) = (-5) + b$

14. $a + (3a + 2) = (a + 3a) + 2$

15. $3 + 2x = 3 + x \cdot 2$

16. $(3b)a = 3(ab)$

17. $s + (-2) + r = r + s - 2$

18. $x(2 + y) = x(y + 2)$

19. $-2(3yx) = -6xy$

20. $(15 + 17) + 85 = (15 + 85) + 17$

Rearrange the terms of each sum and the factors of each product in Exercises 21–30 according to the sum and product conventions stated in this section. Simplify each result if possible.

21. $4 + x - 7$

22. $(-9x)(-3y)$

23. $(y - 8) + (x + 4)$

24. $(-6x)(5xy)$

25. $(2 - x) + (x - 7)$

26. $4 - y + x - 3$

27. $3b(-4) + (-5a)(-6)$

28. $d - b(-5a) - 4c$

29. $6 - n + m - 7$

30. $(ka)(-kb)(-5b)(-7)$

Using the commutative and associative axioms, rearrange the numbers in Exercises 31–35 so that each computation can be performed mentally in the easiest way possible.

31. $8(39 \cdot 125)$

32. $123 + (695 + 877)$

33. $(0.25 \cdot 79)4$

34. $\left(18\frac{9}{16} + 10\frac{5}{12}\right) + 12\frac{7}{16}$

35. $1 + 2 + 3 + \ldots + 98 + 99 + 100$ (the sum of the first 100 natural numbers)

2.2 CLOSURE, IDENTITY, INVERSE AXIOMS

An interesting question which arises when an operation is defined on a set of numbers is whether or not all answers thus obtained are in this specified set. For example, if the operation "addition" is defined on the natural numbers, is the sum of all possible pairs of natural numbers a natural number? We know that $2 + 3 = 5$ and $5 \in N$ (where N = set of natural numbers). Also, $1000 + 2517 = 3517$, and $3517 \in N$. It would be impossible to check all possible combinations, but the sum of two natural numbers is always a natural number. The natural numbers are said to be closed with respect to addition. As a matter of fact, so are the rational numbers, and so is the set of real numbers. All these sets are also closed with respect to multiplication, and the following axioms formalize these facts.

CLOSURE AXIOM FOR ADDITION

For all real numbers a and b, $a + b$ is exactly one real number.

CLOSURE AXIOM FOR MULTIPLICATION

For all real numbers a and b, $a \cdot b$ is exactly one real number.

The closure axioms state that the sum of any two real numbers is always a real number and that the product of any two real numbers is always a real number.

EXAMPLE 2.2.1 Is the set of natural numbers, N, closed with respect to subtraction?

Solution To disprove a statement, or to show that a statement is false, a single counterexample is sufficient. Thus consider $5 - 5 = 0$; $5 \in N$, but $0 \notin N$. Therefore, N is *not* closed with respect to subtraction. Another counterexample is the following: $5 - 7 = -2$, $5 \in N$, $7 \in N$, but $-2 \notin N$.

The set of integers, however, is closed with respect to subtraction. That is, the difference of any two integers is again an integer. The integers are thus an extension of the natural numbers that provide a set which is closed with respect to subtraction.

Similarly, the rational numbers are an extension of the integers that provide a set which is closed with respect to division. (Division by zero is not permitted, however.)

EXAMPLE 2.2.2 Show by counterexample that the set of integers is not closed with respect to division.

Solution Let $a = 3$ and $b = 5$; then $\frac{a}{b} = \frac{3}{5}$, and $\frac{3}{5}$ is not an integer. However, $\frac{3}{5}$ *is* a rational number, as are all ratios of integers except $\frac{a}{0}$.

The preceding discussion leads to the following generalized result.

The set of real numbers is closed with respect to all four operations of arithmetic: addition, subtraction, multiplication, and division (excluding division by zero).

EXAMPLE 2.2.3 Let $S = \{1, 2, 3\}$. Is S closed with respect to addition?

Solution The closure axiom states that for *all* possible combinations of elements in the set, the sum must also be in the set; therefore, since $1 + 3 = 4$, and since 4 is not in set S, S is *not* closed with respect to addition.

The identity and inverse axioms are important not only for their significance as properties of the real numbers but also for solving equations.

THE IDENTITY AXIOM FOR ADDITION

For every real number r there is exactly one real number, 0 (zero), such that

$$r + 0 = r \quad \text{and} \quad 0 + r = r$$

The number 0 is called the **additive identity**.

THE IDENTITY AXIOM FOR MULTIPLICATION

For every real number r there is exactly one real number, 1 (one), such that

$$r \cdot 1 = r \quad \text{and} \quad 1 \cdot r = r$$

The number 1 is called the **multiplicative identity**.

These axioms state that the addition of 0 to a real number does not change the value of the number and that the multiplication of a real number by 1 does not change the value of the number.

THE INVERSE AXIOM FOR ADDITION

For every real number r there exists exactly one real number, $(-r)$, called the additive inverse of r, or the opposite of r, such that

$$r + (-r) = 0 \quad \text{and} \quad -r + r = 0$$

EXAMPLE 2.2.4 Find the additive inverse (or opposite) of each of the following numbers:

a. 3 b. -215 c. $-\frac{1}{2}$ d. $\frac{2}{9}$ e. $\sqrt{2}$

Solution Using the inverse axiom for addition, $r + (-r) = 0$:

a. $3 + (-3) = 0$; thus the inverse of 3 is -3.

b. $-215 + [-(-215)] = 0$; but $-(-215) = 215$; therefore, the inverse of -215 is 215.

c. $-\frac{1}{2} + \left[-\left(-\frac{1}{2}\right)\right] = 0$; thus the inverse of $-\frac{1}{2}$ is $\frac{1}{2}$.

d. $\frac{2}{9} + \left(-\frac{2}{9}\right) = 0$; thus the inverse of $\frac{2}{9}$ is $-\frac{2}{9}$.

e. $\sqrt{2} + (-\sqrt{2}) = 0$; thus the inverse of $\sqrt{2}$ is $-\sqrt{2}$.

THE INVERSE AXIOM FOR MULTIPLICATION

For every nonzero real number r there exists exactly one real number, $\frac{1}{r}$, called the **multiplicative inverse** of r, or the **reciprocal** of r, such that

$$r \cdot \frac{1}{r} = 1 \quad \text{and} \quad \frac{1}{r} \cdot r = 1$$

EXAMPLE 2.2.5 Find the multiplicative inverse (reciprocal) of each of the following numbers:

a. 3 b. -215 c. $\frac{1}{2}$ d. $-\frac{2}{3}$ e. 0

Solution Using the inverse axiom for multiplication, $r \cdot \frac{1}{r} = 1$:

a. $3 \cdot \frac{1}{3} = 1$; therefore, the reciprocal of 3 is $\frac{1}{3}$.

b. $-215 \cdot \left(-\frac{1}{215}\right) = 1$; therefore, the reciprocal of -215 is $-\frac{1}{215}$.

c. $\frac{1}{2}\left(\frac{1}{\frac{1}{2}}\right) = 1$. However, $\frac{1}{2}(2) = 1$. Since each nonzero real number has exactly one reciprocal, $\frac{1}{\frac{1}{2}} = 2$. Therefore, the reciprocal of $\frac{1}{2}$ is 2.

d. $-\frac{2}{3}\left(\frac{1}{-\frac{2}{3}}\right) = 1$. But $-\frac{2}{3}\left(-\frac{3}{2}\right) = 1$. Therefore, $\frac{1}{-\frac{2}{3}} = -\frac{3}{2}$ since the reciprocal of a real number is unique. Thus the reciprocal of $-\frac{2}{3}$ is $-\frac{3}{2}$.

e. 0 has no reciprocal since $\frac{1}{0}$ is undefined.

In Chapter 1 it was seen that the definition of subtraction,

$$r - s = r + (-s)$$

preserves the arithmetical meaning of subtraction—that is, $r - s$ is the number which when added to s produces r, $(r - s) + s = r$. By using the concept of the reciprocal, division may be defined in a similar way.

ALTERNATE DEFINITION OF DIVISION

For all nonzero real numbers r and s,

$$\frac{r}{s} = r \cdot \frac{1}{s} = \frac{1}{s} \cdot r$$

For example, $\frac{10}{2} = 10 \cdot \frac{1}{2} = \frac{1}{2} \cdot 10 = 5$. Similarly, $\frac{x}{2} = \frac{1}{2}x$.

Since multiplication is commutative and associative, this means that an expression such as $\frac{xy}{z}$ can be calculated in several ways without changing the value of the expression. In other words,

$$\frac{xy}{z} = (xy)\left(\frac{1}{z}\right) = \left(\frac{x}{z}\right)y = x\left(\frac{y}{z}\right) = x\left(\frac{1}{z}\right)y = \left(\frac{1}{z}\right)xy$$

As examples,

$$\frac{(6)(7)}{3} = \left(\frac{6}{3}\right)(7) = 2 \cdot 7 = 14$$

$$\frac{8(-10)}{-5} = 8\left(\frac{-10}{-5}\right) = 8(+2) = 16$$

The inverse and identity axioms, combined with the associative and commutative axioms, may be used to simplify certain algebraic expressions —that is, to write these expressions in a shorter or simpler form.

Usually, an algebraic expression is said to be simplified if it is written with as few symbols as possible. There are some exceptions to this general rule, but these exceptions will be stated when they arise.

EXAMPLE 2.2.6 Simplify $(x + 2) - 2$.

Solution

$$\begin{aligned}(x + 2) - 2 &= x + (2 - 2)\\ &= x + 0\\ &= x\end{aligned}$$

Thus x is the simplified form for $(x + 2) - 2$.

EXAMPLE 2.2.7 Simplify $3 + (2p - 3)$.

Solution

$$\begin{aligned}3 + (2p - 3) &= 2p + (3 - 3)\\ &= 2p + 0\\ &= 2p\end{aligned}$$

Thus $2p$ is the simplified form for $3 + (2p - 3)$.

EXAMPLE 2.2.8 Simplify $\frac{2y}{2}$.

Solution

$$\begin{aligned}\frac{2y}{2} &= \left(\frac{1}{2}\cdot 2\right)y\\ &= 1\cdot y\\ &= y\end{aligned}$$

Thus $\frac{2y}{2} = y$.

EXAMPLE 2.2.9 Simplify $\frac{-5x}{-5}$.

Solution

$$\begin{aligned}\frac{-5x}{-5} &= \frac{1}{-5}(-5x)\\ &= 1\cdot x\\ &= x\end{aligned}$$

Thus $\dfrac{-5x}{-5} = x.$

It is often necessary to find the inverse of a number that is expressed in decimal representation.

EXAMPLE 2.2.10 Find the opposite and the reciprocal of 0.25.

Solution The opposite of 0.25 is -0.25, because $0.25 + (-0.25) = 0$ and $(-0.25) + (0.25) = 0$. The reciprocal can be found in two ways:

1. By the axiom, the reciprocal of 0.25 is

$$\frac{1}{0.25} = \frac{1}{\left(\frac{25}{100}\right)} = \frac{100}{25} = 4$$

2. The common fractional representation of 0.25 is $\frac{1}{4}$ and the reciprocal of $\frac{1}{4}$ is

$$\frac{1}{\frac{1}{4}} = 1 \cdot \frac{4}{1} = 4$$

EXERCISES 2.2 A

Determine which of the sets in Exercises 1–10 is closed with respect to the given operation. If the set is closed, write "closed." If it is not closed, give a counterexample.

1. The set of natural numbers, addition
2. The set of natural numbers, division
3. The set of rational numbers, division
4. The set of real numbers, subtraction
5. The set of integers, multiplication
6. The set $\{0, 1\}$, addition
7. The set $\{0, 1\}$, subtraction
8. The set $\{0, 1\}$, multiplication
9. The set $\{0, 1\}$, division
10. The set of *even* natural numbers, $\{2, 4, 6, \ldots\}$
a. Addition b. Subtraction c. Multiplication d. Division

Find the additive inverse (opposite) of each number in Exercises 11–15.

11. 4

12. -3

13. $\frac{1}{5}$

14. $-\frac{2}{7}$

15. 1256

Find the multiplicative inverse (reciprocal) of each number in Exercises 16–25.

16. 12

17. -5

18. $\frac{1}{5}$

19. $-\frac{2}{7}$

20. $\frac{3}{10}$

21. 0.125

22. -2.5

23. $4\frac{1}{3}$

24. 0.0001

25. $-\frac{3}{250}$

Simplify each expression in Exercises 26–45.

26. $(x + 7) - 7$

27. $(4 + y) - 4$

28. $r + (25 - r)$

29. $-9\left(\frac{x}{-9}\right)$

30. $\frac{6b}{6}$

31. $\frac{(5x + 6) - 6}{5}$

32. $(0 - n) + n$

33. $(-n + n) + \frac{-n}{n}$

34. $\frac{2[(5 - x) - 5]}{-2}$

35. $\frac{(-6 - 8x) - (-6)}{-8}$

36. $\frac{-x}{x}(0 - x)$

37. $\frac{(0 - 4y) + 4y}{0 - 4}$

38. $\frac{(3x - 7) + 7}{(3 - 8y) + 8y}$

39. $\frac{-\frac{1}{3}(x + 4)}{3(x + 4)}$

40. $[(-6 + x) + (6 + 8y)] - x$

41. $(n - n)\left(\frac{-n}{n}\right)$

42. $(1 - x - 1)\left(\frac{-x + x}{-x - x}\right)$

43. $\left(4 + \frac{-5}{-5} - 4\right)\left(-5 + \frac{4x}{4} + 5\right)$

44. $\frac{3x + 4 - 4}{3 - 4x + 4x}$

45. $\frac{7 - 4x + 4x - 2}{2 + 4x - 7 - 4x}$

46. What number when added to (-17) yields the additive identity?

47. What number when multiplied by $\frac{2}{3}$ yields the multiplicative identity?

48. The sum of 3 and y equals the additive identity. What must y equal?

49. If the product of x and $-\frac{1}{4}$ equals the multiplicative identity, what must x equal?

50. If the product of 4 and y equals 4, what must y equal?

EXERCISES 2.2 B

Determine which of the sets in Exercises 1–10 is closed with respect to the given operation. If the set is closed, write "closed." If it is not closed, give a counterexample.

1. The set of rational numbers, subtraction
2. The set of integers, division
3. The set of natural numbers, subtraction
4. The set of integers, subtraction
5. The set of real numbers, division
6. The set $\{-1, 0, 1\}$, addition
7. The set $\{-1, 0, 1\}$, subtraction
8. The set $\{-1, 0, 1\}$, multiplication
9. The set $\{-1, 0, 1\}$, division
10. The set of *odd* natural numbers, $\{1, 3, 5, \ldots\}$
a. Addition b. Subtraction c. Multiplication d. Division

Find the additive inverse (opposite) of each number in Exercises 11–15.

11. 7 **12.** -14 **13.** $\frac{1}{12}$

14. $-\frac{4}{9}$ **15.** $-\sqrt{3}$

Find the multiplicative inverse (reciprocal) of each number in Exercises 16–25.

16. 215 **17.** -12 **18.** $\frac{1}{10}$

19. $-\frac{4}{9}$ **20.** $\frac{5}{12}$ **21.** $0.333\ldots$

22. -3.125

23. $5\frac{1}{2}$

24. 1.01

25. $-\frac{5}{17}$

Simplify each expression in Exercises 26–45.

26. $(x - 3) + 3$

27. $(2 + y) - 2$

28. $a + (b - a)$

29. $-3\left(\frac{x}{-3}\right)$

30. $\frac{0.5m}{0.5}$

31. $\frac{(-4x - 9) + 9}{-4}$

32. $m + (0 - m)$

33. $\left((-n + \frac{n}{-n}\right) - (-n)$

34. $\frac{-7[(-3 - x) + 3]}{7}$

35. $\frac{(5 - 8x) - 5}{-8}$

36. $\frac{5x - (0 - 5x)}{5}$

37. $\frac{-xy + [0 - (-xy)]}{0 - 1}$

38. $\frac{5[(2x + 3) - 3]}{-2x + (5 + 2x)}$

39. $\frac{-\frac{1}{2}(x - 6)}{-2(x - 6)}$

40. $(-7x + y) + [(5 - y) + (7x - 5)]$

41. $(-7x + 7x)(5 - x - 5)$

42. $\frac{-5n}{-5}(0 - n + n)$

43. $\left(-6 + \frac{-3y}{-3} + 6\right)\left(\frac{x}{x} + 6 - 6\right)$

44. $\frac{-5n - 2a + 2a}{-5n + 2a - 2a}$

45. $\frac{(a + b - c)(4x + c - c)}{4(a + b - c)}$

46. What number when added to $-\frac{1}{3}$ yields the additive identity?

47. What number when multiplied by $-\frac{3}{5}$ yields the multiplicative identity?

48. The product of 5 and x equals the multiplicative identity. What must x equal?

49. If the sum of 5 and x equals the additive identity, what must x equal?

50. If the sum of -6 and y equals -6, what must y equal?

2.3 THE DISTRIBUTIVE AXIOM

The distributive axiom is significant for many reasons. Three of these reasons are listed below:

1. It links the operations of addition and multiplication.
2. It permits the collecting of "like terms" to simplify expressions, such as $2x + 3x$, which can be written as $5x$.
3. It permits the "removal of parentheses" needed in simplifying expressions and in solving equations.

THE DISTRIBUTIVE AXIOM

For all real numbers r, s, and t,

$$r(s + t) = rs + rt$$

In other words, multiplication "distributes" over addition. The factor r is distributed to *each* term of the sum.

EXAMPLE 2.3.1 Use the distributive axiom to remove the parentheses in $3(x + 2)$.

Solution
$$\begin{aligned} 3(x + 2) &= 3 \cdot x + 3 \cdot 2 \\ &= 3x + 6 \end{aligned}$$

EXAMPLE 2.3.2 Express $5(y + z)$ as a sum by applying the distributive axiom.

Solution $5(y + z) = 5y + 5z$

The following immediate consequences of the distributive axiom are useful:

1. $a(b - c) = ab - ac$
2. $ab + ac = a(b + c)$
3. $ac + bc = (a + b)c$
4. $ac - bc = (a - b)c$
5. $ad + bd + cd = (a + b + c)d$
6. $-(b + c) = (-1)(b + c) = -b - c$
7. $-(b - c) = (-1)(b - c) = -b + c$

EXAMPLE 2.3.3 Express $2(x - 5)$ as a sum, using the distributive axiom or one of its consequences.

Solution
$$\begin{aligned} 2(x - 5) &= 2(x) - 2(5) \qquad \text{(By consequence 1)} \\ &= 2x - 10 \end{aligned}$$

EXAMPLE 2.3.4 Express $2x + 3x$ as a product by using consequence 3.

Solution $2x + 3x = (2 + 3)x = 5x$

EXAMPLE 2.3.5 Simplify $5z - 7z$.

Solution Expressing $5z - 7z$ as a product by using consequence 4,

$$5z - 7z = (5 - 7)z = -2z$$

EXAMPLE 2.3.6 Simplify $3y + 2y - 4y$.

Solution

$$\begin{aligned} 3y + 2y - 4y &= (3 + 2 - 4)y \\ &= 1y \\ &= y \end{aligned}$$

(By consequence 5 and the identity axiom, multiplication)

EXAMPLE 2.3.7 Remove the parentheses: $-(2x + 8)$.

Solution Using consequence 6,

$$-(2x + 8) = -2x - 8$$

EXAMPLE 2.3.8 Remove the parentheses: $-(5 - 3x)$.

Solution Using consequence 7,

$$\begin{aligned} -(5 - 3x) &= -5 + 3x \\ &= 3x + (-5) \\ &= 3x - 5 \end{aligned}$$

EXAMPLE 2.3.9 Remove the parentheses: $-4(2x + 5)$.

Solution

$$-4(2x + 5) = -8x - 20$$

One of the purposes of the distributive axiom, as stated at the beginning of this section, is to "collect *like* terms."

DEFINITIONS

Like terms are terms with identical literal factors.
A literal factor is a factor denoted by a letter.

As examples, $5x$ and $2x$ are like terms; $16x^2$ and $-4x^2$ are like terms; $39xy^2z$ and $-\frac{1}{2}xy^2z$ are like terms.

$3x$ and $3y$ are *not* like terms, because the literal factors, x and y, are not identical. Similarly, $5x^2$ and $5x$ are *not* like terms, and xy and x are *not* like terms. Terms which are not like are called *unlike* terms.

The distributive axiom can be used to collect like terms, as shown in the preceding examples and also in the following examples. For convenience, the distributive axiom and all its consequences will be referred to simply as "the distributive axiom" from now on.

Simplify Examples 2.3.10–15 by collecting like terms.

EXAMPLE 2.3.10 $2x - 3x$

Solution $2x - 3x = (2 - 3)x = -x$

EXAMPLE 2.3.11 $4x^2 + 6x^2 - 2x^2$

Solution $4x^2 + 6x^2 - 2x^2 = (4 + 6 - 2)x^2 = 8x^2$

EXAMPLE 2.3.12 $2x + 3 + 4x$

Solution

$$\begin{aligned} 2x + 3 + 4x &= (2x + 4x) + 3 \\ &= (2 + 4)x + 3 \\ &= 6x + 3 \end{aligned}$$

(Note that $6x$ and 3 are unlike terms)

EXAMPLE 2.3.13 $x^2 + 2x^2$

Solution

$$\begin{aligned} x^2 + 2x^2 &= 1x^2 + 2x^2 \\ &= (1 + 2)x^2 \\ &= 3x^2 \end{aligned}$$

EXAMPLE 2.3.14 $xy - 2y + xy + x$

Solution

$$\begin{aligned} xy - 2y + xy + x &= (xy + xy) + x - 2y \\ &= 2xy + x - 2y \end{aligned}$$

(Note that these three terms are unlike terms)

EXAMPLE 2.3.15 $x^2 + 5x - 6x + 2$

Solution

$$\begin{aligned} x^2 + 5x - 6x + 2 &= x^2 + (5x - 6x) + 2 \\ &= x^2 + (5 - 6)x + 2 \\ &= x^2 + (-1)x + 2 \\ &= x^2 - x + 2 \end{aligned}$$

EXAMPLE 2.3.16 Simplify $3(x + 2y) + 5(x - y)$.

Solution
$$\begin{aligned} 3(x + 2y) + 5(x - y) &= 3x + 6y + 5x - 5y \\ &= (3x + 5x) + (6y - 5y) \\ &= (3 + 5)x + (6 - 5)y \\ &= 8x + y \end{aligned}$$

EXAMPLE 2.3.17 Simplify $2(x + y) - (x - y)$.

Solution
$$\begin{aligned} 2(x + y) - (x - y) &= 2x + 2y - x + y \\ &= (2x - x) + (2y + y) \\ &= x + 3y \end{aligned}$$

With practice, the student should be able to omit many of the steps shown in the examples.

HISTORICAL NOTE

The discovery that there was structure in the number system of arithmetic (such as the closure, commutative, associative, and distributive axioms) was made at the beginning of the nineteenth century by the British school of algebraists—George Peacock (1791–1858), Duncan Farquharson Gregory (1813–1844), Augustus DeMorgan (1806–1871), and others.

In 1843, the Irish mathematician William Rowan Hamilton (1805–1865) invented his algebra of quaternions, which describe rotations in space, in which multiplication is *not* commutative.

In 1857, the English mathematician Arthur Cayley (1821–1895) developed another noncommutative algebra, the algebra of matrices.

Later were invented the nonassociative Lie and Jordan algebras, named after the Norwegian mathematician Sophus Lie (1842–1899) and the French mathematician Camille Jordan (1838–1922). Today abstract algebra is a major area of research.

EXERCISES 2.3 A

For Exercises 1–20, remove parentheses by applying the distributive axiom (or one of its consequences).

1. $2(x + 5)$

2. $-3(y + 2)$

3. $5(n - 7)$

4. $-6(2x - 3)$

5. $-(x - 5)$

6. $-(3x + 4)$

7. $-(6 - y)$

8. $3(x + y - 2)$

9. $-\frac{1}{2}(2x - 6y)$

10. $(a + b)x$

11. $(2a - b)3y$

12. $a - (b - c)$

13. $x - (y + 5)$

14. $-4(-2x + 3y - 4)$

15. $a - 2[b - (c - 3)]$

16. $y - [4z - 2(x + 5)]$

17. $3(a - 2b) + 4(x + 2y)$

18. $0 - (c - d)$

19. $4x^2(x - 1) - (x - 1)$

20. $a(x + 5) - 2(x + 5)$

Combine like terms and simplify by using the distributive axiom (or one of its consequences) in Exercises 21–35.

21. $3x + 2x$

22. $x + 4x + 5x$

23. $2x - x$

24. $3y - 3y$

25. $-4xy + 2xy$

26. $3z + 5z - 7z$

27. $3x + 5 - x$

28. $10k - 5k + 3k$

29. $6 - 4t - 3t$

30. $5x^2 + 5x - 8x - 2$

31. $3z^2 - z + 6z + 4$

32. $3p - q - 2p$

33. $6x^2 + 12xy - 5xy - 10y^2$

34. $x^2 - xy + xy - y^2$

35. $2ab - 3ac + 4ca - 5ba$

In Exercises 36–50, remove parentheses and collect like terms.

36. $3(x + 2) - 5$

37. $-(y - 3) + 2y$

38. $3x^2 - (5x - 2x) + 3$

39. $y^2 - (y + 3y) + 2(y^2 + 1)$

40. $6(x + y) + 2(x + y)$

41. $-7(c + d) - (c + d)$

42. $5(a - b) - 4(a - b)$

43. $3x(2x - 1) - 2x(3x + 2)$

44. $4x(2y - 3) - 3y(2x - 5)$

45. $10 - (x + 5)$

46. $6k - (4 + 5k)$

47. $2a(b - 8) - b(a - 6)$

48. $2n - [3(n - 2) - 5(n + 3)]$

49. $2x - [2 - (x - 2) - 2(x - 2)]$

50. $3(c + 4d) - 5(c + 4d) - (c + 4d)$

By using the distributive, commutative, and associative axioms, rearrange the numbers in Exercises 51–60 so that each computation can be performed mentally in the easiest way possible.

51. $45(37) + 45(63)$

52. $24\left(\frac{1}{2} + \frac{1}{3}\right)$

53. $\frac{1}{7}(65) + \frac{6}{7}(65)$

54. $58(99) + 58$

55. $\left(\frac{3}{2} + \frac{2}{3}\right)12$

56. $87(450) - 86(450)$

57. $246\left(\frac{7}{8}\right) - \frac{3}{8}(246)$

58. $60\left(\frac{5}{12} - \frac{1}{15} + \frac{1}{12}\right)$

59. $18\left(\frac{9}{7}\right) - \left(\frac{5}{6} + \frac{2}{7}\right)18$

60. $497(8) - 497(5) + 7(497)$

EXERCISES 2.3 B

For Exercises 1–20, remove parentheses by applying the distributive axiom (or one of its consequences).

1. $3(x + 2)$

2. $-2(y - 5)$

3. $-4(x + y)$

4. $7(5x - 4)$

5. $-(x + 2)$

6. $-(2y - 9)$

7. $-(8 - 3y)$

8. $-3(2x + 3y - 1)$

9. $\frac{1}{2}(2x + 4y)$

10. $(a - b)y$

11. $(5n + 2)6a$

12. $x - (z + y)$

13. $a - (2b - 3c)$

14. $-2(3x - y - 5)$

15. $c + 5[a - 2(b + 4)]$

16. $6z - [y - (x - 8)]$

17. $5(m - 3y) - 2(n - 4x)$

18. $0 - (c + d)$

19. $9n^2(y + 4) - (y + 4)$

20. $c(x - 6) - d(x - 6)$

Combine like terms and simplify by using the distributive axiom (or one of its consequences) in Exercises 21–35.

21. $5y + 7y$

22. $2x + x - 3x$

23. $5y - y$

24. $-xy + 3xy$

25. $3z + 3z^2 - z$

26. $2k + 7 - 5k$

27. $x^2 + 2x - 3x + 5$

28. $2x + 5 + 3x - 7$

29. $4 - x + 2$

30. $5n + 4m + n$

31. $2s - t + 3s + 2t$

32. $2p + 3q + 3p$

33. $3x^2 - 6xy + 5xy - 10y^2$

34. $5xy - 6xz - 4xy - 2zx$

35. $9x^2 + 12xy - 12xy - 16y^2$

In Exercises 36–50, remove parentheses and collect like terms.

36. $2(x + x) + 5$

37. $-(4 - 2k) + 3k$

38. $4y^2 - (3y + 5y) + 9$

39. $x^2 - (x - 2) + 3(x^2 + 5)$

40. $7(x + y) - 4(x + y)$

41. $-8(a - b) - (a - b)$

42. $2(x + y) - 2(x - y)$

43. $6x(5y - 2) - 3y(5x - 1)$

44. $7y(5y - 2) - 5y(7y + 2)$

45. $8x - (3x - 4)$

46. $6x - (6 + 7x)$

47. $a(b + 1) - 3b(5 - a)$

48. $5k - 2[4(2 - k) - (6 - k)]$

49. $x - 3[x - 3(x - 3)]$

50. $2(m - 4n) - 6(m - 4n) + (m - 4n)$

By using the distributive, commutative, and associative axioms, rearrange the numbers in Exercises 51–60 so that each computation can be performed mentally in the easiest way possible.

51. $37(65) + 35(37)$

52. $123(79) - 23(79)$

53. $\left(\frac{1}{12} + \frac{1}{18}\right)36$

54. $117\left(\frac{11}{9}\right) - \frac{2}{9}(117)$

55. $4(25 + 20)$

56. $8(125 - 50)$

57. $0.25(89) + 89(0.75)$

58. $96(125) + (875)96$

59. $26(98) - 98(13 + 12)$

60. $0.06(7500) + 0.03(7500) - 7500(0.08)$

2.4 EQUATIONS AND SOLUTIONS

2.4.1 Basic Concepts

An equation is a statement having the form $A = B$. However, an equation does not have to be a true statement. Consider the following examples of equations:

$2 + 4 = 6$	(True)
$7 - 3 = 8$	(False)
$x + 5 = 5 + x$	(Open, true for all values of x)
$x + 5 = x + 1$	(Open, false for all values of x)
$x + 5 = 9$	(Open, true for some value of x and false for some value of x)

An **open equation** is an equation containing one or more variables. An open equation is not classified as true or false, but it becomes true or false when each variable is replaced by a numerical value.

The equation $2 + 4 = 6$ is called a *true equation* since it does not contain any variables and since it is known that the sum of 2 and 4 is 6.

The equation $7 - 3 = 8$ is called a *false equation* since it contains no variables and since it is known that the difference between 7 and 3 is 4 and not 8.

The open equation $x + 5 = 5 + x$ becomes true for all replacements of the variable. Such an equation is also called an **identity**.

The open equation $x + 5 = x + 1$ becomes false for all replacements of the variable. Equations such as these are sometimes called *contradictions*.

The open equation $x + 5 = 9$ is neither true nor false, but it becomes either true or false when the variable is replaced by a numerical value.

For example, if $x = 2$, then $x + 5 = 9$ becomes $2 + 5 = 9$, a false equation.

If $x = 4$, then $x + 5 = 9$ becomes $4 + 5 = 9$, a true equation.

An open equation that is sometimes true and sometimes false is also called a **conditional equation** because the equation becomes true on the condition that the variable is replaced by a certain numerical value. The conditional equation $x + 5 = 9$ is true on the condition that x is replaced by 4.

The number 4 is called a **solution** or **root** of the equation $x + 5 = 9$ because $x + 5 = 9$ becomes true when x is replaced by 4. The number 4 is also said to **satisfy** the equation $x + 5 = 9$.

An important concern of algebra is the process of finding all the solutions of an open equation. This process is called **solving the equation.**

DEFINITIONS

A **solution** or **root** of an open equation in one variable is a number that makes the equation true when the variable is replaced by the name of this number.

The **solution set** of an open equation is the set of all solutions of the equation.

To solve an equation means to find the solution set of the equation.

One primitive method for solving an equation is the trial and error method. This method is illustrated in the example below.

EXAMPLE 2.4.1 Solve $2x^2 + 5x = 3$.

Trial and Error Solution First, try positive integers.

If $x = 1$, then $2x^2 + 5x = 2\cdot 1^2 + 5\cdot 1 = 2 + 5 = 7$; $7 = 3$ is false.
If $x = 2$, then $2x^2 + 5x = 2\cdot 2^2 + 5\cdot 2 = 8 + 10 = 18$; $18 = 3$ is false.
If $x = 3$, then $2x^2 + 5x = 2\cdot 3^2 + 5\cdot 3 = 18 + 15 = 33$; $33 = 3$ is false.

Since the resulting numbers are getting larger and larger, it seems useless to continue trying positive integers. Thus negative integers are tried next.

If $x = -1$, then $2x^2 + 5x = 2(-1)^2 + 5(-1) = 2 - 5 = -3$; $-3 = 3$ is false.
If $x = -2$, then $2x^2 + 5x = 2(-2)^2 + 5(-2) = 8 - 10 = -2$; $-2 = 3$ is false.
If $x = -3$, then $2x^2 + 5x = 2(-3)^2 + 5(-3) = 18 - 15 = 3$; $3 = 3$ is true.
At last a solution has been found! $x = -3$.

Have all the solutions been found? There is no way to know for certain by this method because all the integers have not been tried nor have any of the nonintegral rational numbers been tried.

As a matter of fact, it may be shown that $x = \frac{1}{2}$ is a solution.

If $x = \frac{1}{2}$, then $2x^2 + 5x = 2\left(\frac{1}{2}\right)^2 + 5\left(\frac{1}{2}\right) = \frac{1}{2} + \frac{5}{2} = \frac{6}{2} = 3$; $3 = 3$ is true.

The solution set of $2x^2 + 5x = 3$ is $\left\{\frac{1}{2}, -3\right\}$. However, at this point, there is no guarantee that the solution set has been found, that is, that $x = \frac{1}{2}$ and $x = -3$ are the *only* replacements that make the equation $2x^2 + 5x = 3$ true.

Finally, it seems reasonable that there must be a better method than the trial and error method for solving equations. Indeed, the next sections are devoted to the development of procedures for solving linear equations—procedures that will be less time-consuming than the trial and error method and that will give assurance that the solution set has indeed been found.

2.4.2 Equivalence Theorems

DEFINITION

Equivalent equations are equations that have the same solution set.

For example, all the equations listed below are equivalent because each has {5} for its solution set—that is, 5 is the only solution of each equation:

$$x + 2 = 7$$

$$2(x - 3) = 4$$

$$\frac{x}{3} = \frac{5}{3}$$

$$x = 5$$

The process of solving an equation involves the replacement of one equation by a simpler equivalent equation.

For example, a procedure is developed so that $2x - 6 = 4$ is replaced by the simpler equivalent equation $x = 5$. Not only does this technique find the solution but it also guarantees that all the solutions have been found, since it is known that 5 is the only replacement for x that makes $x = 5$ a true statement.

The following equivalence theorems are fundamental to the process of solving equations and are stated without proof.

THE EQUIVALENCE THEOREMS

Let $A = B$ be an open equation, and let C be any real number.

1. **Addition theorem** $A = B$, **if and only if** $A + C = B + C$
   If the same number is added to each side of an equation, the resulting sums are again equal, and the two equations are equivalent.

2. **Subtraction theorem** $A = B$, **if and only if** $A - C = B - C$
   If the same number is subtracted from each side of an equation, the resulting differences are equal, and the two equations are equivalent.

3. **Multiplication theorem** $A = B$ **if and only if** $AC = BC$ **and** $C \neq 0$
   If each side of an equation is multiplied by the same nonzero real number, the resulting products are equal, and the two equations are equivalent.

4. Division theorem $A = B$ if and only if $\frac{A}{C} = \frac{B}{C}$ and $C \neq 0$

If each side of an equation is divided by the same nonzero real number, the resulting quotients are equal, and the two equations are equivalent.

The following examples illustrate the use of the equivalence theorems.

EXAMPLE 2.4.2 Solve $x - 4 = 3$.

Solution $x - 4 = 3$ if and only if

$x - 4 + 4 = 3 + 4$ (Addition theorem—add 4 to both sides)

$x + 0 = 7$ (Inverse axiom, addition)

$x = 7$ (Identity axiom, addition)

Thus 7 is the solution and $\{7\}$ is the solution set.

Check: For $x = 7$, $x - 4 = 3$ becomes $7 - 4 = 3$, a true statement.

EXAMPLE 2.4.3 Solve $x + 2 = 4$.

Solution $x + 2 = 4$ if and only if

$x + 2 - 2 = 4 - 2$ (Subtraction theorem—subtract 2 from each side)

$x + 0 = 2$ (Inverse axiom, addition)

$x = 2$ (Identity axiom, addition)

Thus 2 is the solution. The solution set is $\{2\}$.

Check: For $x = 2$, $x + 2 = 4$ becomes $2 + 2 = 4$, true.

EXAMPLE 2.4.4 Solve $\frac{x}{2} = 3$.

Solution $\frac{x}{2} = 3$ if and only if

$2\left(\frac{x}{2}\right) = 2(3)$ (Multiplication theorem—multiply both sides by 2)

$x = 6$ (Inverse and identity axioms, multiplication)

Thus 6 is the solution.

Check: For $x = 6$, $\frac{x}{2} = 3$ becomes $\frac{6}{2} = 3$, true.

EXAMPLE 2.4.5 Solve $4x = 20$.

Solution $4x = 20$ if and only if

$$\frac{4x}{4} = \frac{20}{4}$$ (Division theorem—divide both sides by 4)

$$x = 5$$ (Inverse and identity axioms, multiplication)

Thus 5 is the solution.

Check: For $x = 5$, $4x = 20$ becomes $4(5) = 20$, true.

Equations such as $2x - 6 = 4$ and $5x + 10 = 0$ are called *linear equations.*

DEFINITION

A linear equation in one variable, x, is an equation that can be expressed in the form $ax + b = 0$ where $a \neq 0$.

By using the equivalence theorems, any linear equation can be solved by transforming the given equation into a simpler equation.

EXAMPLE 2.4.6 Solve $2x - 6 = 4$.

Solution $2x - 6 = 4$

$$2x - 6 + 6 = 4 + 6$$ (Addition theorem—add 6 to both sides)

$$2x = 10$$

$$\frac{2x}{2} = \frac{10}{2}$$ (Division theorem—divide each side by 2)

$$x = 5$$

Thus 5 is the solution.

Check: For $x = 5$, $2x - 6 = 4$ becomes $2(5) - 6 = 4$

$$10 - 6 = 4$$

$$4 = 4, \text{ true}$$

Since $2x - 6 = 4$ is true for $x = 5$, 5 is the solution of $2x - 6 = 4$.

EXAMPLE 2.4.7 Determine if the indicated value of the variable is a solution of the given equation: $3(x - 2) = 5x + 2$; $x = 4$.

Solution For $x = 4$,

$$3(x - 2) = 3(4 - 2) = 3(2) = 6$$
$$5x + 2 = 5(4) + 2 = 20 + 2 = 22$$

Since $6 \neq 22$, 4 is *not* a solution.

2.4.3 Solving Linear Equations

The examples of the previous section have shown how to find the solution of a linear equation systematically. The technique involves the use of the equivalence theorems to transform the given equation into one having the form $x = a$ where a is the solution desired. It should be noted that the solution process also involves the use of the commutative, associative, distributive, identity, and inverse axioms as well as the axioms of the equal relation.

The *operations* that are to be used in transforming the equation are determined by analyzing the original equation. The symbols indicate the order in which the operations were performed on the number named by the variable. The solution requires that the *inverse* operations be performed in the *reverse* order.

EXAMPLE 2.4.8 Solve

$$\frac{3(x - 5)}{2} + 4 = 10$$

Analysis (This work is usually done mentally.):

Order of Operations on x	*Inverse Operations*	*Reversed Order*
1. 5 was subtracted from x.	1. Add 5.	1. Subtract 4.
2. Result multiplied by 3.	2. Divide by 3.	2. Multiply by 2.
3. Result divided by 2.	3. Multiply by 2.	3. Divide by 3.
4. 4 was added.	4. Subtract 4.	4. Add 5.

Solution (This work is written.)

$$\frac{3(x - 5)}{2} + 4 = 10$$

$$\frac{3(x - 5)}{2} = 10 - 4 = 6 \qquad \text{(Subtract 4 from each side)}$$

$$3(x - 5) = 2(6) = 12 \qquad \text{(Multiply each side by 2)}$$

$$x - 5 = \frac{12}{3} = 4 \qquad \text{(Divide each side by 3)}$$

$$x = 4 + 5 = 9 \qquad \text{(Add 5 to each side)}$$

The solution is 9.

Check: If $x = 9$, then

$$\begin{aligned}\frac{3(x-5)}{2} + 4 &= \frac{3(9-5)}{2} + 4 \\ &= \frac{3(4)}{2} + 4 \\ &= \frac{12}{2} + 4 \\ &= 6 + 4 = 10\end{aligned}$$

EXAMPLE 2.4.9 Solve $3 - y = 6$.

Solution $3 - y = 6$

$$\begin{aligned} -3 + 3 - y &= -3 + 6 && \text{(Add } -3 \text{ to each side)} \\ -y &= 3 \\ (-1)(-y) &= (-1)(3) && \text{(Multiply each side by } -1) \\ y &= -3 \end{aligned}$$

The solution is -3.

Check: If $y = -3$, then $3 - y = 3 - (-3) = 3 + 3 = 6$. Thus $3 - y = 6$ is true for $y = -3$.

Note in this example that both sides were multiplied by -1 so that the final equivalent equation would have the form $y = a$ and *not* $-y = a$.

EXAMPLE 2.4.10 Solve $4(k + 3) = 10$.

Solution $4(k + 3) = 10$

$$\begin{aligned} 4k + 12 &= 10 && \text{(Distributive axiom)} \\ 4k + 12 - 12 &= 10 - 12 && \text{(Subtraction theorem—subtract 12 from each side)} \\ 4k &= -2 \\ \frac{4k}{4} &= \frac{-2}{4} && \text{(Division theorem—divide each side by 4)} \\ k &= -\frac{1}{2} \end{aligned}$$

The solution is $-\frac{1}{2}$.

Check:

$$\begin{aligned} 4(k + 3) &= 10 \\ 4\left(-\frac{1}{2} + 3\right) &= 10 \\ 4\left(\frac{5}{2}\right) &= 10 \\ 10 &= 10, \text{ true} \end{aligned}$$

EXAMPLE 2.4.11 Solve $8 = 8 - 5p$.

Solution $8 = 8 - 5p$

$8 + 5p = 8 - 5p + 5p$ (Addition theorem—add $5p$ to each side)

$8 + 5p = 8$

$5p = 0$ (Subtraction theorem—subtract 8 from each side)

$\frac{5p}{5} = \frac{0}{5}$ (Division theorem—divide each side by 5)

$p = 0$

The solution is 0. The solution set is $\{0\}$. Note that this is *not* the empty set.

Check:

$$8 = 8 - 5p$$
$$8 = 8 - 5(0)$$
$$8 = 8 - 0$$
$$8 = 8, \text{ true}$$

EXAMPLE 2.4.12 Solve $x + 2 = x + 5$.

Solution $x + 2 = x + 5$

$$x + 2 - 2 = x + 5 - 2$$
$$x = x + 3$$
$$x - x = x - x + 3$$
$$0 = 3$$

But 0 can never equal 3. Therefore, there is no value of x which makes the equation true, and the solution set is the empty set, $\varnothing$.

EXAMPLE 2.4.13 Solve $5(x - 4) = 5x - 20$.

Solution

$$5(x - 4) = 5x - 20$$
$$5x - 20 = 5x - 20$$

Since the left side and the right side of the equation are identical, the statement is true regardless of the value of x; in other words, all real numbers satisfy the equation and the solution set is R, the set of real numbers.

EXERCISES 2.4 A

For Exercises 1–10, determine whether the indicated value of the variable is a solution of the open equation.

1. $5x - 12 = 3;\ x = 3$

2. $\frac{3y}{2} + \frac{7}{2} = 2;\ y = -1$

3. $x^2 - 3x = 0;\ x = -3$

4. $(3t - 1)(3t + 5) = 7;\ t = \frac{2}{3}$

5. $2x^2 - 5x = 3;\ x = \frac{-1}{2}$

6. $y(y + 4) - 4 = 0;\ y = 2$

7. $\frac{2}{x} + \frac{x}{2} = 0;\ x = -2$

8. $a^2 + 5a + 1 = a^2;\ a = -\frac{1}{5}$

9. $\frac{p}{3} - 7 = 1;\ p = 24$

10. $(z + 5) - 5 = \frac{2}{3};\ z = -\frac{2}{3}$

Solve and check the equations in Exercises 11–45.

11. $x + 5 = 6$

12. $x - 2 = 5$

13. $y + 3 = -4$

14. $z - 1 = -7$

15. $2 - k = 3$

16. $4 - x = -1$

17. $2m = 6$

18. $5x = 12$

19. $2p + 3 = 6$

20. $2y - 1 = 7$

21. $2 - 3x = 4$

22. $1 = t + 9$

23. $2 = 2s + 3$

24. $x - 7 = 0$

25. $q + 2 = 0$

26. $2 - x = 0$

27. $4 - 3y = 0 - 3y$

28. $5(t + 2) = 15$

29. $2(n - 1) = 8$

30. $3(x + 4) = -12$

31. $2 + 3(x + 1) = 11$

32. $\frac{1}{2}r = 7$

33. $\frac{x}{4} = 5$

34. $2(2 + x) = 2(2 - x)$

35. $\frac{x + 2}{3} = -2$

36. $\frac{x - 1}{2} = 6$

37. $2x - 5 = -(5 - 2x)$

38. $\frac{x}{2} + 4 = 7$

39. $\frac{2(x + 5)}{3} = 4$

40. $2 - 2x = 2x - 2$

41. $-3(x - 1) = 12$

42. $\frac{-2(x + 1)}{5} + 1 = 3$

43. $2x + 2 = 2x - 2$

44. $2(x + 2) = 2x + 4$

45. $\frac{3(x - 2) + 6}{5} = 0$

EXERCISES 2.4 B

For Exercises 1–10, determine whether the indicated value of the variable is a solution of the open equation.

1. $\frac{3}{x} = 12; x = \frac{1}{4}$

2. $x + \frac{1}{x} + \frac{26}{5} = 0; x = 5$

3. $\frac{5a - 1}{5a + 1} = \frac{2}{3}; a = 1$

4. $(5x - 1)(2x + 1) = 1; x = \frac{1}{5}$

5. $x - \frac{x}{2} = 2x + 3; x = -2$

6. $5 + (k - 5) = -\frac{2}{3}; k = -\frac{2}{3}$

7. $2(x - 3) = 2x - 3; x = 5$

8. $5t^2 - 10t = 0; t = 0$

9. $y - \frac{1}{y} = 0; y = -1$

10. $3x - 2[x - (x - 2)] = 8; x = 4$

Solve and check the equations in Exercises 11–45.

11. $x + 7 = 12$

12. $y - 5 = 1$

13. $z + 4 = -2$

14. $x - 6 = -4$

15. $4 - k = 3$

16. $5 - q = -2$

17. $3p = 12$

18. $7x = 105$

19. $3m + 2 = 6$

20. $2x - 3 = 7$

21. $3 - 2n = 5$

22. $-2 = s + 3$

23. $14 = 3x - 2$

24. $x - 5 = 0$

25. $t + 7 = 0$

26. $4 - x = 0$

27. $6 - 2p = 2p - 6$

28. $2(x + 3) = 16$

29. $3(k - 4) = 12$

30. $-2(y + 1) + 2 = 6$

31. $3 - 2(x + 4) = 11$

32. $\frac{1}{3}z = 5$

33. $\frac{x}{5} = 1$

34. $\frac{x + 4}{2} = -1$

35. $2(4 - x) = 8 - 2x$

36. $\frac{x}{3} - 2 = 7$

37. $\frac{x + 3}{2} - 6 = 1$

38. $5 - x = x - 5$

39. $\frac{x-2}{3} = 4$

40. $\frac{x+4}{2} = 2$

41. $\frac{2x-1}{3} = -3$

42. $-(x-6) = 6 - x$

43. $5 - x = 4 - x$

44. $0 = \frac{2(3-x)}{5} - 2$

45. $2(3x-5) = 6x - 10$

2.5 MORE LINEAR EQUATIONS

The following examples illustrate the solution of more complicated linear equations in one variable.

EXAMPLE 2.5.1 Solve $2x + 9 = 5x - 12$.

Solution

$$
\begin{array}{rcll}
2x + 9 & = & 5x - 12 & \\
-9 & & -9 & \\
2x & = & 5x - 21 & \text{(Subtract 9 from both sides)} \\
-5x & & -5x & \\
-3x & = & -21 & \text{(Subtract } 5x \text{ from both sides)} \\
\frac{-3x}{-3} & = & \frac{-21}{-3} & \text{(Divide both sides by } -3\text{)} \\
x & = & 7 &
\end{array}
$$

Check: Left side. $2x + 9 = 2 \cdot 7 + 9 = 14 + 9 = 23$

Right side. $5x - 12 = 5 \cdot 7 - 12 = 35 - 12 = 23$

Thus $23 = 23$, and $\{7\}$ is the solution set.

EXAMPLE 2.5.2 Solve $5(2 - y) = 4y - (y + 4)$.

Solution

$$
\begin{array}{rcll}
5(2 - y) & = & 4y - (y + 4) & \\
10 - 5y & = & 4y - y - 4 & \text{(Using the distributive axiom)} \\
10 - 5y & = & 3y - 4 & \\
3y - 4 & = & 10 - 5y & \text{(Using the symmetric axiom)} \\
+4 & & +4 & \\
3y & = & 14 - 5y & \text{(Add 4 to both sides)} \\
+5y & & +5y &
\end{array}
$$

$$8y = 14 \qquad \text{(Add } 5y \text{ to both sides)}$$

$$y = \frac{14}{8} \qquad \text{(Divide both sides by 8)}$$

$$y = \frac{7}{4}$$

Check: Left side. $5(2 - y) = 5\left(2 - \frac{7}{4}\right) = 5\left(\frac{8}{4} - \frac{7}{4}\right) = 5 \cdot \frac{1}{4} = \frac{5}{4}$

Right side. $4y - (y + 4) = 4 \cdot \frac{7}{4} - \left(\frac{7}{4} + 4\right)$

$$= \frac{28}{4} - \frac{23}{4} = \frac{5}{4}$$

Thus $\frac{5}{4} = \frac{5}{4}$, and $\left\{\frac{7}{4}\right\}$ is the solution set.

EXAMPLE 2.5.3 Solve $3x + 25 + 4x = 1 + x$.

Solution It is usually desirable to simplify each side as much as possible before attempting the solution.

The left side of the equation is $3x + 25 + 4x$.

Combine like terms to obtain $7x + 25$. Thus

$$3x + 25 + 4x = 1 + x$$

$$7x + 25 = 1 + x \qquad \text{(Combine like terms of left side)}$$

$$7x + 25 - 25 = 1 + x - 25 \qquad \text{(Subtract 25 from each side)}$$

$$7x = -24 + x \qquad \text{(Simplify right side)}$$

$$7x - x = -24 + x - x \qquad \text{(Subtract } x \text{ from each side)}$$

$$6x = -24$$

$$\frac{6x}{6} = \frac{-24}{6} \qquad \text{(Divide each side by 6)}$$

$$x = -4$$

Check:

$$3x + 25 + 4x = 1 + x$$

$$3(-4) + 25 + 4(-4) = 1 + (-4)$$

$$-12 + 25 - 16 = 1 - 4$$

$$-28 + 25 = -3$$

$$-3 = -3, \text{ true}$$

The solution set is $\{-4\}$.

EXAMPLE 2.5.4 Solve $y + 7 - 4(y - 8) = 2(y + 2)$.

Solution

$$y + 7 - 4(y - 8) = 2(y + 2)$$

$$y + 7 - 4y + 32 = 2y + 4 \qquad \text{(Distributive axiom to remove parentheses)}$$

$$-3y + 39 = 2y + 4 \qquad \text{(Combine like terms on left side)}$$

$$-3y + 39 - 39 = 2y + 4 - 39 \qquad \text{(Subtract 39 from each side)}$$

$$-3y = 2y - 35 \qquad \text{(Simplify)}$$

$$-3y - 2y = 2y - 2y - 35 \qquad \text{(Subtract } 2y \text{ from each side)}$$

$$-5y = -35 \qquad \text{(Simplify)}$$

$$\frac{-5y}{-5} = \frac{-35}{-5} \qquad \text{(Divide each side by } -5\text{)}$$

$$y = 7$$

The check is left for the student.

EXAMPLE 2.5.5 Solve

$$\frac{2t - 4}{4} + 5 = t - (3 + t)$$

Solution

$$\frac{2t - 4}{4} + 5 = t - (3 + t)$$

$$\frac{2t - 4}{4} + 5 = t - 3 - t \qquad \text{(Remove parentheses on right side)}$$

$$\frac{2t - 4}{4} + 5 = -3 \qquad \text{(Simplify right side)}$$

$$\frac{2t - 4}{4} + 5 - 5 = -3 - 5 \qquad \text{(Subtract 5 from each side)}$$

$$\frac{2t - 4}{4} = -8 \qquad \text{(Simplify)}$$

$$4\left(\frac{2t - 4}{4}\right) = 4(-8) \qquad \text{(Multiply each side by 4)}$$

$$2t - 4 = -32 \qquad \text{(Simplify)}$$

$$2t - 4 + 4 = -32 + 4 \qquad \text{(Add 4 to each side)}$$

$$2t = -28 \qquad \text{(Simplify)}$$

$$\frac{2t}{2} = \frac{-28}{2} \qquad \text{(Divide each side by 2)}$$

$$t = -14$$

The solution set is $\{-14\}$, and the student should verify this solution.

With a little practice, several of the above steps can be done mentally.

EXERCISES 2.5 A

Solve and check the equations in Exercises 1–30.

1. $2x + 1 = x - 3$

2. $3x + 10 = 4 - 3x$

3. $7y + 7 = 4y - 2$

4. $2(m + 3) = 3(2 - m)$

5. $4p - 3(p - 5) = 25$

6. $10y - 275 = 25(y + 1)$

7. $x - 20 = 5 + x$

8. $3t + 1 - 2t = 0$

9. $5(y - 4) - 6(y + 1) = 4$

10. $10 + 2(3x + 4) = 3x + 19$

11. $3x - 6 = 3(x - 2)$

12. $\dfrac{2 - 3z}{2} = 3 - (z + 2)$

13. $x + 2 = \dfrac{3x - 2}{5}$

14. $\dfrac{8y + 7}{5} = 2y - 3$

15. $2(x - 1) + 4x - 3 = 6x - 5$

16. $8 - \dfrac{7t + 3}{4} - 2(3 - t) = 0$

17. $(2x - 5) - (4 - 3x) = 11$

18. $3(x + 2y) - 2(x + 3y) = 2x + 5$

19. $x - 2[x - 2(x - 2)] = 4$

20. $x(x - 2) - x(x + 2) = x + 20$

21. $4x - 1 = 2x - 5$

22. $13y - 6 = 19y + 42$

23. $4t - 5 = 7 - 2t$

24. $x + 4 = \dfrac{4(x - 1)}{4}$

25. $\dfrac{4(y - 2)}{5} = y - 3$

26. $10t - 2(3t - 1) = 23 - 3t$

27. $4(3x - 4) = 5(x - 2) + x$

28. $5(3 - y) = 121 - 6(25 - y)$

29. $30 - 3(2x - 3) = 5(3 - x)$

30. $12 = 8x - 3(7 - x)$

EXERCISES 2.5 B

Solve and check the equations in Exercises 1–30.

1. $3x + 2 = 2x - 5$

2. $y - 7 = 7 - y$

3. $y - 7 = y + 7$

4. $-2(t - 1) = 4(1 - 2t)$

5. $x - 2(x + 1) = -1$

6. $3k + 3(k - 1) = 21$

7. $x + (x + 2) + (x - 1) = 0$

8. $4p + 9 - 5p = 0$

9. $2(x + 7) - 5 = -1 + 4(x - 2)$

10. $5 + 3(2y + 1) = 6y + 8$

11. $\dfrac{5x + 2}{3} = 3x - 2$

12. $\dfrac{2 - 5x}{4} = 1 - (x + 1)$

13. $2x - 4 = 2(x - 2)$

14. $4(t - 20) - 5(t - 22) = 2t$

15. $5(4p - 3) + 3(2p + 2) = 8p$

16. $8x + (5 - x) = -30$

17. $(7 - 3x) - (5 + 2x) = 12$

18. $5(2x^2 - x) = 7 - 2(6 - 5x^2)$

19. $x - [1 - (x - 1)] = 3x$

20. $y(2y - 5) - 2y(y + 5) = 75$

21. $3x + 1 = 5x - 2$

22. $2y = 7 - 3(y - 1)$

23. $5(t - 6) = 3(23 - 2t)$

24. $3(1 - x) = 2(1 + x)$

25. $2x - 16 = 3(13 - 3x)$

26. $6(y - 3) = 5(2y - 5)$

27. $12 - 5(t - 2) = 2 - t$

28. $2n - (6 - n) = 4n - 13$

29. $x + 2(3x - 5) = 4$

30. $6(3w + 4) - 3(10w + 16) = 8(2w - 17)$

2.6 LITERAL EQUATIONS

A formula is a general equation for the solution of a specific problem. For example, a formula for finding the area of a rectangle is $A = LW$, where A represents area, and L and W the length and width of the rectangle. The circumference of a circle can be expressed as $C = 2\pi r$, where C represents the circumference, and r the radius of the circle. The reader is familiar with many formulas, and some were presented in Chapter 1.

Often a formula is not in the form desired or is not convenient for the solution of a problem. Since a formula is an equation, it can be treated as such and can be changed to a desired equivalent equation by the methods shown in the preceding sections.

EXAMPLE 2.6.1 The formula for the perimeter of a rectangle is

$$P = 2L + 2W$$

where P stands for the perimeter, L stands for the length, and W for the width. Solve this formula for W in terms of the other letters.

Solution The technique for solution is to treat W as a variable and all other numerals and letters as constants.

$$P = 2L + 2W$$

$$2W + 2L = P \qquad \text{(Symmetric axiom)}$$

$$2W + 2L - 2L = P - 2L \qquad \text{(Subtract } 2L \text{ from each side)}$$

$$2W = P - 2L \qquad \text{(Simplify)}$$

$$\frac{2W}{2} = \frac{P - 2L}{2} \qquad \text{(Divide each side by 2)}$$

$$W = \frac{P - 2L}{2}$$

Now the equation is solved for W.

An equation involving various letters, some of which are constant, is called a **literal equation.**

EXAMPLE 2.6.2 Solve the formula for the circumference of a circle, $C = 2\pi r$, for r.

Solution $C = 2\pi r$

$$2\pi r = C \qquad \text{(Symmetric axiom)}$$

$$\frac{2\pi r}{2\pi} = \frac{C}{2\pi} \qquad \text{(Divide each side by } 2\pi)$$

$$r = \frac{C}{2\pi}$$

Thus the radius of any circle is equal to the circumference of the circle divided by 2π.

EXAMPLE 2.6.3 $C = \frac{5}{9}(F - 32)$ is the formula for converting Fahrenheit temperature to degrees Celsius or centigrade. Rewrite the formula so that it can be readily used to convert centigrade temperature to Fahrenheit —that is, solve for F.

Solution $C = \frac{5}{9}(F - 32)$

$$9C = 9\left(\frac{5}{9}\right)(F - 32) \qquad \text{(Multiply each side by 9)}$$

$$9C = 5(F - 32) \qquad \text{(Simplify)}$$

$$9C = 5F - 160 \qquad \text{(Distributive axiom)}$$

$$9C + 160 = 5F - 160 + 160 \qquad \text{(Add 160 to each side)}$$

$$5F = 9C + 160 \qquad \text{(Symmetric axiom and simplification)}$$

$$F = \frac{9C + 160}{5} \qquad \text{(Divide each side by 5)}$$

$$F = \frac{9}{5}C + 32 \qquad \text{(Further simplification of right side)}$$

Thus $F = \frac{9}{5}C + 32$.

EXAMPLE 2.6.4 If $mx - y + b = 0$, solve for each of the following:

a. x b. y c. m d. b

Solution In each instance, consider all letters constant except the letter for which you are solving:

a. Solve for x:

$mx - y + b = 0$	
$mx - y + y + b = 0 + y$	(Add y to each side)
$mx + b - b = y - b$	(Subtract b from each side)
$mx = y - b$	(Simplify)
$x = \dfrac{y - b}{m}$	(Divide each side by m [$m \neq 0$])

b. Solve for y:

$mx - y + b = 0$	
$mx - y + y + b = 0 + y$	(Add y to each side)
$mx + b = y$	(Simplify)
$y = mx + b$	(Symmetric axiom)

c. Solve for m:

$mx - y + b = 0$	
$mx - y + b - b = 0 - b$	(Subtract b from each side)
$mx - y + y = -b + y$	(Add y to each side)
$mx = y - b$	(Commutative axiom, addition)
$m = \dfrac{y - b}{x}$	(Divide each side by x [$x \neq 0$])

d. Solve for b:

$mx - y + b = 0$	
$mx + b = y$	(Add y to each side)
$b = y - mx$	(Subtract mx from each side)

EXAMPLE 2.6.5 Solve $2x + y = 10$ for y.

Solution $2x + y = 10$

$$-2x + 2x + y = 10 - 2x$$

$$y = 10 - 2x$$

EXAMPLE 2.6.6 Solve $4x - y = 12$ for y.

Solution

$$4x - y = 12$$
$$-4x + 4x - y = -4x + 12$$
$$-y = -4x + 12$$
$$(-1)(-y) = (-1)(-4x + 12)$$
$$y = 4x - 12$$

EXERCISES 2.6 A

Solve the formulas in Exercises 1–25 for the indicated variable.

1. $A = LW$ (Area of rectangle); L
2. $V = LWH$ (Volume of a parallelepiped—box); W
3. $V = \pi r^2 h$ (Volume of a circular cylinder); h
4. $P = 2L + 2W$ (Perimeter of a rectangle); L
5. $D = rt$ (Uniform motion formula); t
6. $y = mx + b$ (Slope-intercept equation of a line); x
7. $A = \frac{bh}{2}$ (Area of a triangle); h
8. $A = \frac{1}{2}h(a + b)$ (Area of a trapezoid); a
9. $A = P + Prt$ (Simple interest); r
10. $\frac{a}{b} = \frac{c}{d}$ (Proportion); b
11. $A + B + C = 180$ (Sum of angles of a triangle); B
12. $S = \frac{n}{2}(a + l)$ (Sum of n terms of an arithmetic progression); a
13. $y = \frac{x - \bar{x}}{s}$ (Statistics); s
14. $P = a + b + c$ (Perimeter of a triangle); b
15. $D = A(n - 1)$ (Physics: law of small prisms); n
16. $S = \frac{f}{H - h}$ (Surveying: photogrammetry); H
17. $f = \frac{ab}{a + b}$ (Photography); a

18. $Q = \frac{100M}{C}$ (Psychology); C

19. $E = \frac{I - O}{I}$ (Engineering); O

20. $R = \frac{C - S}{n}$ (Economics: depreciation); n

21. $c^2 = a^2 + b^2$ (Theorem of Pythagoras); a^2

22. $D = \frac{1}{2} dACV^2$ (Rockets, missiles); C

23. $C = \frac{WA}{150}$ (Medical dosage); W

24. $H = 17 - \frac{A}{2}$ (Age-sleep relation of growing child); A

25. $r = \frac{R(E - G)}{E}$ (Radio, television); G

Solve the equations in Exercises 26–35 for the specified variable.

26. $x + y = 8$; y
27. $y - 5x = 10$; y
28. $3x - y = 6$; y
29. $x + 4y = 12$; x
30. $x - 2y = 9$; x
31. $x + y + 1 = 0$; x
32. $2y - x + 4 = 0$; x
33. $3x + 2y = 6$; y
34. $2x - 7y = 8$; x
35. $ax + by + c = 0$; y

EXERCISES 2.6 B

Solve the formulas in Exercises 1–25 for the indicated variable.

1. $W = Fd$ (Work); d
2. $C = 2\pi r$ (Circumference of a circle); r
3. $V = \frac{1}{3} bh$ (Volume of a cone); b
4. $A = a(a + 2s)$ (Surface area of a square pyramid); s
5. $P = I^2 r$ (Electric power); r
6. $S = \frac{180(n - 2)}{n}$ (Size of an angle of a regular polygon); n
7. $A = \frac{1}{2} h(a + b)$ (Area of a trapezoid); h

8. $A = P + Prt$ (Simple interest); P

9. $\frac{S}{s} = \frac{d}{D}$ (Pulleys and gears); d

10. $\frac{PV}{T} = \frac{pv}{t}$ (Boyle's gas law); t

11. $S = \frac{n}{2}(a + l)$ (Arithmetic progression); n

12. $K = \frac{wv^2}{2g}$ (Kinetic energy); w

13. $v = nxy$ (Statistical variance); x

14. $R = r + s + t$ (Resistors in series); t

15. $p = \frac{s}{s + f}$ (Probability—life insurance); f

16. $E = KI$ (Electricity); I

17. $f = \frac{pq}{p + q}$ (Mirrors); q

18. $K = \frac{a^2c}{1 - a}$ (Chemistry: ionization constant); c

19. $F = 6(C - 14) + 470$ (Television channel number); C

20. $d = \frac{r}{1 + nr}$ (Discount rate); r

21. $C = \frac{100W}{L}$ (Anthropology: cephalic index); W

22. $r = c(x + a)$ (Biology and chemistry: growth rate); a

23. $C = \frac{yA}{y + 12}$ (Medical dosage); A

24. $W = 5H - 190$ (Height–weight relation); H

25. $F = \frac{fV}{V + v}$ (Acoustics); v

Solve the equations in Exercises 26–35 for the specified variable.

26. $y - 2x = 5$; y

27. $4x + y = 12$; y

28. $5x - y = 20$; y

29. $x - 6y = 18$; x

30. $x + 7y = 4$; x

31. $x - 2y + 6 = 0$; x

32. $8 - x - 4y = 0$; x

33. $2x - 5y = 10$; x

34. $5x + 4y = 20$; y

35. $ax + by + c = 0$; x

2.7 APPLICATIONS: NUMERICAL PROBLEMS, AGE PROBLEMS, CONSECUTIVE INTEGERS

Numerical problems are very often stated in words. Besides mathematics, these problems may come from many other areas such as business or science. If the verbal problem can be translated into a linear equation in one variable, then it can be solved by the methods of the preceding sections.

Techniques that are useful in translating a verbal problem into an equation are summarized below.

Verbal-problem Techniques

1. Read the problem slowly at least two times.
2. Illustrate the problem with a simple picture or diagram, if possible.
3. Designate one of the unknown numbers by a variable, such as x.
4. Represent the other unknown numbers, if any, in terms of the variable selected.
5. Summarize the numerical information in a chart, if possible.
6. List any formula or formulas that might apply.
7. Find an equation relating the numbers of the problem.
8. Solve the equation.
9. Check the solution.

Certain expressions occur quite frequently in verbal problems. Some of these are summarized below together with their symbolic translations.

$a + b$	a plus b the sum of a and b a added to b a more than b a greater than b a increased by b a augmented by b	$a - b$	a minus b the difference between a and b (when b is subtracted from a) b subtracted from a b less than a b smaller than a a decreased by b a diminished by b the excess of a over b
ab	a times b the product of a and b a multiplied by b	$\frac{a}{b}$	a divided by b the quotient when a is divided by b
$2x$	twice x, the double of x	$\frac{x}{2}$	one half of x
$\frac{3x}{5}$	three fifths of x	$\frac{x}{3}$	one third of x
$=$	equals, equal, is, is equal to, are equal to, is the same as, results in, becomes, was, will be		

EXAMPLE 2.7.1 Five less than three times a certain number is the same as four more than one third of this number. Find the number.

Solution Let $x =$ the unknown number. Then

$$3x - 5 = \frac{x}{3} + 4$$
$$3x = \frac{x}{3} + 9$$
$$3(3x) = 3\left(\frac{x}{3} + 9\right)$$
$$9x = x + 27$$
$$8x = 27$$
$$x = \frac{27}{8}$$

EXAMPLE 2.7.2 Twice the difference obtained when seven is subtracted from a certain number results in twenty. Find the number.

Solution Let $x =$ the unknown number. Then

$$2(x - 7) = 20$$
$$2x - 14 = 20$$
$$2x = 34$$
$$x = 17$$

EXAMPLE 2.7.3 A man invests twice as much money in stock A as in stock B, and \$740 more in stock C than in stocks A and B combined. If his total investment is \$5000, how much did he invest in each stock?

Solution

Let $x =$ the number of dollars invested in stock B
then $2x =$ the number of dollars invested in stock A
and $x + 2x + 740 =$ the number of dollars invested in stock C

The total investment, \$5000, $= x + 2x + (x + 2x + 740)$.

$$x + 2x + (x + 2x + 740) = 5000$$
$$6x + 740 = 5000$$
$$6x = 4260$$
$$x = 710$$

He invested \$710 in stock B.
He invested $2(710) =$ \$1420 in stock A.
He invested \$1420 + \$710 + \$740 = \$2870 in stock C.

Check: $710 + 1420 + 2870 = 5000$.

Another type of application involves consecutive numbers—specifically, integers. For example, if x stands for an integer, then $x + 1$ is the next or consecutive integer, followed by $x + 2$, $x + 3$, ... So if $x = 3$, then $x + 1 = 4$, $x + 2 = 5$, and $x + 3 = 6$, and 3, 4, 5, 6 are consecutive integers. Consecutive *even* integers are numbers such as 8, 10, 12, 14, whereas consecutive *odd* integers are numbers such as 9, 11, 13, 15, 17. The following chart is a guide, assuming that x is an integer.

Consecutive Integers

$x, x + 1$	Two consecutive integers
$x, x + 1, x + 2$	Three consecutive integers
$x, x + 2$	Two consecutive even integers
$x, x + 2$	Two consecutive odd integers

EXAMPLE 2.7.4 Find three consecutive integers whose sum is 48.

Solution

Let

$$x = \text{first integer}$$
$$x + 1 = \text{second integer}$$
$$x + 2 = \text{third integer}$$

$$x + (x + 1) + (x + 2) = 48$$
$$3x + 3 = 48$$
$$3x = 45$$
$$x = 15$$

Therefore, $x = 15$, $x + 1 = 16$, and $x + 2 = 17$; the three consecutive integers are 15, 16, 17.

EXAMPLE 2.7.5 If three times the smallest of three consecutive *even* integers is sixteen less than five times the largest, find the three integers.

Solution

Let $\quad x = \text{smallest integer}$

Then $\quad x + 2 = \text{next consecutive } \textit{even} \text{ integer}$

and $\qquad x + 4 =$ third consecutive *even* integer (the largest)

$$\begin{aligned} 3x &= 5(x + 4) - 16 \\ 3x &= 5x + 20 - 16 \\ -2x &= 4 \\ x &= -2 \\ x + 2 &= -2 + 2 = 0 \\ x + 4 &= -2 + 4 = 2 \end{aligned}$$

Thus the three consecutive even integers are $-2, 0, 2$.

Problems involving ages of people or objects are also interesting and follow a pattern similar to that of consecutive integers.

EXAMPLE 2.7.6 Four years ago a father was twice as old as his son. The difference in their ages is 18 years. How old is each today?

Solution

	Now	*4 Years Ago*
Son	x	$x - 4$
Father	$x + 18$	$(x + 18) - 4$

$$\begin{aligned} (x + 18) - 4 &= 2(x - 4) \\ x + 14 &= 2x - 8 \\ x &= 22 \\ x + 18 &= 40 \end{aligned}$$

The father is 40 years old and the son is 22.

Again, the following chart may be helpful.

Age Problems

$x + k$	k years older than x
$x - k$	k years younger than x
$x + a$	Age a years from now, present age x
$x - a$	Age a years ago, present age x

EXERCISES 2.7 A

Solve and check Exercises 1–17.

1. Three more than twice a certain number is 7 less than 3 times the number. Find the number.

2. One half the sum of a given number and 6 is the same as 10 decreased by the difference obtained when 5 is subtracted from the number. Find the number.

3. A resort owner has 100 rooms for rent. At the end of one day he finds that he has rented 3 more than twice as many double rooms as single rooms, and half as many twin rooms as single rooms. He has 6 rooms unrented. If all the unrented rooms are singles, how many rooms of each type are there for rent in the resort?

4. A television repairman, preparing his order of tubes for the next month, finds his needs to be as follows: for every picture tube he orders, he needs 25 tubes of type A, 43 tubes of type B, 18 tubes of type C, and 36 tubes of type D. If he orders a total of 984 tubes, how many picture tubes did he order?

5. A small rental agency has bicycles, motor scooters, and dune buggies for rent. The total storage area allows for 165 vehicles for maximum profit in the following manner: 4 less than 3 times as many motor scooters as bicycles, and one third as many dune buggies as bicycles. How many vehicles of each type does the agency have?

6. A survey of the foreign language department of a certain college revealed the following facts about student enrollment in French, Spanish, Japanese, and Swahili: 6 less than twice as many students took Spanish as French. The enrollment in Swahili topped the enrollment in French by 9, and half as many students took Japanese as Spanish. If the total class enrollment is 460, how many students were enrolled in each language?

7. The sum of three consecutive integers is 42. Find the integers.

8. Find four consecutive integers so that the largest subtracted from 4 times the second is 53 more than the smallest.

9. Find three consecutive odd integers so that the sum of twice the largest and 3 times the smallest is 73.

10. Find three consecutive even integers so that the sum of twice the largest and 3 times the smallest is 8.

11. The average of four consecutive odd integers is 34. Find the integers.

12. The average of six consecutive even integers is -9. Find the integers.

13. Fourteen years ago, Jack was one fourth as old as he will be 7 years from now. How old is Jack now?

14. In 12 years a building will be twice as old as it is now. In how many years will the building be 3 times as old as it is now?

15. When Mr. and Mrs. Anderson were married, Mr. Anderson was 3 years older than his wife. On their 25th wedding anniversary, the sum of their ages was 97 years. How old was each on his wedding day?

16. Seven years ago John was 3 times as old as Mary. Now John is twice as old as Mary. Find the present age of each.

17. The sum of the ages of Bill and Betty is 44 years. Four years ago Bill was 5 times as old as Betty. Find the present age of each.

EXERCISES 2.7 B

Solve and check Exercises 1–17.

1. Three times a number added to 19 is 16 more than 4 times the number. Find the number.

2. Four times a number decreased by 24 is 18 less than 5 times the number. Find the number.

3. The difference between two thirds of a number and twice the number is 16. Find the number.

4. A photographer plans to spend a vacation in Hawaii. He will take twice as many rolls of color slide film as color print film, and as many rolls of black and white film as the other two types combined. If he takes a total of 120 rolls of film, how many rolls of each kind will he take?

5. In a supermarket three types of refrigerators are used to store bottles. If the first refrigerator holds 24 more bottles than the second, and the third holds three fourths as many bottles as the first, find the storage capacity of each if their total capacity is 317 bottles.

6. In a football game twice the number of points scored by the winning team was 14 less than 3 times the number of points scored by the losing team. If the winners won by 4 points, what was the final score in the game?

7. A jar of 62 coins contains pennies, nickels, and dimes. If there are one third as many nickels as dimes, and 6 more than 3 times as many pennies as nickels, how many coins of each kind are in the jar? How much money is in the jar?

8. If x is the middle integer of three consecutive integers, write algebraic expressions involving x for the smallest and the largest of the three integers.

9. If the sum of the three consecutive integers in Exercise 8 is 159, what are the integers?

10. The largest of four consecutive odd integers is 5 less than twice the smallest. Find the integers.

11. Find three consecutive integers whose sum is the difference between 5 times the smallest and the largest.

12. The average of four consecutive even integers is -1. Find the integers.

13. Of five increasing numbers, the three smallest are consecutive integers, and the second, fourth, and fifth numbers are consecutive odd integers. If the sum of the numbers is 51, find the numbers.

14. If a roof on a house is $2y - 6$ years old, what was the age of the roof 5 years ago?

15. If the roof on a house will be twice as old in 6 years as it is now, how old is the roof now?

16. John's sister is 3 years younger than John, and John's father is 28 years older than John. Five years ago the father's age was 1 year more than twice the sum of his children's ages then. How old is each now?

17. Mike has a brother who is one half his age and a sister who is twice his age. If the combined ages of the three siblings is 5 less than 4 times Mike's age, how old is Mike's sister?

2.8 APPLICATIONS: GEOMETRIC

The solution of geometric problems requires the application of formulas concerning geometric figures. In this type of problem, it is useful to draw the figure discussed and mark it with the numerical information. The equation is obtained by substituting the numerical information into the appropriate formula.

Some geometric figures with their formulas are supplied in Figure 2.8.1.

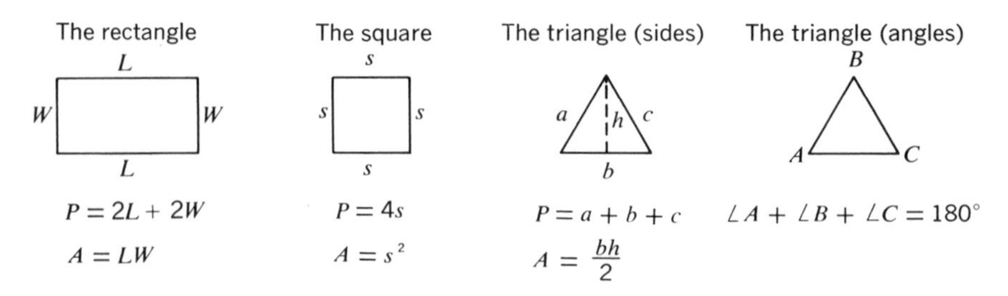

FIGURE 2.8.1

EXAMPLE 2.8.1 The length of a rectangle is 4 times its width. The perimeter is unchanged if the width is doubled and the length is decreased by 9 inches. What are the original dimensions of the rectangle?

Solution

Let $\qquad W =$ the original width

Then $\qquad 4W =$ the original length

Equation: $2 \cdot 2W + 2(4W - 9) = 2W + 2 \cdot 4W$; $W = 9$ inches, $L = 36$ inches.

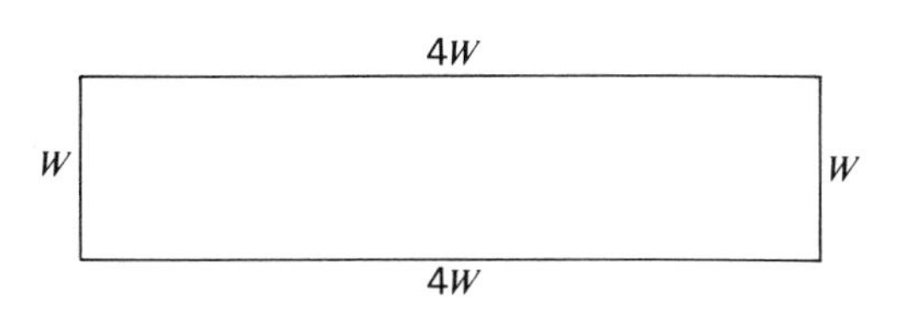

FIGURE 2.8.2

EXAMPLE 2.8.2 One angle of a triangle is 10 degrees more than the second angle, and the third angle is 10 degrees less than the second angle. How many degrees are in each angle of the triangle?

Solution

Let $\qquad x =$ the number of degrees in the second angle

Then $\qquad x + 10 =$ measure of first angle

$\qquad x - 10 =$ measure of third angle

Equation: $x + (x + 10) + (x - 10) = 180$; $x = 60$ degrees, $x + 10 = 70$ degrees, $x - 10 = 50$ degrees.

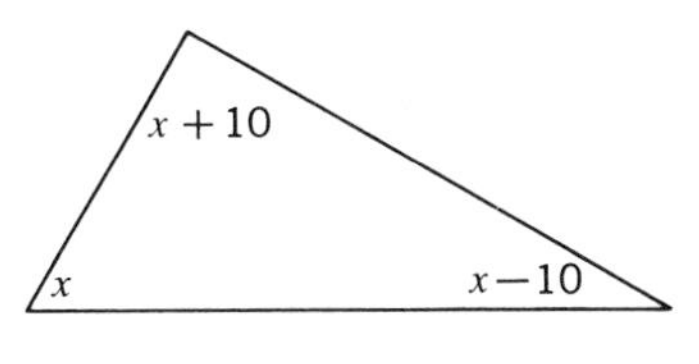

FIGURE 2.8.3

EXERCISES 2.8 A

Solve and check Exercises 1–10.

1. The width of a rectangular garden is 2 feet more than one third of its length. The perimeter is 52 feet. Find the dimensions of the rectangle.

2. The measurements of the angles of a triangle can be obtained by multiplying each of three consecutive integers by 10. Find the number of degrees in each angle.

3. What will be the cost of installing a hardwood floor to cover a rectangular area whose length is 6 feet less than twice its width if the cost is 80 cents a square foot and the perimeter is 78 feet?

4. In the diagram, the large rectangle is divided into four smaller rectangles. If the sum of the areas of the rectangles labelled I, II, and III is 5240, find the area of rectangle IV.

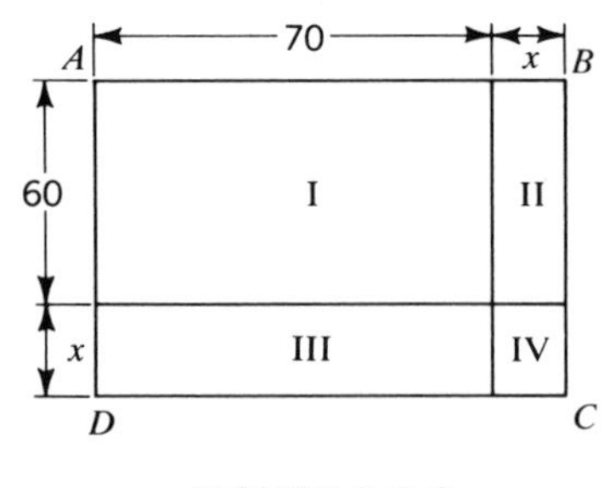

FIGURE 2.8.4

5. Find the area of a square whose perimeter is 40 inches.

6. Two rectangular carpets have widths of 8 feet and 10 feet, respectively. If the length of the first carpet is 3 feet longer than the length of the second carpet, and if the sum of their areas is 258 square feet, find the length of each carpet.

7. A 12-inch-by-18-inch rectangular picture is to have a frame of uniform width. The perimeter of the framed picture determines the length of wood needed for the frame. Find the width of the frame if the perimeter of the framed picture is 84 inches.

8. A piece of copper tubing is to be bent into the shape of a triangle in such a way that one side of the triangle is 3 inches more than twice the second side, and the third side is 1.5 times the second side. If the piece of tubing is 48 inches long, find the dimensions of the triangle.

9. A 4-by-5 portrait camera produces negatives that are 4 inches by 5 inches, rectangular in shape. A standard print from this negative is 8 inches by 10 inches.
   a. How many prints can be made from a piece of paper 112 inches by 110 inches if no paper is wasted?
   b. How many 11-inch-by-14-inch prints can be made from a piece of paper 112 inches by 110 inches if no paper is wasted?

10. An owner building his home specified a triangular window whose height was 2 feet shorter than its base. However, in order to meet the city lighting requirements, it was necessary to increase the height by 4 feet to permit an increase in area of 6 square feet. Find the dimensions of the original window and the dimensions of the new window.

EXERCISES 2.8 B

Solve and check Exercises 1–10.

1. The cross section of an irrigation ditch has the shape of a rectangle whose depth is 1 foot less than its width. By digging the ditch 2 feet deeper, the area of the cross section can be increased by 6 square feet. Find the width and the depth of the ditch before digging.
2. One side of a triangle is 2 inches less than a second side. The third side is 3 times the shortest side. If the perimeter is 7 inches, find the length of each side.
3. If the length of a Ping-Pong table is 2 feet less than twice its width, find the length and width if the perimeter is 29 feet.
4. Equilateral triangles (three equal sides) are cut off the three corners of a larger equilateral triangle to form a regular hexagon (six equal sides). If the perimeter of the hexagon is 54 inches, find the length of a side of the larger triangle.

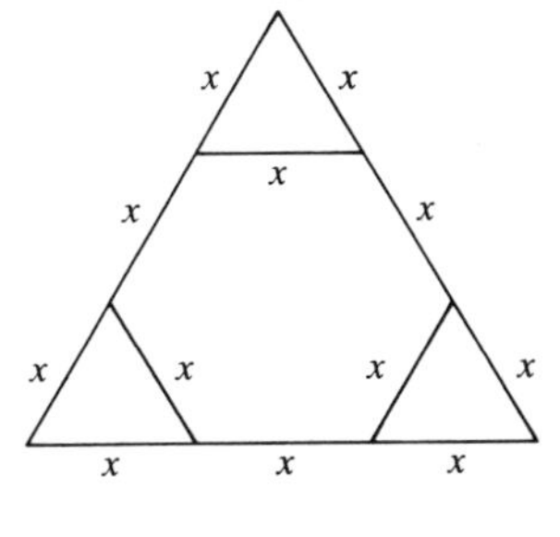

FIGURE 2.8.5

5. The area of a square whose side is 8 feet is the same as the area of a rectangle with a 16-foot length. Find the width of the rectangle.
6. A man wants to build a 4-foot-high brick wall at a uniform distance from the two 24-foot sides of his house and from a 50-foot length of the house. If he uses 4-inch-by-8-inch bricks costing 10 cents apiece, how far from his house can he build the wall and keep the cost of the bricks at $212.40?

7. A wood template has the shape of a rectangle surmounted by an equilateral triangle (three equal sides). The height of the rectangle is 3 inches less than the base of the rectangle (also a side of the triangle). It takes a total length of 34 inches of metal stripping to go around the template. Find the length of a side of the triangle.

8. An isosceles triangle is a triangle with two equal sides and two equal angles. If each of the equal angles of an isosceles triangle is 5 degrees less than twice the third angle, how many degrees are there in each angle of the triangle?

9. A farmer planned to fence three sides of a rectangular area, the fourth side (a length) being along a river. He planned a rectangle whose length was 5 feet longer than the width. His son, who had studied algebra, told him that the area would be the same but less fencing would be needed if he increased the length by 15 feet and decreased the width by 10 feet. Following his son's advice, the farmer found that he used 120 feet of fencing. Find the amount of fencing that would have been needed for the original plan.

10. Each page of a book is 9 inches by 7 inches and is to have a 1-inch margin along the two 9-inch sides. The margins at the other two sides of the page are equal in width. How wide should each of these margins be in order to have 40 square inches of printing?

2.9 APPLICATIONS: MIXTURE PROBLEMS

Many problems can be classified as mixture problems. The essential feature of this type of problem is that two or more items each having a specified unit value are combined to form a mixture. The sum of the values of the components must equal the value of the final mixture.

In this type of problem, the chart is a very useful device for summarizing the information. From the chart, the equation may be obtained very easily. The final entry in the last column is equal to the sum of the other entries in this column.

EXAMPLE 2.9.1 A piggy bank contains 23 coins, consisting of dimes and quarters. The total value is \$3.35. How many coins of each kind are there?

Solution

Let $\qquad d =$ the number of dimes

Then $\qquad 23 - d =$ the number of quarters

Item	*Unit Value* ·	*Number of Items* =	*Value*
Dimes	10 cents	d	$10d$
Quarters	25 cents	$23 - d$	$25(23 - d)$
Mixture		23	335 cents

Equation: $10d + 25(23 - d) = 335$; $d = 16$ dimes, $23 - d = 7$ quarters.

EXAMPLE 2.9.2 How many pounds of candy worth 75 cents a pound should be mixed with 20 pounds of candy worth $1.30 a pound to obtain a mixture worth $1.00 a pound?

Solution Let $x =$ the number of pounds of 75-cent candy.

Item	*Unit Value* ·	*Number of Items* =	*Value*
Candy	75 cents	x pounds	$75x$
Candy	130 cents	20 pounds	$20(130)$
Mixture	100 cents	$x + 20$	$100(x + 20)$

Equation: $75x + 20(130) = 100(x + 20)$, $x = 24$ pounds.

EXAMPLE 2.9.3 Tickets to a certain event cost 75 cents for children and $2.00 for adults. A total of $2430 was collected. If there were 3 times as many adult tickets sold as children's tickets, how many adult tickets were sold?

Solution

Let $x =$ the number of children's tickets

Then $3x =$ the number of adult tickets

Item	*Unit Value* ·	*Number of Items* =	*Value*
Children's tickets	75 cents	x	$75x$
Adult tickets	200 cents	$3x$	$200(3x)$
Mixture			243,000

Equation: $75x + 200(3x) = 243{,}000$; $x = 360$, $3x = 1080$.

EXAMPLE 2.9.4 A businessman makes two types of investments: Type A, which yields a 5 percent annual return, and type B, which yields a 4 percent return. If his total investment is \$10,000, and his income from the investment at the end of one year is \$440, how much money did he invest in type A?

Solution Let d = amount invested in type A.

Item	*Unit Value* · *(Percent Return)*	*Number of Items (Amount Invested) in Dollars*	= *Value (Income)*
Type A	0.05	d	$0.05d$
Type B	0.04	$10{,}000 - d$	$0.04(10{,}000 - d)$
Mixture (Totals)	—	10,000	440

Equation: $0.05d + 0.04(10{,}000 - d) = 440$, $d = 4000$ dollars.

EXERCISES 2.9 A

Solve and check Exercises 1–10.

1. How many gallons of high-octane gasoline worth 40 cents a gallon should be blended with 300 gallons of low-octane gasoline worth 35 cents a gallon to produce a mixture worth 37 cents a gallon?
2. A farmer combines seed worth 12 cents a pound with seed worth 20 cents a pound to obtain a 50-pound mixture worth 15 cents a pound. How many pounds of each type of seed does he use?
3. A student attending a college which gives only A, B, and C grades on a 4, 3, 2 scale, respectively, is trying to compute his grade point average. For example, 3 units of A yield $3 \cdot 4 = 12$ grade points, and 4 units of C yield $4 \cdot 2 = 8$ grade points. The grade point average is the quotient of the number of grade points achieved, divided by the number of units taken. The student receives twice as many units of B as A, and 3 more units of C than of A. If his grade points totaled 54, how many units of A, B, and C did he receive? What was his grade point average?
4. Using the general information from Exercise 3, another student finds he has 4 more units of B than of A, and 1 more unit of B than he has of C. If his total number of grade points is 45, how many units of A, B, and C did this student receive, and what was his grade point average?

5. A cashier in a ticket booth starts the day with $100 worth of change. The change is in quarters, dimes, and dollar bills. If he has 50 times as many dimes as dollar bills and 60 more than 10 times as many quarters as dollar bills, how many dimes, quarters, and dollar bills did the cashier start with?
6. The student association of a certain college decides to invest its building fund, since it cannot start construction of a student union for 3 more years. Some of the money is invested at 6 percent per year, and twice this amount is invested at $5\frac{1}{4}$ percent per year, yielding a total income at the end of the first year of $825. How much money is invested at each rate?
7. A businessman wishes to invest a part of $7500 in stocks earning 5 percent dividends and the remainder in tax-exempt bonds earning $3\frac{1}{2}$ percent. How much must he invest in the stocks to receive an average return of 4 percent on the total amount of money?
8. At a movie theater on a Friday evening there were 420 paid admissions. If loge seats sold for $2.50 each and general admission seats sold for $1.75 each, how many tickets of each kind were sold if the day's receipts totaled $817.50?
9. How many liters of water must be added to a 75 percent acid solution (75 percent acid and 25 percent water by volume) to obtain 100 liters of a solution that is 30 percent acid?
10. How many grams of sodium hydroxide consisting of 88 percent pure sodium hydroxide and 12 percent water are needed to make $5\frac{1}{2}$ liters of a 1 normal solution? (A 1 normal solution consists of 40 grams of sodium hydroxide per liter.)

EXERCISES 2.9 B

Solve and check Exercises 1–10.

1. A young couple (husband and wife both working) managed to save $8000. They decided to invest $5000, part at $5\frac{1}{2}$ percent annual interest and part at 4 percent. If the $5\frac{1}{2}$ percent investment yielded an income of $104 more than the 4 percent investment, how much did the couple invest at each rate?
2. A corporation invested $100,000 in the following manner: 6 percent bonds, 8 percent stocks, and 9 percent mortgages. If the amount invested in bonds is the same as the amount invested in stocks, and if the total income per year from all three investments is $7500, how much is invested in each of the three types of investment?

3. Six ounces of a tranquilizer in powdered form and of normal strength is mixed with a double-strength tranquilizer to form a mixture which is 1.5 times normal strength. How many ounces of the double-strength powder must be used to yield this new mixture?
4. A jar at the checkout counter of a grocery store has a sign asking for contributions to a charitable cause. At the end of the week, the jar contains 62 coins worth a total of $7.40. If there was 1 less than half as many dimes as nickels, and 4 more quarters than dimes, and these were the only types of coins, how many quarters were in the jar?
5. A grocer mixes 45 pounds of ordinary coffee beans worth 70 cents a pound with rare coffee beans worth 90 cents a pound. How many pounds of rare beans should he use to obtain a mixture worth 75 cents a pound?
6. Concrete is a mixture of cement and sand. How many pounds of cement worth 24 cents a pound and how many pounds of sand worth 4 cents a pound have been mixed to obtain 120 pounds of concrete worth 12 cents a pound?
7. How many doughnuts worth 12 cents each and how many worth 7 cents each are combined to produce an assortment of three dozen doughnuts costing $3.22?
8. In constructing a state road, the county pays 40 percent as much as the state, and the city pays 65 percent as much as the county. What is the share of each for a road costing $3071?
9. How much water must be evaporated from a 5 percent salt solution weighing 50 pounds to obtain an 8 percent solution?
10. In football, a touchdown counts 6 points, a conversion 1 point, a field goal 3 points, and a safety 2 points. A team scored 49 points in a game by making the same number of safeties as field goals, 3 times as many touchdowns as field goals, and half as many conversions as touchdowns. How many plays of each kind did the team make?

2.10 APPLICATIONS: UNIFORM MOTION

If an object moves with a constant rate of speed, then the motion of the object is called *uniform motion*. The basic principle involved in such motion is the formula

$$\text{rate} \times \text{time} = \text{distance}$$
$$r \cdot t = d$$

In solving problems of this type, it is useful to make a sketch illustrating the facts of the problem and also a chart organizing the facts.

EXAMPLE 2.10.1 At a certain time two trains start from the same depot and travel in opposite directions. If one travels 35 mph and the other travels 60 mph, in how many hours will they be 285 miles apart?

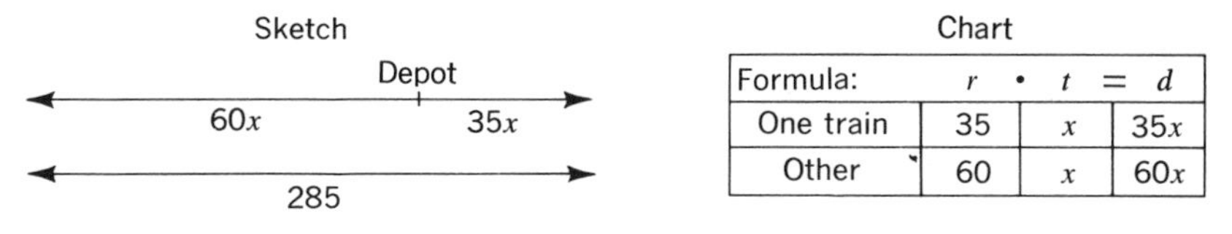

Chart

Formula:	r	$\cdot\ t$	$= d$
One train	35	x	$35x$
Other	60	x	$60x$

FIGURE 2.10.1

Solution

Equation:

$$\begin{aligned} \text{distance of one train} + \text{distance of other} &= \text{total distance} \\ 35x + 60x &= 285 \\ x &= 3 \text{ hours} \end{aligned}$$

EXAMPLE 2.10.2 A plane leaves an airport at 9 a.m. and travels east at a uniform speed of 280 mph. At 10:30 a.m. a jet flying the same course left the same airport and overtook the first plane at noon. At what uniform rate did the jet fly?

Sketch

9 A.M. $280 \cdot 3$ 12:00

10:30 A.M. $x \cdot 1\frac{1}{2}$ 12:00

Chart

Formula:	r	$\cdot\ t$	$= d$
Plane	280	3	840
Jet	x	$1\frac{1}{2}$	$\frac{3x}{2}$

FIGURE 2.10.2

Solution

Equation: distance of plane = distance of jet

$$\frac{3x}{2} = 840$$

$$x = 560 \text{ mph}$$

It is important when working this type of problem to keep the *units* consistent. For example, if the distance is in *miles* and the time in *hours*, then the rate must be in miles per hour. If the distance is in feet and the rate in feet per second, then the time must be expressed in seconds.

EXAMPLE 2.10.3 A snail and a rabbit are having a race. The rabbit gives the snail a 30-foot head start on the 35-foot racecourse. If the snail proceeds at 1 foot per minute and the rabbit hops at 3 feet per second, how long is it before the rabbit catches the snail? Does the rabbit win?

Solution First, the units in the snail's rate should be converted to feet per second (or the rabbit's rate to feet per minute).

$$1 \text{ foot per minute} = \frac{1}{60} \text{ foot per second}$$

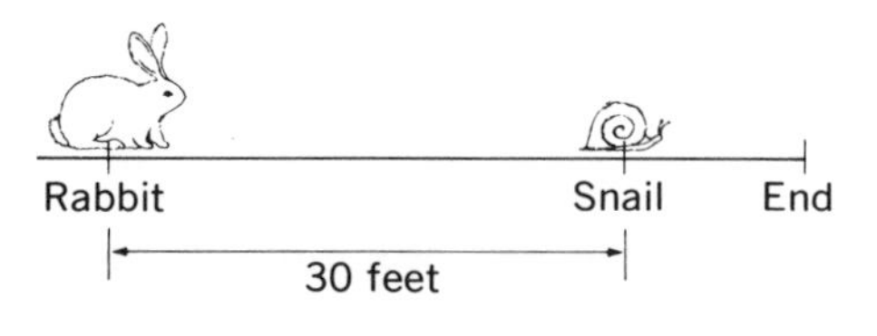

FIGURE 2.10.3

The distance to be caught up is 30 feet. The rate at which the rabbit catches the snail is the difference between their respective rates.

Let $t =$ the time in seconds:

$$d = r \cdot t$$

$$30 = \left(3 - \frac{1}{60}\right)t$$

$$\frac{1800}{179} = t$$

$t \approx 10$ seconds. In 10 seconds the snail has progressed an additional

$$10\left(\frac{1}{60}\right) \text{ feet} = \frac{1}{6} \text{ foot}$$

and the rabbit wins the race. (Why?)

EXERCISES 2.10 A

Solve and check Exercises 1–10.

1. A car and a motorcycle stop at a rest stop on the highway. The car travels north at 60 mph and the motorcycle goes south at 45 mph. If the two vehicles leave the rest stop at the same time, how long will it

take them to be 210 miles apart? (It is assumed that the highway is perfectly straight.)

2. How far apart will the car and the motorcycle from Exercise 1 be at the end of 40 minutes?

3. Two trains, 465 miles apart, travel toward each other, one traveling 15 mph faster than the other. They pass each other in 3 hours. What is the rate of the slower train?

4. At 8 a.m. a truck leaves a depot, traveling west at 45 mph. Two hours later a car starts from the same depot and travels at 60 mph until it overtakes the truck. At what clock time does the car overtake the truck? (It is assumed that the truck driver does not stop for lunch.)

5. Two planes leave an airport at the same time and travel in opposite directions. One plane travels 80 mph faster than the other plane. What is the rate of each plane if the planes are 2190 miles apart at the end of 3 hours?

6. At 6 a.m. two trains leave the same depot and travel in opposite directions. They are 648 miles apart at the end of 6 hours. Find the rate of each train if one travels 12 mph slower than the other.

7. A police car moving at 60 mph pursues a traffic offender driving at 40 mph. If the offender was 0.5 mile ahead when the police car started, how long does it take for the police car to overtake the offender?

8. Mr. Lately has a business appointment in Redding, California. He leaves Sacramento, a town 150 miles south of Redding, and drives at a rate of 50 mph. Suddenly he realizes that he will be late if he does not speed up, so he increases his rate to 60 mph. If his total driving time is 2 hours and 50 minutes, how far did he drive at 50 mph?

9. A ship is heading due west out of San Francisco at a rate of 35 mph (*not* knots). A passenger on board realizes he has forgotten to bring his medicine. A helicopter, flying at a rate of 95 mph, is sent to deliver the medicine to the ship. If the helicopter starts out when the ship has been underway for 6 hours, how far from San Francisco is the ship when the helicopter reaches it?

10. Jim's home town is 340 miles from his college. Jim leaves college at 8 a.m. to drive home for Christmas vacation. His girlfriend, Debbie, who lives in his home town, decides to drive to meet him. Debbie leaves at 11 a.m. and drives along the same highway as Jim. If Jim drives 10 mph faster than Debbie and they meet at noon, what was Jim's speed?

EXERCISES 2.10 B

Solve and check Exercises 1–10.

1. At noon two cars leave Sacramento. One car, southbound for Fresno, travels 15 mph faster than the other car, northbound for Portland, Oregon. If the two cars are 476 miles apart at the end of 4 hours, find the rate of speed of each car.
2. A plane took 4 hours to fly to Chicago from San Francisco. The return trip from Chicago to San Francisco took 1 hour longer because strong headwinds reduced the plane's speed by 96 mph. Find the rate of the plane going to Chicago.
3. A car speeding at 80 mph is 3 miles beyond a police car when the police car starts in pursuit at 90 mph. How long does it take the police car to overtake the speeding car?
4. Two men drive from Los Angeles to San Francisco by different routes. The journey is 420 miles by one route and 390 miles by the other route. The man taking the shorter route drives 10 miles an hour faster than the second man, and arrives in San Francisco $6\frac{1}{2}$ hours after leaving Los Angeles. How long does it take the second man to get from Los Angeles to San Francisco?
5. A fisherman drove from his home to a mountain lake in 4 hours. Returning by the same route, he took 1 hour longer because heavy traffic forced him to drive 8 mph slower. Find his average rate going to the lake.
6. A passenger train and a freight train leave the same station at the same time and travel in opposite directions. The average rate of the passenger train is 20 mph faster than the freight train. The trains are 520 miles apart at the end of 5 hours. Find the average rate of each train.
7. A Coast Guard boat averaging 30 mph starts one half hour later from the same dock to overtake another boat, moving at 18 mph. How long does it take the Coast Guard boat?
8. How far can a boat travel out to sea at 30 mph and return at 20 mph if the boat has just enough fuel for a 6-hour trip?
9. A man walks from his home into the countryside at 4 mph. He visits with friends for an hour and they drive him home, at a rate of 36 mph. If the man returns 6 hours after he started, how far from his home is his friend's house?
10. A couple attends an out-of-town wedding. The trip to the wedding takes 3 hours, the wedding and reception last 2 hours, and as a result of the celebration, the trip home takes 4 hours, at a rate of 15 mph less than the going rate. What was the couple's rate of travel on the way home?

SUMMARY

The Axioms of Equality

1. Reflexive axiom	$r = r$	For every r in the replacement set
2. Symmetric axiom	If $r = s$, then $s = r$	For every r and s in the replacement set
3. Transitive axiom	If $r = s$ and $s = t$, then $r = t$	For every r, s, and t in the replacement set

Properties of the Set of Real Numbers

Let r, s, and t be real numbers.

Axioms

1. Closure axiom	$r + s$ is a real number rs is a real number	
2. Commutative axiom	$r + s = s + r$ $rs = sr$	(Addition and multiplication are commutative, but subtraction and division are not)
3. Associative axiom	$(r + s) + t = r + (s + t)$ $(rs)t = r(st)$	(Addition and multiplication are associative, but subtraction and division are not)
4. Distributive axiom	$r(s + t) = rs + rt$	(Multiplication distributes over addition)
5. Identity axiom	There exists a unique real number, called 0, so that $r + 0 = r$.	(0 is the addition identity)
	There exists a unique real number, called 1, so that $r \cdot 1 = r$.	(1 is the multiplication identity)

6. Inverse axiom	For each real number r there exists a unique real number, $-r$, called the opposite of r, so that $r + (-r) = 0$.	(Addition)
	For each real number r, except $r = 0$, there exists a unique real number, $\frac{1}{r}$, called the reciprocal of r, so that $r\left(\frac{1}{r}\right) = 1$.	(Multiplication)

The Equivalence Theorems

Let $A = B$ be an open equation, and let C be any real number.

1. Addition theorem

$$A = B, \text{ if and only if } A + C = B + C$$

If the same number is added to each side of an equation, the resulting sums are again equal, and the two equations are equivalent.

2. Subtraction theorem

$$A = B, \text{ if and only if } A - C = B - C$$

If the same number is subtracted from each side of an equation, the resulting differences are equal, and the two equations are equivalent.

3. Multiplication theorem

$$A = B, \text{ if and only if } AC = BC \text{ and } C \neq 0$$

If each side of an equation is multiplied by the same nonzero number, the resulting products are equal, and the two equations are equivalent.

4. Division theorem

$$A = B \text{ if and only if } \frac{A}{C} = \frac{B}{C} \text{ and } C \neq 0$$

If each side of an equation is divided by the same nonzero number, the resulting quotients are equal, and the two equations are equivalent.

- ☐ The **solution** or **root** of an open equation is a number that makes the equation true when the variable is replaced by the name of this number.
- ☐ The **solution set** of an open equation is the set of all solutions of the equation.
- ☐ To **solve an equation** means to find the solution set of the equation.
- ☐ Two **equations** are **equivalent** if and only if they have the same solution set.
- ☐ A **linear equation in one variable**, x, is an equation which can be expressed in the form $ax + b = 0$, $a \neq 0$.

REVIEW EXERCISES

Name the axioms that are illustrated by Exercises 1–10.

1. $\frac{-1}{2} + \frac{1}{3} = \frac{1}{3} + \frac{-1}{2}$

2. $\frac{-1}{2} + \left[-\left(\frac{-1}{2}\right)\right] = 0$

3. $\left(\frac{-1}{2}\right)(-2) = 1$

4. $\frac{-2}{3}\left(3 + \frac{-1}{5}\right) = \frac{-2}{3}(3) + \frac{-2}{3}\left(\frac{-1}{5}\right)$

5. $\left(\frac{-1}{2} + \frac{1}{3}\right) + \frac{1}{4} = \frac{-1}{2} + \left(\frac{1}{3} + \frac{1}{4}\right)$

6. $\left(\frac{-3}{5}\right)\left(\frac{1}{2}\right) = \frac{1}{2}\left(\frac{-3}{5}\right)$

7. $\frac{-3}{5} \cdot \frac{1}{1} = \frac{-3}{5}$

8. $-10\left(\frac{-1}{2} \cdot \frac{-1}{3}\right) = 5\left(\frac{-1}{3}\right)$

9. $-2(-4 + 5)$ is a real number.

10. $\frac{3}{5} + 0 = \frac{3}{5}$

State which equivalence theorem or theorems (addition, subtraction, multiplication, division) are needed to solve Exercises 11–20.

11. $2x - 5 = 8$

12. $\frac{x + 6}{2} = 7$

13. $\frac{x}{2} + 9 = 15$

14. $3x + 8 = 2$

15. $5x = 2x + 6$

16. $3x = 8x - 20$

17. $4x - 5 = 7 - 2x$

18. $6x + 7 = 5x + 2$

19. $\frac{5x - (x - 2)}{3} = 6$

20. $3(x - 2) - (x + 2) = 4$

Solve and check the equations in Exercises 21–50.

21. $x + 5 = -7$

22. $3x + 2 = x + 4$

23. $5 - 2y = 15$

24. $3 + 4x - 2 = 4 + 2x$

25. $2(x - 4) = 3x - 16$

26. $3(y + 2) + 4 = 5y - 2$

27. $4 - (3x + 8) = 17$

28. $5(z - 1) - (1 - 4z) = 30$

29. $5t + 7t - 2 = 5 + 7 - 2t$

30. $x + 2 = x + 3$

31. $2(x - 5) = 2x - 10$

32. $6 - \frac{x + 2}{3} = 4$

33. $4 = 2x + 3$

34. $4x - (x - 5) = 4 - (x + 7)$

35. $0 = \frac{x + 5}{2}$

36. $7x - 5(x - 2) = 20$

37. $9 - 6(2 - x) = 7x$

38. $x + 4 = 2x - 3$

39. $z + 1 = z$

40. $3 - x = x - 3$

41. $4(x + 3) = 12$

42. $2(x - 3) - 3(x + 1) = 0$

43. $y - 2 = 2 - y$

44. $4 - 3(x + 2) = 7$

45. $3(x + 2) = 3x + 6$

46. $0 = 2(x + 5)$

47. $\frac{t + 7}{2} = 5$

48. $5 - \frac{x + 3}{2} = 6$

49. $2(x - 3) + 3(x + 2) = x + 8$

50. $x - 3[x - 3(x - 3)] = 1$

Write an algebraic expression to represent each verbal expression in Exercises 51–60.

51. 4 less than 6 times a certain number, x

52. 9 more than one half a certain number, x

53. The number of cents in x dimes and twice as many nickels

54. The amount of money invested at 4 percent if x dollars is invested at 5 percent and a total of \$5000 is invested

55. The distance apart two cars are at the end of 5 hours if they travel in opposite directions and if the rate of one is 3 times the rate, x, of the other

56. The width of a rectangle whose perimeter is 48 and whose length is x

57. The weight of a mixture when x pounds of one kind of rice is mixed with 40 pounds of another type

58. A overtakes B at the end of x hours. If A started 30 minutes after B, how many hours did B travel?

59. The total income from x dollars invested at 5 percent and \$2000 invested at $6\frac{1}{2}$ percent

60. The sum of three consecutive integers if the smallest is x

Solve and check Exercises 61–70.

61. Six more than the double of a number is equal to 5 times the difference obtained when 3 is subtracted from the number. Find the number.

62. A bus and a train leave the same station at the same time and both travel north. The rate of the train is 15 mph less than twice the rate of the bus. At the end of 3 hours, the train is 90 miles farther north than the bus. Find the rate of each.

63. How many ounces of copper must be added to 18 ounces of an alloy composed of 40 percent copper to make an alloy composed of 50 percent copper?

64. How much paint is required to paint 3 walls of a room if each wall is a rectangle with a length 2 feet less than twice the height? The perimeter of each wall is 50 feet and 1 gallon of paint covers about 400 square feet.

65. A woman paid $2.21 for some fruit consisting of peaches costing 29 cents a pound, plums at 25 cents a pound, and bananas at 10 cents a pound. If she bought twice as many pounds of plums as peaches, and 1 pound less bananas than plums, how many pounds of each kind of fruit did she buy?

66. A corner lot has the shape of a right triangle (one angle is 90 degrees). The sides forming the 90-degree angle are along the streets, and one of these sides is 11 feet longer than the other side. To widen the street along the longer side, a strip 10 feet wide is removed from the lot, thus reducing the length of this side by 11 feet. The area of the lot was decreased by 2310 square feet. Find the original length of the shorter side along the street.

67. During an average life of 70 years, it is estimated that a person sleeps 3 years more than he works; that he spends one half the years he works on recreation and church functions; 7 years eating and drinking; 5 years traveling, 2 years dressing, and 3 years being ill. How many years does he work?

68. A person traveling on a plane is allowed 44 pounds of baggage free but must pay $3.20 per pound for all excess baggage. A wife had 5 pounds of baggage more than her husband had. Together they paid $67.20 for excess baggage. How much baggage did each have?

69. A rush order is received in Boomtown for some machine parts that must be delivered from Bolttown, 280 miles away. To speed the delivery, a station wagon traveling 65 mph leaves Boomtown at 11 a.m. to intercept the delivery truck traveling 50 mph. If the delivery truck left Bolttown at 10 a.m., at what clock time is the delivery truck intercepted?

70. A dillar, a dollar,
Say, algebra scholar,
Can you tell me my age?

Twice nine years ago,
I have eight years to go.
Use x, mathematical sage.

CHAPTER 3

OPERATIONS ON POLYNOMIALS

A fundamental concept in the study of algebra is the class of algebraic expressions known as polynomials. The applications of polynomials to real-world situations are too numerous to list. For instance, the distance a missile travels when thrown into the air and allowed to fall back to earth can be expressed in terms of a polynomial; so can the velocity of the same missile, and its acceleration. In business, revenue obtained from the sale of a commodity is often expressed by a polynomial, and so are marginal revenue, supply and demand, efficiency of investment, and so forth. A study of any of the fields of application of mathematics will surely convince the reader of the importance of polynomials.

It is the object of this chapter to introduce polynomials, to define operations on polynomials, to present some basic factoring techniques, and to show the solution of factorable quadratic equations in one variable.

3.1 PRODUCTS OF MONOMIALS

DEFINITION

A monomial is:

1. A constant, or
2. A term of the form cx^n,
   where c is a constant, x is a variable, and n is a natural number, or
3. A product of terms as described in (2).

Each of the following is an example of a monomial:

$$3,\ x^2,\ 5x^3,\ \sqrt{2}xy,\ x^2y^9,\ \frac{1}{2}xyz^5$$

$\sqrt{x}$ is *not* a monomial; neither is $\frac{1}{x}, \frac{5}{x^2}$, or any other term which does not fit the description of being a constant or of the form cx^n, where c is constant and n is a natural number.

A monomial that has only one variable is called a **monomial in one variable**. For example, $3x$, $15x^3$, and $-2y^2$ are each a monomial in one variable. The monomial $4x^2y^3$ is not a monomial in one variable but a **monomial in two variables.**

A definition is needed here to explain what is meant by x^n, where n is a natural number. In Chapter 1 the operation "raising to a power" was discussed. For example,

$$x^2 = x \cdot x$$
$$x^3 = x \cdot x \cdot x$$

DEFINITION OF x^n

If x is a real number and n is a natural number, then

$$x^1 = x \text{ and } x^n = \overbrace{x \cdot x \cdot x \cdot \ldots \cdot x}^{n \text{ factors}} \text{ if } n > 1$$

EXAMPLE 3.1.1

$$\begin{aligned} x^2 \cdot x^3 &= (x \cdot x) \cdot (x \cdot x \cdot x) \\ &= x \cdot x \cdot x \cdot x \cdot x \\ &= x^5 \end{aligned}$$

EXAMPLE 3.1.2

$$\begin{aligned} 3x^2 \cdot 2x^3 &= (3 \cdot x \cdot x) \cdot (2 \cdot x \cdot x \cdot x) \\ &= (3 \cdot 2) \cdot (x \cdot x \cdot x \cdot x \cdot x) \\ &= 6x^5 \end{aligned}$$

(Associative and commutative axioms, multiplication)

The preceding examples lead to the following theorem.

THE FIRST THEOREM OF EXPONENTS

If x is a real number and m and n are natural numbers, then

$$x^m \cdot x^n = x^{m+n}$$

EXAMPLE 3.1.3

$$x^2 \cdot x^3 = x^{2+3} = x^5$$

EXAMPLE 3.1.4

$$3x^2 \cdot 2x^3 = (3 \cdot 2) \cdot (x^2 \cdot x^3) = 6(x^{2+3}) = 6x^5$$

EXAMPLE 3.1.5

$$3^2 \cdot 3^3 = 3^{2+3} = 3^5 = 243$$

EXAMPLE 3.1.6

$$\begin{aligned} 2^3 \cdot 3^2 \cdot 2^4 = 2^{3+4} \cdot 3^2 &= 2^7 \cdot 3^2 \\ &= 128 \cdot 9 \\ &= 1152 \end{aligned}$$

It should be emphasized that the variable need not be the letter x but can be any letter that designates a real number.

EXAMPLE 3.1.7

$$y^5 \cdot y^9 = y^{5+9} = y^{14}$$

EXAMPLE 3.1.8

$$5z^2 \cdot 6z = 5 \cdot 6 \cdot z^2 \cdot z = 30z^{2+1} = 30z^3$$

Now consider raising a power to a power. For example,

$$(x^3)^2$$

By the definition of x^n,

$$(x^3)^2 = x^3 \cdot x^3$$

and by the first theorem of exponents,

$$x^3 \cdot x^3 = x^{3+3} = x^6$$

Therefore, $(x^3)^2 = x^6$. Also,

$$(x^2)^4 = x^2 \cdot x^2 \cdot x^2 \cdot x^2 = x^{2+2+2+2} = x^{4 \cdot 2} = x^8$$

The problem of raising a power to a power can be generalized by the following theorem of exponents:

THE SECOND THEOREM OF EXPONENTS

If x is a real number and m and n are natural numbers, then

$$(x^m)^n = x^{mn}$$

EXAMPLE 3.1.9

$$(x^3)^4 = x^{3\cdot 4} = x^{12}$$

Note: Because multiplication is commutative, $x^{mn} = x^{nm}$; therefore, $(x^m)^n = (x^n)^m$.

EXAMPLE 3.1.10

$$(y^5)^m = y^{5m}, \ m \text{ is a natural number}$$

The student should use caution when applying the first theorem of exponents. For example,

$$5x^2 \cdot 7y^3 = 35x^2y^3$$

The theorem does not apply to the variables x and y, since they are different. However, the following example illustrates that the theorem *may* be used for monomials in more than one variable when it applies.

EXAMPLE 3.1.11

$$\begin{aligned}(5x^2yz^3)(3xy^4) &= 15x^2xyy^4z^3 \\ &= 15x^3y^5z^3\end{aligned}$$

EXAMPLE 3.1.12

$$(2^3)^2 = 2^6 = 64$$

EXERCISES 3.1 A

Multiply in Exercises 1–20.

1. $x^3 \cdot x^2$
2. $3x \cdot x^4$
3. $2y^2 \cdot 3y^4$
4. $5z^2 \cdot 7z^9$
5. $(-2x^4)(3x^2)$
6. $(-x^{11})(-5x^2)$
7. $\left(\frac{1}{2}y^3\right)(4y^5)$
8. $(x^2)(3x)(-x^3)$
9. $(-z)(-z^2)(-z^3)$
10. $(xy^2)(3xy)$
11. $(5xz)(3x^2y)$
12. $(abc)(a^2b^2c^2)$
13. $(x^2)^4$
14. $(x^4)^2$
15. $(y^3)^4$
16. $(x^3)^2 \cdot x^2$
17. $xy(y^2)^5$
18. $(-x)^2$
19. $(-x^2)^3$
20. $(2xy)(xz^2)(-yz)$

★ *Use the first or second theorem of exponents to simplify the expressions in Exercises 21–30. Assume all exponents to be natural numbers.*

21. $x^2 \cdot x^m$
22. $2x^p \cdot x^q$
23. $(x^3)^n$
24. $(x^n)^3$
25. $x^n y^m \cdot xy^{2n}$
26. $y^{n+1}y^{n-1}$
27. $(rt^k)(r^k t^k)$
28. $(r^{n-1}t^n)(rt^{n-2})$
29. $(x^n)^2(y^2)^n$
30. $[(t^a)^b]^c$

In Exercises 31–40, use the first or second theorem of exponents to simplify each expression.

31. $2^3 \cdot 2^2$
32. $3^3 \cdot 3^3$
33. $(2^2)^3$
34. $(3^2)^4$
35. $2^3 \cdot (3^2)^3 \cdot 2^4$
36. $(-2)^3(-2)^2$
37. $[(-10)^2]^3$
38. $(-2)^3(-2)^4(-5)^6$
39. $[(-2)^4]^2 \cdot [(-5)^2]^4$
40. $(-2)^{3^2}(-5)^{2^3}$

EXERCISES 3.1 B

Multiply in Exercises 1–20.

1. $z^3 \cdot z^4$
2. $5y \cdot y^3$
3. $3x^2 \cdot 4x^4$
4. $7a \cdot 5a^5$
5. $(-4x^3)(6x^2)$
6. $(-y^5)(-3y^2)$
7. $(-a^2)(-a)(-a^3)$
8. $(ab)(-a^2b)$
9. $(x^2)(-3x)(5x^4)$
10. $(x^2y)(2xy^2)$
11. $(3ab)(-2a)(-4b)$
12. $(rst)(rs)(st)$
13. $(x^3)^2$
14. $(x^2)^3$
15. $5(z^2)^3$
16. $x(x^2)^3$
17. $ab(b^4)^5$
18. $(-y^2)^2$
19. $-(y^2)^2$
20. $2(-3x)(xy)(y^2)^4$

★ *Use the first or second theorem of exponents to simplify the expressions in Exercises 21–30. Assume all exponents to be natural numbers.*

21. $x^3 \cdot x^a$
22. $5y \cdot y^m$
23. $(x^a)^b$
24. $x^{2a}y^a \cdot x^a y^2$

25. $x^b(xy^2)$

26. $(rs^2)(r^a s^a)$

27. $t^{n+2}t^{n-3}$

28. $(y^3)^b(z^b)^2$

29. $[(y^p)^q]^2$

30. $(x^2)^{3n}(x^{2n})^3$

In Exercises 31–40, use the first or second theorem of exponents to simplify each expression.

31. $4^2 \cdot 4$

32. $3 \cdot 3^3$

33. $(5^2)^2$

34. $5^2 \cdot 5^2$

35. $3^2 \cdot 4^3 \cdot (2^2)^3$

36. $(-3)^4(-3)$

37. $((-10)^3)^3$

38. $(-5)^4(-5)^2(-2)^5$

39. $[(-5)^3]^3[(-2)^3]^3$

40. $(-2)^{23}[(-5)^3]^2$

3.2 SUMS AND DIFFERENCES OF POLYNOMIALS

DEFINITION

A **polynomial** is a monomial or an algebraic sum of monomials.

EXAMPLE 3.2.1 Each of the following is an example of a polynomial:

a. $x^2 + 3x + 2$ is a polynomial in one variable.
b. $15y^9 - y$ is a polynomial in one variable.
c. $x^2 - 3xy + 2y^2$ is a polynomial in two variables.
d. $9x^5 - 3x^4 + 2x^2 + 5x - 1$ is a polynomial in one variable.
e. $x^2 + y^2 + z^2 - 4xyz$ is a polynomial in three variables.

Some common types of polynomials have special names. A **monomial** is a polynomial of one term, a **binomial** is a polynomial of two terms, and a **trinomial** is a polynomial of three terms.

Each of the following is an example of a binomial:

$$x + 1,\ 3y^2 + y,\ -5x^2 - 2,\ a + b,\ x^2 - y^2$$

Each of the following is an example of a trinomial:

$$x^2 + 2x + 1,\ 5y^3 - 3y + 2,\ a + b + c,\ x^2 - 6xy + 9y^2$$

Polynomials are added or subtracted by using the commutative and associative axioms to rearrange the terms and by using the distributive axiom to combine similar terms.

EXAMPLE 3.2.2 Add $8x^2 - 5x + 7$ to $x^2 + 3x - 4$.

Solution

$$\begin{aligned}(8x^2 - 5x + 7) + (x^2 + 3x - 4) &= (8x^2 + x^2) + (-5x + 3x) + (7 - 4)\\ &= (8 + 1)x^2 + (-5 + 3)x + (7 - 4)\\ &= 9x^2 - 2x + 3\end{aligned}$$

Most of these steps can be performed mentally, and as a rule, the final result can be written immediately after inspection of the problem.

EXAMPLE 3.2.3 Simplify by performing the indicated operations: $(4x + 8y + 9) + (3x - 9y - 2)$.

Solution $(4x + 8y + 9) + (3x - 9y - 2) = 7x - y + 7$

To subtract one polynomial from another, the subtraction problem must be changed to an addition problem by using the definition of subtraction: $r - s = r + (-s)$. This is necessary because the rearrangement of terms is justified by the associative and commutative axioms, and these axioms are valid for the addition operation, but they are not valid for the subtraction operation.

EXAMPLE 3.2.4 Subtract $4x^2 + xy - 6y^2$ from $2x^2 + 3xy - 9y^2$.

Solution

$(2x^2 + 3xy - 9y^2) - (4x^2 + xy - 6y^2)$ (Recalling that "subtract a from b" is translated as "$b - a$")

$= (2x^2 + 3xy - 9y^2) + (-4x^2 - xy + 6y^2)$ (Using the definition of subtraction)

$= (2x^2 - 4x^2) + (3xy - xy) + (-9y^2 + 6y^2)$ (Using the associative and commutative axioms several times)

$= (2 - 4)x^2 + (3 - 1)xy + (-9 + 6)y^2$ (Using the distributive axiom)

$= -2x^2 + 2xy - 3y^2$

Again, it is usually possible to write the final result immediately after inspection of the problem. Thus, by performing much of the work mentally, the solution would be written as illustrated in Example 3.2.5.

EXAMPLE 3.2.5 Subtract $5x - 2y + 4$ from $-3x + 2y + 4$.

Solution

$$\begin{aligned}(-3x + 2y + 4) - (5x - 2y + 4) \\ &= (-3x + 2y + 4) + (-5x + 2y - 4) \\ &= -8x + 4y\end{aligned}$$

It is quite simple to check the addition or subtraction of polynomials by assigning a value to each variable.

EXAMPLE 3.2.6 Check the result of Example 3.2.5 by letting $x = 3$ and $y = 4$.

Solution If $x = 3$ and $y = 4$,

$$\begin{aligned}5x - 2y + 4 &= 5(3) - 2(4) + 4 \\ &= 15 - 8 + 4 \\ &= 11\end{aligned}$$

$$\begin{aligned}-3x + 2y + 4 &= -3(3) + 2(4) + 4 \\ &= -9 + 8 + 4 \\ &= 3\end{aligned}$$

Now $$(-3x + 2y + 4) - (5x - 2y + 4) = 3 - 11 = -8$$

and $$-8x + 4y = -8(3) + 4(4) = -24 + 16 = -8$$

Therefore, for $x = 3$ and $y = 4$,

$$(-3x + 2y + 4) - (5x - 2y + 4) = -8x + 4y$$

It is reasonable to conclude that $-8x + 4y$ is the correct difference. However, there is a danger involved. Special values of the variables may produce a "check" even though the answer is wrong. This can usually be prevented by avoiding the use of the special numbers 0, 1, and -1 as replacement values.

It is conventional to express a polynomial in **descending powers** of a variable. This means that the term containing the highest power of the variable is written first, at the left; the next highest power is written second; and so on, with the lowest power written last, at the right. For example, $5x^4 - 6x^3 - 2x^2 + 8x - 4$ is written in descending powers of the variable x. As another example, $y^3 + 3y^2z - z^4$ is written in descending powers of the variable y.

If the order just described is reversed, then the polynomial is said to be written in **ascending powers** of the variable. For example, $1 - 2x - x^2$ is written in ascending powers of the variable x.

Some problems are more easily done if each polynomial in the problem is arranged in either descending powers of a variable or ascending powers of a variable.

EXAMPLE 3.2.7 Simplify $(3y - 6 - 2y^2) + (5y^2 + 9 - 2y)$.

Solution First rearrange each polynomial in descending powers of the variable y:

$$(3y - 6 - 2y^2) + (5y^2 + 9 - 2y)$$
$$= (-2y^2 + 3y - 6) + (5y^2 - 2y + 9)$$

and then combine like terms:

$$= (-2y^2 + 5y^2) + (3y - 2y) + (-6 + 9)$$
$$= 3y^2 + y + 3$$

EXERCISES 3.2 A

Simplify and check the expressions in Exercises 1–20.

1. $(3x^2 + 2x) + (4x^2 + 3x + 1)$
2. $(y^2 + 3) - (y^3 - 1)$
3. $(x^3 + 2x - 1) + (x^2 + 3x + 2)$
4. $(4z^2 + 2z + 6) + (3z^2 + 5z + 9)$
5. $(2x^2 + y - z) + (3x - y + 2z)$
6. $(7x^2 - 5x + 2) + (x^2 + x - 6)$
7. $(3a + 5b - 7) + (2a - 5b + 7)$
8. $(-2x - 3y - 4) + (2x - 3y - 4)$
9. $3(2a - 5) + 4(7 - 3a)$
10. $(5y^2 - 6y + 9) + (5y - 8 - 4y^2)$
11. $(2a - b - c) - (a + b + c)$
12. $(-r - s + t) - (r + s - t)$
13. $(8 - 2x + y) - (9 - 2x + y)$
14. $(1 + x - x^2) - (2 - x^2 + x^4)$
15. $3(2a - b) - 4(b - 2a)$
16. $(y^2 + 6y + 9) - (y^2 - 4y + 4)$
17. $(25x^4 - 10x^2 + 1) - (9x^2 + 6x + 1)$
18. $(x^3 - x^2 + x) + (x^2 - x + 1)$
19. $[(x^2 + xy + y^2) - (x^2 - xy - y^2)] - (3x^2 + 2xy - y^2)$
20. $(a - b + c) - [(a + b - c) - (a - b - c)]$

In Exercises 21–30, perform the operations as directed.

21. Add $3x^2 + 5 - x$, $x - 8$, $x^2 + 4$. Check, using $x = 6$.
22. Subtract $2a^2 - 5b^2 - 4$ from $a^2 - b^2 + 1$. Check, using $a = 3$, $b = 2$.

23. From $-t^2 - 3t - 5$ subtract $-t^2 - 5t + 4$. Check, using $t = 5$.

24. Find the sum of $c^3 - 3c^2 - c - 7$ and $4c - 3c^2 - 6c^3 + 6$. Check, using $c = 1$.

25. Find the sum of $x^3 - 3x^2 + x - 5$ and $x^2 - 3x + 4$. Check, using $x = 2$.

26. From the sum of $3x^2 + 5x + 9$ and $x^2 - 7x + 14$ subtract $2x^2 - 3x + 5$.

27. Subtract $3a^2 + 5ab - 6b^2$ from the sum of $2a^2 - 3ab + 4b^2$ and $5ab - 6b^2$.

28. From $x^3 - x^2 - 3x + 2$ subtract the sum of $x^4 + 5x - 1$ and $x^3 + 3x^2 - 6x + 2$.

29. Subtract the sum of $x^3 + 2x^2y - xy^2 + y^3$ and $5x^2y + 6xy^2 - 2y^3$ from $4x^3 - 3x^2y + 3xy^2 + 4$.

30. Find the sum of $a^3 - 3a^2b + 5ab^2 - b^3$, $2a^3 + 5a^2b - 2ab^2$, and $6a^2b + 15ab^2 - 9b^3$.

EXERCISES 3.2 B

Simplify and check the expressions in Exercises 1–20.

1. $(5y^2 + 3y + 1) + (3y^2 - y - 4)$

2. $(x^3 + 2) - (x^2 + 3)$

3. $(-2x^3 + 3x + 4) + (-x^3 + x^2 - 2x)$

4. $(a^3 + 5a^2 + 2) - (3a^3 + a^2 - 4)$

5. $(r^2 + s + 1) + (r^2 - s - 1)$

6. $(2x - 3y + 4z) + (5x + 7y - 9z)$

7. $(x^2 - xy + y^2) + (x^2 + xy - y^2)$

8. $7(x - 2) - 3(2 - x)$

9. $-5(2y + 1) + 6(2y - 3)$

10. $(a^2 - a^2b^2 + 2b^2) - (-a^2 + a^2b^2 - 2b^2)$

11. $(3t^2 + 4t - 7) + (-5t^2 + t - 1)$

12. $(x + y - 1) - (x - y + 1)$

13. $(3x - 1) - (1 - 3x)$

14. $(x^2 - x + 1) - (x^2 - x + 1)$

15. $-5(t + 5) + 2(5t + 1)$

16. $(x^4 + 2x^2 + 1) - (x^2 + 10x + 25)$

17. $(4x^2 - 4xy + y^2) - (9y^2 - 6xy + x^2)$

18. $(x^3 + x^2 + x) - (x^2 + x + 1)$

19. $(3x^2 - 2xy - 2y^2) - [(2x^2 + 4xy - y^2) - (x^2 + 2xy + y^2)]$

20. $[(a^2 + ab - ac) - (b^2 - ac - c^2)] - (c^2 - ac + ab)$

In Exercises 21–30, perform the operations as directed.

21. Add $4 - 3x^2$, $2 + 3x + 8x^2$, $3 - 2x^2 - 5x$. Check, using $x = 3$.

22. From $4a + 2b - 4c$ subtract $3a - 3b - 3c$. Check, using $a = 2$, $b = 3$, $c = 5$.

23. Subtract $2y^3 - 5y^2 - 3y + 9$ from $-4y^3 - 3y^2 - 4$. Check, using $y = 10$.

24. Subtract $2x^2 - 6$ from the sum of $-3x^2 + 5x$ and $4 - 7x$. Check, using $x = 1$.

25. Find the number that must be added to $2x^2 - 3xy - y^2$ to obtain $y^2 + xy + x^2$. Check, using $x = 5$, $y = 1$.

26. Subtract $x^3 - 17x + 1$ from the sum of $x^4 + 3x^3 - x^2 + 2$ and $5x^4 - x^3 + 4x - 7$.

27. From the sum of $a^2 + 3ab + 4b^2$ and $2a^2 - 4ab + 5b^2$ subtract $3a^2 - 7ab + 8b^2$.

28. Subtract the sum of $x^4 + 3x^2 + 2$ and $2x^4 - 5x^3 + x^2 - 1$ from $6x^3 - 7x^2 + x$.

29. From $2x^3 - 4x^2y + 8xy^2 + 11y^3$ subtract the sum of $x^3 + y^3$ and $3x^2y - 5xy^2 - 7y^3$.

30. Find the sum of $a^3 + b^3 - a^2b + ab^2$, $a^2b + b^3 - 3a^3$, and $b^3 - 5ab^2 + a^2b$.

3.3 PRODUCTS OF POLYNOMIALS

In order to find the products of polynomials, the distributive axiom is used. It is restated below for emphasis.

THE DISTRIBUTIVE AXIOM

If r, s, and t are real numbers, then

$$r(s + t) = rs + rt$$

This axiom can be applied for multiplying a monomial by a binomial, and it can be extended to multiplying a binomial by a binomial, a monomial by a trinomial, and so on.

EXAMPLE 3.3.1 Multiply $x(2x^2 + 1)$.

Solution

$x(2x^2 + 1)$	$= x \cdot 2x^2 + x \cdot 1$	(Distributive axiom)
	$= 2 \cdot x \cdot x^2 + 1 \cdot x$	(Associative and commutative axioms, multiplication)
	$= 2 \cdot x \cdot x^2 + x$	(Identity axiom, multiplication)
	$= 2x^3 + x$	(First theorem of exponents)

Therefore

$$x(2x^2 + 1) = 2x^3 + x$$ (Transitive axiom of equality)

Of course, these steps can be done mentally by simply remembering the application of the distributive axiom.

EXAMPLE 3.3.2 Multiply $3x^3(2x^2 + 4x)$.

Solution

$$\begin{aligned} 3x^3(2x^2 + 4x) &= 3x^3 \cdot 2x^2 + 3x^3 \cdot 4x \\ &= 6x^5 + 12x^4 \end{aligned}$$

To multiply a binomial by a binomial, the distributive axiom is used, as well as the fact that the real numbers are closed with respect to addition.

EXAMPLE 3.3.3 Multiply $(2x + 5)(3x + 2)$.

Solution

$(2x + 5)(3x + 2)$	$= (2x + 5)(3x) + (2x + 5)(2)$	(Closure and distributive axioms)
	$= 3x(2x + 5) + 2(2x + 5)$	(Commutative axiom, multiplication)
	$= 6x^2 + 15x + 4x + 10$	(Distributive axiom)
	$= 6x^2 + 19x + 10$	(Distributive axiom used to combine like terms)

EXAMPLE 3.3.4 Write as a single polynomial (multiply)

$$(x^2 + 2)(3x^2 + x + 1)$$

Solution

$$\begin{aligned}(x^2 + 2)(3x^2 + x + 1) &= (x^2 + 2)(3x^2) + (x^2 + 2)(x) + (x^2 + 2)(1)\\ &= 3x^4 + 6x^2 + x^3 + 2x + x^2 + 2\\ &= 3x^4 + x^3 + 7x^2 + 2x + 2\end{aligned}$$

Note that the answer was stated in descending powers of the variable—that is, the variable was arranged so that the first term had the largest exponent, the second term the next largest exponent, and so on, with the constant term last. This rearrangement can be done because of the commutative and associative properties of addition of real numbers, and it makes the answer easier to check.

Sometimes it is convenient to arrange the polynomials to be multiplied in vertical order rather than horizontal order, as shown in the following example.

EXAMPLE 3.3.5 Multiply $3x^2 + 2x + 4$ by $4x^2 - x + 1$.

Solution

$$\begin{array}{rrrrrl} & & 3x^2 & +\,2x & +\,4 & \\ & & 4x^2 & -\,x & +\,1 & \\ \hline 12x^4 & +\,8x^3 & +\,16x^2 & & & \text{(Multiply top row by } 4x^2\text{)}\\ & -\,3x^3 & -\,2x^2 & -\,4x & & \text{(Multiply top row by } -x\text{)}\\ & & +\,3x^2 & +\,2x & +\,4 & \text{(Multiply top row by } +1\text{)}\\ \hline 12x^4 & +\,5x^3 & +\,17x^2 & -\,2x & +\,4 & \text{(Combine like terms)}\end{array}$$

EXERCISES 3.3 A

Multiply, simplify, and check the expressions in Exercises 1–10.

1. $x(x + 3)$

2. $x^2(2x + 1)$

3. $y(3y^3 + y + 2)$

4. $x^3(2x^4 + 3x + 1)$

5. $3xy^2(x^2y + xy + y^2)$

6. $-x^3(2x^2 - 3x - 14)$

7. $x^2(3x^2 + 2)$
8. $2ab(a^2 + 3ab + b^2)$
9. $-3z^3(2z^2 - 4z - 1)$
10. $5xz(x^2 + 2xz - 4z^2)$

In Exercises 11–35, multiply and arrange the answer in descending powers of x. Check each answer.

11. $(x + 2)(x + 3)$
12. $(x + 4)(x - 2)$
13. $(x - 1)(x - 4)$
14. $(x - 3)(x + 5)$
15. $(2x + 1)(3x + 5)$
16. $(3x - 2)(5x + 2)$
17. $(4x - 1)(5x - 3)$
18. $(2x + 5)(x - 3)$
19. $(x + y)(2x - y)$
20. $(2x + y)(3x - 2y)$
21. $x(x + 2)(x + 3)$
22. $(x^2 + 3x + 1)(x + 2)$
23. $(3x^2 - x + 1)(2x - 1)$
24. $x^2(x + 1)(x - 2)$
25. $-x(x - 2)(x^2 + x - 1)$
26. $(x + 2)^2$
27. $(x - 3)^2$
28. $x(x + 1)^2$
29. $x(x - 2)^2$
30. $(x^2 - 3x + 2)(x^2 + 2x + 1)$
31. $(3x^2 + 5x + 1)(x^2 - x - 2)$
32. $(x - 3)(x + 3)$
33. $(x + 2)(x - 2)$
34. $(2x + 1)(2x - 1)$
35. $(x^2 - 3)(x^2 + 3)$

In Exercises 36–60, multiply and express each product as a simplified polynomial. Check each answer.

36. $2x(x^2 - 7)(x^2 + 7)$
37. $-5y(y + 9)(y - 9)$
38. $-xy(10x - 3y)(10x + 3y)$
39. $(x + 2)(x + 2)^2$
40. $(y - 5)^3$
41. $(4x + 3y)^3$
42. $(x - 6)(x^2 + 6x + 36)$
43. $(2x + 5y)(4x^2 - 10xy + 25y^2)$
44. $(x + y)(x - y)(x^2 + y^2)$
45. $(2 - x^2)(x^2 + 2)(x^4 + 4)$
46. $(ab + 1)^3$
47. $[(x + 1) + y][(x + 1) - y]$
48. $(a - b - 4)(a - b + 4)$
49. $(x + y - 5)(5 + x - y)$
50. $(x + y + z + 3)(x - z + y - 3)$
51. $(x + y + z)^2$
52. $(a - b + c)^2$
★ 53. $(x^n + 1)(x^n - 1)$
★ 54. $(y^{2n} - 1)(y^n + 1)$
★ 55. $a^n(a^2 - 2ab - b^2)$
★ 56. $x(x^{2a} + x^a + 1)$
★ 57. $x^n(x^{2n+1} - x^n)$
★ 58. $(x^n + 1)(x^{2n} - x^n - 3)$
★ 59. $(x^n + 2)(x^n - 2)$
★ 60. $(x^a + 3)^2$

EXERCISES 3.3 B

Multiply, simplify, and check the expressions in Exercises 1–10.

1. $x(x - 2)$

2. $y^2(3y + 4)$

3. $z(4z^2 - 3z + 1)$

4. $x^2(2x^3 + 9x^2 - 3x + 2)$

5. $2ab^2(a^2b + 2ab - b^2)$

6. $-y^2(3y^2 - 2y + 9)$

7. $x^2y(4x^2y + xy^2 - y^3)$

8. $-xy^2(-2x^2 + xy - 4xy^2)$

9. $b^3(a^3 + ab^2c - b^3c^2)$

10. $3xyz(x^2 - 3xy + 2xz)$

In Exercises 11–35, multiply and arrange the answer in descending powers of x. Check each answer.

11. $(x + 4)(x + 5)$

12. $(x + 2)(x - 3)$

13. $(x - 3)(x - 6)$

14. $(x - 2)(x + 5)$

15. $(2x + 3)(3x + 2)$

16. $(5x - 1)(3x + 2)$

17. $(4x - 3)(2x - 5)$

18. $(6x + 1)(x - 6)$

19. $(2x + y)(x - y)$

20. $(3x + 2y)(2x - y)$

21. $x(x + 1)(x + 2)$

22. $(2x^2 + 3x + 1)(x + 3)$

23. $(x^2 + 4x - 2)(2x - 3)$

24. $x^2(x - 3)(2x + 1)$

25. $-x(2x + 3)(x^2 - x + 2)$

26. $(x + 3)^2$

27. $(2x + 1)^2$

28. $x(x - 1)^2$

29. $x(3x + 2)^2$

30. $(x^2 + 5x - 1)(x^2 - 2x + 1)$

31. $(4x^2 - x + 2)(2x^2 + 3x - 4)$

32. $(x + 5)(x - 5)$

33. $(4x + 1)(1 - 4x)$

34. $(3x + 2)(2 - 3x)$

35. $(x^2 + 4)(x^2 - 4)$

In Exercises 36–60, multiply and express each product as a simplified polynomial. Check each answer.

36. $3y(y^2 + 8)(y^2 - 8)$

37. $-4x(x - 6)(x + 6)$

38. $-ab(4x + 9y)(4x - 9y)$

39. $(y - 3)(y - 3)^2$

40. $(x + 4)^3$

41. $(5x - 6y)^3$

42. $(x + 7)(x^2 - 7x + 49)$

43. $(9x - 10y)(81x^2 + 90xy + 100y^2)$

44. $(a^2 + 4b^2)(a + 2b)(a - 2b)$

45. $(5 + x^2)(5 - x^2)(x^4 + 25)$

46. $(1 - cd)^3$

47. $[(x - y) + 10][(x - y) - 10]$

48. $(x - 6 + y)(x - 6 - y)$

49. $(a + 2 - b)(b - 2 + a)$

50. $(x - y + z + 5)(x + z + y - 5)$

51. $(a - b - c)^2$

52. $(x + y - z)^2$

★ 53. $(1 - c^{2n})(1 + c^{2n})$

★ 54. $(k^{2n} + 1)(k^{3n} - 1)$

★ 55. $c^n(cx^2 + c^2xy - c^nx^2y^2)$

★ 56. $x^a(x^{2a} + x^a - 1)$

★ 57. $x(x^{2n+1} - x^n)$

★ 58. $(x^a + 2)(x^{2a} - 3x^a + 1)$

★ 59. $(x^6 + 3)(x^6 - 3)$

★ 60. $(x^{2a} + 1)^2$

3.4 SPECIAL PRODUCTS

The student may have made some interesting observations concerning certain products in the preceding exercise set. Some of these products occur frequently enough to deserve special attention.

1. **Square of binomial:** $$(A + B)^2 = (A + B)(A + B) = A^2 + 2AB + B^2$$

2. **Cube of binomial:** $$\begin{aligned}(A + B)^3 &= (A + B)(A + B)(A + B)\\ &= (A + B)(A + B)^2\\ &= (A + B)(A^2 + 2AB + B^2)\\ &= A^3 + 3A^2B + 3AB^2 + B^3\end{aligned}$$

3. **Difference of squares:** $(A - B)(A + B) = A^2 - B^2$

EXAMPLE 3.4.1 Find the product $(x + 5)^2$.

Solution This fits the form of a square of a binomial

$$(A + B)^2 = A^2 + 2AB + B^2$$

where $A = x$ and $B = 5$. Thus

$$\begin{aligned}(A + B)^2 &= A^2 + 2AB + B^2\\ (x + 5)^2 &= x^2 + 2(5)x + 5^2\\ &= x^2 + 10x + 25\end{aligned}$$

EXAMPLE 3.4.2 Expand $(3x - 1)^2$.

Solution To "expand" means to multiply. This time the form of a square of a binomial must be examined more closely. Consider

$$(A + B)^2 = A^2 + 2AB + B^2$$

The expanded result is a trinomial. The first term, A^2, is the square of the first term of the binomial, $A + B$. The middle term of the trinomial, $2AB$, is equal to twice the product of the first and second terms of the binomial,

$A + B$. The third term, B^2, is the square of the last term of the binomial, $A + B$. Thus

$$(3x - 1)^2 = (3x + (-1))^2 = (3x)^2 + 2(3x)(-1) + (-1)^2$$
$$= 9x^2 - 6x + 1$$

This result can be checked by multiplying

$$(3x - 1)(3x - 1) = 3x(3x - 1) - 1(3x - 1)$$
$$= 9x^2 - 3x - 3x + 1$$
$$= 9x^2 - 6x + 1$$

EXAMPLE 3.4.3 Expand $(y - 2)^3$.

Solution $(y - 2)^3 = (y + (-2))^3$

Using the form of a cube of a binomial,

$$(A + B)^3 = A^3 + 3A^2B + 3AB^2 + B^3$$

where $A = y$ and $B = -2$,

$$(y - 2)^3 = y^3 + 3(-2)y^2 + 3(-2)^2y + (-2)^3$$
$$= y^3 - 6y^2 + 12y - 8$$

EXAMPLE 3.4.4 Expand $(2b + 1)^3$.

Solution This time the form of a cube of a binomial is used with $A = 2b$ and $B = 1$:

$$(2b + 1)^3 = (2b)^3 + 3(2b)^2(1) + 3(2b)(1)^2 + (1)^3$$
$$= 8b^3 + 12b^2 + 6b + 1$$

EXAMPLE 3.4.5 Expand $(x - 3)(x + 3)$.

Solution The binomials to be multiplied are identical except for the middle sign. The form to recognize is the difference of squares form, $A^2 - B^2$,

$$(A + B)(A - B) = A^2 - B^2$$

with $A = x$, $B = 3$. Thus

$$(x - 3)(x + 3) = x^2 - (3)^2$$
$$= x^2 - 9$$

EXAMPLE 3.4.6 Expand $(2y - 5)(2y + 5)$.

Solution Again the difference of squares form may be recognized, with $A = 2y$ and $B = 5$. Thus

$$\begin{aligned}(2y - 5)(2y + 5) &= (2y)^2 - (5)^2 \\ &= 4y^2 - 25\end{aligned}$$

By learning these forms and applying them, special products can be obtained more rapidly than by the longer multiplication method used previously.

A resulting product can be quickly checked by assigning a value to each variable.

EXAMPLE 3.4.7 Check that $(2x + 5)^2 = 4x^2 + 20x + 25$ by letting $x = 3$.

Solution For $x = 3$,

$$(2x + 5)^2 = (2[3] + 5)^2 = (6 + 5)^2 = (11)^2 = 121$$

Check:

$$4x^2 + 20x + 25 = 4(3)^2 + 20(3) + 25 = 36 + 60 + 25 = 121$$

EXERCISES 3.4 A

Expand the products in Exercises 1–40 by using one of the special product forms. (DO NOT MULTIPLY.) Check by assigning a value to the variable or variables. (For Exercises 1–24, let the variable = 5.)

1. $(x + 1)^2$
2. $(x - 1)^2$
3. $(y + 2)^2$
4. $(y - 2)^2$
5. $(2y + 1)^2$
6. $(2y - 1)^2$
7. $(3x + 4)^2$
8. $(4x - 3)^2$
9. $(3 + 2x)^2$
10. $(4 - x)^2$
11. $(x + 1)^3$
12. $(x - 1)^3$
13. $(y + 2)^3$
14. $(y - 2)^3$
15. $(5y + 1)^3$
16. $(2y - 1)^3$
17. $(3x + 4)^3$
18. $(4x - 3)^3$
19. $(3 + 2x)^3$
20. $(10 - x)^3$
21. $(x + 2)(x - 2)$
22. $(y - 3)(y + 3)$

23. $(5x + 1)(5x - 1)$

24. $(3x + 2)(3x - 2)$

25. $(x + y)(x - y)$

26. $(6x - 7y)(6x + 7y)$

27. $(a + 8b)(a - 8b)$

28. $(x + y)^2$

29. $(x + y)^3$

30. $(x - y)^2$

31. $(x - y)^3$

32. $(10x + 9y)^2$

33. $(10x + 9y)^3$

34. $(x^2 - 6)(x^2 + 6)$

35. $(x^2 - y^2)(x^2 + y^2)$

36. $(7 - x)(7 + x)$

37. $(2 - 3x)(2 + 3x)$

★ **38.** $(x^n + 1)^2$, (n is a natural number)

★ **39.** $(x^n + 1)^3$, (n is a natural number)

★ **40.** $(x^n + 1)(x^n - 1)$, (n is a natural number)

EXERCISES 3.4 B

Expand the products in Exercises 1–40 by using one of the special product forms. (DO NOT MULTIPLY.) Check by assigning a value to the variable or variables. (For Exercises 1–24, let the variable = 5.)

1. $(y + 3)^2$

2. $(y - 3)^2$

3. $(x + 4)^2$

4. $(x - 4)^2$

5. $(3x + 1)^2$

6. $(3x - 1)^2$

7. $(7x + 8)^2$

8. $(8x - 7)^2$

9. $(4 + 2a)^2$

10. $(8 - b)^2$

11. $(y + 8)^3$

12. $(y - 8)^3$

13. $(x + 4)^3$

14. $(x - 4)^3$

15. $(7x + 1)^3$

16. $(7x - 1)^3$

17. $(10x + 5)^3$

18. $(5x - 10)^3$

19. $(9 + 2a)^3$

20. $(6 - b)^3$

21. $(y + 1)(y - 1)$

22. $(x - 4)(x + 4)$

23. $(4x + 1)(4x - 1)$

24. $(3y + 5)(3y - 5)$

25. $(a + b)(a - b)$

26. $(2a - 5b)(2a + 5b)$

27. $(x + 2y)(x - 2y)$

28. $(2a + b)^2$

29. $(2a - b)^2$

30. $(2a + b)(2a - b)$

31. $(2a + b)^3$

32. $(2a - b)^3$

33. $(x^2 + 1)^2$

34. $(x^2 + 1)^3$

35. $(x^2 - 1)^2$

36. $(x^2 - 1)(x^2 + 1)$

37. $(x^2 - 1)^3$

★ **38.** $(x^a + 2)^2$, (a is a natural number)

★ **39.** $(x^a + 2)^3$, (a is a natural number)

★ **40.** $(x^a + 2)(x^a - 2)$, (a is a natural number)

3.5 MONOMIAL FACTORS

The symmetric axiom of equality states:

$$\text{If } r = s, \text{ then } s = r$$

Applying this property to the distributive principle for real numbers, we find that

$$\text{if } r(s + t) = rs + rt, \text{ then } rs + rt = r(s + t)$$

$r(s + t)$ is the factored form of $rs + rt$—that is, the sum $rs + rt$ is expressed as a product of factors. For example, the expression

$$ax + bx$$

contains a common factor, x; each term contains x as a factor. Remembering that the commutative property for multiplication holds for all real numbers, this expression can be written immediately in factored form:

$$ax + bx = x(a + b)$$

(*Note:* It is customary to write the common monomial factor at the left in the factored form.)

It is desirable to express a polynomial as a product of the greatest monomial factor and another polynomial. Thus, although

$$6x^4 + 12x^2 = 2x(3x^3 + 6x)$$

$2x$ is not the greatest common factor. The greatest common factor is $6x^2$, so the desired factored form is

$$6x^4 + 12x^2 = 6x^2(x^2 + 2)$$

Factoring polynomials can be thought of as the inverse or opposite process of multiplying polynomials. For example, the instructions "simplify," "multiply," "express as a single polynomial," or "expand"

$$x(2x + 1)(3x - 1)$$

means the following:

First multiply $(2x + 1)(3x - 1)$ and then multiply the product by x:

$$\begin{aligned} &x[(2x + 1)(3x - 1)] \\ &= x[6x^2 + x - 1] \\ &= 6x^3 + x^2 - x \end{aligned}$$

To factor the polynomial $6x^3 + x^2 - x$, reverse the steps, observing that x is the greatest common factor:

$$6x^3 + x^2 - x = x(6x^2 + x - 1)$$

It is evident from the problem that $6x^2 + x - 1$ is the product of two linear binomials, $2x + 1$ and $3x - 1$, but it is rather difficult to tell by just looking at it. This problem will be explored in greater detail in the next sections. It is presented at this time to develop an awareness of the inverse relationship involved in multiplying and factoring polynomials.

The example below illustrates the procedure for finding the greatest common monomial factor of a polynomial.

EXAMPLE 3.5.1 Factor $abx^2 - ab^2x^2 - a^2bx^2$.

Solution Writing each term of the polynomial in factored form, it reads

$$abxx - abbxx - aabxx$$

Now it is easy to identify the greatest common factor:

$$\begin{aligned} abx^2 - ab^2x^2 - a^2bx^2 &= abxx - abbxx - aabxx \\ &= abxx(1 - b - a) \\ &= abx^2(1 - b - a) \end{aligned}$$

Note the necessity for writing 1 as the first term of the second factor. If this term had been omitted, the product of these two expressions would not have yielded the original polynomial.

EXAMPLE 3.5.2 Express $16ax^2 + 40a^2x - 8ax$ as a product of polynomials.

Solution Again, it is helpful to express each term in factored form:

$$(2^4ax^2) + (5 \cdot 2^3a^2x) - (2^3ax)$$

The common factors are 2^3, a, and x; thus the greatest common factor is $8ax$, and

$$16ax^2 + 40a^2x - 8ax = 8ax(2x + 5a - 1)$$

When the leftmost term of a polynomial written in descending powers of a variable begins with a minus sign, it is customary to include -1 as one of the factors of the greatest common monomial factor. The next example demonstrates this convention.

EXAMPLE 3.5.3 Find the greatest common monomial factor of

$$-x^3 - x^2 + x$$

Solution

$$\begin{aligned} -x^3 - x^2 + x &= (-1)x^3 + (-1)x^2 + (-1)(-x) \\ &= (-x)x^2 + (-x)x + (-x)(-1) \\ &= (-x)(x^2 + x - 1) \\ &= -x(x^2 + x - 1) \end{aligned}$$

A factoring problem can be checked by multiplication.

EXAMPLE 3.5.4 Check that $-x^3 - x^2 + x = -x(x^2 + x - 1)$ by multiplication.

Solution

$$\begin{aligned} -x(x^2 + x - 1) &= (-x)(x^2) + (-x)(x) + (-x)(-1) \\ &= -x^3 - x^2 + x \end{aligned}$$

A factoring problem can also be checked by assigning a value to each variable and showing that the value of the expanded form is equal to the value of the factored form.

EXAMPLE 3.5.5 Factor $8ax^3 - 6ax^2 + 12ax$ and check by letting $a = 5$ and $x = 2$.

Solution

$$\begin{aligned} &8ax^3 - 6ax^2 + 12ax \\ &= 2ax(4x^2) - 2ax(3x) + 2ax(6) \\ &= 2ax(4x^2 - 3x + 6) \end{aligned}$$

Check: For $a = 5$ and $x = 2$,

$$\begin{aligned} 8ax^3 - 6ax^2 + 12ax &= 8(5)(2^3) - 6(5)(2^2) + 12(5)(2) \\ &= 40(8) - 30(4) + 12(10) \\ &= 320 - 120 + 120 = 320 \\ 2ax(4x^2 - 3x + 6) &= 2(5)(2)(4 \cdot 2^2 - 3 \cdot 2 + 6) \\ &= 2(10)(16 - 6 + 6) \\ &= 20(16) = 320 \end{aligned}$$

It is so easy to make a mistake in algebraic work that good algebraic techniques must involve some type of checking.

EXERCISES 3.5 A

In Exercises 1–45, factor and check each result.

1. $4x + 12$

2. $6x - 3$

3. $ay - a$

4. $5x^2 + 25x$

5. $4y^3 - 100y^2$

6. $4x^2 + 2x$

7. $x^4 - x^5$

8. $x^5 - x^4$

9. $c^2n^2 + cn^3$

10. $24x^3 - 30x^2$

11. $6p - 12q + 6$

12. $36x^2 + 36x + 9y$

13. $ax + ay - az$

14. $14x^2 + 21xy + 7$

15. $4an + 16a^2n^2$

16. $6anx + 3abn$

17. $a^2x - ax^2$

18. $ax^3 + a^2x^2 - a^3x$

19. $24x^2 + 12x + 6$

20. $ay^2 + aby + ab$

21. $c^4 - c^3 + c^2 - 2c$

22. $x^2 - x + xy$

23. $-25x^3 - 15x$

24. $a^3b^3 + a^2b^4 - a^2b^3$

25. $-xy^3 - xy^2 - xy$

26. $-xyz - xy - yz$

27. $r^2s - rs^2 - 4rs$

28. $6ax^2y - 12axy^2 + 6axy$

29. $18ab + 27a^2b^2 - 63$

30. $-ab - ac$

31. $-6x^2 + 2x^3 - x^4$

32. $-56mn + 72m^2n^2$

33. $3p^2q + 9p^2q^2 - 12p^2q^3$

34. $15x^2 - 10x^3 - 15x^4$

35. $x^{50} + x^{51}$

36. $x(x + 3) - x^2$

37. $3xy + x(x - 3)$

38. $a(c + d) - ac$

39. $c^2 - c(x - y)$

40. $b^2t - bt(b + c)$

★ **41.** $x^n + x^{n+1}$

★ **42.** $x^{2n} - x^n$

★ **43.** $x^{3n} + x^n$

★ **44.** $y^{4n} - y^nz$

★ **45.** $x^{n+2} + x^{n+1} + x^n$

★ *In Exercises 46–50, factor by first replacing the binomial in parentheses by a single letter as follows:*

$$\begin{aligned} x(x - 3) - y(x - 3) &= xN - yN \text{ where } N = x - 3 \\ &= N(x - y) \\ &= (x - 3)(x - y) \end{aligned}$$

46. $a(x + y) + 3(x + y)$

47. $x^2(x - 4) - 5(x - 4)$

48. $x^2(x + 5) - 2y^2(x + 5)$

49. $y(y^2 - 3) + (y^2 - 3)$

50. $5a^2(t^2 + 4) - (t^2 + 4)$

EXERCISES 3.5 B

In Exercises 1–45, factor and check each result.

1. $5x - 20$

2. $14x + 7$

3. $by + by^2$

4. $6x^2 - 36x$

5. $36y^3 - 144y^2$

6. $6x^3 - 3x^2$

7. $x^3 - x^4$

8. $x^4 - x^3$

9. $a^2n^2 - 3an^2$

10. $75x^2 + 45x^4$

11. $4x^2 + 8x + 16$

12. $15p - 20q - 35$

13. $bx - by + bz$

14. $15mn + 10m^2n^2 - 5$

15. $7an - 14a^2n^2$

16. $3a - 6a^2 + 9a^3$

17. $3x^3 - 9x^2 + 18x$

18. $4an^2 - 2an + 6a^2n$

19. $x^3 - x^2 + x$

20. $3m^2n - 6mn^2 + 12mn$

21. $-xy - x^2y$

22. $-x^3y + 2x^2y^2 + xy^2$

23. $4\pi r^2 + 4\pi R^2 + 4\pi rR$

24. $y^2 - y + xy$

25. $6x^2y + 3x - 8xy^2$

26. $15mn + 20m^2n^2 - 5m^3n^3$

27. $-x^2 - x$

28. $-ab^4 - ab^3 + ab^2$

29. $9m^2n^2p - 3mn^2p + 12m^2np$

30. $x^2y^2z^2 + 2xyz - xz$

31. $-x^3y - y^3x$

32. $42x^2y - 63xy^2$

33. $-6r^2sx - 6rs^2y$

34. $144y - 6y^2 - 150$

35. $y^{20} - y^{19} + y^{18}$

36. $2x(x + 1) - 3xy$

37. $5xy^2 + xy(x + 1)$

38. $a(x - y) - a^3$

39. $a^2n - an(x + y)$

40. $c^2k^2 - ck(c - k)$

★ **41.** $x^{n+1} - x^n$

★ **42.** $ay^n + ay^{2n}$

★ **43.** $x^{3n} - x^{2n}$

★ **44.** $h^2 + 4h^{n+2}$

★ **45.** $x^{n+1} - x^n + x^{n+2}$

★ *In Exercises 46–50, factor each of the following by first replacing the binomial in parentheses by a single letter as follows:*

$$\begin{aligned} x(x - 3) - y(x - 3) &= xN - yN \text{ where } N = x - 3 \\ &= N(x - y) \\ &= (x - 3)(x - y) \end{aligned}$$

46. $c(a - 2) - 4b(a - 2)$

47. $y^2(y + 1) + 4(y + 1)$

48. $t(t^3 + 1) - (t^3 + 1)$

49. $u^2(u - v) + (u - v)$

50. $4(a + b)x^2 + 2(a + b)x + (a + b)$

3.6 FACTORING QUADRATIC POLYNOMIALS: I

A polynomial in x of the form $ax^2 + bx + c$, where $a \neq 0$, is said to be of **second degree**, or **quadratic**, since the highest power of the variable, x, is 2. (a, b, and c are constants.)

Examining the product of the two binomials

$$(x + 2)(x + 3) \tag{1}$$

the distributive axiom permits writing this product as

$$(x + 2)x + (x + 2)3 \tag{2}$$

Applying the distributive axiom again,

$$x^2 + 2x + 3x + 2 \cdot 3 \tag{3}$$

$$= x^2 + (2 + 3)x + 6 \tag{4}$$

$$= x^2 + 5x + 6 \tag{5}$$

The first term of the trinomial (5) is the product of the first terms of the two binomials in (1), and the third term of (5) is the product of the second terms of the binomials in (1):

$$\overbrace{(x + 2)(x}^{x^2} + 3) \quad (x + \underbrace{2)(x + 3}_{+6})$$

The middle term of the trinomial (5), $5x$, is the algebraic sum of the outer product and inner product of the binomials:

$$\overbrace{(x + 2)(x + 3)}^{\text{Outer product}} \quad (x + \underbrace{2)(x}_{\text{Inner product}} + 3)$$

Outer product $= 3x$; inner product $= 2x$; algebraic sum of outer and inner products $= 3x + 2x = 5x$. Thus

$$x^2 + 5x + 6 = x^2 + (2 + 3)x + (2 \cdot 3) = (x + 2)(x + 3)$$

In general, the product of two binomials of the form $X + A$ and $X + B$ is

$$(X + A)(X + B) = X^2 + (A + B)X + A \cdot B$$

EXAMPLE 3.6.1 Express $x^2 + 3x + 2$ in factored form.

Solution If $x^2 + 3x + 2$ is factorable over the integers—that is, if it can be factored into binomials whose coefficients are integers—then the product of the first terms of the binomials must be x^2:

$$(x \quad)(x \quad)$$

and the factors will be in the form

$$(x + A)(x + B)$$

where $A + B = 3$ and $A \cdot B = 2$. Since the only factors of 2 are 2 and 1, and -2 and -1, and the sum $A + B$ is a positive number, $A = 2$ and $B = 1$. (The same result would have been obtained for $A = 1$ and $B = 2$, since addition and multiplication are each commutative.) Thus

$$x^2 + 3x + 2 = (x + 2)(x + 1)$$

Also

$$x^2 + 3x + 2 = (x + 1)(x + 2)$$

The answer can be checked by multiplication or by substitution of a numerical value for x.

EXAMPLE 3.6.2 Factor $x^2 + 3x - 4$ and check by (a) multiplication and (b) substituting 5 for x.

Solution Again, the factors will be in the form $(X + A)(X + B)$ with $A + B = 3$ and $A \cdot B = -4$. The factors of -4 are $(-1)(4)$, $1(-4)$, and $2(-2)$.

If $A = -1$ and $B = 4$, then $A + B = -1 + 4 = 3$. Thus

$$x^2 + 3x - 4 = (x - 1)(x + 4)$$

Check: a. Multiplying,

$$\begin{aligned} (x - 1)(x + 4) &= x(x + 4) + (-1)(x + 4) \\ &= x^2 + 4x - x - 4 \\ &= x^2 + 3x - 4 \end{aligned}$$

b. For $x = 5$,

$$x^2 + 3x - 4 = 5^2 + 3(5) - 4 = 25 + 15 - 4 = 36$$

$$(x - 1)(x + 4) = (5 - 1)(5 + 4) = (4)(9) = 36$$

If all the numerical coefficients and constants of a factored expression are integers, the polynomial is said to be **factored over the integers**. All the examples worked thus far have been factored over the integers.

EXAMPLE 3.6.3 Factor $x^2 - 10x - 24$ over the integers and check.

Solution The integral factors of -24 are $-1(24)$, $1(-24)$, $-2(12)$, $2(-12)$, $-3(8)$, $3(-8)$, $-4(6)$, $4(-6)$. The only pair of factors whose sum is -10 is 2 and -12. Therefore

$$x^2 - 10x - 24 = (x + 2)(x - 12)$$

Also $$x^2 - 10x - 24 = (x - 12)(x + 2)$$

Check: Let $x = 7$.

$$x^2 - 10x - 24 = 7^2 - 10(7) - 24 = 49 - 70 - 24$$
$$= -21 - 24 = -45$$

Then $$(x + 2)(x - 12) = (7 + 2)(7 - 12) = (9)(-5) = -45$$

EXAMPLE 3.6.4 Factor $x^2 + 3xy - 10y^2$ over the integers and check.

Solution

1. Comparing the forms:

$$X^2 + (A + B)X + AB = (X + A)(X + B)$$
$$x^2 + 3xy - 10y^2 = (x \ ? \ y)(x \ ? \ y)$$

It is seen that A and B must have the forms py and qy where $p + q = 3$ and $pq = -10$.

2. The pairs of factors of -10 are $(-1)(10)$, $(10)(-1)$, $(-2)(5)$, and $(2)(-5)$. Since $(-2) + (5) = 3$, select $p = -2$ and $q = 5$.

3. Then $x^2 + 3xy - 10y^2 = (x - 2y)(x + 5y)$.
Also, by the commutative axiom for multiplication,

$$x^2 + 3xy - 10y^2 = (x + 5y)(x - 2y)$$

Check: Let $x = 3$ and $y = 2$. Then

$$x^2 + 3xy - 10y^2 = 3^2 + 3(3)(2) - 10(2^2)$$
$$= 9 + 18 - 40 = 27 - 40 = -13$$

$$(x - 2y)(x + 5y) = (3 - 2\cdot 2)(3 + 5\cdot 2)$$
$$= (3 - 4)(3 + 10) = (-1)(13) = -13$$

EXAMPLE 3.6.5 Factor $x^2 + x + 1$ over the integers, if possible.

Solution Comparing the forms:

$$X^2 + (A + B)X + AB = (X + A)(X + B)$$

it is seen that $A + B = 1$ and $AB = 1$. The only pairs of factors of 1 which are integers are (1)(1) and $(-1)(-1)$; in other words, $A = 1$ and $B = 1$, or $A = -1$ and $B = -1$. But $A + B = 1 + 1 = 2$ or $A + B = -1 + (-1) = -2$, and according to the problem, $A + B = 1$. Therefore, $x^2 + x + 1$ cannot be factored over the integers.

A polynomial that cannot be factored over the integers is said to be **prime** over the integers.

EXERCISES 3.6 A

Supply the missing factor in Exercises 1–5.

1. $x^2 + 9x + 14 = (x + 7)(\quad)$
2. $x^2 - 14x + 24 = (x - 2)(\quad)$
3. $x^2 - 11x - 26 = (x + 2)(\quad)$
4. $x^2 + 4x + 3 = (x + 1)(\quad)$
5. $x^2 + 2x - 8 = (x - 2)(\quad)$

Supply the missing signs in Exercises 6–10.

6. $x^2 + 11x + 30 = (x \quad 6)(x \quad 5)$
7. $x^2 + 5x - 6 = (x \quad 1)(x \quad 6)$
8. $x^2 - 5x - 6 = (x \quad 1)(x \quad 6)$
9. $x^2 - 7x + 6 = (x \quad 1)(x \quad 6)$
10. $x^2 + 11x + 10 = (x \quad 10)(x \quad 1)$

Factor over the integers, if possible, in Exercises 11–30.

11. $x^2 - 9x + 20$
12. $x^2 - 10x + 21$
13. $r^2 + 4r - 32$
14. $p^2 - 6p - 40$
15. $x^2 + 5x + 6$
16. $p^2 - 5p + 6$
17. $y^2 - 7y + 12$
18. $x^2 - 2x - 3$
19. $x^2 - 2x - 4$
20. $y^2 - y - 2$

21. $x^2 + 2xy + y^2$

22. $x^2 - 2xy - 15y^2$

23. $r^2 + rs - 2s^2$

24. $x^2 + 3xz - 70z^2$

25. $a^2 + 7ab + 10b^2$

26. $a^2 - 2ab - 10b^2$

27. $x^2 + 3ax - 10a^2$

28. $64 - 16x + x^2$

29. $-10 - 3x + x^2$

30. $24 - 10y - y^2$

In Exercises 31–40, factor over the integers by first removing the greatest common monomial factor and then factoring the general trinomial factor, if possible.

31. $4x^2 + 32x + 60$

32. $5x^2 + 30x - 35$

33. $100x^3 - 200x^2 - 800x$

34. $6y^4 - 24y^3 - 72y^2$

35. $15y^2 + 30y + 45$

36. $84 + 7a - 7a^2$

37. $20cd - cd^2 - cd^3$

38. $30u^2 + 150u - 120$

39. $2x^3 + 8x^2y - 120xy^2$

40. $5uv - uv^3 - u^3v$

EXERCISES 3.6 B

Supply the missing factor in Exercises 1–5.

1. $x^2 + 9x + 8 = (x + 8)(\quad)$

2. $x^2 + 7x - 8 = (x + 8)(\quad)$

3. $x^2 - 7x - 8 = (x - 8)(\quad)$

4. $x^2 - x - 6 = (x + 2)(\quad)$

5. $x^2 + 5x + 6 = (x + 2)(\quad)$

Supply the missing signs in Exercises 6–10.

6. $x^2 - 15x - 16 = (x \quad 16)(x \quad 1)$

7. $x^2 + 15x - 16 = (x \quad 16)(x \quad 1)$

8. $x^2 + 7x + 10 = (x \quad 2)(x \quad 5)$

9. $x^2 - 3x - 10 = (x \quad 2)(x \quad 5)$

10. $x^2 - x - 42 = (x \quad 7)(x \quad 6)$

Factor over the integers, if possible, in Exercises 11–30.

11. $x^2 - 6x + 8$

12. $x^2 - 2x - 8$

13. $x^2 - x - 6$

14. $x^2 - 5x + 6$

15. $y^2 + 9y + 8$

16. $x^2 - 8x + 12$

17. $z^2 + 3z + 4$

18. $a^2 + 8a - 20$

19. $a^2 - 9a + 20$

20. $a^2 + 9a + 20$

21. $x^2 + xy - 20y^2$

22. $x^2 + 21xy + 20y^2$

23. $x^2 - 19xt - 20t^2$

24. $x^2 + 19xz - 20z^2$

25. $p^2 + 10pq + 25q^2$

26. $r^2 + 6rs + 9s^2$

27. $x^2 + 11xy + 10y^2$

28. $45 + 14y + y^2$

29. $21 - 4a - a^2$

30. $-45 - 44x + x^2$

In Exercises 31–40, factor over the integers by first removing the greatest common monomial factor and then factoring the general trinomial factor, if possible.

31. $3x^2 - 9x + 6$

32. $10x^2 - 20x - 350$

33. $150x^4 - 1050x^3 + 1500x^2$

34. $4a^3y + 4a^2y - 48ay$

35. $20y^2 + 140y - 120$

36. $35b^2 - 30b - 5b^3$

37. $28nk^2 - 3nk^3 - nk^4$

38. $x^3y - 2x^2y^2 - 35xy^3$

39. $u^4 + 2u^3v - 3u^2v^2$

40. $18d^2 + 18dx + 36x^2$

3.7 FACTORING QUADRATIC POLYNOMIALS: 2

The reader may have noticed that all the polynomials of the form $ax^2 + bx + c$ in the preceding section were such that $a = 1$—that is, they were of the form $x^2 + bx + c$.

What happens if $a \neq 1$? If $a \neq 1$, then the trinomial $ax^2 + bx + c$, if it is factorable over the integers, is the product of two binomials

$$(rx + p)(sx + q)$$

so that $ax^2 + bx + c = (rx + p)(sx + q) = (rs)x^2 + (rq + ps)x + pq$ and $a = rs$, $b = rq + ps$, and $c = pq$.

It is important to note here that unless otherwise specified, the trinomials being considered are of the form $ax^2 + bx + c$, where a, b, and c are integers ($a \neq 0$). To factor such a trinomial means to write it as the product of two linear polynomials with integral coefficients:

$$ax^2 + bx + c = (rx + p)(sx + q)$$

where p, q, r, and s are integers. It is not always possible to do this, but when a given polynomial has been reduced as far as possible to factors with integral coefficients and all common monomial factors have been extracted, it is said that the polynomial is completely factored over the set of integers.

EXAMPLE 3.7.1 Factor $6x^2 + x - 15$.

Solution If this trinomial has two binomial factors, the product of the first terms must be $6x^2$. The possibilities are $2x$ and $3x$, or $6x$ and x.

$$(2x \quad)(3x \quad) \text{ or } (6x \quad)(x \quad)$$

The product of the second terms must be -15. The possible factors of -15 are 5 and -3, 3 and -5, 15 and -1, and 1 and -15:

$$\begin{array}{ll}
(2x + 5)(3x - 3) & (6x + 5)(x - 3) \\
(2x - 5)(3x + 3) & (6x - 5)(x + 3) \\
(2x + 3)(3x - 5) & (6x + 3)(x - 5) \\
(2x - 3)(3x + 5) & (6x - 3)(x + 5) \\
(2x + 15)(3x - 1) & (6x + 15)(x - 1) \\
(2x - 15)(3x + 1) & (6x - 15)(x + 1) \\
(2x + 1)(3x - 15) & (6x + 1)(x - 15) \\
(2x - 1)(3x + 15) & (6x - 1)(x + 15)
\end{array}$$

There are sixteen possibilities, and it is necessary to select the correct one, because **complete factorization is unique and results in one and only one correct set of factors (except for the order in which they are written).** All the above pairs of binomials yield the correct first and third terms of the original trinomial, but only one will produce the correct middle term:

$$\overbrace{(2x - \underbrace{3)(3x}_{-9x} + 5)}^{10x}$$

$$(2 \cdot 5 - 3 \cdot 3)x = 10x - 9x = x$$

Therefore

$$6x^2 + x - 15 = (2x - 3)(3x + 5)$$

Also

$$6x^2 + x - 15 = (3x + 5)(2x - 3)$$

EXAMPLE 3.7.2 Check the answer to Exercise 3.7.1 by (a) multiplication and (b) substituting 2 for x.

Solution

a. $$\begin{aligned} (2x - 3)(3x + 5) &= 2x(3x + 5) - 3(3x + 5) \\ &= 6x^2 + 10x - 9x - 15 \\ &= 6x^2 + x - 15 \end{aligned}$$

b. For $x = 2$,

$$6x^2 + x - 15 = 6 \cdot 2^2 + 2 - 15 = 24 - 13 = 11$$
$$(2x - 3)(3x + 5) = (2 \cdot 2 - 3)(3 \cdot 2 + 5) = (4 - 3)(6 + 5) = 11$$

EXAMPLE 3.7.3 Factor over the integers, if possible, and check the result: $8y^2 - 10y - 7$.

Solution

1. Comparing with the general form,

$$\begin{aligned}(rs)X^2 + (rq + ps)X + pq &= (rX + p)(sX + q)\\ 8y^2 - \quad 10y - 7 &= (ay + b)(cy + d)\end{aligned}$$

where $ac = 8$ and $bd = -7$ and $ad + bc = -10$.

2. The factors of 8 are (1)(8) and (2)(4). Therefore, $8y^2 - 10y - 7$ has the form $(y + b)(8y + d)$ or $(2y + b)(4y + d)$.

3. Since $bd = -7$, the possibilities for bd are $(-1)(7)$, $(7)(-1)$, $(1)(-7)$, and $(-7)(1)$.

4. The possibilities to be tried are:

$$\begin{array}{ll}(y - 1)(8y + 7) & (2y - 1)(4y + 7)\\ (y + 7)(8y - 1) & (2y + 7)(4y - 1)\\ (y + 1)(8y - 7) & (2y + 1)(4y - 7)\\ (y - 7)(8y + 1) & (2y - 7)(4y + 1)\end{array}$$

5. After several trials,

$$(2y + 1)(4y - 7)$$

$4y$ (Inner product)

$-14y$ (Outer product)

$-10y$ (Sum of inner and outer products)

Thus

$$8y^2 - 10y - 7 = (2y + 1)(4y - 7)$$

Check: Let $y = 5$.

$$\begin{aligned}8y^2 - 10y - 7 &= 8(5^2) - 10(5) - 7\\ &= 2 \cdot 4(25) - 50 - 7 = 200 - 57 = 143\\ (2y + 1)(4y - 7) &= (2 \cdot 5 + 1)(4 \cdot 5 - 7) = (11)(13)\\ &= 143\end{aligned}$$

EXAMPLE 3.7.4 Factor over the integers, if possible: $5x^2 + 6x - 1$.

Solution If the trinomial can be factored, the possibilities are

$$(x + 1)(5x - 1) \text{ and } (x - 1)(5x + 1)$$

Trying these,

$$\begin{aligned}(x + 1)(5x - 1) &= x(5x - 1) + (1)(5x - 1)\\ &= 5x^2 - x + 5x - 1\\ &= 5x^2 + 4x - 1\\ (x - 1)(5x + 1) &= x(5x - 1) + (-1)(5x + 1)\\ &= 5x^2 - x - 5x - 1\\ &= 5x^2 - 6x - 1\end{aligned}$$

Since none of these possibilities works and since they are the only possibilities, $5x^2 + 6x - 1$ *cannot* be factored over the integers; it is prime.

EXERCISES 3.7 A

Supply the missing signs in Exercises 1–10.

1. $8x^2 - 14x - 15 = (4x \quad 3)(2x \quad 5)$
2. $8x^2 + 14x - 15 = (4x \quad 3)(2x \quad 5)$
3. $8x^2 - 26x + 15 = (4x \quad 3)(2x \quad 5)$
4. $8x^2 + 26x + 15 = (4x \quad 3)(2x \quad 5)$
5. $8x^2 - 22x + 15 = (4x \quad 5)(2x \quad 3)$
6. $8x^2 + 119x - 15 = (8x \quad 1)(x \quad 15)$
7. $8x^2 - 37x - 15 = (8x \quad 3)(x \quad 5)$
8. $8x^2 - 29x + 15 = (8x \quad 5)(x \quad 3)$
9. $8x^2 - 19x - 15 = (8x \quad 5)(x \quad 3)$
10. $8x^2 + 43x + 15 = (8x \quad 3)(x \quad 5)$

Supply the missing factor in Exercises 11–20.

11. $2x^2 - 9x - 110 = (x - 10)(\quad)$
12. $6x^2 - 7x - 20 = (3x + 4)(\quad)$
13. $6x^2 - x - 2 = (2x + 1)(\quad)$
14. $3y^2 + 25y - 50 = (y + 10)(\quad)$
15. $12y^2 - 17y - 5 = (4y + 1)(\quad)$
16. $8a^2 - 26a + 21 = (2a - 3)(\quad)$

17. $10x^2 + 23x - 5 = (2x + 5)(\quad)$
18. $10x^2 + 27x + 5 = (2x + 5)(\quad)$
19. $10x^2 - 23x - 5 = (2x - 5)(\quad)$
20. $10x^2 - 27x + 5 = (2x - 5)(\quad)$

Factor over the integers, if possible, in Exercises 21–40.

21. $2x^2 - 3x + 1$
22. $4x^2 - 20x + 25$
23. $12x^2 - 23x + 5$
24. $12x^2 - 17x - 5$
25. $12x^2 - 16x + 5$
26. $3a^2 - 4a - 7$
27. $3a^2 + 4a - 7$
28. $3a^2 + 21a + 7$
29. $3a^2 - 20a - 7$
30. $3a^2 + 20a - 7$
31. $5x^2 + 5x - 2$
32. $15x^2 - 7x - 2$
33. $9y^2 + 24y + 16$
34. $25x^2 - 30x + 9$
35. $10x^2 + 33x - 7$
36. $10x^2 - 9x - 7$
37. $3x + 2x^2 + 1$
38. $-4a + 1 + 4a^2$
39. $15 + 16x + 4x^2$
40. $8 - 21a - 9a^2$

★ **41.** For what integers k is $x^2 + kx + 4$ factorable over the integers?

★ **42.** For what integers k is $x^2 + 18x + k$ factorable over the integers?

★ **43.** For what integers k is $5x^2 + kx - 9$ factorable over the integers?

★ **44.** If a and b are positive integers, determine the missing signs in each of the following:
a. $x^2 - (a + b)x + ab = (x \quad a)(x \quad b)$
b. $x^2 + (a - b)x - ab = (x \quad a)(x \quad b)$
c. $x^2 - (a - b)x - ab = (x \quad a)(x \quad b)$
d. $x^2 + (a + b)x + ab = (x \quad a)(x \quad b)$

★ **45.** Which of the following are factorable over the integers if a and b are positive integers?
a. $x^2 + (b - a)x - ab$
b. $x^2 + (a - b)x + ab$
c. $x^2 - (a - b)x - ab$
d. $x^2 - (b - a)x + ab$

EXERCISES 3.7 B

Supply the missing signs in Exercises 1–10.

1. $6x^2 + 19x + 15 = (2x \quad 3)(3x \quad 5)$
2. $6x^2 - x - 15 = (2x \quad 3)(3x \quad 5)$

3. $6x^2 + x - 15 = (2x \quad 3)(3x \quad 5)$

4. $6x^2 - 19x + 15 = (2x \quad 3)(3x \quad 5)$

5. $6x^2 - 9x - 15 = (2x \quad 5)(3x \quad 3)$

6. $6x^2 - 13x - 15 = (6x \quad 5)(x \quad 3)$

7. $6x^2 + 9x - 15 = (6x \quad 15)(x \quad 1)$

8. $6x^2 - 23x + 15 = (6x \quad 5)(x \quad 3)$

9. $6x^2 + 43x - 15 = (2x \quad 15)(3x \quad 1)$

10. $6x^2 - 47x + 15 = (2x \quad 15)(3x \quad 1)$

Supply the missing factor in Exercises 11–20.

11. $6x^2 + 9x - 15 = (2x + 5)(\quad)$

12. $8a^2 - 10a - 3 = (2a - 3)(\quad)$

13. $3y^2 - 9y - 84 = (3y + 12)(\quad)$

14. $6x^2 + x - 2 = (2x - 1)(\quad)$

15. $10p^2 + 21p - 10 = (2p + 5)(\quad)$

16. $6x^2 - 8x - 64 = (3x + 8)(\quad)$

17. $4x^2 - 21x - 18 = (4x + 3)(\quad)$

18. $4x^2 + 27x + 18 = (4x + 3)(\quad)$

19. $4x^2 + 21x - 18 = (4x - 3)(\quad)$

20. $4x^2 - 27x + 18 = (4x - 3)(\quad)$

Factor over the integers, if possible, in Exercises 21–40.

21. $3x^2 - 2x - 1$

22. $3x^2 + 3x - 6$

23. $3x^2 - 17x - 6$

24. $9x^2 + 3x + 2$

25. $6x^2 - 11x - 150$

26. $12x^2 - 11x + 2$

27. $6y^2 + 63y - 150$

28. $8a^2 - 34a + 33$

29. $8a^2 - 35a + 33$

30. $8a^2 - 91a + 33$

31. $8a^2 - 50a + 33$

32. $8a^2 + 38a - 33$

33. $4x^2 + 12x + 9$

34. $9b^2 - 30b + 25$

35. $8p^2 + 2p - 21$

36. $8p^2 + 22p - 21$

37. $5x + 4x^2 + 1$

38. $-3y - 1 + 4y^2$

39. $6 - 11x + 4x^2$

40. $-5 - 11x + 16x^2$

★ **41.** For what integers p is $2x^2 + px + 34$ factorable over the integers?

★ **42.** For what integers p is $x^2 + 14x + p$ factorable over the integers?

★ **43.** For what integers p is $6x^2 + px - 5$ factorable over the integers?

★ **44.** If a, b, and c are positive integers, determine the missing signs in each of the following:

a. $a^2x^2 - a(b + c)x + bc = (ax \quad b)(ax \quad c)$
b. $a^2x^2 + a(b - c)x - bc = (ax \quad b)(ax \quad c)$
c. $a^2x^2 - a(b - c)x - bc = (ax \quad b)(ax \quad c)$
d. $a^2x^2 + a(b + c)x + bc = (ax \quad b)(ax \quad c)$

★ **45.** Which of the following are factorable over the integers if a, b, and c are positive integers?

a. $a^2x^2 + a(c - b)x + bc$
b. $a^2x^2 + a(c - b)x - bc$
c. $a^2x^2 - a(c - b)x - bc$
d. $a^2x^2 - a(c - b)x + bc$

3.8 FACTORING PERFECT SQUARE TRINOMIALS

There are some polynomials whose factors are readily recognized, and the student should familiarize himself with these.

The polynomial $x^2 + 2ax + a^2$ is called a **perfect square trinomial** and is the result of the square of a binomial. Note that

$$\begin{aligned}(x + a)^2 = (x + a)(x + a) &= (x + a)x + (x + a)a \\ &= x^2 + ax + ax + a^2 \\ &= x^2 + 2ax + a^2\end{aligned}$$

When a perfect square trinomial is written in descending powers of the variable x, the first term and the last term are positive and perfect squares, such as x^2, 25, 16, a^2, and so on. The middle term is equal to *twice* the product of the first and second terms of the binomial factor, because the outer product and the inner product are the same.

PERFECT SQUARE TRINOMIAL FORM

$$\boldsymbol{x^2 + 2ax + a^2 = (x + a)^2}$$

EXAMPLE 3.8.1 Find the term which when added to $x^2 + 8x$ will make it a perfect square trinomial.

Solution The factors of this perfect square trinomial must be of the form

$$(x + a)(x + a)$$

and their product is $x^2 + 2ax + a^2$. The middle term of this perfect square trinomial is $2ax$, and the middle term of the trinomial we wish to form is $8x$; therefore, let

$$2a = 8$$
$$a = 4$$

Since $a = 4$, $a^2 = 16$. The missing term is a^2, so we must add 16 to $x^2 + 8x$ to make it a perfect square trinomial:

$$x^2 + 8x + 16 = (x + 4)^2$$

Note: The student should recognize the fact that the solution could have been found very simply and mechanically by taking half of the coefficient of x, $\frac{1}{2}(8) = 4$, and squaring this number, $4^2 = 16$.

EXAMPLE 3.8.2 Find the term which when added to $x^2 + 20x$ will make it a perfect square trinomial and thus complete the square.

Solution The middle term of the desired trinomial is $20x$. The coefficient of x is 20; $\frac{1}{2}(20) = 10$; $10^2 = 100$. Therefore, the missing term is 100 and

$$x^2 + 20x + 100 = (x + 10)^2$$

EXAMPLE 3.8.3 Express $x^2 + 10x + 25$ as the square of a binomial.

Solution Using the form

$$x^2 + 2ax + a^2 = (x + a)^2$$
$$x^2 + 10x + 25 = x^2 + 2(5x) + 5^2 = (x + 5)^2$$

EXAMPLE 3.8.4 Express $4y^2 - 12y + 9$ as the square of a binomial.

Solution

$$\begin{aligned} x^2 + 2ax \quad + a^2 &= (x + a)^2 \\ 4y^2 - 12y + 9 &= (2y)^2 + 2(-6y) + 3^2 \\ &= (2y)^2 + 2(-3)(2y) + (-3)^2 = (2y + [-3])^2 \\ &= (2y - 3)^2 \end{aligned}$$

EXAMPLE 3.8.5 Is $x^2 + 14x - 49$ a perfect square?

Solution No, because the third term is *not* positive.

EXAMPLE 3.8.6 Is $y^2 + 3y + 9$ a perfect square?

Solution The only possibility is $(y + 3)^2$. But

$$(y + 3)^2 = y^2 + 2(3)y + 9 = y^2 + 6y + 9$$

Thus $y^2 + 3y + 9$ is not a perfect square.

EXAMPLE 3.8.7 Is $x^2 + 16 - 8x$ a perfect square?

Solution First rearranging the polynomial in descending powers of x,

$$x^2 - 8x + 16 = x^2 - 8x + 4^2 = (x - 4)^2$$

The answer is yes.

EXERCISES 3.8 A

Fill in the missing terms so that the result is a perfect square trinomial in Exercises 1–10.

1. $x^2 - 6x + (\quad)$
2. $x^2 + 20x + (\quad)$
3. $x^2 - 14x + (\quad)$
4. $x^2 + 40x + (\quad)$
5. $x^2 - 2x + (\quad)$
6. $y^2 + (\quad) + 81$
7. $y^2 - (\quad) + 100$
8. $x^2 - (\quad) + 4y^2$
9. $u^2 + (\quad) + 121v^2$
10. $(\quad) - 6xy + y^2$

Determine which of Exercises 11–20 are perfect square trinomials.

11. $x^2 + 8x + 16$
12. $y^2 + 2y + 1$
13. $a^2 + 2ab - b^2$
14. $9y^2 - 18y + 36$
15. $x^2 + 49y^2 - 14xy$
16. $x^2 + 36 - 12x$
17. $n^2 + 4ny + 16y^2$
18. $x^2 + 4$
19. $u^2 + 81 + 9u$
20. $a^2 - b^2$

In Exercises 21–45, factor over the integers, if possible. Check the answers.

21. $x^2 + 4x + 4$
22. $x^2 - 2x + 1$
23. $x^2 + 18x + 81$
24. $x^2 - 10x + 25$

25. $y^2 + 16y + 64$

26. $9a^2 + 6a + 1$

27. $16p^2 - 8p + 1$

28. $36 - 12x + x^2$

29. $64y^2 - 48y + 9$

30. $x^2 + 14x + 49$

31. $4x^2 + 28x + 49$

32. $9p^2 - 30p + 25$

33. $12a + 36 + a^2$

34. $4x^2 + 4xy + y^2$

35. $9s^2 + 6st + 4t^2$

36. $x^4 + 20x^2 + 100$

37. $121u^2 - 22u + 1$

38. $625a^2 + 25ab + b^2$

39. $4x^2 + 9y^2$

40. $u^2 + 441 - 42u$

★ **41.** $(x + y)^2 + 2(x + y) + 1$

★ **42.** $64(a + b)^2 - 16(a + b) + 1$

★ **43.** $100a^2 + 20a(b - c) + (b - c)^2$

★ **44.** $x^2 - 2x(y + z) + (y + z)^2$

★ **45.** $(x + y)^2 - 4(a + b)(x + y) + 4(a + b)^2$

EXERCISES 3.8 B

Fill in the missing terms so that the result is a perfect square trinomial in Exercises 1–10.

1. $a^2 + 12a + (\quad)$

2. $x^2 - 18x + (\quad)$

3. $p^2 - 36p + (\quad)$

4. $x^2 + 6xy + (\quad)$

5. $x^2 + 2x + (\quad)$

6. $y^2 - (\quad) + 36$

7. $y^2 + (\quad) + 49$

8. $u^2 - (\quad) + 9v^2$

9. $a^2 + (\quad) + 144b^2$

10. $(\quad) - 10xy + y^2$

Determine which of Exercises 11–20 are perfect square trinomials.

11. $x^2 - 4x - 4$

12. $q^2 + 10q + 25$

13. $p^2 + 8p + 9$

14. $9x^2 + 12xy + 4y^2$

15. $x^2 + x + 1$

16. $u^4 - 12u^2 + 36$

17. $y^2 + 1 - 2y$

18. $c^2 - 4cd - 4d^2$

19. $a^2 + b^2$

20. $k^2 + 64y^2 - 8ky$

In Exercises 21–45, factor over the integers, if possible. Check the answers.

21. $x^2 + 10x + 25$

22. $y^2 - 24y + 144$

23. $x^2 - 18x + 81$

24. $r^2 - 16r + 64$

25. $4x^2 - 12x + 9$

26. $25m^2 + 10m + 1$

27. $121n^2 - 22n + 1$

28. $16 - 8x + x^2$

29. $16x^2 - 24x + 9$

30. $x^2 + 40x + 400$

31. $9x^2 + 36x + 36$

32. $4y^2 + 28y + 49$

33. $-24x + 16 + 9x^2$

34. $r^2 + 12rs + 36s^2$

35. $4a^2 - 4ab + b^2$

36. $y^4 - 60y^2 + 900$

37. $144u^2 + 12u + 1$

38. $169a^2 - 26ab + b^2$

39. $x^2 + 400 - 40x$

40. $25x^2 + 16y^2$

★ **41.** $(x - 5y)^2 - 8(x - 5y) + 16$

★ **42.** $16x^2 + 72(a - b) + 81(a - b)^2$

★ **43.** $100a^2 - 140(b - c) + 49(b - c)^2$

★ **44.** $(p^2 - q^2)^2 + 4p^2q^2$

★ **45.** $a^2 + b^2 + c^2 + 2ab + 2ac + 2bc$

3.9 FACTORING SPECIAL BINOMIALS

The product of the two binomials $x + a$ and $x - a$ is not a trinomial, but another binomial.

$$\begin{aligned}(x + a)(x - a) &= (x + a)x - (x + a)a \\ &= x^2 + ax - ax - a^2 \\ &= x^2 - a^2\end{aligned}$$

Notice that the outer and inner products are alike except for sign, which makes them additive inverses whose sum is always zero. Since the resulting binomial is the difference of the squares of a and x, this special product is called the difference of squares.

DIFFERENCE OF SQUARES FORM

$$A^2 - B^2 = (A + B)(A - B)$$

EXAMPLE 3.9.1 Factor $x^2 - 25$.

Solution Since x^2 and 25 are both perfect squares and the binomial is of the form $A^2 - B^2$, it factors as $(A + B)(A - B)$, where $A^2 = x^2$ and $B^2 = 25 = 5^2$.

$$\begin{aligned}A^2 - B^2 &= (A + B)(A - B) \\ x^2 - 25 = x^2 - 5^2 &= (x + 5)(x - 5)\end{aligned}$$

EXAMPLE 3.9.2 Write $9x^2 - 4$ in factored form.

Solution $9x^2$ is the square of $3x$, since $(3x)(3x) = 9x^2$. 4 is the square of 2. Therefore

$$9x^2 - 4 = (3x)^2 - (2)^2 = (3x + 2)(3x - 2)$$

EXAMPLE 3.9.3 Factor $2x^2 - 50$.

Solution The two terms have a common factor, 2, so extract this common factor first:

$$\begin{aligned} 2x^2 - 50 &= 2(x^2 - 25) \\ &= 2(x - 5)(x + 5) \end{aligned}$$

EXAMPLE 3.9.4 Factor $4x^2 - 25y^2$.

Solution $4x^2$ is the square of $2x$. $25y^2$ is the square of $5y$. Therefore

$$4x^2 - 25y^2 = (2x)^2 - (5y)^2 = (2x + 5y)(2x - 5y)$$

EXAMPLE 3.9.5 Factor $x^2 + 1$, if possible.

Solution x^2 is the square of x. 1 is the square of 1. However, these squares are *not* separated by a minus sign, and they are therefore *not the difference* of two squares but their sum.

A binomial of the form $x^2 + a^2$ cannot be expressed as the product of two linear factors.

Two other factorable binomials are useful to know.

DIFFERENCE OF CUBES

$$A^3 - B^3 = (A - B)(A^2 + AB + B^2)$$

SUM OF CUBES

$$A^3 + B^3 = (A + B)(A^2 - AB + B^2)$$

These can be verified by direct multiplication.

EXAMPLE 3.9.6 Factor $x^3 - 8$.

Solution $x^3 - 8$ fits the difference of cubes form, with $A^3 = x^3$ and $B^3 = 8 = 2^3$.

$$\begin{aligned} A^3 - B^3 &= (A - B)(A^2 + AB + B^2) \\ x^3 - 8 = x^3 - 2^3 &= (x - 2)(x^2 + 2x + 4) \end{aligned}$$

EXAMPLE 3.9.7 Factor $27x^3 + 125y^3$.

Solution Since $27x^3 = (3x)^3$ and $125y^3 = (5y)^3$, and since

$$A^3 + B^3 = (A + B)(A^2 - AB + B^2)$$

$$27x^3 + 125y^3 = (3x)^3 + (5y)^3 = (3x + 5y)([3x]^2 - [3x][5y] + [5y]^2)$$
$$= (3x + 5y)(9x^2 - 15xy + 25y^2)$$

EXAMPLE 3.9.8 Check by multiplication that

$$x^3 - 8 = (x - 2)(x^2 + 2x + 4)$$

Solution

$$\begin{array}{l} x^2 + 2x + 4 \\ \underline{\quad x - 2 \qquad\quad} \\ x^3 + 2x^2 + 4x \\ \underline{\qquad - 2x^2 - 4x - 8} \\ x^3 \qquad\qquad\qquad - 8 \end{array}$$

EXAMPLE 3.9.9 Check by multiplication that

$$27x^3 + 125y^3 = (3x + 5)(9x^2 - 15xy + 25y^2).$$

Solution

$$\begin{array}{l} 9x^2 - 15xy + 25y^2 \\ \underline{\quad 3x + 5y \qquad\qquad\quad} \\ 27x^3 - 45x^2y + 75xy^2 \\ \underline{\qquad + 45x^2y - 75xy^2 + 125y^3} \\ 27x^3 \qquad\qquad\qquad\qquad + 125y^3 \end{array}$$

EXERCISES 3.9 A

In Exercises 1–20, factor the polynomials over the integers, if possible. Check by multiplication.

1. $x^2 - 36$
2. $4x^2 - 1$
3. $x^2 + 4$
4. $16 - 9a^2$
5. $9y^2 - 100z^2$
6. $n^4 - 4x^2$
7. $x^2 - y^2$
8. $x^3 - y^3$
9. $x^2 - 2$
10. $x^2 + y^2$
11. $x^3 + y^3$
12. $x^3 - 1$
13. $x^3 + 1$
14. $100a^2 - 1$
15. $8a^3 - 1$
16. $8a^3 + 1$
17. $4a^2 - 121$
18. $49r^2 - 64s^2$
19. $81a^2b^2 - 25c^4$
20. $64x^3 + 125y^6$

In Exercises 21–30, factor the polynomials over the integers, if possible. Remove all common factors first. Check the results.

21. $3x^2 - 147$

22. $a^3 - 16a$

23. $y^4 - y$

24. $2x^3 + 686$

25. $2y^3 - 16y^2 - 24y$

26. $100x^2 - 4y^2$

27. $100x^2 - 4x$

28. $100x^2 - 4x^4$

29. $x^4 + 216x$

30. $2y^4z - 54yz^4$

★ *Factor over the integers in Exercises 31–35.*

31. $x^{2n} - y^{2n}$

32. $x^{3n} - 125$

33. $a^{3n} + 216b^{3n}$

34. $a^{6n} + 1$

35. $(x + y)^3 - (x - y)^3$

★ *In Exercises 36–40, express each as a polynomial.*

36. $\dfrac{x^2 - 25}{x - 5}$

37. $\dfrac{x^3 - 343}{x - 7}$

38. $\dfrac{512x^3 + y^3}{8x + y}$

39. $\dfrac{a^{2n} - b^{2n}}{a^n + b^n}$

40. $\dfrac{a^{3n} + b^{3n}}{a^n + b^n}$

★ *In Exercises 41–42, using the facts*

$$x^2 - 1 = (x - 1)(x + 1)$$
$$x^3 - 1 = (x - 1)(x^2 + x + 1)$$
$$x^4 - 1 = (x - 1)(x^3 + x^2 + x + 1)$$

factor each similarly and check by multiplication.

41. $x^5 - 1$

42. $x^7 - 1$

EXERCISES 3.9 B

In Exercises 1–20, factor the polynomials over the integers, if possible. Check by multiplication.

1. $x^2 - 81$

2. $9x^2 - 1$

3. $4x^2 - 9$

4. $x^2 + 9$

5. $4x^2 - 25y^2$

6. $y^2 - x^2$

7. $y^3 - x^3$

8. $y^3 - 8y^2$

9. $y^2 + x^2$

10. $y^3 + x^3$

11. $1 + y^3$

12. $1 - y^3$

13. $16a^2 - 121b^2$

14. $n^3 - 27$

15. $27n^3 - 1$

16. $125n^3 + 64k^3$

17. $y^2 + 121$

★ **18.** $(n + 2)^2 - 36$

★ **19.** $49a^2 - (x + y)^2$

★ **20.** $(a + b)^2 - (a - b)^2$

In Exercises 21–30, factor the polynomials over the integers, if possible. Remove all common factors first. Check the results.

21. $2a^2 - 200$

22. $ax^2 + 3ax^3$

23. $16x^3 - 4x$

24. $p^2y - q^2y$

25. $216x^4 - x$

26. $16y^2 - 64y$

27. $16y^2 - 64$

28. $16y^2 - 4y^4$

29. $8y^4 - 16y^2$

30. $2y^6 + 1024y^3$

★ *Factor over the integers in Exercises 31–35.*

31. $a^{4n} - b^{2n}$

32. $x^{6n} - 64y^{3n}$

33. $r^{3n} + 343s^{9n}$

34. $(a + b)^6 - (a - b)^6$

35. $729u^{3n} + 1000v^{3n}$

★ *In Exercises 36–40, express each as a polynomial.*

36. $\dfrac{x^2 - 36y^2}{x + 6y}$

37. $\dfrac{1000y^3 + 1}{10y + 1}$

38. $\dfrac{a^{3n} - b^{3n}}{a^n - b^n}$

39. $\dfrac{a^{6n} - b^{6n}}{a^{3n} + b^{3n}}$

40. $\dfrac{a^{6n} + b^{6n}}{a^{2n} + b^{2n}}$

★ *In Exercises 41–42, using the facts*

$$\frac{1 + x^3}{1 + x} = 1 - x + x^2$$

and

$$\frac{1 + x^5}{1 + x} = 1 - x + x^2 - x^3 + x^4$$

express each as a polynomial and check by multiplication.

41. $\dfrac{1 + x^7}{1 + x}$

42. $\dfrac{1 + x^9}{1 + x}$

3.10 COMPLETE FACTORING

A polynomial is said to be **completely factored over the integers** when all common factors have been removed and when no further factoring over the integers is possible. For example, $2x^2 - 8 = 2(x^2 - 4)$. However, $x^2 - 4$ can be factored, since it is the difference of squares. Therefore, in completely factored form,

$$2x^2 - 8 = 2(x + 2)(x - 2)$$

It is often necessary to perform several factoring processes before a polynomial is completely factored.

EXAMPLE 3.10.1 Factor completely $3ax^4 - 3ay^4$.

Solution

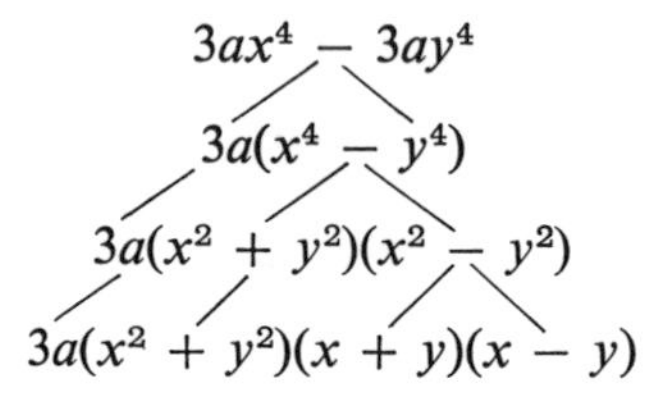

Summarizing these steps:

1. Remove greatest common monomial factor by the distributive principle:

$$3ax^4 - 3ay^4 = 3a(x^4 - y^4)$$

2. Factor the difference of two squares:

$$= 3a(x^2 + y^2)(x^2 - y^2)$$

3. Factor the difference of two squares:

$$= 3a(x^2 + y^2)(x + y)(x - y)$$

4. By the commutative property of multiplication:

$$= 3a(x + y)(x - y)(x^2 + y^2)$$

5. By the transitive property of equality:

$$3ax^4 - 3ay^4 = 3a\,(x + y)(x - y)(x^2 + y^2)$$

It is desirable to arrange the factors in such an order that monomials are written first, then linear factors, followed by quadratic factors, and so on.

PROCEDURE FOR COMPLETE FACTORIZATION

Steps in Order	*Factoring Form*
1. Remove greatest common monomial factor.	$ab + ac = a(b + c)$
2. Factor binomials, if present.	$a^2 - b^2 = (a + b)(a - b)$ $a^3 + b^3 = (a + b)(a^2 - ab + b^2)$ $a^3 - b^3 = (a - b)(a^2 + ab + b^2)$
3. Factor trinomials, if present.	$a^2 + 2ab + b^2 = (a + b)^2$ $x^2 + (p + q)x + pq = (x + p)(x + q)$ $rsx^2 + (rq + ps)x + pq = (rx + p)(sx + q)$

EXAMPLE 3.10.2 Completely factor $6x^2 + 30x - 84$.

Solution
$$\begin{aligned} 6x^2 + 30x - 84 &= 6(x^2 + 5x - 14) \\ &= 6(x + 7)(x - 2) \end{aligned}$$

EXAMPLE 3.10.3 Factor completely over the integers $x^4 - 5x^2 - 36$.

Solution
$$\begin{aligned} x^4 - 5x^2 - 36 &= (x^2 - 9)(x^2 + 4) \\ &= (x + 3)(x - 3)(x^2 + 4) \end{aligned}$$

As before, a factorization can be checked by multiplication or by substitution of a value for each variable.

EXAMPLE 3.10.4 Check Example 3.10.2 by multiplication.

Solution
$$\begin{aligned} 6(x + 7)(x - 2) &= 6(x^2 + 5x - 14) \\ &= 6x^2 + 30x - 84 \end{aligned}$$

EXAMPLE 3.10.5 Check Example 3.10.3 by letting $x = 2$.

Solution For $x = 2$,

$$\begin{aligned} x^4 - 5x^2 - 36 &= 2^4 - 5\cdot 2^2 - 36 \\ &= 16 - 20 - 36 \\ &= 16 - 56 = -40 \end{aligned}$$

$$\begin{aligned} (x + 3)(x - 3)(x^2 + 4) &= (2 + 3)(2 - 3)(2^2 + 4) \\ &= (5)(-1)(8) = -40 \end{aligned}$$

EXERCISES 3.10 A

In Exercises 1–40, factor completely over the integers, if possible. Check each answer.

1. $3y^2 - 12$

2. $2x^2 + 8x - 90$

3. $p^3 - 36p$

4. $100 - 100x^4$

5. $4x^2 - 6x + 2$

6. $36a^2 - 36a + 9$

7. $4x^2 - 6x - 10$

8. $4k + 2kx - 2kx^2$

9. $4x^3 - 20x^2 + 25x$

10. $36x^3 - 69x^2 + 15x$

11. $p^2 - 49p^4$

12. $x^4 + 4x^2$

13. $16x - 81x^3$

14. $81a^5 - 16a$

15. $6a^3 + a^2 - 12a$

16. $x^4 + 3x^2 - 4$

17. $-x^3 - 2x^2 + 24x$

18. $50x^4 - 128y^4$

19. $12x^3 + 14x^2 - 10x$

20. $y^4 - 41y^2 + 400$

21. $4u^4 - 84u^2 - 400$

22. $-8x^4 - 8x^3 - 2x^2$

23. $4x^4 + 16x^3 + 64x^2$

24. $3x^3 + 3x^2 - 126x$

25. $m^5 - m$

26. $5x^3 - 625$

27. $64x^3 - x$

28. $64x^4 + x$

29. $x^4 - y^4$

30. $x^6 - y^6$

31. $m^4 - m^2n^2 - m^2$

32. $x^6 - 64$

33. $3x^4 + 81x$

34. $t^4 - 48t^2 - 49$

35. $t^6 - 999t^3 - 1000$

★ **36.** $5x^{2n} - 5$

★ **37.** $6x^{2n} - 24x^n + 24$

★ **38.** $y^{n+2} + 2y^{n+1} - 15y^n$

★ **39.** $a^2(b + c)^2 - a^2d^2$

★ **40.** $x + 2y - a(x + 2y)$

EXERCISES 3.10 B

In Exercises 1–40, factor completely over the integers, if possible. Check each answer.

1. $9x^2 - 9x - 18$

2. $9x^2y + 9xy - 18y$

3. $3x^3 - 17x^2 - 6x$

4. $144x^3y - 100xy^3$

5. $324x^4 - 4y^4$

6. $27t^3 + 36t^2 + 63t$

7. $49x - x^3$

8. $y^8 - 16y^4$

9. $5x^2 - 15x - 50$

10. $4y^2 - 28y + 49$

11. $6x^2 - 20x + 16$

12. $2c^2 - 20c^3 + 50c^4$

13. $y^5 - y$
14. $y^7 - y^4$
15. $6x^2 - 54x + 84$
16. $-3x^3 - 27x$
17. $5a^5 + 60a^3 - 320a$
18. $6b^4 + b^3 - 12b^2$
19. $x^4 - 16x^2$
20. $x^4 - 13x^2 + 36$
21. $2a^3x + 16x$
22. $4a^4 - 16b^4$
23. $2x^4y^4 - 32$
24. $t^6 + 5t^4 - 36t^2$
25. $4x^4 - 48x^3 + 144x^2$
26. $x^4 - 24x^2 - 25$
27. $16u^4 - 36v^4$
28. $-x^4 + 4x^3 + 5x^2$
29. $-81x^4 + 54x^3 - 9x^2$
30. $n^6 - 64$
31. $n^6 + 64n^3$
32. $a^2 - b^2 - c^2$

★ 33. $y^{2n+2} - y^{2n+1} - 2y^{2n}$
★ 34. $y^{2n} + 20y^n + 100$
★ 35. $x^{3n} - x^n$
★ 36. $x^2(y - 1)^2 - 100(y - 1)^2$
★ 37. $4a^2 - (a - b)^2$
★ 38. $a^2(u - v)^2 - a^2(u + v)^2$
★ 39. $k(a^2 - b^2) + a^2 - b^2$
★ 40. $x^{6n+1} - x$

3.11 APPLICATIONS

3.11.1 Solutions of Equations by Factoring

An equation of the form $ax^2 + bx + c = 0$, where $a \neq 0$, is called a **quadratic equation**. If the quadratic trinomial $ax^2 + bx + c$ is factorable over the set of integers, then there is a simple method for solving the equation. The method depends on the following theorem.

THE ZERO-PRODUCT THEOREM

Let r and s be real numbers. Then $rs = 0$ if and only if $r = 0$ or $s = 0$.

EXAMPLE 3.11.1 If $3x = 0$, then by the zero-product theorem $3 = 0$ or $x = 0$. Since $3 \neq 0$, x *must* $= 0$.

EXAMPLE 3.11.2 If $x(x + 2) = 0$, then by the zero-product theorem $x = 0$ or $x + 2 = 0$. Also, if $x + 2 = 0$, $x = -2$.

Therefore, $x(x + 2) = 0$ if $x = 0$ or if $x = -2$.

Check:

$$x(x + 2) = 0 \qquad x(x + 2) = 0$$
$$0(0 + 2) = 0 \qquad -2(-2 + 2) = 0$$
$$0 = 0 \qquad -2(0) = 0$$
$$0 = 0$$

The zero-product theorem enables one to solve quadratic equations that are equivalent to

$$(x + a)(x + b) = 0$$

By the zero-product theorem,

$$x + a = 0 \quad \text{or} \quad x + b = 0$$

If $x + a = 0$, then $x = -a$.
If $x + b = 0$, then $x = -b$.

EXAMPLE 3.11.3 Solve the equation $x^2 + 15x + 54 = 0$.

Solution In factored form,

$$x^2 + 15x + 54 = (x + 6)(x + 9)$$

and the given equation is equivalent to

$$(x + 6)(x + 9) = 0$$

By the zero-product theorem,

$$\begin{aligned} x + 6 &= 0 \quad \text{or} \quad x + 9 = 0 \\ x &= -6 \quad \text{or} \quad x = -9 \end{aligned}$$

Check:

$$\begin{aligned} x^2 + 15x + 54 &= 0 & x^2 + 15x + 54 &= 0 \\ (-6)^2 + (15)(-6) + 54 &= 0 & (-9)^2 + (15)(-9) + 54 &= 0 \\ 36 - 90 + 54 &= 0 & 81 - 135 + 54 &= 0 \\ 0 &= 0 & 0 &= 0 \end{aligned}$$

The solution set is the union of the solution sets of the two equations, $x + 6 = 0$ and $x + 9 = 0$; $\{-6\} \cup \{-9\}$. Therefore, the solution set is $\{-6, -9\}$.

EXAMPLE 3.11.4 Solve for x: $x^2 + 3x = 0$.

Solution The quadratic polynomial $x^2 + 3x$ has a common factor, x. An equivalent equation is

$$x(x + 3) = 0$$

By the zero-product theorem:

$$\begin{aligned} x &= 0 \quad \text{or} \quad x + 3 = 0 \\ x &= 0 \quad \text{or} \quad x = -3 \end{aligned}$$

and the solution set is $\{0, -3\}$.

EXAMPLE 3.11.5 Solve for x: $2x^2 + 4x = x^2 + 5$.

Solution Before attempting a solution, express the given equation as an equivalent equation whose right side is zero:

$$2x^2 + 4x = x^2 + 5$$
$$x^2 + 4x - 5 = 0$$

Now factor:

$$(x + 5)(x - 1) = 0$$

By the zero-product theorem:

$$x + 5 = 0 \quad \text{or} \quad x - 1 = 0$$
$$x = -5 \quad \text{or} \quad x = 1$$

Check:

For $x = -5$,

$$2x^2 + 4x = x^2 + 5$$
$$(2)(-5)^2 + (4)(-5) = (-5)^2 + 5$$
$$50 + (-20) = 25 + 5$$
$$30 = 30$$

For $x = 1$,

$$2x^2 + 4x = x^2 + 5$$
$$(2)(1)^2 + (4)(1) = (1)^2 + 5$$
$$2 + 4 = 1 + 5$$
$$6 = 6$$

Since both answers check, the solution set is $\{-5, 1\}$.

Note: Since the solution of quadratic equations depends on the quadratic polynomial being equal to zero, the solutions are also called the **zeros** of the polynomial.

3.11.2 Rapid Calculations (Optional)

A very useful application of factoring is to obtain shortcuts in arithmetic. The following examples illustrate the use of several factoring forms.

Sum and Difference Form: $(a + b)(a - b) = a^2 - b^2$

EXAMPLE 3.11.6

$$(29)(31) = (30 - 1)(30 + 1) = (30)^2 - 1 = 900 - 1 = 899$$

EXAMPLE 3.11.7 $(52)(48) = (50 + 2)(50 - 2) = 2500 - 4 = 2496$

Difference of Squares Form: $a^2 - b^2 = (a - b)(a + b)$

EXAMPLE 3.11.8 $(35)^2 - (34)^2 = (35 - 34)(35 + 34) = (1)(69) = 69$

EXAMPLE 3.11.9 $(17)^2 - (15)^2 = (17 - 15)(17 + 15) = (2)(32) = 64$

Perfect Square Form: $(a + b)^2 = a^2 + 2ab + b^2$

EXAMPLE 3.11.10 $(32)^2 = (30 + 2)^2 = 900 + 2(60) + 4 = 1024$

EXAMPLE 3.11.11 $(21)^2 = (20 + 1)^2 = 400 + 40 + 1 = 441$

First and Ten Form: $(10a + b)(10a + [10 - b]) = a(a + 1) \oplus b(10 - b)$ where $\oplus$ means "tack on," as shown in the following examples.

This form is used for numbers that are two-digit numbers having the same tens digit and with the sum of the units digits equal to 10.

EXAMPLE 3.11.12
$$\begin{aligned}(32)(38) &= (3)(4) \oplus (2)(8)\\ &= 12 \oplus 16 = 1216\end{aligned}$$

EXAMPLE 3.11.13
$$\begin{aligned}(65)(65) &= (6)(7) \oplus (5)(5)\\ &= 42 \oplus 25 = 4225\end{aligned}$$

EXAMPLE 3.11.14 $(26)(24) = (2)(3) \oplus (6)(4) = 6 \oplus 24 = 624$

EXERCISES 3.11 A

By factoring, determine the solution set of the equations over the real numbers in Exercises 1–30. Check all answers.

1. $(x + 2)(x - 3) = 0$
2. $(x - 1)(x + 4) = 0$
3. $(2y - 3)(y + 2) = 0$
4. $(3a + 1)(5a - 2) = 0$
5. $x^2 - 3x + 2 = 0$
6. $2p^2 - 3p = 0$
7. $z^2 - 5z = 0$
8. $2n^2 - 32 = 0$
9. $x(x + 2) = 0$
10. $x^2 - 8x + 12 = 0$

11. $y^2 + 7y = -10$

12. $x^2 = 6 + x$

13. $a^2 - 12a - 30 = 15$

14. $x^2 + 8x - 20 = 4 - 2x$

15. $(m + 3)(m + 4) = 0$

16. $(m + 3)(m + 4) = 6$

17. $(x + 3)(x - 3) = 8x$

18. $(x + 2)(x - 1) = x^2 + 1$

19. $(y + 2)(y - 2) = y^2 - y - 3$

20. $x^2 = 16$

21. $x^2 = 25$

22. $y^2 = 36$

23. $u^2 = 49$

24. $2u^2 - 4 = u^2 - 5$

25. $x^2 + 1 = x^2 - 1$

26. $x^2 + 1 = 0$

27. $(x - 5)^2 = (x + 5)^2$

28. $(x + 1)(x^2 - 2x) = (x^2 + x)(x - 2)$

29. $(y + 1)^3 - y^3 = 1$

30. $(y + 1)(y - 1) = 2y^2 + 3$

31. The square of a number is 91 more than 6 times the number. Find the number.

32. The sum of a number and its reciprocal is 2. Find the number.

33. The product of two consecutive odd integers is 255. Find the integers.

34. The product of two consecutive even integers is 528. Find the integers.

35. If 18 is subtracted from 10 times a number, the result is one half the square of the number. Find the number.

36. $A = \pi R^2 - \pi r^2$ is the formula for the area between two concentric circles.

a. Express the right side of the formula in factored form.

b. Using the factored form, calculate the area between two circles, where $R = 43$ and $r = 41$. Use $\pi = \frac{22}{7}$.

37. $M = \frac{4wx^2 - 5wLx + wL^2}{8}$ is the formula for the bending moment of a uniformly loaded beam having length L and load w pounds per unit length where x is the distance from one end.

a. Express the numerator on the right side of the formula in factored form.

b. Using the factored form, calculate M for $L = 12$, $w = 50$, and $x = 4$.

38. $-16t^2 + 2000t - 18{,}400 = 0$ is used to find the time a certain projectile is in flight. Find t.

39. $s^2 = 8rh - 4h^2$ gives the span s of a circular arch whose height is h and whose radius is r.

a. Express the right side of the formula in factored form.

b. Solve for s if $r = 50$ and $h = 20$.

40. $P = EI - RI^2$ is a formula for electric power. Find the amperage I if $P = 360$ watts, $E = 110$ volts, $R = 5$ ohms.

In Exercises 41–50, find the products by using the rapid calculation methods of this section.

41. $(37)(43)$
42. $(84)(86)$
43. $(18)(22)$
44. $(22)^2$
45. $(34)^2$
46. $(31)(39)$
47. $(41)^2 - (40)^2$
48. $(85)^2$
49. $(27)(23)$
50. $(44)^2 - (42)^2$

EXERCISES 3.11 B

By factoring, determine the solution set of the equations over the real numbers in Exercises 1–30. Check all answers.

1. $(x + 5)(x + 2) = 0$
2. $(t - 3)(t - 9) = 0$
3. $(4s - 1)(2s + 3) = 0$
4. $(3x - 4)(6x - 1) = 0$
5. $2y^2 + 7y + 6 = 0$
6. $x^2 + 6x = 0$
7. $x^2 + 6x = 27$
8. $8k - k^2 = 0$
9. $16x + 99 = 3x^2$
10. $15z + 3z^2 = z - 8$
11. $a^2 + 4a + 2 = 7a + a^2$
12. $x^2 - 2x = 0$
13. $x^2 - 2x = 3$
14. $(y + 2)^2 = 3y^2 - 2y + 4$
15. $(x + 1)(x - 2) = 4$
16. $(x - 1)^2 = x^2 - 10$
17. $z^2(2z^2 + z - 1) = 0$
18. $x(x + 2)(x - 3) = 0$
19. $x^2 = 9$
20. $15 - 2x = x^2$
21. $u^2 = 100$
22. $u^2 = 64$
23. $u^2 = 81$
24. $5u^2 + 25 = u^2 + 9$
25. $4 - 3x^2 = 9 - 3x^2$
26. $(x - 3)(x + 4) = (x + 4)(x - 3)$
27. $2x^2 = x^2 - 4$
28. $u^3 - (u - 1)^3 = 7$
29. $(y + 9)^2 - (y - 9)^2 = 36y$
30. $4x^2(x^2 - 25) = (2x - 10)(2x + 10)$

31. If 4 is added to 7 times a certain number, the result is twice the square of the number. Find the number.

32. The square of a number is 20 more than 8 times the number. Find the number.

33. The product of two consecutive odd numbers is 143. Find the two numbers.

34. The product of two consecutive even integers is 18 more than 15 times the larger number. Find the two integers.

35. Five times the square of a number added to 10 times the number equals 315. Find the number.

36. $b^2 = c^2 - a^2$ is a formula for finding side b of a right triangle when the hypotenuse c and side a are given.
a. Rewrite the formula, expressing the right side in factored form.
b. Using the result of (a), calculate b for
1. $c = 85$ and $a = 84$
2. $c = 37$ and $a = 35$

37. $A = 2\pi rh + 2\pi r^2$ is a formula for the total area A of a right circular cylinder having radius r and height h.
a. Rewrite the formula, expressing the right side in factored form.
b. Using the result of (a), calculate A for $r = 15$ and $h = 20$. Use $\pi = \frac{22}{7}$.

38. $D = ckwL^2 - 4c^2kwL + 4c^3kw$ is a formula that gives the maximum deflection for a certain beam of length L.
a. Rewrite the formula, expressing the right side in factored form.
b. Using the result of (a), calculate D for $L = 12$, $c = \frac{1}{4}$, $w = 200$, and $k = \frac{1}{500}$.

39. $1600y = 1200x - x^2$ is an equation of the orbit of a certain projectile initially fired at 200 feet per second.
a. Find x for $y = 0$. (The largest value gives the range.)
b. Find x for $y = 200$. (This gives the horizontal distance the projectile has traveled when its vertical height is 200 feet.

40. In chemistry, $x^2 = 4(1 - x)^2$ gives the number of moles, x, that react when 1 mole of pure ethyl alcohol is mixed with 1 mole of acetic acid. Find x.

In Exercises 41–50, find the products by using the rapid calculation methods of this section.

41. (38)(42)

42. (55)(45)

43. (45)(45)

44. (87)(93)

45. (87)(83)

46. $(61)^2 - (60)^2$

47. $(42)^2$

48. $(65)^2 - (63)^2$

49. (46)(44)

50. $(51)^2$

SUMMARY

- ☐ A **monomial** is a constant, a term of the form cx^n, where c is a constant, x is a variable, and n is a natural number, or a product of terms of the form cx^n.
- ☐ A **polynomial** is a monomial or an algebraic sum of monomials.

Definition of x^n

If n is a natural number, $x^1 = x$ and

$$x^n = \underbrace{x \cdot x \cdot \ldots \cdot x}_{n \text{ factors}}$$

The First Theorem of Exponents

If x is a real number, and m and n are natural numbers, then

$$x^m \cdot x^n = x^{m+n}$$

The Second Theorem of Exponents

If x is a real number, and m and n are natural numbers, then $(x^m)^n = x^{mn}$.

Special Products and Factors

Distributive axiom	$A(B + C) = AB + AC$
General trinomial I	$(X + A)(X + B) = X^2 + (A + B)X + AB$
General trinomial II	$(aX + b)(cX + d) = acX^2 + (ad + bc)X + bd$
Perfect square	$(A + B)^2 = A^2 + 2AB + B^2$
Perfect cube	$(A + B)^3 = A^3 + 3A^2B + 3AB^2 + B^3$
Difference of squares	$A^2 - B^2 = (A + B)(A - B)$
Difference of cubes	$A^3 - B^3 = (A - B)(A^2 + AB + B^2)$
Sum of cubes	$A^3 + B^3 = (A + B)(A^2 - AB + B^2)$

The Zero-Product Theorem

Let a and b be real numbers. Then $ab = 0$ if and only if $a = 0$ or $b = 0$.

REVIEW EXERCISES

Simplify the products in Exercises 1–30.

1. $x^2 \cdot x^3$

2. $r^4 \cdot r^9$

3. $x^3 \cdot x$

4. $p^2 \cdot p^3 \cdot p^4$

5. $(-3x)(4x^3)$

6. $-a^3(2a^5)$

7. $(5n^3x)(-2x^3)$

8. $3(a^2)^3$

9. $3x(-2x)^2$

10. $x^{n+1} \cdot x^{n-1}$

11. $-2a(a - b)$

12. $2ab(ab^2 - ab + a^2b)$

13. $-xy(2x - y + 3z)$

14. $-3p^2(x + p^2)$

15. $(x - 2)(x - 3)$

16. $(y + 1)(y - 4)$

17. $(2a + 3)(3a - 1)$

18. $(a + 5)^2$

19. $(5x + 1)(5x - 1)$

20. $(2y - 3)^2$

21. $x(x + 2)(2x - 7)$

22. $x^3(2x^2 + x - 2)$

23. $-x^4(3x^3 + 2x - 4)$

24. $(2x + y)(3x - 16y)$

25. $x(x + 4)(2x - 5)$

26. $2x(3x + 2)^2$

27. $(x + 2)(x + 3)^2$

28. $(x^2 + 3x + 2)(x - 4)$

29. $(2x^2 - x + 3)(3x + 1)$

30. $(x^2 + 5x + 6)(x^2 - 3x + 1)$

In Exercises 31–55, factor the polynomials as completely as possible over the integers.

31. $3x^2 - 6x + 24xy$

32. $t^2 + 11t + 24$

33. $x^2 + 10x + 25$

34. $49y^4 - 1$

35. $x^3 - 1$

36. $a^3 + 1$

37. $-3x^3 + 27x$

38. $6a^2 - 7a - 3$

39. $36x + 3x^3 - 3x^5$

40. $100p^2 - 25q^2$

41. $64y^3 - 36y$

42. $x^6 - y^6$

43. $12x^2y - 22xy - 20y$

44. $8a^2 + 5ab - 3b^2$

45. $2y^5 - 162y$

46. $15ax^6 + 42ax^5 - 9ax^4$

47. $25r^2 + 10r + 1$

48. $36x^2 - 12x + 1$

49. $4p^2 - 20pq + 25q^2$

50. $x(y + 2) + 3(y + 2)$

51. $5(x - 3) + p(x - 3)$

52. $3x - x^2(x + 4)$

53. $y^2z(a - 1) + yz$

54. $x(a + 2)^2 + y(a + 2)$

55. $3(x + y) - 2(x + y)^2$

In Exercises 56–60, fill in the missing term so that the result is a perfect square trinomial.

56. $x^2 - 6x + (\quad)$

57. $x^2 + 16x + (\quad)$

58. $y^2 - 8y + (\quad)$

59. $a^2 - (\quad) + 36b^2$

60. $y^2 + (\quad) + 49z^2$

Solve the equations over the reals in Exercises 61–70.

61. $2x^2 - 3x - 14 = 0$

62. $x^2 + 7x = 0$

63. $x^2 + 7x = 60$

64. $45y - 15y^2 = 0$

65. $t(t - 11) = -18$

66. $12z^2 + 4z - 1 = 0$

67. $2x - (x + 2) = 3x$

68. $x^2 + 6 = 7x$

69. $x^2 = 3x$

70. $x^2 = 64$

CHAPTER 4

FRACTIONS

In algebra, numbers may be designated by letters or numerals, and polynomials are formed by combining these numbers, using the operations of addition, subtraction, and multiplication. If the numbers used are elements of the set of integers, then the polynomial represents an integer, because the set of integers is closed with respect to addition, subtraction, and multiplication. Thus a polynomial is considered to be an **integral algebraic expression.** However, it is desirable to consider a polynomial in a broader sense, and the numbers used in forming the polynomial may be elements of the set of rational numbers, the set of real numbers, or the set of complex numbers. Until it is stated otherwise, the numbers will be considered as elements of the set of real numbers.

A **rational algebraic expression** is an expression that is obtained by adding, subtracting, multiplying, or dividing polynomials. For example, the expressions

$$t - \frac{1}{t} + 5, \frac{y^2 + 9}{y + 3}, \text{ and } \frac{x - 2}{x^2 - 5x + 6} + \frac{3x + 5}{x^2 - 9}$$

are rational expressions.

The choice of the word "rational" to describe these expressions was influenced by the fact that these expressions designate rational numbers whenever the polynomials designate integers or rational numbers. This is so because the set of rational numbers is closed with respect to the operations of addition, subtraction, multiplication, and division.

Similar to the polynomial, a rational algebraic expression is given a broader meaning. Until stated otherwise, *the numbers involved in forming a rational expression shall be considered as elements of the set of real numbers.*

A rational algebraic expression is also called an **algebraic fraction** or, for simplicity, a **fraction.**

4.1 SIMPLIFICATION OF FRACTIONS

From experience with the arithmetic of fractions, one learns that a fraction can be named in more than one way.

HISTORICAL NOTE

Throughout his existence, man has had difficulty in understanding fractions. The ancient Egyptians limited their fractions by requiring the numerator to be the number 1. They were unable to grasp the concept of a pair of natural numbers representing a single number, such as the fraction $\frac{5}{6}$, which is a single number made from the integers 5 and 6.

Before the time of Archimedes (287–212 B.C.), the Greeks disliked the idea of breaking unity into parts, and so they worked with ratios of integers. Thus if Alpha had 50 coins and Beta had 60 coins, the ratio of their amounts would be 5 to 6.

The Romans avoided fractions by the use of subunits; feet were divided into inches and pounds into ounces. A twelfth part of the Roman unit was called *uncia*, from which is derived our modern "ounce" and "inch." Instead of regarding a measurement as $\frac{5}{6}$ of a unit, the Romans considered this measurement as 10 *uncias* $\left(\frac{5}{6} = \frac{5 \cdot 2}{6 \cdot 2} = \frac{10}{12}\right)$.

The practical necessity of obtaining greater accuracy in measurement and the theoretical need to close the number system with respect to the division operation motivated the extension of the number system to include the fractions. To denote the concept of opposites and to close the number system with respect to the subtraction operation, the number system was extended to include the number 0 and the negatives of the counting numbers and the fractions. Thus finally the set of rational numbers, Q, was exhibited as a set of numbers that is closed with respect to the operations of addition, subtraction, multiplication, and division.

$$Q = \left\{\frac{p}{q} \text{ where } p \text{ and } q \text{ are integers and } q \neq 0\right\}$$

The rational number $\frac{5}{6}$ is interpreted as the quotient obtained when 5 is divided by 6. The number $\frac{5}{6}$ is also interpreted as the ratio of 5 to 6, and this interpretation influenced the choice of the word "rational" to describe numbers that are the quotient of two integers.

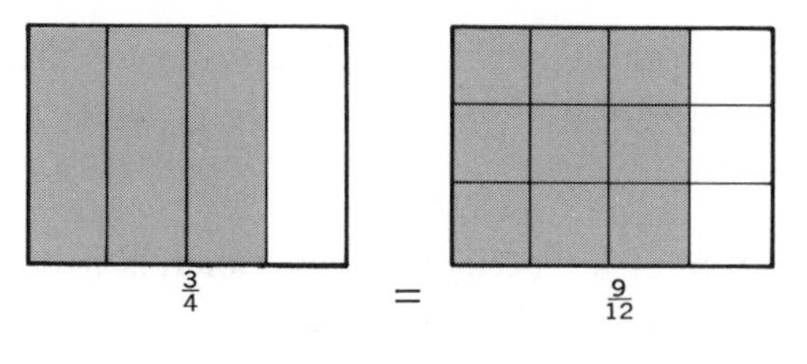

FIGURE 4.1.1

For example, by examining the geometric models illustrated in Figures 4.1.1 and 4.1.2, it can be seen that $\frac{3}{4}$ and $\frac{9}{12}$ are names for the same number.

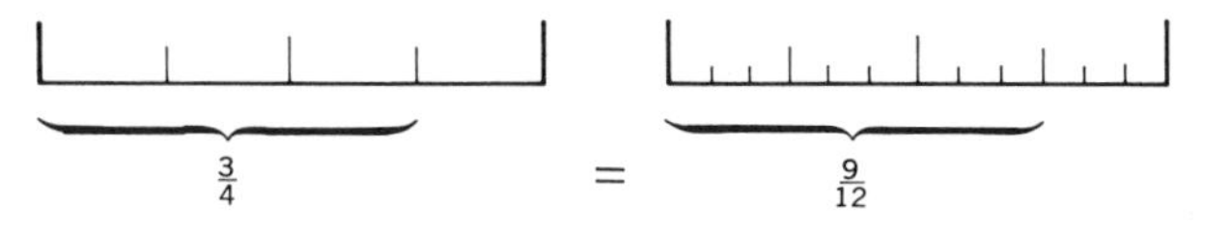

FIGURE 4.1.2

Figures 4.1.1 and 4.1.2 illustrate that further subdivisions of a basic unit cause a subdivision of the original parts of the unit used to represent the fraction.

Before stating a "fundamental theorem," the following definition is in order.

DEFINITION OF EQUAL QUOTIENTS

If a, b, c, and d are real numbers and if $bd \neq 0$, then

$$\frac{a}{b} = \frac{c}{d} \text{ if and only if } ad = bc$$

Thus $\frac{3}{4} = \frac{9}{12}$ because $3(12) = 4(9)$ and $\frac{2}{3} = \frac{4}{x}$ if and only if $2x = 12$ and $x = 6$.

The following theorem is so important in connection with everything pertaining to fractions or quotients that it is called the Fundamental Theorem of Fractions.

THE FUNDAMENTAL THEOREM OF FRACTIONS

If n, d, and k are real numbers and $d \neq 0$ and $k \neq 0$, then

$$\frac{nk}{dk} = \frac{n}{d}$$

This theorem is valid since $(nk)d = (dk)n$ by the associative and commutative axioms for multiplication.

The Fundamental Theorem of Fractions is used to simplify a fraction —that is, to reduce a fraction to lowest terms.

The **simplification of a fraction, or the reduction of a fraction to lowest terms, means the renaming of a fraction so that the numerator and denominator do not have a factor in common.**

Since all common factors of the numerator and denominator must be removed, to simplify a fraction requires that the numerator and denominator be completely factored.

EXAMPLE 4.1.1 Simplify

$$\frac{60}{84}$$

Solution

$$\frac{60}{84} = \frac{2\cdot 2\cdot 3\cdot 5}{2\cdot 2\cdot 3\cdot 7} = \frac{5\cdot 12}{7\cdot 12} = \frac{5}{7}$$

EXAMPLE 4.1.2 Reduce

$$\frac{3x^2}{15x^5}$$

to lowest terms.

Solution

$$\frac{3x^2}{15x^5} = \frac{3\cdot x\cdot x}{3\cdot 5\cdot x\cdot x\cdot x\cdot x\cdot x} = \frac{1\ \cdot 3x^2}{5x^3\cdot 3x^2} = \frac{1}{5x^3}$$

EXAMPLE 4.1.3 Simplify

$$\frac{35x^2y^3}{7xy}$$

Solution

$$\frac{35x^2y^3}{7xy} = \frac{5\cdot 7\cdot xxyyy}{7xy} = \frac{5xy^2\cdot 7xy}{1\cdot 7xy} = \frac{5xy^2}{1} = 5xy^2$$

Note in the preceding example that $\frac{5xy^2}{1}$ names the same number as $5xy^2$, just as $\frac{5}{1}$ names the same number as 5. The integral forms $5xy^2$ and 5 are considered to be the simplified forms of $\frac{5xy^2}{1}$ and $\frac{5}{1}$, respectively.

Since the number $\frac{n}{d}$ can be considered as the quotient obtained when the numerator, n, is divided by the denominator, d, there are three numbers involved in this operation, and accordingly, there are three signs associated with the number $\frac{n}{d}$:

the sign of the numerator, n

the sign of the denominator, d

and the sign of the quotient, $\frac{n}{d}$

As a result, there are four different ways to name a negative number such as $\frac{-3}{7}$:

$$\frac{-3}{7} = -\frac{3}{7} = \frac{3}{-7} = -\frac{-3}{-7}$$

The preferred form is $\frac{-3}{7}$—that is, the minus sign is placed in the numerator. It is very easy to "drop" a minus sign in a calculation, and this convention helps to prevent this error.

EXAMPLE 4.1.4 Simplify

$$\frac{21xy}{-14xz}$$

Solution

$$\frac{21xy}{-14xz} = \frac{-21xy}{14xz} = \frac{(-3y)(7x)}{2z(7x)} = \frac{-3y}{2z}$$

The factoring technique is also used to reduce a fraction with a polynominal numerator or a polynomial denominator.

EXAMPLE 4.1.5 Simplify

$$\frac{x^2 - 9}{x^2 - 3x}$$

Solution Factoring numerator and denominator,

$$\frac{x^2 - 9}{x^2 - 3x} = \frac{(x + 3)(x - 3)}{x(x - 3)} = \frac{x + 3}{x}$$

EXAMPLE 4.1.6 Reduce to lowest terms

$$\frac{x^2 - 1}{x - 1}$$

Solution

$$\frac{x^2 - 1}{x - 1} = \frac{(x + 1)(x - 1)}{x - 1} = \frac{(x + 1)(x - 1)}{1(x - 1)} = \frac{x + 1}{1} = x + 1$$

EXAMPLE 4.1.7 Simplify

$$\frac{3x^2 - 75y^2}{3x^2 - 21xy + 30y^2}$$

Solution

$$\begin{aligned}\frac{3x^2 - 75y^2}{3x^2 - 21xy + 30y^2} &= \frac{3(x^2 - 25y^2)}{3(x^2 - 7xy + 10y^2)} \\ &= \frac{3(x - 5y)(x + 5y)}{3(x - 5y)(x - 2y)} \\ &= \frac{x + 5y}{x - 2y}\end{aligned}$$

Special attention is called to the forms $a - b$ and $b - a$, which are additive inverses of each other.

$$\begin{aligned}-(b - a) = -(b + (-a)) &= -1(b) + (-1(-a)) \\ &= -b + a = a + (-b) = a - b\end{aligned}$$

Thus $\qquad -(b - a) = a - b \quad \text{and} \quad a - b = -(b - a).$

This property is needed sometimes in the process of renaming fractions. Its use is illustrated in the examples below.

EXAMPLE 4.1.8 Simplify

$$\frac{x - 1}{1 - x}$$

Solution Since $1 - x = -(x - 1)$,

$$\frac{x - 1}{1 - x} = \frac{x - 1}{-(x - 1)} = \frac{1(x - 1)}{(-1)(x - 1)} = \frac{1}{-1} = -1$$

EXAMPLE 4.1.9 Simplify

$$\frac{15 - 5y}{y^2 - 3y}$$

Solution

$$\frac{15 - 5y}{y^2 - 3y} = \frac{5(3 - y)}{y(y - 3)} = \frac{5 \cdot -(y - 3)}{y \cdot (y - 3)} = \frac{-5 \cdot (y - 3)}{y \cdot (y - 3)}$$

$$= \frac{-5}{y}$$

EXERCISES 4.1 A

Simplify in Exercises 1–52.

1. $\dfrac{4}{12}$

2. $\dfrac{-12}{36}$

3. $\dfrac{15}{-25}$

4. $\dfrac{-42}{-105}$

5. $-\dfrac{150}{200}$

6. $\dfrac{60x}{90x}$

7. $\dfrac{6xy}{6xy}$

8. $\dfrac{7a^2b}{7ab}$

9. $\dfrac{25x^3}{75x^5}$

10. $\dfrac{75x^5}{25x^3}$

11. $-\dfrac{m}{m^3n}$

12. $\dfrac{3xy}{-6y}$

13. $\dfrac{-300x^3y^2}{-75xy}$

14. $\dfrac{80a^2b^2}{16ab}$

15. $-\dfrac{-a^3b}{-a^3b}$

16. $\dfrac{3x^2y^2}{xy}$

17. $\dfrac{xy}{3x^2y^2}$

18. $\dfrac{xy^2z}{-x^2y^2z^2}$

19. $\dfrac{-ab^2c}{-a^2b^2c^2}$

20. $\dfrac{3x(a + b)}{12x^2}$

21. $\dfrac{7x + 21}{7x}$

22. $\dfrac{4y}{4y^2 - 8y}$

23. $\dfrac{x^2 - 4x}{x^2 - 2x}$

24. $\dfrac{x^2 - 4}{x^2 - 2x}$

25. $\dfrac{y + y^2}{3y + 3}$

26. $\dfrac{3a^2 + 3ab}{4b^2 + 4ab}$

27. $\dfrac{(c - d)^2}{6c - 6d}$

28. $\dfrac{2u + 2v}{6u + 6v}$

29. $\dfrac{3y + y^2}{3y + 9}$

30. $\dfrac{75n^2 - 3}{5n + 1}$

31. $\dfrac{7x - 7y}{7x + 7y}$

32. $\dfrac{7x - 7y}{7y - 7x}$

33. $\dfrac{3a^2 + 3}{3a + 3}$

34. $\dfrac{3t + 1}{4 - 36t^2}$

35. $\dfrac{3x^2 + 6x}{3x}$

36. $\dfrac{(a - b)^3}{(a - b)^2}$

37. $\dfrac{x^2 - 9}{9 - x^2}$

38. $\dfrac{3x + 15}{x^2 + 3x - 10}$

39. $\dfrac{x^2 - 1}{x^2 + 3x - 4}$

40. $\dfrac{x^2 + 2x - 8}{x^2 - 4}$

41. $\dfrac{4t^2 - 4t - 24}{6t^2 - 36t + 54}$

42. $\dfrac{z^2 - 10z + 25}{z^2 - z - 20}$

43. $\dfrac{m^2n - mn^2 + mn}{m^2n + mn^2 - mn}$

44. $\dfrac{x^2y - xy^2 + xyz}{yz^2 - y^2z + xyz}$

45. $\dfrac{6x^2 - 6}{6x^2 + 36x - 42}$

46. $\dfrac{5 - x}{x^2 - 25}$

47. $\dfrac{5a^2 + 15a - 140}{10a^2 + 110a + 280}$

48. $\dfrac{5n^2 - 17n + 6}{5n^2 - 13n - 6}$

49. $\dfrac{9x^2 - 4y^2}{2y - 3x}$

50. $\dfrac{x^3 - 8}{x^2 + 2x - 8}$

51. $\dfrac{x^2 - 25}{x^3 + 125}$

52. $\dfrac{(r - s)(r - t)}{(t - r)(t - s)}$

EXERCISES 4.1 B

Simplify in Exercises 1–52.

1. $\dfrac{6}{9}$

2. $\dfrac{-8}{12}$

3. $\dfrac{16}{-36}$

4. $\dfrac{-15}{-45}$

5. $-\dfrac{120}{280}$

6. $\dfrac{40y^2}{80y^2}$

7. $\dfrac{9xy}{45yz}$

8. $\dfrac{24a^2b^2c^2}{24abc}$

9. $\dfrac{52y^4}{26y}$

10. $\dfrac{26y}{52y^4}$

11. $-\dfrac{a^2}{a^3b}$

12. $\dfrac{15xy}{-60x}$

13. $\dfrac{-24x^4y^2}{-36x^2y}$

14. $\dfrac{5ab^2}{40a^2b}$

15. $-\dfrac{-xyz}{-xyz}$

16. $\dfrac{4a^3b^3}{ab}$

17. $\dfrac{ab}{4a^3b^3}$

18. $\dfrac{5x^2y^2z}{-15xy^2z^2}$

19. $\dfrac{-5a^2bc}{-15ab^2c^2}$

20. $\dfrac{5y(x + y)}{10y^2}$

21. $\dfrac{5y - 40}{5y}$

22. $\dfrac{6xy}{2x^2y + 6xy^2}$

23. $\dfrac{5x - 5y}{5x + 5y}$

24. $\dfrac{5x - 5y}{x^2 - xy}$

25. $\dfrac{y^2 + 5y}{25 + 5y}$

26. $\dfrac{u^2v + uv^2}{(u + v)^2}$

27. $\dfrac{4cd + 2c^2}{4cd + 8d^2}$

28. $\dfrac{3a + 3b}{9a + 9b}$

29. $\dfrac{4r + r^2}{4r + 16}$

30. $\dfrac{9t^2 - 81}{t + 3}$

31. $\dfrac{8a + 8b}{8a - 8b}$

32. $\dfrac{8a - 8b}{8b - 8a}$

33. $\dfrac{4m^2 + 4}{4m + 4}$

34. $\dfrac{t - 3}{2t^2 - 5t - 3}$

35. $\dfrac{8y^2 - 12y}{4y}$

36. $\dfrac{(r + 2s)^2}{(r + 2s)^5}$

37. $\dfrac{2y + 4}{y^2 - 2y - 8}$

38. $\dfrac{r^2 - 2s^2}{2s^2 - r^2}$

39. $\dfrac{a^2 - b^2}{(a - b)^2}$

40. $\dfrac{x^2 - 3x - 10}{x^2 - 2x - 15}$

41. $\dfrac{4a^2 - 4ab}{a^2 - b^2}$

42. $\dfrac{8t^2 - 12t - 20}{8t^2 - 28t + 20}$

43. $\dfrac{2n^3 + 4n^2 + 2n}{n^3 - n^2 - 2n}$

44. $\dfrac{x^4y^2 - 4x^3y^3}{x^3y^5 - 4x^2y^6}$

45. $\dfrac{7 - x}{x^2 - 49}$

46. $\dfrac{75n^2 - 3}{1 + 5n}$

47. $\dfrac{3n^3 - 21n^2 - 24n}{3n^3 - 3n}$

48. $\dfrac{49n^2 - 21n - 10}{49n^2 - 25}$

49. $\dfrac{6y^4 + 15y^3 - 9y^2}{9y^2 + 27y}$

50. $\dfrac{x^3 + 64}{x^2 - 16}$

51. $\dfrac{x^2 + 3x + 9}{x^3 - 27}$

52. $\dfrac{(a - b)(a - c)(b - c)}{(c - a)(c - b)(b - a)}$

4.2 RAISING A FRACTION TO HIGHER TERMS

In order to add or subtract certain fractions, the fractions must be renamed so that they have the same denominator. The process of renaming a fraction by multiplying the numerator and denominator by the same number is called **raising the fraction to higher terms**. This process is justified by the Fundamental Theorem of Fractions used in its symmetric form:

$$\frac{n}{d} = \frac{nk}{dk}$$

In practice, it is useful to consider the renaming of a fraction as a multiplication of the original fraction by the number 1, with 1 renamed as $\frac{k}{k}$. Thus

$$\frac{n}{d} = \frac{n}{d} \cdot 1 = \frac{n}{d} \cdot \frac{k}{k} = \frac{nk}{dk}$$

EXAMPLE 4.2.1 Express the fraction $\frac{3}{5}$ in higher terms with denominator 20.

Solution

$$\frac{3}{5} = \frac{3k}{5k} = \frac{3k}{20}$$

Since $5k = 20$, then $k = 4$ and $3k = 3 \cdot 4 = 12$. Thus

$$\frac{3}{5} = \frac{3 \cdot 4}{5 \cdot 4} = \frac{12}{20}$$

In other words, the question is: "What number multiplied by 5 equals 20?" Clearly, the answer is 4. Thus 4 is the multiplier of the numerator and the denominator.

EXAMPLE 4.2.2 Raise $\frac{7}{9y}$ to higher terms with denominator $18y^3$.

Solution

$$\frac{7}{9y} = \frac{7 \cdot k}{9y \cdot k} = \frac{7k}{18y^3}$$

Since $18y^3 = 9y \cdot 2y^2$, the multiplier k is $2y^2$.

In other words, the multiplier can be found by dividing the new denominator by the original denominator:

$$\frac{18y^3}{9y} = 2y^2$$

Therefore $$\frac{7}{9y} = \frac{7(2y^2)}{9y(2y^2)} = \frac{14y^2}{18y^3}$$

EXAMPLE 4.2.3 Raise

$$\frac{8}{15xy^2}$$

to higher terms with denominator $45x^3y^3$.

Solution Find the multiplier:

$$\frac{45x^3y^3}{15xy^2} = 3x^2y$$

Multiply the numerator and denominator by $3x^2y$:

$$\frac{8}{15xy^2} = \frac{8(3x^2y)}{15xy^2(3x^2y)} = \frac{24x^2y}{45x^3y^3}$$

Another type of problem uses the property that $x = \frac{x}{1}$ to build up a fraction or to change a polynomial into a rational expression with a designated denominator.

EXAMPLE 4.2.4 Rename 6 as a fraction with denominator 5.

Solution

$$6 = \frac{6}{1} = \frac{6(5)}{1(5)} = \frac{30}{5}$$

EXAMPLE 4.2.5 Express the polynomial $x - 2$ as a rational expression with denominator $x + 3$.

Solution

$$x - 2 = \frac{x - 2}{1} = \frac{(x - 2)(x + 3)}{1(x + 3)} = \frac{x^2 + x - 6}{x + 3}$$

Some problems involve the signs of a fraction and the special property that $b - a = -(a - b)$.

EXAMPLE 4.2.6 Express

$$\frac{3}{7 - x}$$

with denominator $x - 7$.

Solution Since $7 - x = -(x - 7)$, then

$$\frac{3}{7 - x} = \frac{3}{-(x - 7)} = \frac{-3}{x - 7}$$

EXAMPLE 4.2.7 Rename

$$\frac{x}{x + 3}$$

as an equal fraction with denominator $x^2 - 9$.

Solution First find the multiplier:

$$\frac{x^2 - 9}{x + 3} = \frac{(x - 3)(x + 3)}{x + 3} = x - 3$$

Thus $$\frac{x}{x + 3} = \frac{x(x - 3)}{(x + 3)(x - 3)} = \frac{x^2 - 3x}{x^2 - 9}$$

EXAMPLE 4.2.8 Express

$$\frac{3}{x + 2}$$

in higher terms having the denominator $x^2 - 5x - 14$.

Solution First find the multiplier:

$$\frac{x^2 - 5x - 14}{x + 2} = \frac{(x - 7)(x + 2)}{x + 2} = x - 7$$

then $$\frac{3}{x + 2} = \frac{3(x - 7)}{(x + 2)(x - 7)} = \frac{3x - 21}{x^2 - 5x - 14}$$

EXAMPLE 4.2.9 Find n so that

$$\frac{6}{6-x} = \frac{n}{x^2-36}$$

Solution

$$\frac{x^2-36}{6-x} = \frac{(x+6)(x-6)}{(-1)(x-6)} = \frac{x+6}{-1} = (-1)(x+6)$$

$$\frac{6}{6-x} = \frac{6(-1)(x+6)}{(-1)(x-6)(-1)(x+6)} = \frac{-6(x+6)}{x^2-36}$$

Thus $n = -6(x+6)$.

EXERCISES 4.2 A

Express the fractions in Exercises 1–38 in higher terms having the denominators as indicated.

1. $\frac{4}{7} = \frac{\quad}{21}$
2. $\frac{3}{5} = \frac{\quad}{25}$
3. $\frac{2}{-3} = \frac{\quad}{42}$
4. $-5 = \frac{\quad}{35}$
5. $\frac{x}{12} = \frac{\quad}{36}$
6. $\frac{2}{5} = \frac{\quad}{15x}$
7. $\frac{3}{4} = \frac{\quad}{48xy}$
8. $\frac{2}{7x} = \frac{\quad}{21x^2}$
9. $\frac{3x}{-8} = \frac{\quad}{40x^2}$
10. $-\frac{2x}{5y^2} = \frac{\quad}{30xy^3}$
11. $-\frac{3xy}{7z} = \frac{\quad}{42xyz}$
12. $\frac{a}{b^2} = \frac{\quad}{5ab^3}$
13. $\frac{5}{x+1} = \frac{\quad}{2x+2}$
14. $\frac{x}{2x-5} = \frac{\quad}{6x-15}$
15. $\frac{x-3}{x+4} = \frac{\quad}{5x+20}$
16. $\frac{2x}{x-2} = \frac{\quad}{x^2-2x}$
17. $\frac{x}{x-5} = \frac{\quad}{2x^2-10x}$
18. $\frac{2}{y} = \frac{\quad}{y^2+2y}$
19. $\frac{3y}{y+3} = \frac{\quad}{y^2-9}$
20. $\frac{t+3}{t-2} = \frac{\quad}{t^2+t-6}$
21. $x+2 = \frac{\quad}{x+1}$
22. $\frac{1}{4x-3} = \frac{\quad}{3-4x}$

23. $\frac{-x}{7 - x} = \frac{}{x - 7}$

24. $\frac{1}{5 - 2x} = \frac{}{2x^2 - 5x}$

25. $\frac{y}{4 - y} = \frac{}{y^2 - 16}$

26. $\frac{x + 3}{x + 4} = \frac{}{x^2 + 6x + 8}$

27. $\frac{a + 1}{a - 3} = \frac{}{a^2 - 2a - 3}$

28. $x = \frac{}{6x^2 - 6x}$

29. $\frac{t - 2}{t + 2} = \frac{}{14 + 9t + t^2}$

30. $\frac{a - b}{a + b} = \frac{}{(a + b)^2}$

31. $\frac{6a^2}{5a - 1} = \frac{}{10a^2 + 13a - 3}$

32. $\frac{r + 1}{r + 2} = \frac{}{7r^4 - 28r^2}$

33. $\frac{-a}{2} = \frac{}{4a^2 - 10a}$

34. $\frac{9}{x^2 - 9} = \frac{}{(x - 3)(x + 3)^2}$

35. $\frac{5n + 5}{n - 1} = \frac{}{3n^3 - 3n}$

36. $\frac{ab}{(a - b)^2} = \frac{}{(b - a)^2}$

37. $\frac{x}{xy + y} = \frac{}{xy(x + 1)^2}$

38. $\frac{3}{y(x + y)} = \frac{}{y^2(x + y)^3}$

EXERCISES 4.2 B

Express the fractions in Exercises 1–38 in higher terms having the denominators as indicated.

1. $\frac{4}{9} = \frac{}{36}$

2. $\frac{-2}{5} = \frac{}{20}$

3. $\frac{3}{-7} = \frac{}{42}$

4. $4 = \frac{}{-10}$

5. $\frac{x}{15} = \frac{}{45}$

6. $\frac{3}{4} = \frac{}{16x^2}$

7. $\frac{2}{5} = \frac{}{15xy}$

8. $\frac{3}{-5x} = \frac{}{45x^2}$

9. $\frac{2x}{15} = \frac{}{45x^2}$

10. $-\frac{5x}{7y^2} = \frac{}{21xy^3}$

11. $-\frac{2xy}{3z} = \frac{}{42x^2z^2}$

12. $\frac{a^2}{b^3} = \frac{}{4a^2b^4}$

13. $\frac{7}{3y - 3} = \frac{}{6y - 6}$

14. $\frac{2y}{4y - 1} = \frac{}{8y^2 - 2y}$

15. $\frac{x + 1}{x - 3} = \frac{}{x^2 - 3x}$

16. $\frac{x - 2}{2x - 1} = \frac{}{12x^2 - 6x}$

17. $\dfrac{2x}{2x-1} = \dfrac{}{4x^2-1}$

18. $\dfrac{1+t}{1+3t} = \dfrac{}{18t^2+6t}$

19. $\dfrac{r}{r+4} = \dfrac{}{r^2+r-12}$

20. $\dfrac{u-5}{u-6} = \dfrac{}{u^2-4u-12}$

21. $6x - 5 = \dfrac{}{x-2}$

22. $\dfrac{1}{5-6y} = \dfrac{}{6y-5}$

23. $\dfrac{-t}{1-t} = \dfrac{}{t-1}$

24. $\dfrac{x}{5-x} = \dfrac{}{x^2-25}$

25. $\dfrac{-1}{7-2r} = \dfrac{}{4r^2-49}$

26. $\dfrac{x+2}{x+4} = \dfrac{}{2x^3+5x^2-12x}$

27. $\dfrac{y-1}{y-2} = \dfrac{}{12-8y+y^2}$

28. $\dfrac{6x}{x+6} = \dfrac{}{2x^2+2x-60}$

29. $x = \dfrac{}{5x+10}$

30. $\dfrac{2x+3y}{2x-3y} = \dfrac{}{(2x-3y)^2}$

31. $\dfrac{x}{3} = \dfrac{}{3x^2-75x}$

32. $\dfrac{3}{3y+1} = \dfrac{}{1-9y^2}$

33. $\dfrac{-2}{n+5} = \dfrac{}{n^3+2n^2-15n}$

34. $\dfrac{4}{16x^2-1} = \dfrac{}{(4x-1)^2(4x+1)}$

35. $\dfrac{4n^2}{4n+1} = \dfrac{}{12n^3-n^2-n}$

36. $\dfrac{1+x}{1-x} = \dfrac{}{(x-1)^2}$

37. $\dfrac{a}{b(a+b)} = \dfrac{}{a^3b^3(a+b)^3}$

38. $\dfrac{a}{1-a} = \dfrac{}{a^2-1}$

4.3 ADDITION AND SUBTRACTION

The procedures for adding and subtracting algebraic fractions are based on the theorems for the addition and subtraction of quotients of real numbers and on the Fundamental Theorem of Fractions.

Let n, m, and d be any real numbers with $d \neq 0$.

THE ADDITION OF QUOTIENTS THEOREM

$$\frac{n}{d} + \frac{m}{d} = \frac{n+m}{d}$$

THE SUBTRACTION OF QUOTIENTS THEOREM

$$\frac{n}{d} - \frac{m}{d} = \frac{n-m}{d}$$

An outline of the proof of the addition theorem is given below to help the student understand the theorems.

$$\frac{n}{d} + \frac{m}{d} = n\left(\frac{1}{d}\right) + m\left(\frac{1}{d}\right)$$ (By the alternate definition of division)

and $$n\left(\frac{1}{d}\right) + m\left(\frac{1}{d}\right) = (n + m)\frac{1}{d}$$ (Distributive axiom)

$$(n + m)\frac{1}{d} = \frac{n + m}{d}$$ (Alternate definition of division)

Thus $$\frac{n}{d} + \frac{m}{d} = \frac{n + m}{d}$$

Since multiplication is distributive over subtraction and $n - m = n + (-m)$, it can similarly be shown that

$$\frac{n}{d} - \frac{m}{d} = \frac{n - m}{d}$$

Examination of the addition and subtraction theorems for the quotients of real numbers reveals that fractions cannot be combined into a single fraction by the operations of addition and subtraction unless the denominators of the fractions being combined are the same. The Fundamental Theorem of Fractions provides a technique for renaming fractions so that they will have the same denominator, called a common denominator. A common denominator can always be found by multiplying the denominators of the fractions that are to be added or subtracted.

To provide insight into this process, a visual model for the addition of two fractions of arithmetic is provided in Figure 4.3.1.

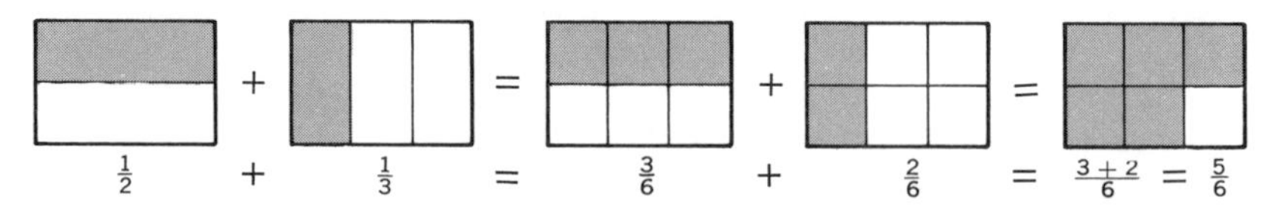

FIGURE 4.3.1

Now, consider the addition problem from arithmetic,

$$\frac{5}{12} + \frac{7}{18}$$

By multiplying the denominators, a common denominator can be found: $12 \cdot 18 = 216$:

$$\frac{5}{12} = \frac{5 \cdot 18}{12 \cdot 18} = \frac{90}{216} \quad \text{and} \quad \frac{7}{18} = \frac{7 \cdot 12}{18 \cdot 12} = \frac{84}{216}$$

Thus $$\frac{5}{12} + \frac{7}{18} = \frac{90}{216} + \frac{84}{216} = \frac{90 + 84}{216} = \frac{174}{216}$$

However, $\frac{174}{216}$ is not in its lowest terms and must be simplified:

$$\frac{174}{216} = \frac{2 \cdot 87}{2 \cdot 108} = \frac{2 \cdot 3 \cdot 29}{2 \cdot 3 \cdot 36} = \frac{29}{36}$$

Thus

$$\frac{5}{12} + \frac{7}{18} = \frac{29}{36}$$

There is another procedure for finding a common denominator which simplifies the arithmetical calculations involved and sometimes, but not always, avoids the final simplification. This procedure is called finding the **least common denominator,** the **L.C.D.** The objective is to find the smallest common denominator—that is, the smallest number which has each of the original denominators for a factor. Thus, to find the least common denominator, the original denominators must be factored into prime factors. Then the L.C.D. is the product of all the primes that occur in each factorization, with each prime taken the greatest number of times it occurs in any denominator.

For example, to find the L.C.D. of $\frac{5}{12}$ and $\frac{7}{18}$, factor 12 and 18:

$$12 = 2 \cdot 2 \cdot 3 \quad \text{and} \quad 18 = 2 \cdot 3 \cdot 3$$

Then the L.C.D. is $2 \cdot 2 \cdot 3 \cdot 3 = 36$.

Now the sum $\frac{5}{12} + \frac{7}{18}$ is obtained as follows:

$$\frac{5}{12} + \frac{7}{18} = \frac{15}{36} + \frac{14}{36} = \frac{29}{36}$$

This procedure is similar to the one that is used in adding or subtracting algebraic fractions, with the objective of finding the smallest number for the common denominator replaced by the objective of finding the polynomial of smallest degree for the common denominator.

EXAMPLE 4.3.1 Express as a single fraction

$$\frac{3}{16x} + \frac{5}{16x}$$

Solution The denominators of the two fractions are the same; therefore, the addition of quotients theorem can be applied directly:

$$\frac{3}{16x} + \frac{5}{16x} = \frac{3 + 5}{16x} = \frac{8}{16x}$$

but

$$\frac{8}{16x} = \frac{8}{8 \cdot 2x} = \frac{1}{2x}$$

Therefore, in lowest terms,

$$\frac{3}{16x} + \frac{5}{16x} = \frac{1}{2x}$$

EXAMPLE 4.3.2 Express as a single fraction

$$\frac{5}{6x^2} + \frac{1}{9x}$$

Solution The two fractions have unlike denominators.

1. Find the L.C.D. by factoring $6x^2$ and $9x$:

$$6x^2 = 2\cdot 3\cdot x\cdot x \quad \text{and} \quad 9x = 3\cdot 3\cdot x$$

Thus the L.C.D. is $2\cdot 3\cdot 3\cdot x\cdot x = 18x^2$.

2. Rename the fractions by using the Fundamental Theorem of Fractions:

$$\frac{5}{6x^2} = \frac{5\cdot 3}{6x^2\cdot 3} = \frac{15}{18x^2} \quad \text{and} \quad \frac{1}{9x} = \frac{1\cdot 2x}{9x\cdot 2x} = \frac{2x}{18x^2}$$

3. Apply the theorem for the addition of quotients:

$$\frac{5}{6x^2} + \frac{1}{9x} = \frac{15}{18x^2} + \frac{2x}{18x^2} = \frac{15 + 2x}{18x^2}$$

$$= \frac{2x + 15}{18x^2} \text{ (Conventional form)}$$

EXAMPLE 4.3.3 Express as a single fraction

$$\frac{3}{y} + \frac{2}{3} - \frac{y}{y + 3}$$

Solution

1. The L.C.D. is $3y(y + 3)$.

2. Rename the fractions so that each has the denominator $3y(y + 3)$:

$$\frac{3}{y} = \frac{3\cdot 3(y + 3)}{y\cdot 3(y + 3)} = \frac{9y + 27}{3y(y + 3)}$$

$$\frac{2}{3} = \frac{2\cdot y(y + 3)}{3\cdot y(y + 3)} = \frac{2y^2 + 6y}{3y(y + 3)}$$

$$\frac{y}{y + 3} = \frac{y\cdot 3y}{(y + 3).3y} = \frac{3y^2}{3y(y + 3)}$$

3. Apply the theorem for the addition and subtraction of quotients and combine like terms:

$$\frac{9y + 27}{3y(y + 3)} + \frac{2y^2 + 6y}{3y(y + 3)} - \frac{3y^2}{3y(y + 3)}$$

$$= \frac{(9y + 27) + (2y^2 + 6y) - (3y^2)}{3y(y + 3)}$$

$$= \frac{-y^2 + 15y + 27}{3y(y + 3)}$$

EXAMPLE 4.3.4 Simplify

$$\frac{5}{x^2 - 6x + 9} - \frac{4}{x^2 - 9}$$

Solution

1. $x^2 - 6x + 9 = (x - 3)(x - 3)$

   $x^2 - 9 = (x - 3)(x + 3)$

   Thus the L.C.D. $= (x - 3)^2(x + 3)$.

2. $$\frac{5}{(x - 3)^2} = \frac{5(x + 3)}{(x - 3)^2(x + 3)} = \frac{5x + 15}{(x - 3)^2(x + 3)}$$

   $$\frac{4}{x^2 - 9} = \frac{4(x - 3)}{(x - 3)(x + 3)(x - 3)} = \frac{4x - 12}{(x - 3)^2(x + 3)}$$

3. $$\frac{5}{(x - 3)^2} - \frac{4}{x^2 - 9} = \frac{(5x + 15)}{(x - 3)^2(x + 3)} - \frac{(4x - 12)}{(x - 3)^2(x + 3)}$$

   $$= \frac{(5x + 15) - (4x - 12)}{(x - 3)^2(x + 3)}$$

   $$= \frac{x + 27}{(x - 3)^2(x + 3)}$$

In subtracting fractions, it is especially important to remember that the bar, the horizontal line separating the numerator and denominator, is a grouping symbol indicating that the numerator is to be considered as a single number and that the denominator is to be considered as a single number.

In Example 4.3.4, the parentheses are used to enclose the numerator of each fraction in order to emphasize this fact.

EXAMPLE 4.3.5 Simplify

$$\frac{2}{x^2 - 4x + 4} + \frac{3x}{4 - x^2}$$

Solution

1. $x^2 - 4x + 4 = (x - 2)(x - 2)$

$$4 - x^2 = -(x^2 - 4) = -(x - 2)(x + 2)$$

It is advisable at this time to rewrite the second fraction

$$\frac{3x}{4 - x^2} = \frac{3x}{-(x^2 - 4)} = \frac{-3x}{x^2 - 4}$$

Thus the L.C.D. $= (x - 2)(x - 2)(x + 2) = (x - 2)^2(x + 2)$.

2. $$\frac{2}{x^2 - 4x + 4} = \frac{2(x + 2)}{(x - 2)(x - 2)(x + 2)} = \frac{2x + 4}{(x - 2)^2(x + 2)}$$

$$\frac{-3x}{x^2 - 4} = \frac{-3x(x - 2)}{(x - 2)(x - 2)(x + 2)} = \frac{-3x^2 + 6x}{(x - 2)^2(x + 2)}$$

3. $$\frac{2}{x^2 - 4x + 4} + \frac{3x}{4 - x^2} = \frac{2}{x^2 - 4x + 4} + \frac{-3x}{x^2 - 4}$$

$$= \frac{(2x + 4) + (-3x^2 + 6x)}{(x - 2)^2(x + 2)}$$

$$= \frac{-3x^2 + 8x + 4}{(x - 2)^2(x + 2)}$$

EXAMPLE 4.3.6 Simplify

$$\frac{3}{x-2} - \frac{2}{3x} - \frac{18}{3x^2 - 6x}$$

Solution

1. $x - 2 = x - 2$ } These expressions are already in factored form.

$3x = 3x$

$3x^2 - 6x = 3x(x - 2)$

Thus the L.C.D. $= 3x(x - 2)$.

2. $$\frac{3}{x - 2} = \frac{3 \cdot 3x}{3x(x - 2)} = \frac{9x}{3x(x - 2)}$$

$$\frac{2}{3x} = \frac{2(x - 2)}{3x(x - 2)} = \frac{2x - 4}{3x(x - 2)}$$

$$\frac{18}{3x^2 - 6x} = \frac{18}{3x(x - 2)}$$

3. $\dfrac{3}{x-2} - \dfrac{2}{3x} - \dfrac{18}{3x^2 - 6x}$

$$= \frac{9x}{3x(x-2)} - \frac{2x-4}{3x(x-2)} - \frac{18}{3x(x-2)}$$

$$= \frac{9x - (2x-4) - 18}{3x(x-2)}$$

$$= \frac{9x - 2x + 4 - 18}{3x(x-2)}$$

$$= \frac{7x - 14}{3x(x-2)}$$

$$= \frac{7(x-2)}{3x(x-2)}$$

$$= \frac{7}{3x}$$

EXERCISES 4.3 A

Rewrite each of Exercises 1–40 as a single fraction in lowest terms.

1. $\dfrac{1}{3} + \dfrac{4}{3}$

2. $\dfrac{2}{7} + \dfrac{5}{7}$

3. $\dfrac{1}{x} + \dfrac{3}{x} - \dfrac{5y}{x}$

4. $\dfrac{2}{x+3} + \dfrac{x}{x+3} + \dfrac{5}{x+3}$

5. $\dfrac{y}{z-1} + \dfrac{2y}{z-1} - \dfrac{y+1}{z-1}$

6. $\dfrac{2}{3} + \dfrac{3}{4}$

7. $\dfrac{4}{9} + \dfrac{5}{12}$

8. $\dfrac{5}{14} + \dfrac{3}{28} - \dfrac{1}{21}$

9. $\dfrac{1}{3} + \dfrac{1}{4} + \dfrac{1}{5}$

10. $\dfrac{2}{3y} + \dfrac{5}{12y} + \dfrac{1}{18y}$

11. $\dfrac{1}{2x} + \dfrac{1}{6x} - \dfrac{1}{3x}$

12. $\dfrac{2}{x} + \dfrac{3}{x^2}$

13. $\dfrac{1}{2x} + \dfrac{4}{3x^2} - \dfrac{1}{6x^3}$

14. $\dfrac{3}{x} + \dfrac{2}{x+1}$

15. $\dfrac{1}{b-1} - \dfrac{1}{b}$

16. $\dfrac{x}{x+2} + \dfrac{1}{x} + \dfrac{x}{2}$

17. $\dfrac{4}{y+1} - \dfrac{3}{y+2}$

18. $\dfrac{2}{a+3} + \dfrac{5}{a} - \dfrac{1}{3}$

19. $\dfrac{1}{x} + \dfrac{1}{y} + \dfrac{1}{z}$

20. $\dfrac{3x}{(x+2)^2} + \dfrac{2}{(x+2)}$

21. $\dfrac{y}{(2y-3)^2} + \dfrac{5}{2y-3} - \dfrac{3}{2y+3}$

22. $2x + \dfrac{3}{x}$

23. $2x - \dfrac{x^2}{x+1}$

24. $2 - \dfrac{x-1}{x+1}$

25. $\dfrac{x}{x^2+5x+4} + \dfrac{3}{x^2+4x+3}$

26. $\dfrac{x+2}{3x^2+5x+2} - \dfrac{x+3}{3x^2-16x-12}$

27. $\dfrac{2y}{y^2-5y+6} + \dfrac{3}{2+3y-2y^2}$

28. $\dfrac{2}{x-2} + \dfrac{3}{2-x}$

29. $\dfrac{a-2}{3a+3} + \dfrac{a-3}{2a+2}$

30. $\dfrac{x}{5x-5} - \dfrac{2}{3-3x}$

31. $\dfrac{x^2+3}{24-2x-x^2} + \dfrac{2x+1}{2x-8}$

32. $\dfrac{2}{c-5} + \dfrac{3}{c+5} - 1$

33. $\dfrac{3}{y^2-5y} + \dfrac{2}{y^2+5y} - \dfrac{4}{y^2-25}$

34. $\dfrac{1}{x^2-49} + \dfrac{1}{(x+7)^2} - \dfrac{1}{(x-7)^2}$

35. $1 + \dfrac{2}{x} - \dfrac{x-1}{x^2-x}$

36. $2a - 5 + \dfrac{25}{2a+5}$

37. $\dfrac{x}{x+y} + \dfrac{y}{x-y} - \dfrac{2xy}{x^2-y^2}$

38. $\dfrac{7y}{2y^2+5y-3} - \dfrac{10y}{3y^2+8y-3}$

39. $\dfrac{1}{a^3+1} + \dfrac{1}{a^2-1}$

40. $\dfrac{10}{45x^2-5} - \dfrac{1}{9x^2-3x-2} + \dfrac{1}{9x^2-9x+2}$

EXERCISES 4.3 B

Rewrite each of Exercises 1–40 as a single fraction in lowest terms.

1. $\dfrac{2}{5} + \dfrac{6}{5}$

2. $\dfrac{1}{9} - \dfrac{5}{9}$

3. $\dfrac{2}{a} + \dfrac{5}{a} - \dfrac{2b}{a}$

4. $\dfrac{p}{p+1} + \dfrac{2}{p+1} - \dfrac{3}{p+1}$

5. $\dfrac{3x+2y}{5} - \dfrac{2x+3y}{5}$

6. $\dfrac{1}{6} + \dfrac{2}{15}$

7. $\dfrac{3}{4} - \dfrac{7}{12}$

8. $\dfrac{2}{3} + \dfrac{3}{5} + \dfrac{5}{2}$

9. $\dfrac{1}{2a} + \dfrac{3}{6a} + \dfrac{5}{14a}$

10. $\dfrac{x}{6y} - \dfrac{2}{8y} + \dfrac{5x}{24y}$

11. $\dfrac{2x}{9y^2} - \dfrac{3x}{8y^2} + \dfrac{5x}{18y^2}$

12. $\dfrac{5x-2}{x} - \dfrac{2x-5}{x}$

13. $\frac{3}{5y} + \frac{2}{15y^2} - \frac{1}{10y^3}$

14. $\frac{1}{a} + \frac{2}{a+1}$

15. $\frac{x}{y+1} - \frac{x}{y}$

16. $\frac{m}{m+4} + \frac{1}{m} + \frac{3}{4}$

17. $\frac{2}{x-1} - \frac{x}{x-2}$

18. $\frac{3}{a} + \frac{2}{a+5} - \frac{1}{5}$

19. $\frac{a}{b} + \frac{b}{a}$

20. $\frac{5x}{(3x-1)^2} + \frac{4}{3x-1}$

21. $\frac{2}{(x+5)^2} - \frac{1}{x-5} + \frac{x}{x+5}$

22. $\frac{x^2-2}{x} - x$

23. $\frac{x-2}{x+1} - x$

24. $x - \frac{2x+1}{x+2}$

25. $\frac{a+b}{1-a^2b^2} - \frac{a-b}{(1-ab)^2}$

26. $\frac{a+4}{5a-5} - \frac{a+5}{4a-4}$

27. $\frac{3x+2}{3x-6} + \frac{x^2+3x+1}{3x^2-5x-2}$

28. $1 - \frac{2}{n-3} - \frac{3}{n+3}$

29. $\frac{1}{3a^2+11a-4} + \frac{1}{3a^2-13a+4}$

30. $\frac{5x}{(3x-1)^2} + \frac{4}{1-3x}$

31. $\frac{a}{b-a} + \frac{b}{a-b}$

32. $\frac{x+7}{4-6x} + \frac{3x^2+14}{9x^2-4}$

33. $\frac{y+9}{4y^2-5y-6} - \frac{y+7}{5y^2-11y+2}$

34. $\frac{5}{2x+8} - \frac{7}{3x-12} + \frac{20}{x^2-16}$

35. $\frac{1}{a^2-1} - \frac{1}{a^2+2a+1}$

36. $\frac{2}{2x+x^2} - \frac{3}{2x-x^2} - \frac{4}{x^2-4}$

37. $\frac{t^2-2t-21}{t^2-4t-5} + \frac{t-6}{5-t}$

38. $\frac{x^2-2x+4}{x-2} - \frac{x^2+2x+4}{x+2} - \frac{8x}{x^2-4}$

39. $\frac{4a^2}{x^2-a^2} - \frac{x-a}{x+a} - \frac{x+a}{x-a}$

40. $\frac{x+a}{x^3-a^3} - \frac{1}{x^2-a^2}$

4.4 MULTIPLICATION

The product of two algebraic fractions may be expressed in a simpler form by applying the theorem for the product of quotients and by applying the Fundamental Theorem of Fractions.

The product of quotients involves the reciprocal of a real number. Recall that for any nonzero real number r the reciprocal of r is $\frac{1}{r}$ and $r \cdot \frac{1}{r} = 1$.

THE PRODUCT OF RECIPROCALS THEOREM

For any nonzero real numbers r and s,

$$\frac{1}{r}\cdot\frac{1}{s} = \frac{1}{rs}$$

As examples, $\frac{1}{2}\cdot\frac{1}{5} = \frac{1}{2\cdot 5} = \frac{1}{10}$ and $\frac{1}{-3}\cdot\frac{1}{7} = \frac{1}{(-3)(7)} = \frac{1}{-21} = \frac{-1}{21}$. Now the theorem for multiplying fractions can be stated.

THE PRODUCT OF QUOTIENTS THEOREM

Let n, d, r, s be any real numbers with $d \neq 0$, $s \neq 0$. Then

$$\frac{n}{d}\cdot\frac{r}{s} = \frac{nr}{ds}$$

This theorem is verified as follows:

$$\frac{n}{d}\cdot\frac{r}{s} = n\left(\frac{1}{d}\right) r\left(\frac{1}{s}\right) \qquad \text{(Definition of division)}$$

$$= n\cdot r\cdot\frac{1}{d}\cdot\frac{1}{s} \qquad \text{(Commutative and associative axioms)}$$

$$= nr\cdot\frac{1}{ds} \qquad \text{(Product of reciprocals theorem)}$$

$$= \frac{nr}{ds} \qquad \text{(Definition of division)}$$

In other words, the product of quotients theorem states that the product of two fractions is obtained by multiplying the numerators to obtain the new numerator and by multiplying the denominators to obtain the new denominator.

EXAMPLE 4.4.1 Express as a single fraction

$$\frac{3}{5}\cdot\frac{2}{7}$$

Solution

$$\frac{3}{5}\cdot\frac{2}{7} = \frac{3\cdot 2}{5\cdot 7} = \frac{6}{35}$$

EXAMPLE 4.4.2 Simplify

$$\frac{5}{8}\cdot\frac{4}{15}$$

Solution

$$\frac{5}{8}\cdot\frac{4}{15} = \frac{5\cdot 4}{8\cdot 15} = \frac{5\cdot 4}{(5\cdot 3)(4\cdot 2)} = \frac{1}{6}$$

EXAMPLE 4.4.3 Express as a simplified single fraction

$$\frac{7x^2}{3y}\cdot\frac{12y^2}{35x^3}$$

Solution

$$\frac{7x^2}{3y}\cdot\frac{12y^2}{35x^3} = \frac{7\cdot 12x^2y^2}{3\cdot 35x^3y}$$

(Using the commutative and associative axioms to rearrange the factors so that the numerals occur first and the letters are arranged in alphabetical order)

$$= \frac{4y(3\cdot 7x^2y)}{5x(3\cdot 7x^2y)}$$

(Rearranging the factors to form the pattern $\frac{ak}{bk}$)

$$= \frac{4y}{5x}$$

(Since $(3\cdot 7x^2y)$ is a common factor, the fraction can be reduced)

EXAMPLE 4.4.4 Simplify

$$\frac{t^2 + 3t}{6t - 2}\cdot\frac{9t^2 - 1}{t^2 - 9}$$

Solution

$$\frac{t^2 + 3t}{6t - 2}\cdot\frac{9t^2 - 1}{t^2 - 9} = \frac{t(t + 3)}{2(3t - 1)}\cdot\frac{(3t - 1)(3t + 1)}{(t - 3)(t + 3)}$$ (Factoring)

$$= \frac{t(3t + 1)(t + 3)(3t - 1)}{2(t - 3)(t + 3)(3t - 1)}$$ (Rearranging factors)

$$= \frac{t(3t + 1)}{2(t - 3)}$$ (Eliminating the common factors)

The final simplified fraction is left in factored form.

Note that each fraction was expressed in factored form, and the terms were rearranged to simplify reducing.

EXAMPLE 4.4.5 Simplify

$$\frac{x^2 + 3x + 2}{x^2 - 4x - 12} \cdot \frac{x - 6}{x + 1}$$

Solution First factor the trinomials, if possible:

$$\frac{x^2 + 3x + 2}{x^2 - 4x - 12} = \frac{(x + 2)(x + 1)}{(x + 2)(x - 6)}$$

Thus

$$\begin{aligned}
\frac{x^2 + 3x + 2}{x^2 - 4x - 12} \cdot \frac{x - 6}{x + 1} &= \frac{(x + 2)(x + 1)}{(x + 2)(x - 6)} \cdot \frac{(x - 6)}{(x + 1)} \\
&= \frac{(x + 1)}{(x - 6)} \cdot \frac{(x - 6)}{(x + 1)} \\
&= \frac{(x - 6)(x + 1)}{(x - 6)(x + 1)} \\
&= 1
\end{aligned}$$

Note that the first fraction was simplified before the two fractions were multiplied.

It is usually desirable to reduce a fraction as soon as possible to avoid any cumbersome arithmetic or algebra.

EXAMPLE 4.4.6 Simplify

$$\frac{3}{10} \cdot \frac{5}{12} \cdot \frac{4}{24}$$

Solution First method:

$$\begin{aligned}
\frac{3}{10} \cdot \frac{5}{12} \cdot \frac{4}{24} &= \frac{3 \cdot 5 \cdot 4}{10 \cdot 12 \cdot 24} \\
&= \frac{60}{2880} \\
&= \frac{2 \cdot 2 \cdot 3 \cdot 5}{2 \cdot 2 \cdot 2 \cdot 2 \cdot 2 \cdot 2 \cdot 3 \cdot 3 \cdot 5} \\
&= \frac{1}{48}
\end{aligned}$$

Second method:

$$\frac{3}{10}\cdot\frac{5}{12}\cdot\frac{4}{24}$$

$$= \frac{3}{10}\cdot\frac{5}{12}\cdot\frac{1}{6}$$

$$= \frac{3\cdot 5}{(2\cdot 5)(3\cdot 4)\cdot 6}$$

$$= \frac{1}{48}$$

Notice how the arithmetic was much simpler when the fractions were reduced as soon as possible rather than multiplying everything first.

EXERCISES 4.4 A

Simplify in Exercises 1–30.

1. $\frac{1}{3}\cdot\frac{1}{4}$

2. $\frac{2}{3}\cdot\frac{6}{5}$

3. $\frac{2}{3}\cdot\frac{5}{7}\cdot\frac{3}{15}$

4. $\frac{x}{3}\cdot\frac{x}{7}$

5. $\frac{x+y}{4}\cdot\frac{x-y}{4}$

6. $\frac{2}{x}\cdot\frac{x^2}{4}$

7. $\frac{5}{3x^2y}\cdot\frac{21y^2}{7x}$

8. $\frac{3a+3}{25a}\cdot\frac{5a}{9a+9}$

9. $\frac{-6xyz}{4a^2b}\cdot\frac{10ab^2}{15xyz^2}$

10. $\frac{r+s}{rs}\cdot\frac{r^2s^2}{(r+s)^2}$

11. $\frac{a^2-25}{5}\cdot\frac{10a}{a^2+4a-5}$

12. $\frac{x^2-4xy+4y^2}{9x^3}\cdot\frac{18x}{4x-8y}$

13. $\frac{x^2+6x+9}{3x^2+6x}\cdot\frac{x^2-4}{x^2+x-6}$

14. $\frac{3x^2+3}{x^2-x-6}\cdot\frac{x^2-5x+6}{x^2+1}$

15. $\frac{a^2-b^2}{3}\cdot\frac{3a-3b}{a^2+ab}$

16. $\frac{5y^2-5y}{4y-40}\cdot\frac{y^3}{3}\cdot\frac{y^2-9y-10}{2-2y}$

17. $\frac{2n^2-3n-2}{n^3+2n^2-3n}\cdot\frac{6n^2-6n}{6n^2+3n}$

18. $\frac{x^2-5x}{x^2-3x+2}\cdot\frac{x^2-x}{5x-25}$

19. $\frac{a^2-3a+2}{a-1}\cdot\frac{a+5}{a^2+3a-10}$

20. $\frac{-48a^3bc^2}{18ab^2}\cdot\frac{-9c}{24a^2c^3}\cdot 4a$

21. $\dfrac{x^2 - 25}{x^2 - 49} \cdot \dfrac{49 - x^2}{25 - x^2}$

22. $\dfrac{x^3 + y^3}{x^2 - xy + y^2} \cdot \dfrac{x^2 - 2xy + y^2}{x^2 - y^2}$

23. $\dfrac{a^3 - 8}{ab + a} \cdot \dfrac{(b + 1)^2}{a^2 - 4}$

24. $\dfrac{xy - xz}{xy + xz} \cdot \dfrac{y}{z - y} \cdot \dfrac{y + z}{y}$

★ 25. $\dfrac{(a + b)^2 - 36}{a + b - 6} \cdot \dfrac{a^2 - ab + 6a}{(a + 6)^2 - b^2}$

26. $6(x - 1) \cdot \dfrac{x}{3x - 3}$

27. $(x + 2)(x - 5) \cdot \dfrac{x - 2}{x - 5}$

28. $6x(x + 6)\left(\dfrac{5}{x}\right)$

29. $3(x + 3)(x - 3)\left(\dfrac{2}{x - 3}\right)$

30. $(2x - 1)(x + 4)(x - 1)\left(\dfrac{x - 2}{(2x - 1)(x - 1)}\right)$

EXERCISES 4.4 B

Simplify in Exercises 1–30.

1. $\dfrac{1}{5} \cdot \dfrac{1}{6}$

2. $\dfrac{5}{8} \cdot \dfrac{16}{30}$

3. $\dfrac{20}{50} \cdot \dfrac{5}{6} \cdot \dfrac{14}{7}$

4. $\dfrac{x}{5} \cdot \dfrac{y}{8}$

5. $\dfrac{3x}{y - 2} \cdot \dfrac{2y}{x - 3}$

6. $\dfrac{3}{y} \cdot \dfrac{y^2}{6xy}$

7. $\dfrac{2}{5ab^2} \cdot \dfrac{15ab}{22}$

8. $\dfrac{4c - 8}{14c} \cdot \dfrac{35c}{6c - 12}$

9. $\dfrac{4abc}{-6xy^2} \cdot \dfrac{-8x^2y}{12a^2bc^2}$

10. $\dfrac{(u - v)^2}{2uv} \cdot \dfrac{8u^3v^3}{(v - u)^3}$

11. $\dfrac{25x^2 - 1}{25} \cdot \dfrac{75x}{5x^2 + 6x + 1}$

12. $\dfrac{7a + 14}{a^2 + 4a + 4} \cdot \dfrac{a^2 - 4}{5a - 10}$

13. $\dfrac{x^2 + 6x + 9}{x^2 + 4x + 3} \cdot \dfrac{2x^2 + 2x}{x^2 - 9}$

14. $\dfrac{10y - 21}{y^2 - 4y + 4} \cdot \dfrac{y^2 + y - 6}{10y^2 + 9y - 63}$

15. $\dfrac{3x^2 - 3x}{x^2 - 81} \cdot \dfrac{x^2 + 4x - 45}{3x - 15}$

16. $\dfrac{3x^2 - 3x}{5x - 40} \cdot \dfrac{x^2}{3} \cdot \dfrac{x^2 - 7x - 8}{4 - 4x}$

17. $\dfrac{2x^2 + 13x + 15}{2x^2 - x - 6} \cdot \dfrac{x^2 - x - 2}{x^2 + 6x + 5}$

18. $\dfrac{10y^2 - 11y - 6}{3y^2 - 6y} \cdot \dfrac{y^2 - y - 2}{6y^2 - 13y + 6}$

19. $3xy^2 \cdot \dfrac{5x - 15}{15xy - 45x^2y^2}$

20. $\dfrac{y^2 + 5y}{y^2 - 5y - 6} \cdot \dfrac{y^2 - 7y + 6}{y^2 + 4y - 5} \cdot \dfrac{1}{y + 1}$

21. $\dfrac{x-y}{z-x}\cdot\dfrac{x-z}{z-y}\cdot\dfrac{y-z}{y-x}$

★ 22. $\dfrac{x^2y^2+xy-42}{x^2y^2-xy-42}\cdot\dfrac{x^2y^2-49}{x^2y^2-36}$

★ 23. $\dfrac{y^3+1}{y^2+1}\cdot\dfrac{y^4-1}{y+1}$

★ 24. $\dfrac{t^3-64}{t^2+4t+16}\cdot\dfrac{t}{t^2-16}$

★ 25. $\dfrac{(x-y)^2-49}{x-y-7}\cdot\dfrac{x^2-xy+7x}{(x+7)^2-y^2}$

26. $(x-3)(x+3)^2\cdot\dfrac{2}{x^2-9}$

27. $x(x-6)(x+6)\cdot\dfrac{6x}{x^2+6x}$

28. $8xy(x+y)\left(\dfrac{1}{4x+4y}\right)$

29. $x^3(x+5)(x-2)\left(\dfrac{x-5}{x^3+5x^2}\right)$

30. $(3x+5)(x-1)(x-2)\left(\dfrac{x+2}{3x^2+2x-5}\right)$

4.5 DIVISION

The quotient of two algebraic fractions may be simplified by applying the definition of division to change the division into a multiplication of two fractions. Since division is defined as multiplication by the reciprocal of the divisor, the procedure involves the use of an expression for the reciprocal of the quotient of two real numbers.

THE RECIPROCAL OF A QUOTIENT THEOREM

Let n and d be any nonzero real numbers. Then

$$\frac{1}{\frac{n}{d}} = \frac{d}{n}$$

Applying this theorem to the division of quotients results in the following procedure:

If n, d, and s are any nonzero real numbers and r is any real number,

$$\frac{r}{s} \div \frac{n}{d} = \frac{r}{s}\cdot\frac{1}{\frac{n}{d}} \qquad \text{(Definition of division)}$$

$$= \frac{r}{s}\cdot\frac{d}{n} \qquad \text{(Reciprocal of a quotient theorem)}$$

$$= \frac{rd}{sn} \qquad \text{(Product of quotients theorem)}$$

Stated formally, the above procedure is called

THE QUOTIENT OF QUOTIENTS THEOREM

$$\frac{r}{s} \div \frac{n}{d} = \frac{r}{s} \cdot \frac{d}{n} = \frac{rd}{sn}$$

provided s, d, and n are nonzero real numbers, and r is any real number.

Sometimes this theorem is stated informally by saying "to divide by a fraction, invert the divisor and multiply."

EXAMPLE 4.5.1 Express as a single fraction

$$\frac{3}{5} \div \frac{1}{4}$$

Solution

$$\frac{3}{5} \div \frac{1}{4} = \frac{3}{5} \cdot \frac{4}{1} = \frac{12}{5}$$

EXAMPLE 4.5.2 Simplify

$$\frac{3a^2}{5b^3} \div \frac{6a}{10b}$$

Solution

$$\begin{aligned} \frac{3a^2}{5b^3} \div \frac{6a}{10b} &= \frac{3a^2}{5b^3} \cdot \frac{10b}{6a} \\ &= \frac{(3a^2)(10b)}{(6a)(5b^3)} \\ &= \frac{(a)(2)}{(2)(b^2)} \\ &= \frac{a}{b^2} \end{aligned}$$

EXAMPLE 4.5.3 Simplify by expressing the indicated quotient as a single fraction in lowest terms:

$$\frac{x+1}{x} \div \frac{x^2+3x+2}{x^2+x}$$

Solution

$$\begin{aligned} \frac{x+1}{x} \div \frac{x^2+3x+2}{x^2+x} &= \frac{x+1}{x} \cdot \frac{x^2+x}{x^2+3x+2} \\ &= \frac{x+1}{x} \cdot \frac{x(x+1)}{(x+2)(x+1)} \\ &= \frac{x(x+1)(x+1)}{x(x+2)(x+1)} \\ &= \frac{x+1}{x+2} \end{aligned}$$

EXAMPLE 4.5.4 Simplify by expressing the indicated quotient as a single fraction in lowest terms:

$$\frac{x^2 + 4x}{x^3 + 4x^2 + 4x} \div \frac{5x + 20}{x^3 - 4x^2 - 12x}$$

Solution

$$\begin{aligned}
\frac{x^2 + 4x}{x^3 + 4x^2 + 4x} \cdot \frac{x^3 - 4x^2 - 12x}{5x + 20} &= \frac{x(x + 4)(x)(x + 2)(x - 6)}{x(x + 2)(x + 2)5(x + 4)} \\
&= \frac{x(x - 6)(x)(x + 2)(x + 4)}{5(x + 2)(x)(x + 2)(x + 4)} \\
&= \frac{x(x - 6)}{5(x + 2)}
\end{aligned}$$

EXAMPLE 4.5.5 Express as a single fraction in lowest terms:

$$\left(x + \frac{1}{y}\right) \div \left(x^2 - \frac{1}{y^2}\right)$$

Solution Express the divisor as a single fraction and the dividend as a single fraction:

$$x + \frac{1}{y} = \frac{xy + 1}{y}$$

$$x^2 - \frac{1}{y^2} = \frac{x^2y^2 - 1}{y^2}$$

Now

$$\left(x + \frac{1}{y}\right) \div \left(x^2 - \frac{1}{y^2}\right) = \frac{xy + 1}{y} \div \frac{x^2y^2 - 1}{y^2}$$

Applying the quotient of quotients theorem,

$$\begin{aligned}
\frac{xy + 1}{y} \div \frac{x^2y^2 - 1}{y^2} &= \frac{xy + 1}{y} \cdot \frac{y^2}{x^2y^2 - 1} \\
&= \frac{xy + 1}{y} \cdot \frac{y^2}{(xy + 1)(xy - 1)} \\
&= \frac{y \cdot y(xy + 1)}{y(xy - 1)(xy + 1)} \\
&= \frac{y}{xy - 1}
\end{aligned}$$

EXERCISES 4.5 A

Simplify in Exercises 1–20.

1. $\dfrac{2}{3} \div \dfrac{4}{3}$

2. $\dfrac{1}{5} \div \dfrac{3}{8}$

3. $\dfrac{1}{3} \div \dfrac{1}{9}$

4. $\dfrac{2x^2y}{3yz} \div \dfrac{6xy^2}{yz^2}$

5. $\dfrac{14x^2}{9y^2} \div \dfrac{35x^2}{36y^2}$

6. $\dfrac{-5a^2b^3}{7cd^2} \div \dfrac{10a^3b^2}{21c^2d}$

7. $\dfrac{7x}{x^2-49} \div \dfrac{x^2-14x+49}{x^2+14x+49}$

8. $\dfrac{4b^2-1}{b^2-4} \div \dfrac{12b^3-6b^2}{6b^3-12b^2}$

9. $\dfrac{y^2-y-12}{y^2-16} \div \dfrac{y^2-9}{12y^2-36y}$

10. $\dfrac{a^2+a-2}{a^2+2a-3} \div \dfrac{a^2+7a+10}{a^2-2a-15}$

11. $\dfrac{x^2+16x+64}{x^2-16x+64} \div \dfrac{256-4x^2}{6x^2-384}$

12. $\dfrac{a^3-8}{a^2+3a+2} \div \dfrac{a^2-4}{a+1}$

13. $\dfrac{b^2-25}{6b^3} \div \dfrac{b^2-2b-35}{2b-14}$

14. $\dfrac{12x^2-3}{15} \div (2x+1)$

15. $\left(1+\dfrac{1}{y}\right) \div \left(1-\dfrac{1}{y^2}\right)$

16. $\left(1+\dfrac{3}{2t-3}\right) \div \left(\dfrac{1}{2t-3}\right)$

17. $\left(1-\dfrac{x^2}{9}\right) \div \left(1+\dfrac{x}{3}\right)$

18. $\left(\dfrac{a^2+a-2}{a^2-6a+5} \div \dfrac{a^2+7a+10}{a^2-10a+25}\right)\cdot\dfrac{a+5}{a-5}$

19. $\dfrac{a^2+a-2}{a^2-6a+5} \div \left(\dfrac{a^2+7a+10}{a^2-10a+25}\cdot\dfrac{a+5}{a-5}\right)$

20. $\left(\dfrac{x^2-x}{x^2+2x-3}\cdot\dfrac{x^2+2x+1}{x^2+4x}\right) \div \dfrac{x^2-3x-4}{x^2-16}$

EXERCISES 4.5 B

Simplify in Exercises 1–20.

1. $\dfrac{3}{5} \div \dfrac{3}{7}$

2. $\dfrac{2}{3} \div \dfrac{5}{12}$

3. $\dfrac{5}{6} \div \dfrac{7}{24}$

4. $\dfrac{4xy^2}{5a^2b} \div \dfrac{14x^2yz}{25a^3b}$

5. $\dfrac{-10ab^2}{4x^2yz} \div \dfrac{-15a^2b^3}{12xy}$

6. $\dfrac{8r^5t^7}{13x^2y} \div \dfrac{40r^5t^6}{26xy}$

7. $\dfrac{(x-2)(x^2-6)}{8x^3+24x^2} \div \dfrac{x^4-36}{2x^9+6x^8}$

8. $\dfrac{4a^2-a-3}{12a^2-7a-12} \div \dfrac{2a^2-a-1}{6a^2-5a-4}$

9. $\dfrac{x^2+12x+36}{x^2-12x+36} \div \dfrac{x^2+5x-6}{x^2-5x-6}$

10. $\dfrac{x^2+x-2}{x^2-x-12} \div \dfrac{x^2+3x+2}{x^2-7x+12}$

11. $\dfrac{a^2+6a-16}{a^2-64} \div (a-2)$

12. $\dfrac{x^3+27}{x^2+3x} \div \dfrac{x^2+9x}{x^2+7x-18}$

13. $\dfrac{1+3x-18x^2}{6x^2-17x-3} \div \dfrac{x-3}{3x-1}$

14. $(a-b) \div \left(\dfrac{a}{b} - \dfrac{b}{a}\right)$

15. $\left(\dfrac{t}{2} + \dfrac{t}{3}\right) \div \left(t^2 - \dfrac{t}{2}\cdot\dfrac{t}{3}\right)$

16. $\left(6 + \dfrac{7}{x} - \dfrac{3}{x^2}\right) \div \left(9 + \dfrac{3}{x} - \dfrac{2}{x^2}\right)$

17. $\left(\dfrac{a^2-3a}{25-a^2} \div \dfrac{9a-6a^2+a^3}{30-11a+a^2}\right) \div \left(\dfrac{a^2+2a-15}{a^2-9}\right)$

18. $\left(\dfrac{2x+3y}{x-2y}\cdot\dfrac{3x-6y}{x-2y}\right) \div \dfrac{6x+9y}{x-2y}$

19. $\left(\dfrac{x^2+x-2}{x^2+4x+4} \div \dfrac{x^2-x}{x^2+x-6}\right)\cdot\dfrac{x^2+3x+2}{x^2-x-2}$

20. $\dfrac{x^2+x-2}{x^2+4x+4} \div \left(\dfrac{x^2-x}{x^2+x-6}\cdot\dfrac{x^2+3x+2}{x^2-x-2}\right)$

4.6 COMPLEX FRACTIONS

A **complex fraction** is a fraction that contains a fraction in the numerator or in the denominator or in both numerator and denominator.

There are two basic methods for simplifying a complex fraction. The method which usually involves the least number of calculations is based on the Fundamental Theorem of Fractions. The numerator and denominator are both multiplied by the least common denominator of all the fractions that appear in the numerator and denominator.

The other method for simplifying a complex fraction is based on the consideration that the complex fraction is an indicated quotient. First the numerator and denominator are expressed as single fractions, then the resulting fractions are divided.

EXAMPLE 4.6.1 Simplify

$$\frac{3 + \frac{1}{3}}{2 - \frac{3}{5}}$$

Solution Multiplying numerator and denominator by 15, the L.C.D.,

$$\frac{(3 + \frac{1}{3})(15)}{(2 - \frac{3}{5})(15)} = \frac{45 + 5}{30 - 9} = \frac{50}{21}$$

Alternate Solution First rewriting the numerator and the denominator each as single fractions,

$$\frac{3 + \frac{1}{3}}{2 - \frac{3}{5}} = \frac{\frac{9}{3} + \frac{1}{3}}{\frac{10}{5} - \frac{3}{5}} = \frac{\frac{10}{3}}{\frac{7}{5}} = \frac{10}{3} \cdot \frac{5}{7} = \frac{50}{21}$$

EXAMPLE 4.6.2 Simplify

$$\frac{\frac{1}{x} + \frac{1}{2}}{\frac{1}{4}}$$

Solution Multiplying numerator and denominator by $4x$, the L.C.D.,

$$\frac{\frac{1}{x} + \frac{1}{2}}{\frac{1}{4}} \cdot \frac{4x}{4x} = \frac{\left(\frac{1}{x} + \frac{1}{2}\right) 4x}{\frac{1}{4} \cdot 4x} = \frac{\frac{4x}{x} + \frac{4x}{2}}{\frac{4x}{4}} = \frac{4 + 2x}{x}$$

Alternate Solution

$$\frac{\frac{1}{x} + \frac{1}{2}}{\frac{1}{4}} = \frac{\frac{2 + x}{2x}}{\frac{1}{4}} = \frac{2 + x}{2x} \div \frac{1}{4}$$

$$= \frac{2 + x}{2x} \cdot \frac{4}{1}$$

$$= \frac{2(2 + x)}{x} = \frac{4 + 2x}{x}$$

A complex fraction may contain a complex fraction in either its numerator or denominator or in both. If this is the case, the complex fraction in the numerator or denominator is simplified first by either of the two methods described above, and then the resulting fraction is simplified.

EXAMPLE 4.6.3 Simplify

$$\frac{1}{1 + \frac{1}{1 + \frac{1}{2}}}$$

Solution First simplify the complex fraction that occurs in the denominator:

$$\frac{1}{1+\frac{1}{2}} = \frac{1 \cdot 2}{(1+\frac{1}{2})2} = \frac{2}{2+1} = \frac{2}{3}$$

Thus

$$\frac{1}{1+\left(\frac{1}{1+\frac{1}{2}}\right)} = \frac{1}{1+\frac{2}{3}}$$

$$= \frac{1 \cdot 3}{(1+\frac{2}{3})3} = \frac{3}{3+2} = \frac{3}{5}$$

Alternate Solution

$$\frac{1}{1+\frac{1}{1+\frac{1}{2}}} = \frac{1}{1+\frac{1}{\frac{3}{2}}}$$

$$= \frac{1}{1+\frac{2}{3}} = \frac{1}{\frac{5}{3}} = \frac{3}{5}$$

EXAMPLE 4.6.4 Simplify

$$\frac{\frac{x}{x-y} - \frac{y}{x+y}}{\frac{x}{x+y} - \frac{y}{y-x}}$$

Solution The denominators within the fraction are $x - y$ and $x + y$ in the numerator and $x + y$ and $y - x$ in the denominator.

Since $y - x = -(x - y)$, it is useful to rename

$$\frac{y}{y-x} \text{ as } \frac{-y}{x-y}$$

$$\frac{\frac{x}{x-y} - \frac{y}{x+y}}{\frac{x}{x+y} - \frac{y}{y-x}} = \frac{\frac{x}{x-y} - \frac{y}{x+y}}{\frac{x}{x+y} - \frac{-y}{x-y}}$$

$$= \frac{(x-y)(x+y)\left(\frac{x}{x-y} - \frac{y}{x+y}\right)}{(x-y)(x+y)\left(\frac{x}{x+y} + \frac{y}{x-y}\right)}$$

$$= \frac{x(x+y) - y(x-y)}{x(x-y) + y(x+y)}$$

$$= \frac{x^2 + xy - xy + y^2}{x^2 - xy + xy + y^2}$$

$$= \frac{x^2 + y^2}{x^2 + y^2} = 1$$

EXERCISES 4.6 A

Simplify in Exercises 1–20.

1. $\dfrac{2 + \frac{3}{4}}{3 + \frac{7}{8}}$

2. $\dfrac{1 + \frac{1}{3}}{1 + \frac{1}{9}}$

3. $\dfrac{\frac{5}{2}}{\frac{3}{4} + \frac{7}{8}}$

4. $\dfrac{1}{\frac{2}{3} + \frac{3}{4}}$

5. $\dfrac{\frac{1}{2} + \frac{1}{3}}{1 - (\frac{1}{2})(\frac{1}{3})}$

6. $\dfrac{\dfrac{1}{x} + \dfrac{1}{2}}{\dfrac{x}{2} - \dfrac{2}{x}}$

7. $\dfrac{\dfrac{x}{y} + \dfrac{y}{5}}{\dfrac{xy}{10}}$

8. $\dfrac{a + \dfrac{3}{a}}{a^2 - \dfrac{9}{a^2}}$

9. $\dfrac{y - \dfrac{9}{y}}{y - 7 + \dfrac{12}{y}}$

10. $\dfrac{\dfrac{1}{3} + \dfrac{1}{n}}{1 - \dfrac{1}{3n}}$

11. $\dfrac{1}{\dfrac{1}{x} + \dfrac{1}{y}}$

12. $\dfrac{2x}{\dfrac{x}{30} + \dfrac{x}{50}}$

13. $\dfrac{1 + \dfrac{x}{y}}{\dfrac{1}{x} + \dfrac{1}{y}}$

14. $\dfrac{1 - \dfrac{1}{x + 1}}{1 + \dfrac{1}{x - 1}}$

15. $\dfrac{3 - \dfrac{2}{\frac{1}{5} + \frac{1}{3}}}{5 - \dfrac{2}{\frac{1}{5} + \frac{1}{3}}}$

16. $\dfrac{\dfrac{1}{a} - \dfrac{2}{a^2} - \dfrac{3}{a^3}}{1 - \dfrac{9}{a^2}}$

17. $\dfrac{1}{1 - \dfrac{1}{1 - \dfrac{1}{x}}}$

18. $\dfrac{1}{y - \dfrac{1}{y + \dfrac{1}{y}}}$

19. $\dfrac{\dfrac{1}{x - 2} - \dfrac{1}{x + 2}}{1 + \dfrac{1}{x^2 - 4}}$

20. $\dfrac{1}{2 - \dfrac{1}{2 + \dfrac{1}{2 - \frac{1}{2}}}}$

EXERCISES 4.6 B

Simplify in Exercises 1–20.

1. $\dfrac{3 - \frac{5}{7}}{5 + \frac{3}{14}}$

2. $\dfrac{1 - \frac{1}{5}}{1 - \frac{1}{25}}$

3. $\dfrac{1}{\frac{1}{4} - \frac{1}{5}}$

4. $\dfrac{\frac{2}{3}}{\frac{1}{4} + \frac{2}{5}}$

5. $\dfrac{\frac{1}{4} - \frac{1}{3}}{1 + (\frac{1}{4})(\frac{1}{3})}$

6. $\dfrac{x - \dfrac{y}{5}}{x + \dfrac{y}{5}}$

7. $\dfrac{\dfrac{y}{3} - \dfrac{3}{y}}{\dfrac{1}{y} + \dfrac{1}{3}}$

8. $\dfrac{3 - \dfrac{1}{y}}{9 - \dfrac{1}{y^2}}$

9. $\dfrac{x + 3 - \dfrac{10}{x}}{x - \dfrac{25}{x}}$

10. $\dfrac{x + 6}{\dfrac{1}{x} + \dfrac{1}{6}}$

11. $\dfrac{1}{\dfrac{2}{a} + \dfrac{3}{b}}$

12. $\dfrac{\dfrac{x}{a} - \dfrac{x}{b}}{\dfrac{x}{ab}}$

13. $\dfrac{1}{1 + \dfrac{3}{x}}$

14. $\dfrac{1 + \dfrac{5}{x - 5}}{1 - \dfrac{5}{x + 5}}$

15. $\dfrac{2 + \dfrac{1}{\frac{1}{4} + \frac{1}{3}}}{4 - \dfrac{1}{\frac{1}{4} + \frac{1}{3}}}$

16. $\dfrac{\dfrac{a}{6b} - \dfrac{3b}{2a}}{\dfrac{a}{2b} + \dfrac{3b}{2a} - 2}$

17. $\dfrac{1}{1 + \dfrac{1}{2 + \frac{1}{3}}}$

18. $\dfrac{1}{1 + \dfrac{1}{1 + \frac{1}{2}}}$

19. $\dfrac{1}{x + \dfrac{1}{x - \dfrac{1}{x}}}$

20. $\dfrac{1}{x - 1 + \dfrac{1}{1 + \dfrac{x}{4 - x}}}$

4.7 CHANGING QUOTIENTS INTO SUMS: LONG DIVISION

It is sometimes desirable to replace a single fraction by a sum or difference of two or more rational expressions. There are several applications (for example, problems in calculus) that require that a quotient of two polynomials be changed into the sum of fractions in which the numerator of each has a degree smaller than the degree of the denominator.

The method used to replace a quotient of two polynomials by a sum or difference is similar to the long-division algorithm used in arithmetic. For example, the division of 865 by 23 is done as follows:

$$\begin{array}{r} 37 \\ 23\overline{)865} \\ 69 \\ \hline 175 \\ 161 \\ \hline 14 \end{array}$$

Thus

$$\frac{865}{23} = 37 + \frac{14}{23} = 37\frac{14}{23}$$

Check: $23(37) + 14 = 851 + 14 = 865$.

The long-division algorithm used to obtain the quotient of two polynomials is similar to the algorithm used in arithmetic and is justified by the same basic properties.

In arithmetic, the division process is stopped when the remainder is less than the divisor.

In algebra, the division process is stopped when the degree of the remainder polynomial is less than the degree of the divisor polynomial.

EXAMPLE 4.7.1 Express

$$\frac{x^3 - 5x^2 + 4x + 7}{x^2 + 2}$$

as a sum of rational expressions with the degree of the polynomial in the numerator of any fraction smaller than the degree of the polynomial in its denominator.

Solution

1. Divide x^3 by x^2 to obtain the partial quotient, x:

$$\begin{array}{r} x\phantom{{}- 5x^2 + 4x + 7} \\ x^2 + 2\overline{)x^3 - 5x^2 + 4x + 7} \end{array}$$

2. Subtract the product of x and $x^2 + 2$ from the dividend:

$$\begin{array}{r} x^3 \qquad\quad + 2x \qquad \\ \hline -5x^2 + 2x + 7 \end{array}$$

3. Repeat the process: Divide $-5x^2$ by x^2 to obtain -5:
4. Subtract the product of -5 and $x^2 + 2$ from the remainder polynomial, $-5x^2 + 2x + 7$:

$$\begin{array}{r} x - 5 \qquad\qquad\quad \\ x^2 + 2\overline{)x^3 - 5x^2 + 4x + 7} \\ x^3 \qquad\quad + 2x \qquad \\ \hline - 5x^2 + 2x + 7 \\ - 5x^2 \qquad\; - 10 \\ \hline 2x + 17 \end{array}$$

5. Since the degree of the remainder, $2x + 17$, is less than the degree of the divisor, $x^2 + 2$, the process is stopped at this point, and the quotient is expressed as the following sum:

$$\frac{x^3 - 5x^2 + 4x + 7}{x^2 + 2} = x - 5 + \frac{2x + 17}{x^2 + 2}$$

Check:

$$\begin{aligned} (x^2 + 2)\left(x - 5 + \frac{2x + 17}{x^2 + 2}\right) \\ &= (x^2 + 2)(x - 5) + (x^2 + 2)\left(\frac{2x + 17}{x^2 + 2}\right) \\ &= (x^3 - 5x^2 + 2x - 10) + (2x + 17) \\ &= x^3 - 5x^2 + 4x + 7 \end{aligned}$$

A quick check could also be made numerically by assigning a value to x. For instance, let $x = 2$. Then

$$x^3 - 5x^2 + 4x + 7 = 2^3 - 5(2^2) + 4(2) + 7 = 3$$

and

$$x^2 + 2 = 2^2 + 2 = 6$$

For $x = 2$,

$$\frac{x^3 - 5x^2 + 4x + 7}{x^2 + 2} = \frac{3}{6} = \frac{1}{2}$$

Also, for $x = 2$,

$$\begin{aligned} x - 5 + \left(\frac{2x + 17}{x^2 + 2}\right) &= 2 - 5 + \left(\frac{4 + 17}{4 + 2}\right) \\ &= -3 + \frac{21}{6} \\ &= \frac{-18 + 21}{6} \\ &= \frac{3}{6} = \frac{1}{2} \end{aligned}$$

Since

$$\frac{x^3 - 5x^2 + 4x + 7}{x^2 + 2} = x - 5 + \frac{2x + 17}{x^2 + 2} \quad \text{for} \quad x = 2$$

there is assurance that the problem has been worked correctly.

EXAMPLE 4.7.2 Divide by using the long-division algorithm:

$$\frac{x^3 + 1}{x + 1}$$

Solution The polynomial $x^3 + 1$ can also be considered as

$$x^3 + 0 \cdot x^2 + 0 \cdot x + 1$$

so places must be provided for the missing x^2 and x terms:

$$\begin{array}{r|l} & x^2 - x + 1 \\ \hline x + 1 & x^3 \qquad\qquad + 1 \\ & x^3 + x^2 \\ \hline & \quad - x^2 \qquad + 1 \\ & \quad - x^2 - x \\ \hline & \qquad\quad + x + 1 \\ & \qquad\qquad x + 1 \\ \hline \end{array}$$

Thus

$$\frac{x^3 + 1}{x + 1} = x^2 - x + 1$$

Check:

$$\begin{aligned} &(x + 1)(x^2 - x + 1) \\ &\quad = (x + 1)(x^2 - x) + (x + 1)(1) \\ &\quad = x^3 + x^2 - x^2 - x + x + 1 = x^3 + 1 \end{aligned}$$

In Example 4.7.2, the remainder is zero. Thus $x + 1$ is a factor of $x^3 + 1$.

Note that

$$\frac{x^3 + 1}{x + 1} = \frac{(x + 1)(x^2 - x + 1)}{(x + 1)} \quad \text{(Factoring the sum of cubes)}$$

$$= x^2 - x + 1$$

For cases such as this, when the divisor is an exact factor of the dividend, it is much easier to use the method of reducing a fraction presented earlier than to use the long-division algorithm.

It is important when using the long-division algorithm that both dividend and divisor be arranged in the same order—that is, the terms of both must be in descending order or the terms of both must be in ascending order.

EXAMPLE 4.7.3 Divide and check

$$\frac{7y + y^4 - 7 + y^3}{3 - y + y^2}$$

Solution First arrange the polynomials in descending powers of y:

$$\frac{7y + y^4 - 7 + y^3}{3 - y + y^2} = \frac{y^4 + y^3 + 7y - 7}{y^2 - y + 3}$$

$$\begin{array}{r l}
 & y^2 + 2y - 1 \\
y^2 - y + 3 \big) & \overline{y^4 + y^3 \qquad\quad + 7y - 7} \\
 & \underline{y^4 - y^3 + 3y^2} \\
 & \quad 2y^3 - 3y^2 + 7y - 7 \\
 & \quad \underline{2y^3 - 2y^2 + 6y} \\
 & \qquad\quad - y^2 + y - 7 \\
 & \qquad\quad \underline{- y^2 + y - 3} \\
 & \qquad\qquad\qquad\quad - 4
\end{array}$$

Therefore

$$\frac{y^4 + y^3 + 7y - 7}{y^2 - y + 3} = y^2 + 2y - 1 + \frac{-4}{y^2 - y + 3}$$

Quick Check: Letting $y = 2$,

$$\frac{y^4 + y^3 + 7y - 7}{y^2 - y + 3} = \frac{16 + 8 + 14 - 7}{4 - 2 + 3} = \frac{31}{5}$$

$$y^2 + 2y - 1 + \frac{-4}{y^2 - y + 3} = 4 + 4 - 1 + \frac{-4}{4 - 2 + 3}$$

$$= 7 - \frac{4}{5} = \frac{31}{5}$$

EXERCISES 4.7 A

In Exercises 1–10, divide by using the long-division algorithm. Check by multiplication.

1. $\frac{2x^3 + 5x^2 - 3x + 7}{x + 4}$

2. $\frac{y^5 + 4y^3 + 4y + 1}{y^2 + 2}$

3. $\frac{2x^3 - 13x^2 + 13x + 10}{x - 5}$

4. $\frac{27y^3 - 8}{3y - 2}$

5. $\dfrac{x^4 + 4}{x^2 + 2x + 2}$

6. $\dfrac{4x^3 + 10x^2 - 16x - 14}{2x + 7}$

7. $\dfrac{x^3 + 4x^2 + 5x + 12}{x + 4}$

8. $\dfrac{5x^4 - 30x^2 + 2x - 1}{x^2 - 6}$

9. $\dfrac{a^3 + 64}{a + 4}$

10. $\dfrac{2x^3 - 3x^2 + 5}{x - 1}$

In Exercises 11–20, divide and check as indicated.

11. $\dfrac{2x^3 - 10x^2 + 3x - 14}{x - 5}$

    Check, using $x = 2$.

12. $\dfrac{t^2 - 10t + 8 + t^3}{t + 4}$

    Check, using $t = 1$.

13. $\dfrac{x^4 - 1}{x - 1}$

    Check, using $x = 3$.

14. $\dfrac{4x^2 + 7}{2x - 1}$

    Check, using $x = 1$.

15. $\dfrac{2x^5 + 4x^2 - 6x^3 - 5}{x^2 - 3}$

    Check, using $x = 2$.

16. $\dfrac{y^4 + 64}{y^2 - 4y + 8}$

    Check, using $y = 2$.

17. $\dfrac{25a^3 + ab^2 + b^3}{5a + 2b}$

    Check, using $a = 1, b = -2$.

18. $\dfrac{x^6 + x^4 + x^2 + 1}{x^2 + 1}$

    Check, using $x = 2$.

19. $\dfrac{x^2 - y^2 - 6y - 9}{x - y - 3}$

    Check, using $x = 4$, y $= 2$.

20. $\dfrac{x^2 - y^2 + 4x - 6y - 5}{x + y + 5}$

    Check, using $x = 3, y = 2$.

EXERCISES 4.7 B

In Exercises 1–10, divide by using the long-division algorithm. Check by multiplication.

1. $\dfrac{5x^4 + x^3 - 4x^2 + 2x - 3}{x + 2}$

2. $\dfrac{y^5 - 3y^4 + 2y^2 - 3y + 2}{y^2 + 3}$

3. $\dfrac{6x^3 - 25x^2 + 3x + 4}{2x - 1}$

4. $\dfrac{y^3 - 125}{y - 5}$

5. $\dfrac{x^4 - 7x^2 + 9}{x^2 - x - 3}$

6. $\dfrac{6x^4 - 32x^2 - 9}{3x^2 - 1}$

7. $\dfrac{x^3 + 3x^2 + 2x + 7}{x + 3}$

8. $\dfrac{2x^4 - 14x^2 - 5}{x^2 - 7}$

9. $\dfrac{a^3 - 729}{a - 9}$

10. $\dfrac{4x^4 - 3x^2 + 7x - 1}{x - 1}$

In Exercises 11–20, divide and check as indicated.

11. $\dfrac{y^3 - 2y^2 - 9y + 18}{y - 3}$
Check, using $y = 4$.

12. $\dfrac{3x^3 - 3x - 10 + 14x^2}{x + 5}$
Check, using $x = -3$.

13. $\dfrac{x^3 + 8}{x + 2}$
Check, using $x = 5$.

14. $\dfrac{9x^3 - 3x^2 + 4x}{3x - 2}$
Check, using $x = 1$.

15. $\dfrac{5y^4 + 12y^2 - 30}{y^2 + 4}$
Check, using $y = 2$.

16. $\dfrac{x^2 + 25 + x^4}{5 + 3x + x^2}$
Check, using $x = 2$.

17. $\dfrac{16x^4 + 7x^2y^2 - 50y^4}{4x - 5y}$
Check, using $x = 1$, $y = 1$.

18. $\dfrac{x^4 - 29x^2y^2 + 4y^4}{x^2 + 5xy - 2y^2}$
Check, using $x = 1$, $y = 1$.

19. $\dfrac{a^2 - 4b^2 + 20b - 25}{a + 2b - 5}$
Check, using $a = 2$, $b = 2$.

20. $\dfrac{4x^2 - 16y^2 - 4x + 24y - 8}{2x - 4y + 2}$
Check, using $x = 3$, $y = 1$.

4.8 FRACTIONAL EQUATIONS

A **fractional equation** is an equation whose terms are rational expressions.

For example, $\dfrac{1}{x} = 3$ and $\dfrac{2}{x + 2} + \dfrac{3}{x - 3} = 15$ are fractional equations.

To transform a fractional equation into an equivalent equation whose terms are integral expressions, both sides of the equation must be multiplied by the least common denominator of all the fractions involved in the equation. Since this may require multiplication by an expression involving the variable, an equivalent equation is obtained only when the variable is restricted to designate those numbers that do not make the multiplier zero.

For example, to solve $\dfrac{1}{x} = 3$, both sides of the equation are multiplied by x, and x is restricted so that $x \neq 0$. Then, if $x \neq 0$,

$$\frac{1}{x}(x) = 3(x)$$

$$1 = 3x$$

$$3x = 1$$

$$x = \frac{1}{3}$$

To solve $\frac{x}{x-3} = \frac{3}{x-3}$, both sides are multiplied by $x - 3$, and x is restricted so that $x - 3 \neq 0$—that is, $x \neq 3$.

Thus, if $x \neq 3$,

$$\frac{x}{x-3}(x-3) = \frac{3}{x-3}(x-3) \quad \text{and} \quad x = 3$$

The solution set of

$$\frac{x}{x-3} = \frac{3}{x-3}$$

is the set of real numbers, x, such that $x \neq 3$ and $x = 3$. There is no value for x that makes both of these statements true, so there is no solution. In other words, the solution set is the empty set, $\varnothing$.

The equivalence theorem for multiplication, introduced earlier, is restated below for convenience.

THE EQUIVALENCE THEOREM FOR MULTIPLICATION

If A, B, and C are any real numbers, then

$$A = B \text{ if and only if } AC = BC \text{ and } C \neq 0$$

This theorem is used in solving fractional equations, and care must be taken to exclude any values of a variable that might yield a zero value for C. The values for which $C \neq 0$ are called **restricted values of the variable.**

EXAMPLE 4.8.1 State the restricted values of y and solve:

$$\frac{2}{y-5} + \frac{1}{y+5} = \frac{11}{y^2 - 25}$$

Solution Since it is required that $y - 5 \neq 0$ and $y + 5 \neq 0$, the restricted values are $y \neq 5$ and $y \neq -5$.

Multiplying both sides by the L.C.D. $(y - 5)(y + 5)$,

$$\left(\frac{2}{y-5} + \frac{1}{y+5}\right)(y-5)(y+5) = \frac{11(y-5)(y+5)}{(y-5)(y+5)}$$

$$2(y+5) + (y-5) = 11$$

$$3y + 5 = 11$$

$$3y = 6$$

$$y = 2$$

Since the common solution to $y \neq 5$, $y \neq -5$, and $y = 2$ is 2, $\{2\}$ is the solution set of the original equation.

Check:

$$\frac{2}{y-5} + \frac{1}{y+5} = \frac{2}{2-5} + \frac{1}{2+5} = \frac{-2}{3} + \frac{1}{7} = \frac{-11}{21}$$

$$\frac{11}{y^2 - 25} = \frac{11}{(2)^2 - 25} = \frac{11}{4 - 25} = \frac{-11}{21}$$

EXAMPLE 4.8.2 State the restricted values of the variable and solve:

$$\frac{x+1}{x+2} - \frac{x+1}{x-3} = \frac{5}{x^2 - x - 6}$$

Solution Restricted values: $x \neq -2$ and $x \neq 3$.

$$\left(\frac{x+1}{x+2}\right)(x+2)(x-3) - \left(\frac{x+1}{x-3}\right)(x+2)(x-3)$$
$$= \frac{5}{x^2 - x - 6}(x+2)(x-3)$$

$$(x+1)(x-3) - (x+1)(x+2) = 5$$
$$(x^2 - 2x - 3) - (x^2 + 3x + 2) = 5$$
$$-5x - 5 = 5$$
$$-5x = 10$$
$$x = -2$$

Thus $x \neq -2$ and $x \neq 3$ and $x = -2$. Since there is no common solution, the solution set is the empty set, $\varnothing$.

EXAMPLE 4.8.3 State the restricted values of the variable and solve:

$$\frac{x}{(x-4)^2} + \frac{2}{x-4} = \frac{3x-8}{(x-4)^2}$$

Solution Restricted values: $x \neq 4$.

$$(x-4)^2 \frac{x}{(x-4)^2} + (x-4)^2 \frac{2}{x-4} = \frac{3x-8}{(x-4)^2}(x-4)^2$$
$$x + 2(x-4) = 3x - 8$$
$$3x - 8 = 3x - 8$$
$$3x = 3x$$
$$x = x$$

Since $x = x$ is true for all real numbers, the solution set is the set of all real numbers except 4; $R, x \neq 4$.

EXAMPLE 4.8.4 State the restricted values and solve for x:

$$\frac{a}{x} - \frac{b}{a} = \frac{1}{3x} \quad \text{where} \quad a \neq 0 \quad \text{and} \quad b \neq 0$$

Solution Restricted values: $x \neq 0$; L.C.D. $= 3ax$.

$$\frac{a}{x} - \frac{b}{a} = \frac{1}{3x}$$

$$3ax\left(\frac{a}{x}\right) - 3ax\left(\frac{b}{a}\right) = 3ax\left(\frac{1}{3x}\right)$$

$$3a^2 - 3bx = a$$

$$-3bx = a - 3a^2$$

$$x = \frac{a - 3a^2}{-3b}$$

or

$$x = \frac{a(3a - 1)}{3b}$$

provided $a \neq 0$, $b \neq 0$, and $a \neq \frac{1}{3}$.

EXAMPLE 4.8.5 Solve

$$\frac{t}{5 - t} - \frac{4 - t}{5 + t} = \frac{t - 25}{t^2 - 25}$$

Solution Restricted values: $t \neq 5$ and $t \neq -5$. Finding the L.C.D.,

$$t^2 - 25 = (t + 5)(t - 5)$$

$$5 + t = t + 5$$

$$5 - t = -(t - 5)$$

Renaming $\frac{t}{5 - t}$ as $\frac{-t}{t - 5}$, then the L.C.D. $= (t + 5)(t - 5)$.

$$\frac{-t}{t - 5} - \frac{4 - t}{5 + t} = \frac{t - 25}{t^2 - 25}$$

$$(t + 5)(t - 5)\left(\frac{-t}{t - 5}\right) - (t + 5)(t - 5)\left(\frac{4 - t}{5 + t}\right) = (t^2 - 25)\left(\frac{t - 25}{t^2 - 25}\right)$$

$$-t(t + 5) - (4 - t)(t - 5) = t - 25$$

$$-t^2 - 5t - (4t - 20 - t^2 + 5t) = t - 25$$

$$-t^2 - 5t - 9t + 20 + t^2 = t - 25$$

$$-14t + 20 = t - 25$$

$$-15t = -45$$

$$t = 3$$

Check: For $t = 3$,

$$\frac{t}{5-t} - \frac{4-t}{5+t} = \frac{3}{5-3} - \frac{4-3}{5+3} = \frac{3}{2} - \frac{1}{8} = \frac{11}{8}$$

$$\frac{t-25}{t^2-25} = \frac{3-25}{9-25} = \frac{-22}{-16} = \frac{11}{8}$$

EXERCISES 4.8 A

In Exercises 1–25, state the restricted values of the variable and solve. Check each solution.

1. $\frac{2}{3x} - \frac{1}{4x} = \frac{1}{12}$

2. $\frac{2}{5x} - \frac{x+2}{10x} = \frac{1}{2x}$

3. $\frac{1}{x} + \frac{1}{2x} + \frac{1}{3x} = \frac{1}{x+5}$

4. $\frac{6}{y} - \frac{1}{y-2} = \frac{3}{y^2-2y}$

5. $\frac{3}{y} + \frac{5}{1-y} + \frac{2y+3}{y^2-1} = 0$

6. $\frac{x}{5} - \frac{x}{6} = 3$

7. $\frac{5}{x} = 20$

8. $\frac{3}{2x-1} = \frac{7}{3x+1}$

9. $\frac{20+y}{10y+5} = \frac{2}{5}$

10. $\frac{3x+1}{25} - \frac{1}{10} = \frac{3x-1}{30}$

11. $\frac{5}{x} = 0$

12. $\frac{x+1}{x-5} - \frac{x+3}{x+2} = \frac{2}{x^2-3x-10}$

13. $\frac{x-6}{x+6} - \frac{x-2}{x+1} = \frac{15}{x^2+7x+6}$

14. $\frac{4t}{9t+18} + \frac{t+2}{3t-6} = \frac{7}{9}$

15. $\frac{3}{y-7} - \frac{2}{y+7} = \frac{y+35}{y^2-49}$

16. $\frac{2x+10}{3x^2-5x-2} + \frac{2}{x-2} = \frac{-6}{3x+1}$

17. $1 + \frac{2}{t-1} = \frac{2}{t(t-1)}$

18. $6 + \frac{4}{x-3} = \frac{x+1}{x-3}$

19. $\frac{x+4}{x+5} + \frac{x+5}{x+3} = 2$

20. $\frac{x}{x^2-3x} - \frac{x+12}{x^2-x-6} = \frac{2}{x^2+2x}$

21. $\frac{5}{5x+3} + \frac{2}{1-2x} = \frac{4-6x}{10x^2+x-3}$

22. $\frac{a-2x}{b-x} = \frac{3}{2}$
Solve for x.

23. $\frac{x}{a} - \frac{a}{b} = \frac{b}{c}$
Solve for x.

24. $\frac{1}{R} = \frac{1}{A} + \frac{1}{B}$
Solve for A.

25. $\frac{F}{V} = \frac{f}{V - v}$
Solve for V.

EXERCISES 4.8 B

In Exercises 1–25, state the restricted values of the variables and solve. Check each solution.

1. $\frac{y}{3} - \frac{y}{2} = 4$

2. $\frac{3}{x} = 15$

3. $\frac{2}{3x - 7} = \frac{5}{x + 2}$

4. $3 - \frac{1}{t} = \frac{7}{5t} - \frac{9}{5}$

5. $\frac{2x - 3}{5x} - \frac{x + 4}{3x} = 2$

6. $\frac{1}{x} = \frac{1}{7x} + \frac{1}{49}$

7. $\frac{y + 12}{y^2 - 16} + \frac{1}{4 - y} = \frac{1}{4 + y}$

8. $\frac{8y - 1}{5} = \frac{16y + 3}{10} - \frac{2y - 5}{5y - 1}$

9. $\frac{5 + x}{x} = 0$

10. $\frac{9y + 2}{12} + \frac{17}{9} = \frac{21y - 8}{18} - \frac{2y - 3}{3}$

11. $\frac{3}{2x - 6} + \frac{1}{4x + 2} = \frac{2x - 3}{2x^2 - 5x - 3}$

12. $\frac{44}{2x^2 - 9x - 5} - \frac{6}{2x + 1} = \frac{4}{x - 5}$

13. $\frac{5}{t} + \frac{1}{2t} - \frac{2}{3t} = 29$

14. $\frac{2x - 29}{x^2 + 7x - 8} = \frac{5}{x + 8} - \frac{3}{x - 1}$

15. $\frac{5t^2}{t^2 - t - 20} = \frac{3t + 2}{t + 4} + \frac{2t + 3}{t - 5}$

16. $\frac{y - 8}{y - 3} - \frac{y + 8}{y + 3} = \frac{y}{9 - y^2}$

17. $\frac{2}{x + 1} + \frac{1}{3x + 3} = \frac{1}{6}$

18. $\frac{x}{x + 2} - \frac{3}{x - 2} = \frac{x^2 - 8}{x^2 - 4}$

19. $\frac{1}{2x - 3} + \frac{x}{4x^2 - 9} = \frac{1}{8x + 12}$

20. $\frac{3x - 2}{x^2 + 4x} - \frac{2x + 5}{5x - x^2} = \frac{5x^2 + x - 5}{x^3 - x^2 - 20x}$

21. $\frac{3x + 2}{3x^2 + 13x - 10} - \frac{2x + 1}{2x^2 + 7x - 15} = 0$

22. $\frac{3}{x + a} - \frac{2}{x - a} = \frac{1}{x}$
Solve for x.

23. $\frac{a-2}{b} + \frac{3}{2b} = \frac{2}{x}$

Solve for x.

24. $E = \frac{I-O}{I}$

Solve for I.

25. $\frac{P}{N} = \frac{p}{N+n}$

Solve for N.

4.9 RATIO AND PROPORTION

At the beginning of this chapter it was stated that a fraction, such as $\frac{5}{6}$, is also called the **ratio** of 5 to 6, or the ratio $\frac{5}{6}$. An equation which states that two ratios are equal is called a **proportion.** Thus a proportion has the form

$$\frac{a}{b} = \frac{c}{d}$$

The statement $\frac{a}{b} = \frac{c}{d}$ is expressed sometimes by saying, "a, b, c, and d are in proportion."

A proportion is a simple fractional equation.

If two numbers are in the ratio $\frac{a}{b}$, then the numbers may be represented as ax and bx ($x \neq 0$) by applying the Fundamental Theorem of Fractions: $\frac{ax}{bx} = \frac{a}{b}$.

EXAMPLE 4.9.1. A sum of \$350 is to be divided between two partners in the ratio $\frac{3}{4}$. How much does each partner receive?

Solution

Let $\qquad x =$ the share of one partner

Then $\qquad 350 - x =$ the share of the other partner

Then

$$\frac{x}{350 - x} = \frac{3}{4}$$

$$4x = 3(350 - x)$$

$$7x = 3(350)$$

$$x = \$150$$

$$350 - x = \$200$$

EXAMPLE 4.9.2 The ratio of women students to men students at a certain college is $\frac{7}{9}$. If there are 2135 women students, how many men students are there?

Solution Let $x =$ the number of men students. Then

$$\frac{7}{9} = \frac{2135}{x}$$

$$7x = 2135 \cdot 9 \qquad \text{(Multiplying both sides by the L.C.D., } 9x\text{)}$$

$$x = 305 \cdot 9$$

$$x = 2745$$

The terms a, b, c, and d of the proportion

$$\frac{a}{b} = \frac{c}{d}$$

also have special names. In the order in which they are stated above, a is the first term, b the second, c the third, and d the fourth. The first and fourth terms are called the **extremes** of a proportion, and the second and third terms are called the **means** of the proportion.

THEOREM

The product of the means of a proportion is equal to the product of the extremes.

In symbols, if $\frac{a}{b} = \frac{c}{d}$, then $bc = ad$.

The proof of this theorem is very simple.

If $$\frac{a}{b} = \frac{c}{d}, \text{ then } bd \cdot \frac{a}{b} = bd \cdot \frac{c}{d}$$

$$ad = bc$$

$$\text{or } bc = ad \quad \text{(By the symmetric axiom)}$$

Other consequences of this theorem are, for instance, that

$$\text{if } \frac{a}{b} = \frac{c}{d}, \text{ then } \frac{a}{c} = \frac{b}{d}$$

$$\text{and } \frac{d}{b} = \frac{c}{a}$$

In words, the extremes and means can be interchanged without changing the proportion.

EXAMPLE 4.9.3 If 3, 4, and 5 are the first three terms, in that order, of a proportion, find the fourth term.

Solution Let d = the fourth term. Then

$$\frac{3}{4} = \frac{5}{d}$$

$$3d = 20$$

$$d = \frac{20}{3} = 6\frac{2}{3}$$

Thus the fourth term is $6\frac{2}{3}$.

DEFINITION

Two variables are said to **vary directly** or to be **directly proportional** if they are related by an equation having the form

$$y = kx$$

The variables are x and y. The number k is called the **constant of proportionality**, or the **constant of variation.**

EXAMPLE 4.9.4 The work W done by a constant force f is directly proportional to the distance s the object is moved. If it requires 60 foot-pounds of work to move an object 4 feet, how much work is required to move the same object 20 feet?

Solution Since $W = fs$, then $\frac{W}{s} = f$, a constant.

Thus $\frac{60}{4} = \frac{W}{20}$ or $60 \cdot 20 = 4W$ or $W = 300$ foot-pounds.

DEFINITION

If one variable is equal to a constant times the reciprocal of the other variable, then the variables are said to **vary inversely,** or they are said to be **inversely proportional** to each other.

This relationship is expressed by the form

$$y = \frac{k}{x}$$

where x and y are the variables, and k is the constant of variation.

EXAMPLE 4.9.5 Under constant temperature, the volume V of a gas varies inversely as the pressure p. If the volume of a certain gas is 30 cubic feet under 4 pounds of pressure, what is its volume at the same temperature when the pressure is 10 pounds?

Solution Since $V = \frac{k}{p}$, $Vp = k$, a constant. Thus $V \cdot 10 = 30 \cdot 4$

$$V = \frac{120}{10} = 12 \text{ cubic feet}$$

DEFINITION

Three variables are said to vary jointly if they are related by an equation having the form

$$z = kxy$$

The variables are z, x, and y, and the constant of variation is k.

EXAMPLE 4.9.6 By Newton's law of gravitation, the force F between two particles of mass kept at a constant distance apart varies jointly as their masses m and n.

a. If $m = 5$, $n = 2$, and $F = 1$, find the constant of variation.

b. Find F if $m = 4$ and $n = 3$.

Solution

a. The formula is $F = kmn$. For $F = 1$, $m = 5$, $n = 2$,

$$1 = k(5)(2)$$

$$k = \frac{1}{10}$$

The constant of variation is $\frac{1}{10}$.

b. Using $F = \frac{1}{10}mn$ for $m = 4$ and $n = 3$,

$$F = \frac{1}{10}(4)(3)$$

$$F = \frac{6}{5}$$

EXERCISES 4.9 A

1. The ancient Greeks considered the most beautiful rectangle to be the one whose sides were in the Golden Ratio. Approximating the Golden Ratio by the ratio $\frac{5}{8}$, find the dimensions of the most beautiful rectangle with perimeter of 78 inches.
2. By weight, the ratio of oxygen to hydrogen in pure distilled water is $\frac{8}{1}$. How many grams of each element are in 100 grams of water?
3. Pie crust is made by combining shortening and flour in the ratio $\frac{2}{3}$ by volume. If 2 cups of flour is required per pie, how much shortening is required to make 15 pies?
4. A son and grandson are to divide an inheritance of $5000 in the ratio $\frac{5}{3}$. How much does each one receive if the inheritance tax is $500 and the lawyer receives $2700?
5. The batting average of a baseball player is the ratio of the number of hits he makes to the number of times he comes to bat. If a certain player has a batting average of 0.325, how many hits has he made in 40 times at bat?
6. On a certain map, $1\frac{1}{2}$ inches represents 50 miles. If two cities are $6\frac{1}{4}$ inches apart on the map, find the number of miles from one city to the other.
7. If the resistance is constant, then the voltage E of an electrical circuit varies directly with the amperage I. If $E = 90$ volts when $I = 18$ amperes, find the amperage when $E = 220$ volts.
8. The cost of a certain size of nail varies directly as the number of nails purchased. If 1500 nails cost $8.10, find the price of 100 nails.
9. For a given distance, the speed r of a moving object varies inversely as its time t. If a trip took 12 hours at a speed of 350 mph, find how long the trip would take at a speed of 400 mph.
10. The weight that a certain kind of horizontal beam can bear varies inversely as the length between its supports. If a 12-foot beam can bear 1600 pounds, how many pounds can an 18-foot beam bear?
11. If the pressure p of a gas varies jointly as its absolute temperature t and its density d, and the pressure is 20 pounds when the temperature is 220 degrees and the density is 1.4, find the pressure when the density is 0.80 and the temperature is 280 degrees.
12. Use the information from Exercise 11 to find the absolute temperature of a gas with density 1.54 if the pressure is 30 pounds.

13. If the first three terms of a proportion are 2, 4, and 6, respectively, find the fourth term.
14. The first term of a proportion is 6 and the second term is 8. The fourth term is the sum of the second and third terms. Find the third and fourth terms.
15. If the exchange ratio of Mexican pesos to United States dollars is $12\frac{1}{2}$ to 1, what is the price, in dollars, of an article costing 450 pesos?
16. The profit of an investment is to be divided among the three partners in a business in the ratio of 5:3:2. How much does each one receive if the profit is $48,000?
17. A man earns $4200 in thirty weeks. At this rate, how much could he earn in 1 year if he worked every week except for 2 weeks' vacation without pay?
18. The force required to stretch a spring varies directly as the elongation of the spring. If a force of 30 pounds is required to stretch a certain spring 2 inches, what force will be required to stretch the spring 5 inches?
19. Under constant tension, the frequency (vibrations per second) of a string varies inversely as the length of the string. A frequency of 440 vibrations per second (concert A) is obtained by a certain string 32 inches long. Using the same type of string, what length is needed for a frequency of 256 vibrations per second (middle C)?
20. The intensity of illumination on an object varies inversely as the square of the distance of the object from the source of light. If the intensity is 8 lumens at a distance of 3 feet from the light source, what is the intensity at a distance of 5 feet from the same source?

EXERCISES 4.9 B

1. A sampling revealed that on a certain day, for every 1000 tin cans manufactured, 15 were defective. If 250,000 cans were made on that day, how many were defective?
2. The ratio of sodium to chlorine in common table salt is $\frac{35}{23}$. Find the amount of each element in the salt compound weighing 290 pounds.
3. A piece of wire 84 inches long is cut so that the parts are in the ratio $\frac{4}{3}$. How long is each part?
4. One grocer advertises 16 cans of a beverage for $1.00. Another grocer advertises 12 cans for 69 cents. Which is the better buy and why?

5. Two companies contribute a total of \$26,400 to a certain charity. If the ratio of their contributions is $\frac{5}{6}$, find the amount that each company contributes.
6. A blueprint has the scale 1 inch = 4 feet. What are the dimensions of a rectangular living room that measures $4\frac{1}{2}$ by 5 inches on the blueprint?
7. The cost of a certain chemical varies directly as the weight. If 24 pounds costs \$3.50, how much would 5 pounds cost?
8. Water pressure varies directly as the depth beneath the surface of the water. If the pressure is 1250 pounds per square foot when the depth is 20 feet, find the pressure at a depth of 1 mile (1 mile = 5280 feet).
9. The time required to complete a certain job varies inversely as the number of men working on the job. If 12 men can paint an apartment house in 1 week, in how many days could 15 men paint the house?
10. For a fixed volume, the area of the base of a cylinder varies inversely as its height. If the area of the base of a tin can 8 inches high is 12 square inches, what should be the height of a can with a base of 15 square inches to produce the same volume?
11. If I is the interest from a principal P invested at simple interest for N number of years, then I varies jointly as P and N. If the interest on a \$10,000 investment is \$1100 at the end of 2 years, how much money must be invested at this rate to yield an interest payment of \$1540 at the end of $3\frac{1}{2}$ years?
12. If the potential energy e of an object varies jointly as its mass m and its height h above sea level, and it is known that $e = 200$ when $m = 20$ and $h = 15$, find the potential energy of an object with $m = 35$ and $h = 12$.
13. Find the fourth term of a proportion if its first three terms are 7, 14, and 20, in that order.
14. Find the means of a proportion if the product of the extremes is 36 and if the ratio of one mean to the other is 4 to 1.
15. If 5 pounds of oranges costs 89 cents, what is the cost of 12 pounds of oranges?
16. If the ratio of kilometers to miles is 8 to 5, what speed in miles per hour is equivalent to 60 kilometers per hour?
17. The ratio of the size of a medium egg to a large egg is 2 to 3. If medium eggs cost 34 cents per dozen, what should be the equivalent cost for a dozen large eggs?
18. For a certain type of concrete mix, the ratio of cement to sand to rock is 1:3:5. Find the number of cubic feet of each component in a 108-cubic-foot mix of this concrete.

19. The gravitational attraction between two bodies varies inversely as the square of the distance between the bodies. If a force F of 25 force units results from two bodies that are 6 units apart, find the gravitational force generated by the two bodies if they are 10 units apart.
20. The length L of a pendulum of a clock varies directly as the square of the period P of the pendulum. (The period is the time it takes for the end of the pendulum to make a movement from one extreme position to the other and then return.) If a pendulum 18 inches long has a period of $1\frac{1}{2}$ seconds, how much should the pendulum be shortened so that the new period will be exactly one second?

4.10 VERBAL PROBLEMS

Verbal problems that lead to fractional equations are illustrated in the following examples.

4.10.1 Work Problems

A formula for a work problem is $w = tr$, where w is the amount of work done, t the time spent working, and r the amount of work done per unit time (the rate). If the work done is a complete job, then $w = 1$.

When several persons work together, it is assumed that the total work done is the sum of the amounts of work done by those persons working.

In summary, $w = tr$ and $w(\text{total}) = w(\text{worker A}) + w(\text{worker B})$.

If the work done is the complete job, then $w = 1$ and $r = \frac{1}{t}$.

EXAMPLE 4.10.1 One machine requires 90 minutes to complete a certain job. Another machine completes the same job in 120 minutes. If the machines worked together, how long would it take them to complete the job?

Solution Let x = the number of minutes the machines work together.

	Working Alone			*Working Together*		
Formula:	t ·	r =	w	t ·	r =	w
First machine	90	$\frac{1}{90}$	1	x	$\frac{1}{90}$	$\frac{x}{90}$
Second machine	120	$\frac{1}{120}$	1	x	$\frac{1}{120}$	$\frac{x}{120}$

Equation:

work done by first + work done by second = total work

$$\frac{x}{90} + \frac{x}{120} = 1$$

$$360\left(\frac{x}{90} + \frac{x}{120}\right) = 360$$

$$4x + 3x = 360$$

$$7x = 360$$

$$x = \frac{360}{7} = 51\frac{3}{7} \text{ minutes}$$

EXAMPLE 4.10.2 Working alone, a painter's son requires twice as much time to paint a job as his father does. If the painter works alone for 2 days and then completes the job by working with his son for 3 days, how long would it have taken the father to paint the job alone?

Solution

Let x = the time of the painter working alone

Then $2x$ = the time of the son working alone

	Working Alone			*Working Together*		
Formula:	t ·	r =	w	t ·	r =	w
Painter	x	$\frac{1}{x}$	1	5	$\frac{1}{x}$	$\frac{5}{x}$
Son	$2x$	$\frac{1}{2x}$	1	3	$\frac{1}{2x}$	$\frac{3}{2x}$

Equation:

painter's work + son's work = total work

$$\frac{5}{x} + \frac{3}{2x} = 1$$

$$10 + 3 = 2x$$

$$2x = 13$$

$$x = \frac{13}{2} = 6\frac{1}{2} \text{ days}$$

4.10.2 Motion Problems

Some uniform-motion problems state that two times are equal or that two rates are equal. Since $d = rt$ implies that $t = \frac{d}{r}$ and $r = \frac{d}{t}$, a fractional equation results from a statement of this kind.

EXAMPLE 4.10.3 A boat travels 45 miles downstream in the same time that it takes to travel 30 miles upstream. If the speed of the boat in still water is 12 mph, find the rate of the current.

Solution

Let $x =$ the rate of the current

Then $12 + x =$ the rate downstream

$12 - x =$ the rate upstream

Formula:	r	$\cdot\ t$	$= d$
Downstream	$12 + x$	$\frac{45}{12 + x}$	45
Upstream	$12 - x$	$\frac{30}{12 - x}$	30

Equation:

$$\frac{30}{12 - x} = \frac{45}{12 + x}$$

$$30(12 + x) = 45(12 - x)$$

$$360 + 30x = 540 - 45x$$

$$75x = 180$$

$$x = \frac{12}{5} = 2\frac{2}{5} \text{ mph}$$

EXERCISES 4.10 A

1. The denominator of a certain fraction is twice the numerator. If 2 is subtracted from both the numerator and the denominator, the resulting fraction is $\frac{3}{7}$. Find the original fraction.

2. The denominator of a certain fraction exceeds the numerator by 12. If the numerator is increased by 4 and if the denominator is decreased by 3, the value of the resulting fraction is $\frac{3}{4}$. What is the value of the original fraction?
3. A mason can lay the same amount of brick in 8 days that his helper can in 12 days. If they worked together, how long would it take them to lay this amount of brick?
4. One pipe can fill a tank in 18 minutes, a second in 30 minutes, and a third in 45 minutes. If all three pipes were opened at the same time, how long would it take to fill the tank?
5. One crew of men can sheetrock a certain building in 15 days. With the help of a second crew, the building can be sheetrocked in 6 days. How long would it take the second crew, working alone, to sheetrock the building?
6. A man can mow his lawn in 40 minutes. His son can mow the lawn in 60 minutes. The man starts to mow the lawn and after 10 minutes his son joins him. How long does it take them to complete the job, provided they have two lawnmowers?
7. A boat travels 50 miles downstream in the same time it takes to travel 30 miles upstream. If the rate of the current is 8 mph, find the rate of the boat in still water.
8. A plane with a wind travels 800 miles in the same time it travels 600 miles against the same wind. If the speed of the plane in still air is 175 mph, find the speed of the wind.
9. An experienced typist can type 3 times as fast as a new one. Working together, they complete a certain job in 9 hours. How long would it have taken each one alone?
10. It takes one train 5 hours longer to travel 500 miles than another traveling at the same rate to go 400 miles. Find the time of each train.
11. A boat can go 30 miles down a river in $\frac{3}{5}$ the time it takes to return the same distance upstream. If the rate of the current is 3 mph, what is the rate of the boat in still water?
12. A person averaged 60 mph on a trip from Sacramento to San Francisco. Returning, he averaged 40 mph due to heavy traffic. What was his average rate of speed for the whole trip? (The answer is *not* 50 mph!)
13. For two resistors R_1 and R_2, connected in parallel, the total resistance R can be found by using the following formula:

$$\frac{1}{R} = \frac{1}{R_1} + \frac{1}{R_2}$$

If two resistors, connected in parallel, are such that one has 3 times the resistance of the other, and if the total resistance is 15 ohms, find the number of ohms in each resistor.

14. Applying Kirchoff's Laws to the electric circuit shown in the figure below, the voltage E across the 6-ohm resistor is given by

$$\frac{6 - E}{0.4} - \frac{E - 4}{0.6} - \frac{E}{6} = 0$$

Solve for E.

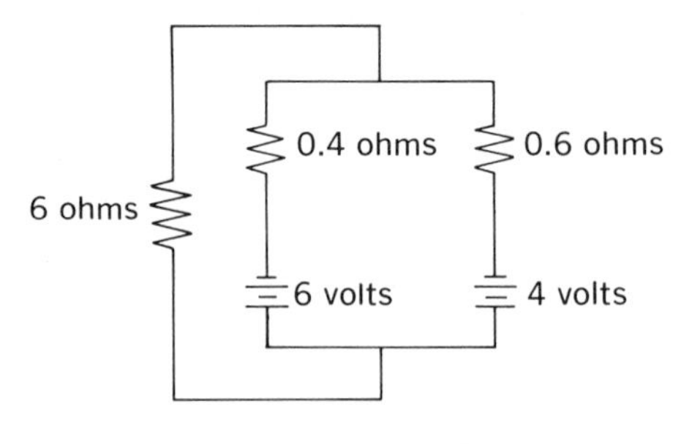

FIGURE 4.10.1

EXERCISES 4.10 B

1. The ratio of the numerator of a certain fraction to the denominator is $\frac{3}{5}$. If 20 is subtracted from the numerator and 10 is added to the denominator, the resulting fraction is equal to $\frac{1}{2}$. Find the original fraction.
2. The sum of two integers is 168. When the larger is divided by the smaller the quotient is 6 and the remainder is 7. Find these two integers.
3. A roofer requires 10 hours to shingle a roof. His helper can do the same work in 15 hours. How long would it take them if they worked together?
4. Three machines can each manufacture a certain article in 12 minutes, 15 minutes, and 20 minutes, respectively. If the three machines worked simultaneously, how long would it take them to manufacture 1000 articles?
5. Three old machines together required 80 minutes to do a certain job. Two new machines were installed, one of which worked 3 times as fast as all the old machines together. The five machines, working together, took 15 minutes to do the same type of job. How long would it take the other new machine, working alone, to do the work?

6. An express train travels 150 miles in the same time that a freight train travels 100 miles. The rate of the express train is 20 mph faster than that of the freight train. Find the rate of each train.
7. A plane with a tail wind completes a 2600-mile trip in the same time that a plane going in the opposite direction completes a trip of 2400 miles. Find the speed of the wind if each plane has a speed in still air of 600 mph.
8. It takes a motorboat the same time to go 5 miles upstream as it does to go 8 miles downstream. If the rate of the current is 6 mph, what is the rate of the boat in still water?
9. A new machine can process checks 5 times as fast as an old one. With both machines operating, 1000 checks can be processed in $\frac{1}{2}$ hour. How long would it take each machine alone to process 1000 checks?
10. Two pipes can fill a tank in 9 and 12 hours, respectively. A third pipe can drain the tank in 18 hours. How long would it take to fill the tank if all three pipes were open?
11. One number is 4 times another. The sum of their reciprocals is $\frac{3}{4}$. Find the numbers.
12. The focal length F of a concave mirror can be found by using the formula:

$$\frac{1}{F} = \frac{1}{a} + \frac{1}{b}$$

where a is the distance of the object from the mirror and b is the distance of the image from the mirror. A certain concave mirror has a focal length of 5 inches. At what distance from this mirror will an object be if the image distance is $\frac{1}{4}$ the object distance?

SUMMARY

The Fundamental Theorem of Fractions

If n, d, and k are real numbers, and $d \neq 0$ and $k \neq 0$, then

$$\frac{nk}{dk} = \frac{n}{d}$$

The Product of Reciprocals Theorem

If d and k are nonzero real numbers, then

$$\frac{1}{d} \cdot \frac{1}{k} = \frac{1}{dk}$$

The Addition of Quotients Theorem

If n, m, and d are any real numbers, and $d \neq 0$, then

$$\frac{n}{d} + \frac{m}{d} = \frac{n + m}{d}$$

The Subtraction of Quotients Theorem

If n, m, and d are any real numbers, and $d \neq 0$, then

$$\frac{n}{d} - \frac{m}{d} = \frac{n - m}{d}$$

The Product of Quotients Theorem

If n, d, r, and s are any real numbers, and $d \neq 0$ and $s \neq 0$, then

$$\frac{n}{d} \cdot \frac{r}{s} = \frac{nr}{ds}$$

The Reciprocal of a Quotient Theorem

If n and d are nonzero real numbers, then

$$\frac{1}{\frac{n}{d}} = \frac{d}{n}$$

The Quotient of Quotients Theorem

If n, d, and s are any nonzero real numbers, and r is any real number, then

$$\frac{r}{s} \div \frac{n}{d} = \frac{rd}{sn}$$

☐ An equation which states that two ratios are equal is called a **proportion.**

☐ **Direct Variation.** Two variables are said to **vary directly** or to be **directly proportional** if they are related by an equation having the form $y = kx$. (x and y are variables, k is a constant.)

☐ **Inverse Variation.** Two variables are said to **vary inversely** or to be **inversely proportional** to each other if they are related by an equation having the form $y = \frac{k}{x}$. (x and y are variables, k is a constant.)

Joint Variation. Three variables are said to **vary jointly** if they are related by an equation $z = kxy$ (z, x, and y are variables, k is a constant.)

REVIEW EXERCISES

Simplify in Exercises 1–10.

1. $\dfrac{156}{390}$

2. $\dfrac{15x^3y^2}{27xy^3z}$

3. $\dfrac{x-7}{7-x}$

4. $\dfrac{9x^2-36y^2}{x^2-4xy+4y^2}$

5. $\dfrac{2x^2+5x-150}{2x^2+17x-30}$

6. $\dfrac{x^2-2x+1}{15x^2}\cdot\dfrac{20x^5}{1-x^2}$

7. $\dfrac{x^3+5x^2+6x}{x^2-x}\cdot\dfrac{x^2-3x+2}{(x^2-4)(x+3)}$

8. $\dfrac{5xy+15y}{15y}\div\dfrac{x^2+4x+3}{x^2-1}$

9. $\dfrac{\dfrac{1}{x}-\dfrac{1}{3}}{x-\dfrac{9}{x}}$

10. $\left(x+\dfrac{3}{y}\right)\div\left(x-\dfrac{3}{y}\right)$

In Exercises 11–20 write each as a single fraction reduced to lowest terms.

11. $\dfrac{5}{x+3}-\dfrac{x}{x+3}$

12. $\dfrac{x}{x+3}+\dfrac{5x^2}{x^2-9}$

13. $\dfrac{x}{x^2-6x+5}+\dfrac{3}{x-1}$

14. $\dfrac{2}{x-1}-\dfrac{4}{x}+\dfrac{2}{x+1}$

15. $\dfrac{1}{x}+\dfrac{1}{2}+\dfrac{5}{x+2}$

16. $\dfrac{3}{x^2+5x+4}-\dfrac{2}{x^2+4x+3}$

17. $\dfrac{1}{(x+2)^2}-\dfrac{2}{x^2-4}+\dfrac{1}{(x-2)^2}$

18. $\left(\dfrac{3x-1}{3x}-\dfrac{3x}{3x+1}\right)\div\left(\dfrac{5x-2}{5x}-\dfrac{5x}{5x+2}\right)$

19. $3x-\dfrac{x}{4x-\dfrac{5x}{2}}$

20. $4x+\dfrac{6}{2x-\dfrac{6x}{3}}$

In Exercises 21–25 divide by using the long-division algorithm. Check by multiplication.

21. $\dfrac{x^3-4x^2+3x-1}{x-3}$

22. $\dfrac{x^4+x^2-2}{x^2+3}$

23. $\dfrac{39+20x^2+15x^3}{5x+10}$

24. $\dfrac{a^3-1}{a-1}$

25. $\dfrac{7x^3 - 3x + x^2 + 3x^3 - 5x - 10}{1 + 2x}$

26. Given the algebraic expression

$$\left(\frac{x - 2}{x^2 + 1}\right) \div \left(\frac{k - 3}{x + 3}\right)$$

a. For what value(s) of x will the expression be undefined?

b. For what value(s) of k will the expression be undefined?

c. For what value(s) of x will the expression equal 0? Why?

Solve and check Exercises 27–35. State the restricted values of the variable.

27. $\dfrac{5}{x - 6} + \dfrac{3}{x + 6} = \dfrac{8}{x^2 - 36}$

28. $\dfrac{x + 3}{x + 2} = \dfrac{3}{2}$

29. $\dfrac{2}{x + 3} - \dfrac{2}{x - 3} = \dfrac{1}{3 - x}$

30. $\dfrac{2x + 3}{x - 1} = \dfrac{2x - 5}{x + 3}$

31. $\dfrac{5}{3x + 1} - \dfrac{x + 2}{4x} + \dfrac{1}{4} = 0$

32. $\dfrac{x}{x + 4} - \dfrac{4}{x - 4} = \dfrac{x^2 + 16}{x^2 - 16}$

33. $\dfrac{1}{1 - x} + \dfrac{x}{x - 1} = 1$

34. $\dfrac{x - a}{x + b} = \dfrac{x + b}{x - a}$
Solve for x.

35. $\dfrac{p - x}{x} = c + \dfrac{1}{x}$
Solve for x.

36. A mechanic can do a certain repair job in $3\frac{1}{2}$ hours. Working together with his assistant, they can complete the job in 2 hours. How long would it take the assistant working alone?

37. A plane travels 1140 miles with a wind of 35 mph in the same time it travels 860 miles against the wind. Find the speed of the plane in still air.

38. The force applied to stretch an elastic spring varies directly as the amount of elongation. If a force of 18 pounds determines an elongation of 2.5 inches, what is the constant of variation?

39. The numerator of a fraction is 3 less than the denominator. If 1 is added to the denominator and 2 is subtracted from the numerator, the value of the new fraction formed is $\frac{1}{4}$. Find the original fraction.

40. If F varies inversely as d^2 and if $F = 6$ when $d = 2$, find the value of F when $d = 6$.

41. A swimming pool can be filled by a pipe in 10 hours and by a hose in 15 hours. How long does it take to fill the pool using both pipe and hose, assuming there is no loss in water pressure?

42. The area A of a triangle varies jointly as the length of the base b and the altitude h upon that base. If $A = 6$ when $b = 3$ and $h = 4$, find the area of a triangle with base 12 units and altitude 7 units.

43. For what value of x is the proportion $\frac{x + 2}{x + 9} = \frac{5}{12}$ true?

44. If a car uses 16 gallons of gasoline on a 720-mile journey, how much gas will be used on a 450-mile trip? (Assume the mileage per gallon to be the same for both trips.)

CHAPTER 5

GRAPHING AND LINEAR SYSTEMS

In geometry, a straight line is thought of as a set of infinitely many points where each point is said to be on the line. It has already been seen how the geometric line can be used to visualize the set of real numbers and their properties.

So far, the discussion has focused on linear equations in *one* variable. In this chapter the concept of linear equation will be extended to include the case of *two* variables.

The geometric plane is used to visualize equations in two variables and thereby to provide a better understanding of such equations.

In this chapter the student will be introduced to a rectangular coordinate system where a one-to-one correspondence is established between the set of points on a plane and the set of ordered pairs of real numbers. He will then study linear equations in two variables and their graphs. Finally, he will be shown how to solve a system of linear equations in two variables and how to use these techniques for solving a variety of practical applications.

5.1 RECTANGULAR COORDINATES

5.1.1 Ordered Pairs and Solutions

An **ordered pair** is an expression having the form **(a, b)** where **a** is called the **first component** (or first member) of the ordered pair and **b** is called the **second component** (or second member) of the ordered pair.

HISTORICAL NOTE

The concept of a coordinate system probably originated with the ancient Egyptian surveyors. The hieroglyphic symbol used to designate the districts into which Egypt was divided was a grid symbol.

Records indicate that the Greeks used the ideas of longitude and latitude to locate points in the sky and on the earth. The Romans, who were noted for their surveying techniques, arranged the streets of their cities on a rectangular coordinate system.

The Arab and Persian mathematicians were the first to use geometric figures for algebraic problems. Examples are found in the works of the Arab al-Khowarizmi (ca. 825) and the Persian Omar Khayyam (ca. 1100). This usage is again found in the writings of Fibonacci (1220), Pacioli (1494), and Cardan (1545).

René Descartes (1596–1650) is credited with the invention of analytic geometry, since he used a rectangular coordinate system to establish a relationship between equations and curves. Descartes' best-known work is a philosophical treatise, published in 1637, called *A Discourse on the Method of Rightly Conducting the Reason and Seeking Truth in the Sciences.* This publication had three appendices, the third of which is the famous *La Géométrie.* In this appendix Descartes deals with such topics as the solution of equations of degree greater than 2 and the use of exponents to indicate powers of numbers. But what Descartes is best remembered for is his introduction of modern analytic geometry in *La Géométrie*, and the Cartesian coordinate system is named in his honor. Although Descartes is credited with the development of analytic geometry, Pierre de Fermat, another great French mathematician, also formulated coordinate geometry at the same time and made a considerable contribution in this field. The modern terms *coordinates*, *abscissa*, and *ordinate* were contributed by the German mathematician, Gottfried Wilhelm Leibniz, in 1692.

The order in which the components of an ordered pair are written is important. For example, the ordered pair (3, 5) is *not* the same as the ordered pair (5, 3).

DEFINITION

A solution of an open equation in two variables x and y is an ordered pair (a, b) such that the equation becomes true when x is replaced by a and y is replaced by b.

EXAMPLE 5.1.1 Which of the following is a solution of $3x - y = 10$?
a. (4, 2) b. (2, 4)

Solution

a. For (4, 2), $x = 4$ and $y = 2$. Then $3x - y = 3(4) - 2 = 12 - 2 = 10$. Since $3x - y = 10$ is true for $x = 4$ and $y = 2$, (4, 2) is a solution.

b. For (2, 4), $x = 2$ and $y = 4$. Then $3x - y = 3(2) - 4 = 6 - 4 = 2$. Since $3x - y = 10$ is false for $x = 2$ and $y = 4$, (2, 4) is not a solution.

EXAMPLE 5.1.2 Find solutions of $y = 2x + 3$ for:
a. $x = 5$ b. $x = 0$ c. $x = -5$ d. $y = 7$

Solution

a. For $x = 5$, $y = 2(5) + 3 = 10 + 3 = 13$, and (5, 13) is a solution.

b. For $x = 0$, $y = 2(0) + 3 = 3$, and (0, 3) is a solution.

c. For $x = -5$, $y = 2(-5) + 3 = -10 + 3 = -7$, and $(-5, -7)$ is a solution.

d. For $y = 7$, $7 = 2x + 3$, $2x + 3 = 7$, $2x = 4$, $x = 2$. Therefore, (2, 7) is a solution.

In general, most equations in two variables have infinitely many solutions that are ordered pairs of real numbers. As an example of an exception, $x^2 + y^2 = -1$ has no solutions in the set of ordered pairs of real numbers since x^2 and y^2 and $x^2 + y^2$ must be positive or zero.

5.1.2 Rectangular Coordinates

In Chapter 1 a one-to-one correspondence was established between the set of real numbers and the set of points on a line: Every point on the number line represents a unique real number, and for every real number there exists a corresponding unique point on the number line.

A similar one-to-one correspondence can be established between the set of ordered pairs of real numbers and the set of points on a plane.

A **number plane** is a plane whose points have been placed in one-to-one correspondence with the set of ordered pairs of real numbers.

A **rectangular coordinate system** is obtained by taking two perpendicular number lines, called the **axes**, intersecting at their origins.

The point of intersection of the axes is called the **origin** of the coordinate system.

It is customary to select one axis horizontal, called the x-axis, with its positive direction to the right, and the other axis vertical, called the y-axis, with its positive direction upward.

Unless it is specified otherwise, the unit segment on the x-axis is selected to be the same length as the unit segment on the y-axis—that is, the same scale is used for both axes.

The axes separate the plane into four regions, called **quadrants**, that are numbered consecutively starting with the upper right quadrant and proceeding counterclockwise (see Figure 5.1.1).

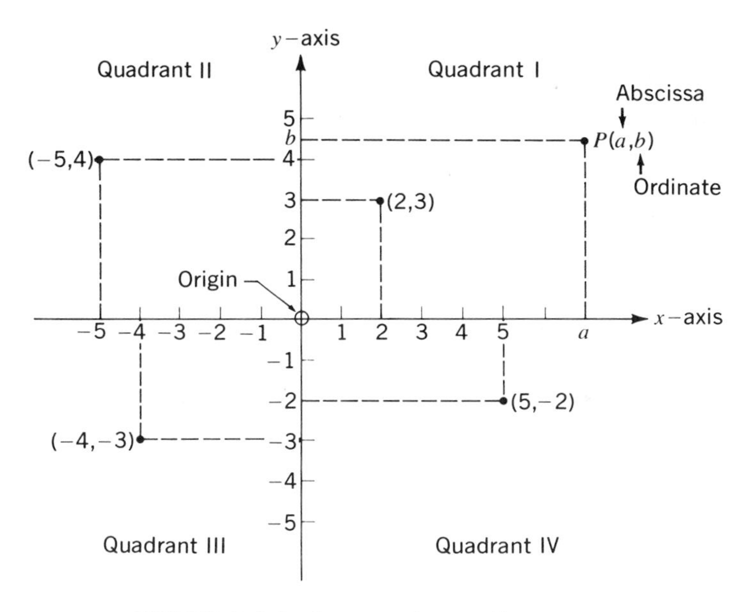

FIGURE 5.1.1 A rectangular coordinate system.

To each ordered pair (a, b) of real numbers is associated a unique point P, the point of intersection of a vertical line through point a on the x-axis and a horizontal line through point b on the y-axis.

Conversely, to each point P in the plane is associated a unique ordered pair (a, b) where a is the coordinate of the point of intersection of the vertical line through P and the x-axis and b is the coordinate of the point of intersection of the horizontal line through P and the y-axis.

A point a on the x-axis is assigned the ordered pair $(a, 0)$, and a point b on the y-axis is assigned the ordered pair $(0, b)$. The origin is assigned the ordered pair $(0, 0)$.

The numbers of the ordered pair (a, b) are called the **coordinates** of P, with the first component a called the **abscissa** and the second component b called the **ordinate** (see Figure 5.1.1). A point will be designated symbolically by $P: (a, b)$.

The association in which two perpendicular coordinate lines are used to establish a one-to-one correspondence between the set of points on a plane and the set of ordered pairs or real numbers is called a **rectangular** (or **Cartesian**) **coordinate system.**

EXAMPLE 5.1.3 Graph $P:(3, 4)$.

Solution To locate the point P, whose coordinates are $(3, 4)$, draw a vertical line through the point 3 on the x-axis and a horizontal line through the point 4 on the y-axis. The point P is the intersection of these two lines (see Figure 5.1.2). This point is located in the first quadrant.

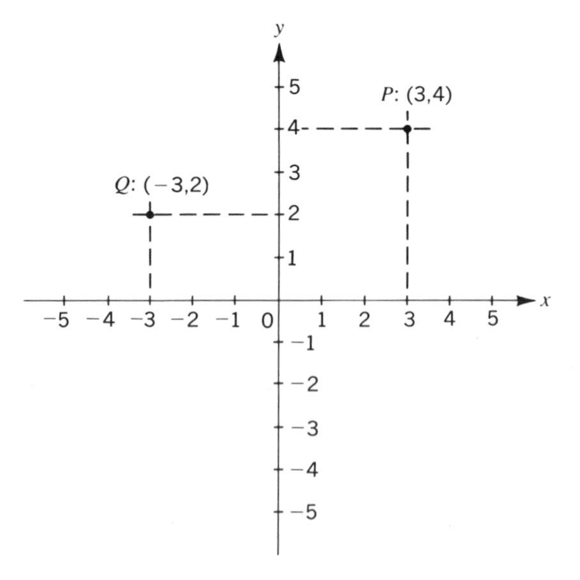

FIGURE 5.1.2

EXAMPLE 5.1.4 Graph Q: $(-3, 2)$.

Solution To locate the point Q, whose coordinates are $(-3, 2)$, draw a vertical line through the point $x = -3$ and a horizontal line through the point $y = 2$. The intersection of these two lines is point Q. This point is in the second quadrant (see Figure 5.1.2).

If vertical lines are drawn through every point which corresponds to an integer on the x-axis and horizontal lines are drawn through every point which corresponds to an integer on the y-axis, the intersections of these lines

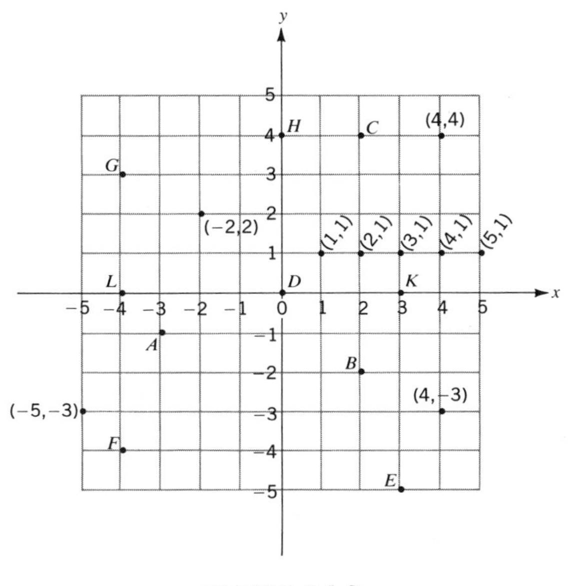

FIGURE 5.1.3

will represent a one-to-one correspondence between the points of intersection of these lines and the set of ordered pairs of integers. The lines form a grid, as shown in Figure 5.1.3.

There are many more points on a number plane other than those corresponding to ordered pairs of integers, just as there are many more points on a number line other than those corresponding to integers.

The set of points on a number plane is in one-to-one correspondence with the set of ordered pairs of real numbers:

$$\{(x, y) \mid x \text{ and } y \text{ are real numbers}\}$$

EXAMPLE 5.1.5 Graph the solutions of the equation $2x + y = 6$ for: a. $x = 0$, b. $x = 1$, c. $y = 0$, d. $y = -2$. Join these points in the order a, b, c, d.

Solution

a. For $x = 0$, $2(0) + y = 6$ and $y = 6$. The solution is $(0, 6)$.

b. For $x = 1$, $2(1) + y = 6$ and $y = 4$. The solution is $(1, 4)$.

c. For $y = 0$, $2x + 0 = 6$ and $x = 3$. The solution is $(3, 0)$.

d. For $y = -2$, $2x + (-2) = 6$, $2x = 8$, and $x = 4$. The solution is $(4, -2)$.

The graph is shown in Figure 5.1.4.

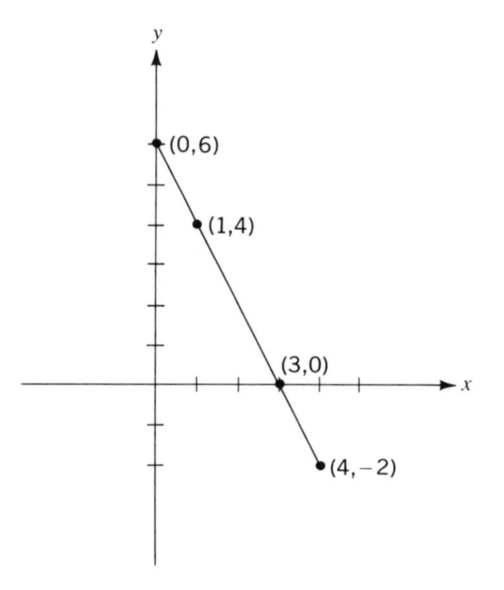

FIGURE 5.1.4

EXERCISES 5.1 A

For Exercises 1–5, state the coordinates of each of the points in Figure 5.1.3.

1. A **2.** B **3.** C **4.** D **5.** E

For Exercises 6–13, locate the points whose coordinates are given on one set of axes. Label each point with its coordinates.

6. $(2, 3)$ **7.** $(3, 2)$ **8.** $(-1, -4)$ **9.** $(1, -3)$
10. $(-1, 3)$ **11.** $(0, 4)$ **12.** $(2, 0)$ **13.** $(0, 0)$

14. Locate the points $(3, 2)$ and $(-1, -2)$ and connect them with a straight line. Locate the points $(2, -2)$ and $(-1, 4)$ and connect them with a straight line. What are the coordinates of the point of intersection?

15. If a point is on the y-axis, what is its abscissa?

16. A point is on the x-axis four units to the left of the origin.
  a. What is its abscissa?
  b. What is its ordinate?

For Exercises 17–19, graph the given points, join them as indicated, and name the figure thus formed.

17. A:(0, 0), B:(0, 4), C:(3, 0); join A to B to C to A.

18. A:(−1, 2), B:(4, 2), C:(4, −3), D:(−1, −3); join A to B to C to D to A.

19. A:(−3, 5), B:(2, 5), C:(−3, −1), D:(2, −1); join A to B to D to C to A.

Which of Exercises 20–25 are solutions of $y = 2x + 3$?

20. (0, 3)
21. (1, 5)
22. (2, 6)
23. (3, 0)
24. (5, 13)
25. (−4, −1)

Which of Exercises 26–31 are solutions of $2x + y = 6$?

26. (3, 0)
27. (0, 3)
28. (−3, 0)
29. (−3, 12)
30. (4, −2)
31. (4, 1)

Which of Exercises 32–36 are solutions of $x = 5$?

32. (0, 5)
33. (5, 0)
34. (5, 2)
35. (5, −4)
36. (3, 5)

From the set {(2, 6), (−2, 6), (6, −2), (−2, 0), (0, 6)} *select one or more solutions for each equation in Exercises 37–42.*

37. $3x + 2y = 6$
38. $y = 3x + 6$
39. $y = 10 - 2x$
40. $3x + y + 6 = 0$
41. $x = -2$
42. $y = 6$

Find solutions of each equation in Exercises 43–56 for:

a. $x = 4$ b. $x = -4$ c. $y = 0$ d. $y = 2$

43. $y = 2x - 8$
44. $x + y = 10$
45. $x - y = 6$
46. $x = 3y - 8$
47. $2x - y = 4$
48. $y - 3x = 5$

49. $x = 4$

50. $y = 2$

51. $(x + 1)(y - 1) = 15$

52. $xy = x - 4$

53. $y = 2x^2$

54. $y^2 = x + 5$

55. $8y = 16 - x^2$

56. $x^2 + 4y^2 = 16$

For Exercises 57–58, copy and complete the table of values, then plot the points on a set of coordinate axes. If the points were connected, what geometric term would describe this graph?

57. $y = 3x - 6$

x	y
0	?
?	0
1	?
-2	?
?	-3

58. $3x + 5y = 15$

x	y
0	?
?	0
-5	?
?	-6
10	?

EXERCISES 5.1 B

For Exercises 1–5, state the coordinates of each of the points in Figure 5.1.3.

1. F **2.** G **3.** H **4.** K **5.** L

For Exercises 6–13, locate the points whose coordinates are given on one set of axes. Label each point with its coordinates.

6. (3, 5)

7. (5, 3)

8. $(-4, -1)$

9. $(2, -4)$

10. $(-4, 2)$

11. $(0, -3)$

12. $(-5, 0)$

13. $(-5, -3)$

14. Locate the points (3, 5) and $(6, -4)$ and connect them with a straight line. Locate the points (7, 1) and $(1, -5)$ and connect them with a straight line. What are the coordinates of the point of intersection?

15. If a point is on the x-axis, what is its ordinate?

16. A point is on the y-axis three units below the origin.

a. What is its abscissa?

b. What is its ordinate?

For Exercises 17–19, graph the given points, join them as indicated, and name the figure thus formed.

17. $A:(-3, 0)$, $B:(0, 5)$, $C:(3, 0)$; join A to B to C to A.
18. $A:(-2, 1)$, $B:(1, 2)$, $C:(4, 1)$, $D:(1, -4)$; join A to B to C to D to A.
19. $A:(-4, 2)$, $B:(-1, -5)$, $C:(2, 2)$, $D:(5, -5)$, $E:(8, 2)$; join A to B to C to D to E.

Which of Exercises 20–25 are solutions of $y = \frac{1}{2}x - 2$?

20. $(4, 0)$
21. $(2, 1)$
22. $(2, -1)$
23. $(10, 8)$
24. $(0, 2)$
25. $(-2, -3)$

Which of Exercises 26–31 are solutions of $3x - y = 12$?

26. $(0, 12)$
27. $(0, -12)$
28. $(4, 0)$
29. $(0, 4)$
30. $(-1, 15)$
31. $(-2, -18)$

Which of Exercises 32–36 are solutions of $y = 4$?

32. $(4, 4)$
33. $(4, 0)$
34. $(0, 4)$
35. $(-4, 4)$
36. $(4, -4)$

From the set $\{(3, 5), (5, 3), (0, 5), (3, 0), (5, -3)\}$ *select one or more solutions for each equation in Exercises 37–42.*

37. $5x + 3y = 15$
38. $2y = 3x - 9$
39. $y = 17 - 4x$
40. $5x + 2y = 25$
41. $y = 5$
42. $x = 3$

Find solutions of each equation in Exercises 43–56 for:

a. $x = 3$ b. $y = 5$ c. $y = -5$ d. $x = 0$

43. $y = 3x + 4$
44. $x + 2y = 3$
45. $2x - y = 5$
46. $x = y - 2$
47. $x + 3y = 6$
48. $5x + 2y = 5$
49. $x = 3$
50. $y = -5$

51. $3xy^2 = 25$

52. $x = \dfrac{12}{y + 1}$

53. $y^2 = 2x + 10$

54. $y = 30x^2 - 25x$

55. $3x = 25 - y^2$

56. $y^2 - x^2 = 16$

For Exercises 57–58, copy and complete the table of values, then plot the points on a set of coordinate axes. If the points were connected, what geometric term would describe this graph?

57. $x - y = 8$

x	y
0	?
?	0
−2	?
?	−4
3	?

58. $3x - 2y = 12$

x	y
0	?
?	0
2	?
−4	?
?	6

5.2 GRAPHS OF LINEAR EQUATIONS

In the previous section, it was seen that a solution of an equation in two variables is an ordered pair of numbers for which the equation becomes a true statement.

DEFINITION

The **solution set** of an equation in two variables is the set of all solutions of the equation.

For an equation having the form

$$y = mx + b \text{ where } m \neq 0$$

there is exactly one real number y for each real number x. Consequently, an equation of the form $y = mx + b$ has infinitely many solutions.

A **linear function** is the set of ordered pairs (x, y) defined by a rule having the form $y = mx + b$ where $m \neq 0$—in other words, the solution set of $y = mx + b$.

The **graph of a linear function** is the graph of the set of ordered pairs belonging to the linear function.

The graph of a linear function is a straight line.

EXAMPLE 5.2.1 Graph $\{(x, y) \mid y = 2x + 1\}$.

Solution Solutions can be found by arbitrarily assigning values for x and obtaining corresponding values for y. It is convenient to make a table of values:

x	y	$2x + 1 = y$
(1	3)	$2(1) + 1 = 3$
(2	5)	$2(2) + 1 = 5$
(3	7)	$2(3) + 1 = 7$
(0	1)	$2(0) + 1 = 1$
(−2	−3)	$2(-2) + 1 = -3$

Thus the ordered pairs (1, 3), (2, 5), (3, 7), (0, 1), and (−2, −3) are solutions for the given equation. Figure 5.2.1 is the graph of these five points. As shown previously, an ordered pair (p, q) is a solution of an equation in two variables, x and y, if the equation is true when $x = p$ and $y = q$.

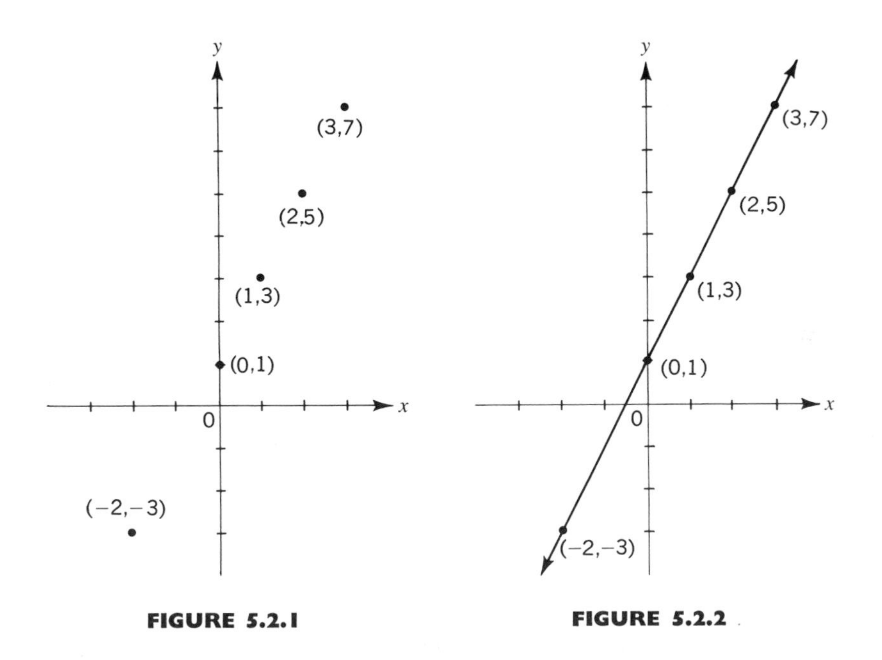

FIGURE 5.2.1 **FIGURE 5.2.2**

The plotted points in Figure 5.2.2 seem to lie in a straight line, and, in fact, they do. There are infinitely many solutions to the equation $y = 2x + 1$, and only part of the graph representing the solutions can be shown. The arrows at each end of the line representing the solution set indicate that the line continues indefinitely in both directions.

DEFINITION

Any equation in two variables x and y that can be expressed in the form

$$Ax + By + C = 0$$

where A and B are real numbers, not both zero, is called a **linear equation in two variables.**

If $B \neq 0$, then this equation can be solved for y.

$$Ax + By + C = 0$$

$$By = -Ax - C$$

$$y = -\frac{A}{B}x - \frac{C}{B}$$

$$y = mx + b \text{ where } m = -\frac{A}{B}$$

$$\text{and} \quad b = -\frac{C}{B}$$

Thus if $B \neq 0$, then $Ax + By + C = 0$ defines a linear function.
If $B = 0$, then the general equation becomes

$$Ax + C = 0$$

Solving for x, $$x = -\frac{C}{A}$$

or $x = c$, a real number.

DEFINITION

The **graph of a linear equation** is the graph of its solution set, and for any linear equation in two variables, this graph is a straight line.

To verify that the graph of a linear equation in two variables actually is a straight line, two things must be shown:

1. The set of points whose coordinates satisfy a given linear equation $Ax + By + C = 0$ (A and B real numbers, not both zero) all lie on a straight line.
2. The coordinates (x, y) of any point lying on a straight line satisfy a linear equation in x and y.

Both these statements can be proved, but for now they will be taken for granted, and it will be assumed that in a Cartesian coordinate system, the graph of a linear equation in two variables is a line and every line in the plane is the graph of a linear equation in two variables.

Since two points uniquely determine a line, it is only necessary to find two solutions (ordered pairs) that satisfy a given linear equation, plot the points corresponding to the coordinates, and draw a line through these points. This line represents the graph of the given equation. It is a good idea, however, to find a third solution as a check.

EXAMPLE 5.2.2 Graph $2x + y = 6$.

Solution Make a table of solutions. Any value may be selected for x, and then the corresponding y value is computed, using the equation.

Similarly, any value may be selected for y and its x value computed.
For this equation, it is convenient to solve for y: $y = 6 - 2x$.
Selecting $x = 0$, $x = 2$, and $x = 5$, the following table is made:

$(x$	$y)$	$6 - 2x = y$
$(0$	$6)$	$6 - 2(0) = 6$
$(2$	$2)$	$6 - 2(2) = 2$
$(5$	$-4)$	$6 - 2(5) = -4$

The points are plotted and a straight line is drawn through them (see Figure 5.2.3).

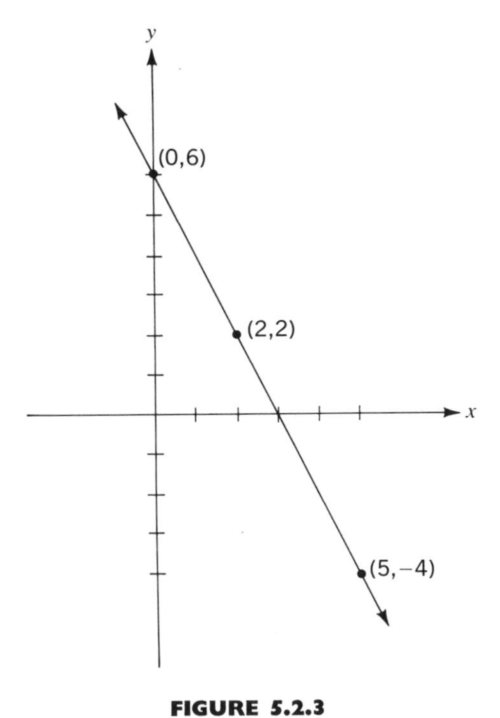

FIGURE 5.2.3

EXAMPLE 5.2.3 Graph $\{(x, y) \mid x + 2y + 4 = 0\}$.

Solution It is a good idea to select zero values for x and y whenever possible, because this results in simple calculations.

Selecting $x = 0$, $y = 0$, and $y = -3$, the following table is made:

$(x$	$y)$	$x + 2y + 4 = 0$
$(0$	$-2)$	$0 + 2y + 4 = 0,\ 2y = -4,\ y = -2$
$(-4$	$0)$	$x + 2(0) + 4 = 0,\ x + 4 = 0,\ x = -4$
$(2$	$-3)$	$x + 2(-3) + 4 = 0,\ x - 6 + 4 = 0,\ x - 2 = 0,\ x = 2$

The plotted points and the line through them are shown in Figure 5.2.4.

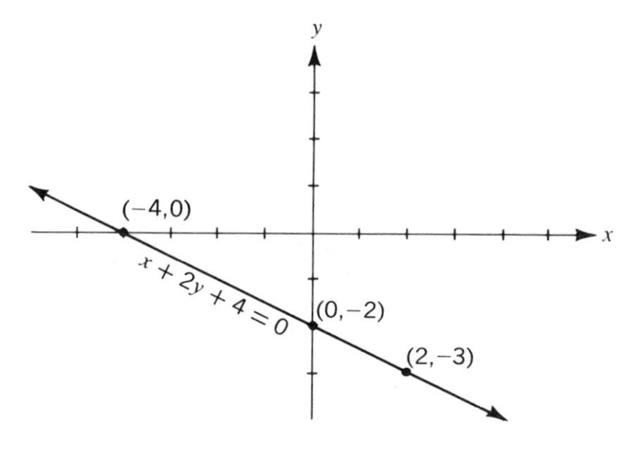

FIGURE 5.2.4

EXAMPLE 5.2.4 Graph $\{(x, y) \mid y = 3\}$.

Solution All ordered pairs whose second component is 3 are solutions of $y = 3$; for example, (2, 3), (0, 3), (−3, 3). The graph is a horizontal line, as shown in Figure 5.2.5.

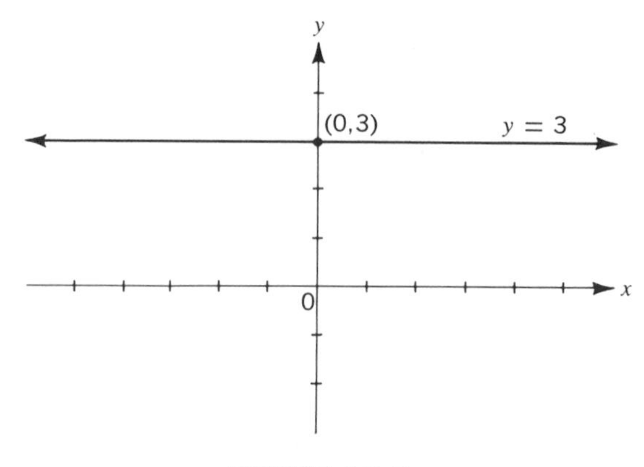

FIGURE 5.2.5

EXAMPLE 5.2.5 Graph $\{(x, y) \mid x = 2\}$.

Solution All ordered pairs whose first component is 2 are solutions of $x = 2$; for example, $(2, 4)$, $(2, 0)$, $(2, -3)$. The graph is a vertical line, as shown in Figure 5.2.6.

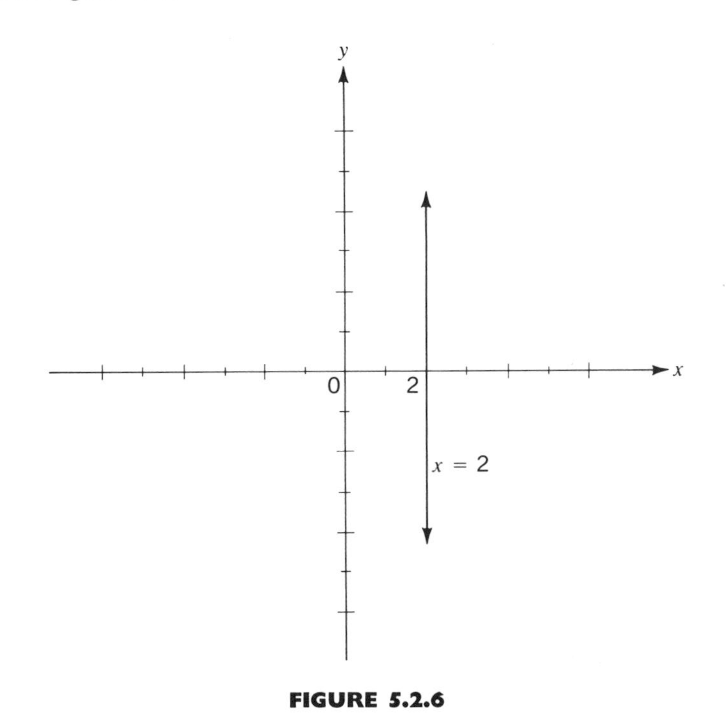

FIGURE 5.2.6

EXERCISES 5.2 A

For Exercises 1–4, draw the graphs of the equations on one set of coordinate axes by first making a table of values and then plotting the points whose coordinates are derived from the table of values. Label the points and draw a line through them. (Use at least three points.)

1. $y = x + 2$

2. $y = x - 1$

3. $y = x$

4. $y = x + 4$

5. What characteristic do the graphs in Exercises 1–4 have in common? How do these graphs differ from one another?

6. Describe the set of points in the plane whose
   a. Abscissa is 5
   b. Ordinate is 6

Graph each of Exercises 7–18 on a number plane.

7. $y = 2x + 6$

8. $y = 3x - 9$

9. $y = 4 - x$

10. $y = 8 - 4x$

11. $x + 2y = 0$

12. $3x - 4y = 0$

13. $x - y = 5$

14. $x + 2y = 8$

15. $2x + 3y = 6$

16. $5x - 2y = 10$

17. $x = -3$

18. $y + 1 = 0$

19. The relation between centigrade and Fahrenheit temperature scales is given by the formula

$$F = \frac{9}{5}C + 32$$

where F = degrees Fahrenheit

and C = degrees centigrade

Represent this relation graphically. (Let the horizontal axis represent C.) Graph from $C = -20$ to $C = 120$.

20. Refer to the graph in Exercise 19 and find the Fahrenheit temperature corresponding to each of the following centigrade temperatures:
a. 0 degrees b. 5 degrees c. 37 degrees d. 100 degrees
e. -15 degrees

21. Refer to the graph in Exercise 19 and find the centigrade temperature for each of the following:
a. 0 degrees F b. 100 degrees F c. 59 degrees F
d. 140 degrees F e. -22 degrees F

22. Plot the following points on the same set of axes. Use ten squares = 1 unit on the y scale.

TABLE OF SQUARE ROOTS

(correct to one decimal place)

Number, x	0	1	2	3	4	5	6
Square root, y	0.0	1.0	1.4	1.7	2.0	2.2	2.4

a. Join the points by a smooth curve.

b. Using the graph, estimate $\sqrt{3.5}$ to the nearest tenth. (This process is called interpolation.)

c. Extending the graph, estimate $\sqrt{7}$ to the nearest tenth. (This process is called extrapolation.)

23. The accompanying table describes the relation between the number x of the sides of a polygon and the sum y of the measures of the angles of the polygon measured in units of 180 degrees.

Number of Sides, x	*Sum of ∠s in Units of 180°, y*
3	1
4	2
5	3
7	5
10	8

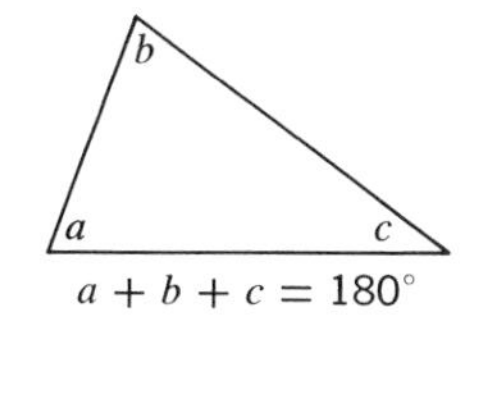

$a + b + c = 180°$

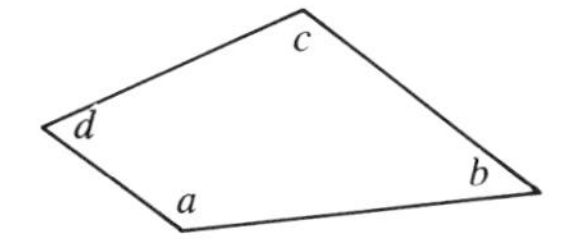

$a + b + c + d = 360°$

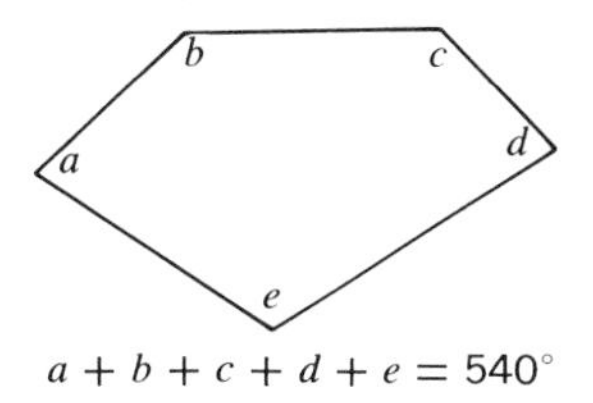

$a + b + c + d + e = 540°$

a. Graph the ordered pairs (x, y) shown in the table.

b. Join the points consecutively by straight lines.

c. From the graph, determine

1. The sum in degrees of the ∠s of a polygon having 8 sides. (Interpolate.)
2. The sum in degrees of the ∠s of a polygon having 12 sides. (Extrapolate by extending the graph.)
3. The number of sides of a polygon whose angles have a sum of 720 degrees.

d. Find an equation for which this is the graph.

24. If a point having coordinates $(a, 0)$ is on a graph, then a is called an x-intercept of the graph. If a point having coordinates $(0, b)$ is on a graph, then b is called a y-intercept of the graph. Graph each of the following equations and determine its x-intercept and its y-intercept.

a. $y = 2x - 6$, b. $x + y = 4$, c. $3x - 2y = 6$

For Exercises 25–30, graph each given pair of lines on the same set of coordinate axes and state how the two lines in each pair seem to be related.

25. $3x - y = 6$
$y = 3x - 6$

26. $y = 8 - 2x$
$2x + y = 4$

27. $x + y = 0$
$x - y = 0$

28. $2x = 3y - 6$
$3y = 2x + 6$

29. $y = x + 4$
$2x - 2y = 8$

30. $2y = -x - 1$
$x + 2y + 1 = 0$

EXERCISES 5.2 B

For Exercises 1–4, draw the graphs of the equations on one set of coordinate axes by first making a table of values and then plotting the points whose coordinates are derived from the table of values. Label the points and draw a line through them. (Use at least three points.)

1. $y = 2x + 1$

2. $y = 3x + 1$

3. $y = -5x + 1$

4. $x + y - 1 = 0$

5. What characteristic do the graphs in Exercises 1–4 have in common? How do these graphs differ from one another?

6. Describe the set of points in the plane whose
a. Abscissa is -2
b. Ordinate is -4

Graph each of Exercises 7–18 on a number plane.

7. $y = 5x - 10$

8. $y = 4x + 12$

9. $y = 6 - 2x$

10. $y = 12 - 3x$

11. $x - 5y = 0$

12. $5x - 2y = 0$

13. $x - 3y = 9$

14. $x + 4y = 12$

15. $3x + 5y = 15$

16. $4x - 5y = 10$

17. $2x + 8 = 0$

18. $2y - 5 = 0$

19. The number of hours H that a growing child should sleep may be related to the age A of the child in years by the formula

$$H = 17 - \frac{A}{2}$$

Represent this relation graphically by selecting the horizontal axis as the A-axis. Graph from $A = 0$ to $A = 18$.

20. The table below shows the changes in temperature in a town on a certain day from 9 a.m. to 3 p.m.

Hour	9	10	11	12	1	2	3
Temperature	62°	66°	71°	74°	70°	69°	67°

a. Draw a graph to show these changes in temperature. Join the points consecutively by straight lines.

b. On the graph, mark the points that indicate the approximate time when the temperature was 68 degrees.

c. Determine from the graph for how long a time the temperature was above 68 degrees that day.

21. The formula $A = 100(1 + 0.06t)$ expresses the amount of money A in dollars to which a sum of \$100 accumulates if invested at 6 percent simple interest for t years.

a. Graph this equation, finding points for $t = 2, 4, 6, 8$, and 10. Join the points by a straight line.

b. From the graph, find how much money has accumulated at the end of 5 years.

c. From the graph, find how long it would take the sum to accumulate to \$142.

22. Using the straight-line depreciation method, a car costing \$3000 with a probable scrap value of \$500 at the end of 10 years will have a book value B dollars at the end of n years where

$$B = 3000 - 250n$$

a. Graph this equation for $0 \leq n \leq 10$.

b. From the graph determine the book value at the end of $5\frac{1}{2}$ years.

c. When will the book value be \$1875?

d. On the same set of axes, graph $A = 250n$, the amount in the depreciation fund.

e. When is the amount in the depreciation fund equal to the book value?

23. a. For what value of k will the point $(k, 3)$ be on the graph of $y = 2x - 7$?
b. For what value of k will the point $(2, k)$ be on the graph of $y = 3x - 1$?

24. Determine the x-intercept and the y-intercept for each of the following. (See Exercises 5.2 A, Number 24.)

a. $y = 3x - 2$, b. $x - y = 4$, c. $2x - 3y = 12$

For Exercises 25–30, graph each given pair of lines on the same set of coordinate axes and state how the two lines in each pair seem to be related.

25. $x = y + 1$
$y = x + 1$

26. $x = 5 - 2y$
$x + 2y = 5$

27. $2x - 3y = 6$
$3x - 2y = 6$

28. $4x + 4y = 8$
$5x + 5y = 15$

29. $4x + 4y = 8$
$5x + 5y = 10$

30. $2y = 10 - 5x$
$5x + 2y - 20 = 0$

5.3 SLOPE AND THE LINEAR EQUATION

5.3.1 Slope

If a roof rises 1 foot for every 4 feet of horizontal distance, the steepness or pitch of the roof is $\frac{1}{4}$ (Figure 5.3.1). The ratio $\frac{1}{4}$ represents the **slope** of the roof, and it is the ratio of the rise to the run. It is often important to

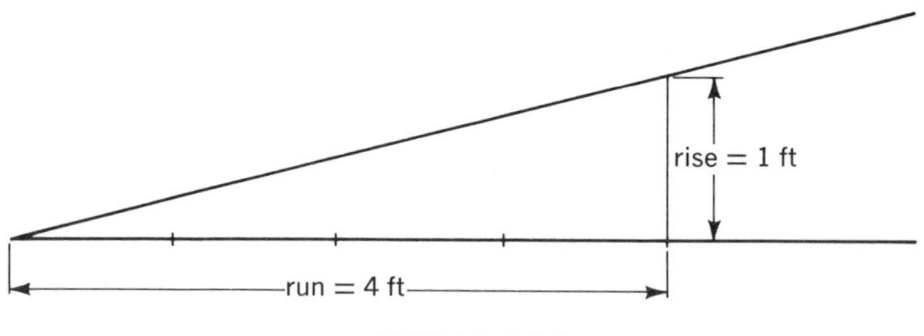

FIGURE 5.3.1

evaluate the slope of a line, and the computation is similar to the roof problem. Consider the graph of the equation $2x - y = 4$ (Figure 5.3.2). Choose any two points on the graph, for example, point P:(5, 6) and point Q:(1, −2):

$$\text{Slope} = \frac{\text{rise}}{\text{run}} = \frac{\text{change in ordinate } (y)}{\text{change in abscissa } (x)}$$

$$= \frac{6 - (-2)}{5 - 1} = \frac{8}{4} = \frac{2}{1}$$

Therefore, the slope is $\frac{2}{1}$, or 2. The slope of a line is the same for all points on the line, not just the two specific points selected as in the above example.

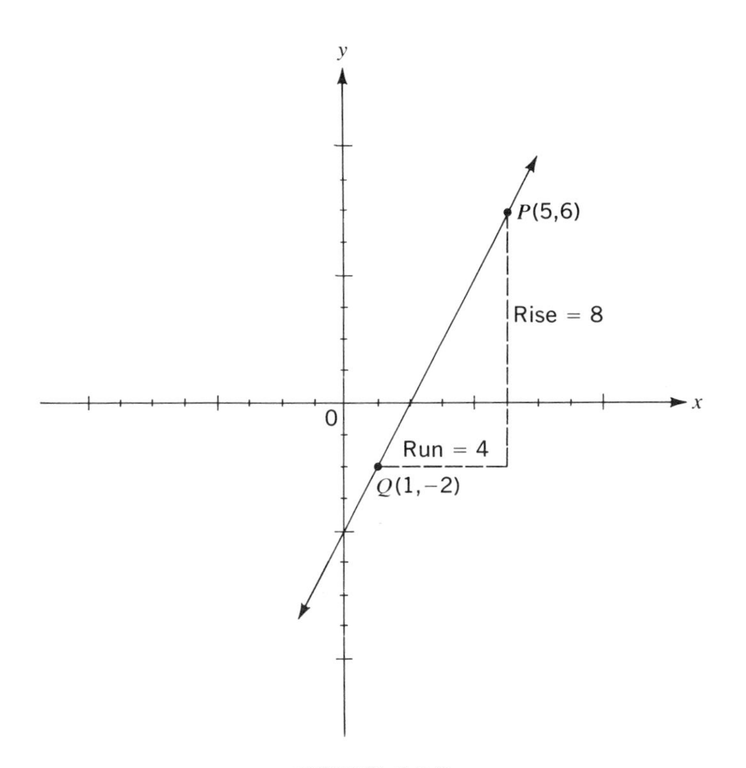

FIGURE 5.3.2

A definition for the slope of a general line can be conveniently expressed by using subscripts to indicate two points on the line. For example, P_1, (read "P sub one") and P_2 (read "P sub two") name two points. The numeral 1 at the lower right of P_1 and the numeral 2 at the lower right of P_2 are subscripts used to indicate that P_1 is the first point and P_2 is the second point.

Similarly, the coordinates of P_1 can be expressed as (x_1, y_1) and those of P_2 as (x_2, y_2).

DEFINITION

If $P_1:(x_1, y_1)$ and $P_2:(x_2, y_2)$ are any two points on a line and if $x_1 \neq x_2$, then the slope m of the line joining P_1 and P_2 is given by

$$m = \frac{y_2 - y_1}{x_2 - x_1}$$

See Figure 5.3.3.

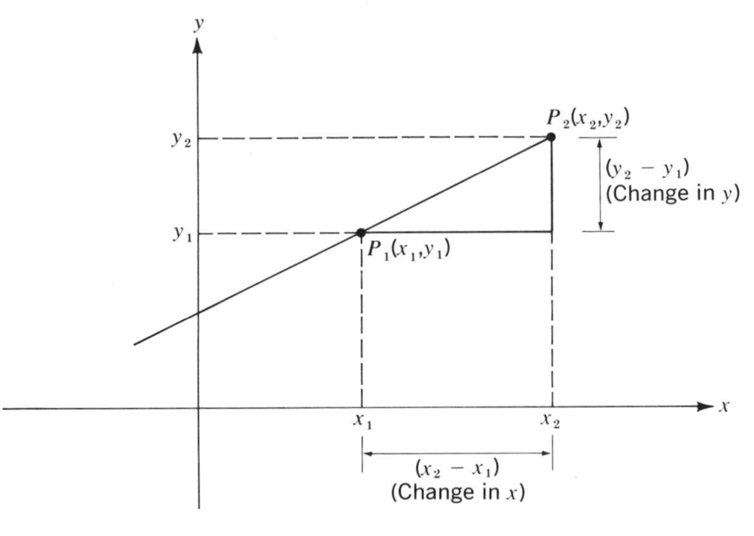

FIGURE 5.3.3

If a line is vertical (parallel to the y-axis), then $x_1 = x_2$ and $x_2 - x_1 = 0$. Since the denominator of the slope ratio is zero, the **slope of a vertical line is undefined** (see Figure 5.3.4).

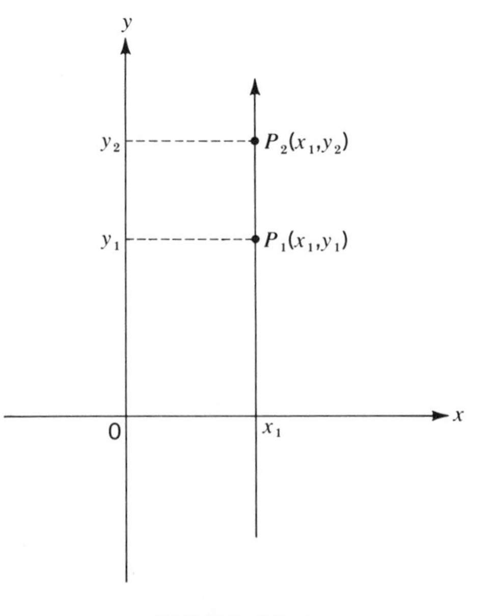

FIGURE 5.3.4

If a line is horizontal (parallel to the x-axis), then $y_1 = y_2$ and $y_2 - y_1 = 0$. In this case, the numerator of the slope ratio is zero, and **the slope of a horizontal line is 0** (see Figure 5.3.5).

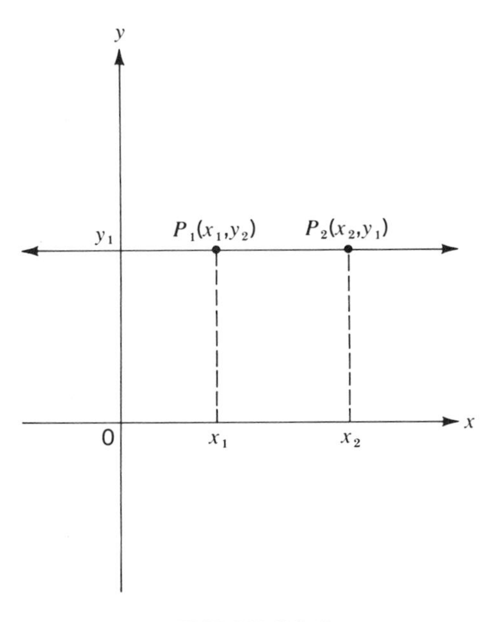

FIGURE 5.3.5

EXAMPLE 5.3.1 Find the slope of the line passing through the points whose coordinates are (3, 2) and (5, 1), respectively.

Solution Using

$$m = \frac{y_2 - y_1}{x_2 - x_1} \text{ with } (x_2, y_2) = (5, 1)$$

$$\text{and } (x_1, y_1) = (3, 2)$$

$$m = \frac{1 - 2}{5 - 3} = -\frac{1}{2}$$

Alternate Solution Using

$$m = \frac{y_2 - y_1}{x_2 - x_1} \text{ with } (x_2, y_2) = (3, 2)$$

$$\text{and } (x_1, y_1) = (5, 1)$$

$$m = \frac{2 - 1}{3 - 5} = -\frac{1}{2}$$

Note in Example 5.3.1 that the value of the slope does not depend on which point is called the first point and which the second.

Note also that the slope is negative, which means that the line "falls to the right" (see Figure 5.3.6).

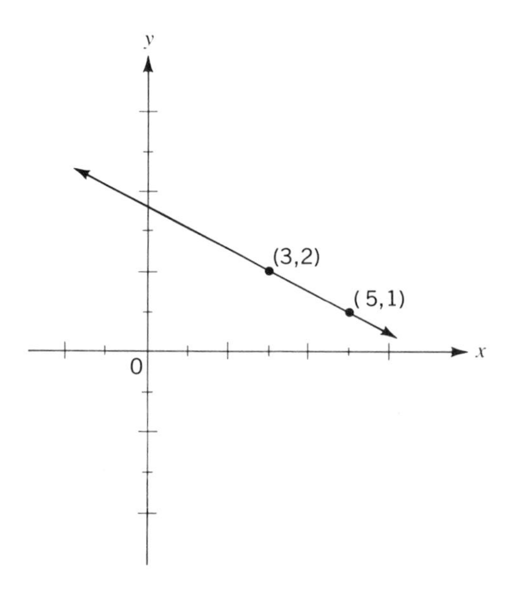

FIGURE 5.3.6

EXAMPLE 5.3.2 Find the slope of the line passing through (2, 1) and (3, 4).

Solution

$$m = \frac{y_2 - y_1}{x_2 - x_1} = \frac{4 - 1}{3 - 2} = 3$$

In this case the slope is positive and the line "rises to the right" (see Figure 5.3.7).

Although the slope of any given nonvertical line is uniquely determined, a given slope does not determine a unique line. Consider the equations

$$y = 2x + 1 \tag{1}$$

and

$$y = 2x \tag{2}$$

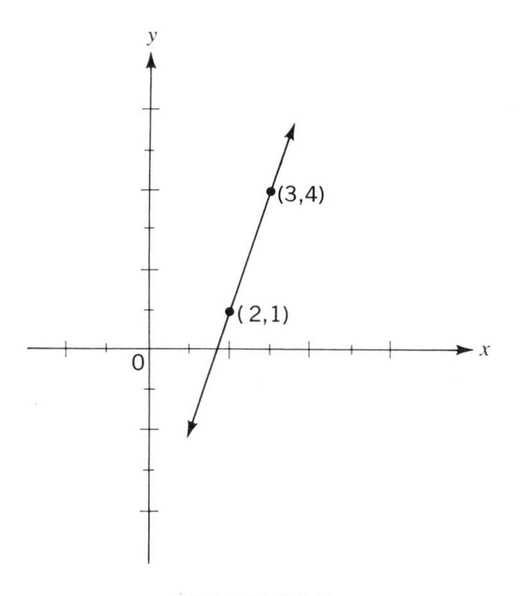

FIGURE 5.3.7

Select two points on the graph of equation (1) and two points on the graph of equation (2). The coordinates (0, 1) and (1, 3) satisfy equation (1). (See Figure 5.3.8.) Slope of line (1) is $\frac{3-1}{1-0} = 2$. The coordinates (0, 0) and (1, 2) satisfy equation (2). Slope of line (2) is $\frac{2-0}{1-0} = 2$. Both lines have the

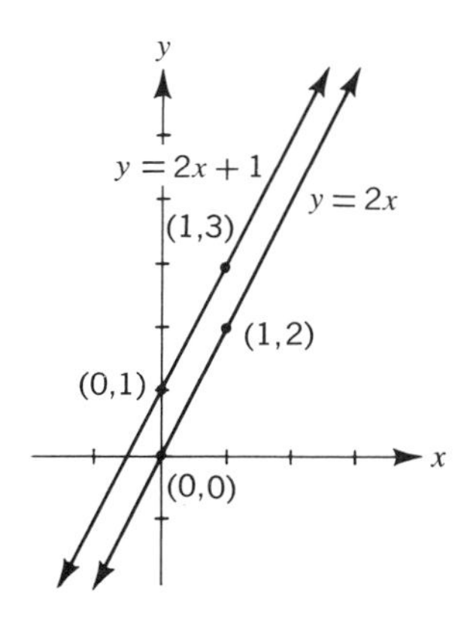

FIGURE 5.3.8

same slope, but from Figure 5.3.8 it is clear that the lines are distinct. They are, in fact, parallel, which leads to the following theorem.

THEOREM

Two distinct lines which are not vertical but are in the same plane have the same slope if and only if they are parallel.

5.3.2 Slope-Intercept Form of a Linear Equation

Every linear equation of the form $Ax + By + C = 0$, where A and B are real numbers not both zero, is equivalent to an equation that is solved explicitly for one variable in terms of the other. For example, the equation

$$2x + 3y + 1 = 0$$

is equivalent to the equation

$$y = -\frac{2}{3}x - \frac{1}{3}$$

To find the slope of the line represented by this equation, two arbitrary points are selected.

For $x = 1$, and $y = -\frac{2}{3}x - \frac{1}{3}$

$$y = -\frac{2}{3}(1) - \frac{1}{3} = -1$$

P_1 is selected as $(1, -1)$.

For $x = -2$, and $y = -\frac{2}{3}x - \frac{1}{3}$

$$y = -\frac{2}{3}(-2) - \frac{1}{3}$$

$$y = \frac{4}{3} - \frac{1}{3} = 1$$

P_2 is selected as $(-2, 1)$.

Calculating the slope,

$$m = \frac{1 - (-1)}{-2 - 1} = \frac{2}{-3} = -\frac{2}{3}$$

Referring to the equation $y = -\frac{2}{3}x - \frac{1}{3}$, it can be seen that the slope $-\frac{2}{3}$ is also the coefficient of x. This is not just a coincidence. In general, if a line is represented by the equation

$$y = mx + b$$

then m, the coefficient of x, represents the slope of the line.

If $x = 0$, then $y = b$, and the ordered pair $(0, b)$ is a solution of $y = mx + b$. The point whose coordinates are $(0, b)$ is located on the y-axis and represents the point where the line crosses the y-axis. This number b in the equation $y = mx + b$ is called the **y-intercept**.

$$\mathbf{y = \underset{\substack{\uparrow \\ \text{slope}}}{m}x + \underset{\substack{\uparrow \\ y\text{-intercept}}}{b}}$$

Since the slope and y-intercept can be determined so easily from the form $y = mx + b$, a special name has been given to this form.

DEFINITIONS

$y = mx + b$ is called the slope-intercept form of a linear equation.

$Ax + By + C = 0$ is called the standard form of a linear equation.

EXAMPLE 5.3.3 Find the slope and the y-intercept of the line whose equation is $5x - 4y = 20$.

Solution Solving the equation for y,

$$\begin{aligned} 5x - 4y &= 20 \\ 5x &= 4y + 20 \\ 4y + 20 &= 5x \\ 4y &= 5x - 20 \\ y &= \frac{5}{4}x - 5 \end{aligned}$$

Comparing with $y = mx + b$, the slope of the line is $m = \frac{5}{4}$, and the y-intercept is $b = -5$.

EXAMPLE 5.3.4 Check the value of the slope found in Example 5.3.3 by selecting any two points on the line and using the slope formula.

Solution Selecting $y = 5$ and using $5x - 4y = 20$,

$$5x - 4(5) = 20$$
$$5x = 40 \text{ and } x = 8$$

Let $P_1 = (8, 5)$. Selecting $y = -10$ and using $5x - 4y = 20$,

$$5x - 4(-10) = 20$$
$$5x = -20 \text{ and } x = -4$$

Let $P_2 = (-4, -10)$.

$$m = \frac{y_2 - y_1}{x_2 - x_1} = \frac{-10 - 5}{-4 - 8} = \frac{-15}{-12} = \frac{5}{4}$$

EXAMPLE 5.3.5 Which of the following lines is parallel to $y = 3x - 5$?
a. $3x - y + 6 = 0$ b. $x + 3y - 5 = 0$

Solution

1. The slope of $y = 3x - 5$ is 3, the coefficient of x, since the equation is in the slope-intercept form.
2. Solving each equation for y to obtain the slope-intercept form,

a. $3x - y + 6 = 0$

$$3x + 6 = y$$
$$y = 3x + 6 \text{ and slope } m = 3$$

Since $y = 3x - 5$ and $3x - y + 6 = 0$ have the same slope, these lines are parallel.

b. $x + 3y - 5 = 0$

$$3y = -x + 5$$
$$y = -\frac{1}{3}x + \frac{5}{3} \text{ and } m = -\frac{1}{3}$$

Since $-\frac{1}{3} \neq 3$, the lines $y = 3x - 5$ and $x + 3y - 5$ are not parallel.

EXAMPLE 5.3.6 Find the value of k so that the line through $(1, k)$ and $(5, 4k)$ has the property stated:

a. Slope -6
b. Slope 0
c. Parallel to $3x - y = 2$

Solution

a. $m = \dfrac{4k - k}{5 - 1} = \dfrac{3k}{4}$ and $m = -6$

$$\frac{3k}{4} = -6,\ 3k = -24,\ k = -8$$

b. $\dfrac{3k}{4} = 0,\ k = 0$

c. Solving $3x - y = 2$ for y,

$$3x = y + 2$$
$$y = 3x - 2$$

and the slope of this line is 3.

Since parallel lines have the same slope,

$$\frac{3k}{4} = 3,\ 3k = 12,\ k = 4$$

EXERCISES 5.3 A

For Exercises 1–5, find the slope of the line segment joining the two given points.

1. (3, 2) and (1, 6)

2. (3, 2) and (4, 1)

3. (1, 0) and (−4, −3)

4. (−2, −1) and (2, 1)

5. (−5, −4) and (−2, −3)

Write each of the equations in Exercises 6–15 in the slope-intercept form, $y = mx + b$, if possible, and determine the slope and the y-intercept of the graph of the equation. Check the value of the slope by selecting any two points on the line and by using the slope formula.

6. $2x + y = 5$

7. $y - 2x + 4 = 0$

8. $2x - y = 8$

9. $3x + 2y = 4$

10. $5x - 2y = 10$

11. $3(x + 1) = 2y$

12. $5(x - y) + 3 = 0$

13. $2x + 8 = 0$

14. $3y - 15 = 0$

15. $y = x$

16. Which of the following lines are parallel?

a. $y = \dfrac{2}{3}x + 1$

b. $y = \dfrac{3}{2}x - 1$

c. $2x - 3y = 2$

d. $3x + 2y + 2 = 0$

17. Which of the lines in Exercise 16 have the same y-intercept?

18. With reference to one set of coordinate axes, draw the following lines through the point $(-2, 4)$:

a. With slope $-\frac{3}{4}$

b. With slope 2

c. With slope undefined

d. With slope zero

19. With reference to one set of coordinate axes, draw the following lines through the point $(0, 0)$:

a. With slope 4

b. With slope $-\frac{1}{4}$

How do these lines seem to be related?

For each pair of lines in Exercises 20–25, state whether the lines are parallel or not.

20. $2x + 2y = 1$
$3x + 3y = 1$

21. $x = y - 2$
$y = x - 2$

22. $2x = 3y$
$3x = 2y$

23. $2(x + 3y) = 3(x + 2y - 1)$
$5x + 10 = 0$

24. $5x + 3y = 15$
$3x + 5y = 30$

25. $x = 3y + 6$
$2x - 6y = 3$

In Exercises 26–30, find the value of k so that the line through $(2, k)$ *and* $(5, 3k)$ *has the property stated.*

26. Slope 4

27. Slope -2

28. Slope 0

29. Parallel to $2x + y = 4$

30. Parallel to $y = x + 5$

In Exercises 31–35, plot the line referred to in each stated exercise.

31. Exercise 26

32. Exercise 27

33. Exercise 28

34. Exercise 29

35. Exercise 30

36. As an example of a slope, a highway grade expressed as a percentage means the number of feet the road changes in elevation for 100 feet measured horizontally.

a. A certain highway has a $2\frac{1}{2}$ percent grade. How many feet does it

rise in a 1-mile stretch (horizontal distance)?
(*Note:* 1 mile = 5280 feet.)

b. How many feet does a $-3\frac{1}{4}$ percent grade highway drop for a $\frac{1}{2}$-mile horizontal stretch?

37. A certain county specification requires that an inclined water pipe must have a slope greater than or equal to $\frac{1}{4}$. Which of the water pipes whose rises and runs are given below meets this specification?
a. Rise = 20 feet, run = 64 feet
b. Rise = 125 feet, run = 500 feet
c. Rise = 60 feet, run = 250 feet

EXERCISES 5.3 B

For Exercises 1–5, find the slope of the line segment joining the two given points.

1. (4, 2) and (2, 1)
2. (0, 1) and (−2, −2)
3. (0, 4) and (3, 0)
4. (−3, 2) and (2, −3)
5. (−1, −2) and (−3, −4)

Write each of the equations in Exercises 6–15 in the slope-intercept form, $y = mx + b$, if possible, and determine the slope and the y-intercept of the graph of the equation. Check the value of the slope by selecting any two points on the line and by using the slope formula.

6. $3x + 2y = 12$
7. $2y - 4x = 5$
8. $\frac{1}{2}(x + y + 10) = 0$
9. $4x + 16 = 2y$
10. $2x - 3y = 9$
11. $x - (y - 1) + 5 = 0$
12. $2(x - y) - 3(x + y) = 5$
13. $4y + 8 = 0$
14. $3x - 9 = 0$
15. $x + y = 0$

16. Which of the following lines are parallel?
a. $x + 9 = 3y$
b. $x + 3y + 9 = 0$
c. $y = \dfrac{x + 1}{3}$
d. $9x - 3y + 1 = 0$

17. Which of the lines in Exercise 16 have the same y-intercept?

18. With reference to one set of coordinate axes, draw the following lines:

a. Through the point $(2, -1)$ with slope $\frac{1}{4}$

b. Through the point $(-2, -3)$ with slope $-\frac{2}{3}$

c. Through the point $(4, -2)$ with slope 0

d. Through the point $(-1, -3)$ with an undefined slope

19. With reference to one set of coordinate axes, draw the following lines through the point $(4, 6)$:

a. With slope $\frac{2}{5}$

b. With slope $-\frac{5}{2}$

How do these lines seem to be related?

For each pair of lines in Exercises 20–25, state whether the lines are parallel or not.

20. $x = 2y - 6$
$y = 2x - 6$

21. $5x - 2y = 3$
$10x - 4y = 5$

22. $4x = 2y + 1$
$2x - y = 4$

23. $3x - 4y = 12$
$4x - 3y = 12$

24. $x + y = 0$
$x - y = 0$

25. $(x + y) - (x - y) = 6$
$2y - 8 = 0$

In Exercises 26–30, find the value of k so that the line through $(2k, 3)$ and $(k, 6)$ has the property stated.

26. Slope $\frac{1}{2}$

27. Slope -1

28. Slope undefined

29. Parallel to $x - 3y = 6$

30. Parallel to $y = x - 4$

In Exercises 31–35, plot the line referred to in each stated exercise.

31. Exercise 26

32. Exercise 27

33. Exercise 28

34. Exercise 29

35. Exercise 30

36. In construction, the pitch of a roof (this is a slope) is the ratio of the vertical rise to the horizontal half-span.

a. For a half-span of 18 feet, how high must the roof rise if the pitch of the roof is 4 to 12?

b. For a roof whose pitch is 3 to 12, find the half-span if the roof rises 7 feet.

5.4 DETERMINING THE EQUATION OF A LINE

In the preceding discussion, the emphasis was on graphing the line whose equation was given. Now consider the problem of finding the equation of a line whose geometric conditions are given.

In order to accomplish this task, one of the following conditions must be given:

1. One point on the line and the slope of the line
2. Two points on the line

If the coordinates of a point on a line are given and if the slope of the line is known, then the equation of the line can be found by using the definition of the slope. Since (x, y) is to be a general point on the line, then

$$m = \frac{y - y_1}{x - x_1}$$

where (x_1, y_1) and m are the given point and slope, respectively.

Multiplying both sides of this equation by $x - x_1$ yields the point-slope form of a line, $y - y_1 = m(x - x_1)$.

DEFINITION

The **point-slope form** of a line with slope m passing through the point (x_1, y_1) is

$$y - y_1 = m(x - x_1)$$

EXAMPLE 5.4.1 Find an equation of a line whose slope is -2 and that passes through the point $(3, 4)$.

Solution $m = -2$ and $(x_1, y_1) = (3, 4)$

Using the point-slope form of a line,

$$\begin{aligned} y - y_1 &= m(x - x_1) \\ y - 4 &= -2(x - 3) \\ y - 4 &= -2x + 6 \\ y &= -2x + 10 && \text{(Slope-intercept form)} \\ 2x + y - 10 &= 0 && \text{(Standard form)} \end{aligned}$$

It is conventional to express the answer in standard form—that is, the linear equation is written in the form

$$Ax + By + C = 0 \text{ where } A \geq 0$$

EXAMPLE 5.4.2 Find an equation of a line passing through the points (3, 4) and (2, 1).

Solution

1. Find the slope:

$$m = \frac{y_2 - y_1}{x_2 - x_1} = \frac{4 - 1}{3 - 2} = 3$$

2. Selecting (x_1, y_1) as (2, 1), and using the point-slope form,

$$\begin{aligned} y - y_1 &= m(x - x_1) \\ y - 1 &= 3(x - 2) \\ y - 1 &= 3x - 6 \\ y &= 3x - 5 && \text{(Slope-intercept form)} \\ 3x - y - 5 &= 0 && \text{(Standard form)} \end{aligned}$$

Alternate Solution and Check: Selecting (x_1, y_1) as (3, 4) and (x_2, y_2) as (2, 1), the slope,

$$m = \frac{y_2 - y_1}{x^2 - x_1} = \frac{1 - 4}{2 - 3} = \frac{-3}{-1} = 3$$

Using the point-slope form,

$$\begin{aligned} y - y_1 &= m(x - x_1) \\ y - 4 &= 3(x - 3) \\ y - 4 &= 3x - 9 \\ y &= 3x - 5 && \text{(Slope-intercept form)} \\ 3x - y - 5 &= 0 && \text{(Standard form)} \end{aligned}$$

Note in the preceding example that when two points on the line are given, it makes no difference which is called the first and which is called the second in finding the equation of the line. Since two possibilities are available, one choice can serve as a check for the other choice.

EXERCISES 5.4 A

1. The line L has slope $-\frac{3}{4}$ and passes through the point (3, 4). Find an equation of the line.

2. Find an equation of the line that passes through the points $(-2, -3)$ and (2, 1).

For Exercises 3–8, find an equation of the form $Ax + By + C = 0$ for the line satisfying the stated conditions.

3. Parallel to $3x + 2y = 6$ and passing through the point (1, 2)

4. Slope $-\frac{1}{2}$, y-intercept 3

5. Slope $-\frac{1}{2}$, passing through (3, 0)

6. Slope $\frac{3}{4}$, passing through the origin

7. Parallel to the x-axis, passing through (1, 2)

8. Parallel to the y-axis, passing through (1, 2)

In Exercises 9–13, find the slope, if it exists, and write an equation of the line containing the two points.

9. (2, 1) and (4, 2)

10. (3, 0) and (0, 3)

11. (−2, −3) and (4, −1)

12. (2, 3) and (−5, 3)

13. (4, 6) and $(4, -\frac{1}{2})$

14. Using the graph, select two points, find the slope of the line, and write its equation.

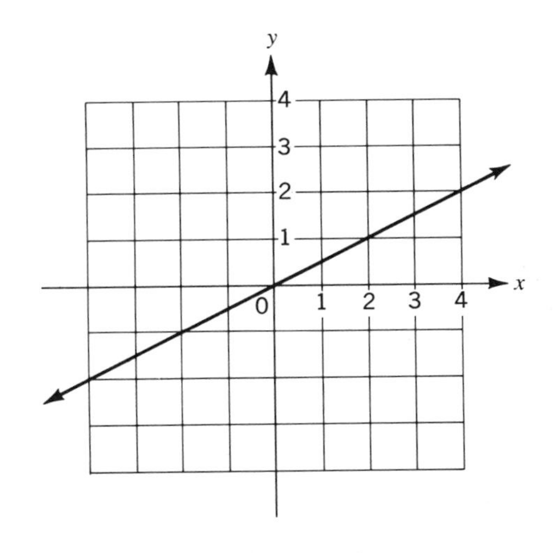

FIGURE 5.4.1

For Exercises 15–16, assuming that each given set of ordered pairs lies on a straight line, find an equation of the line and complete the table.

15.

x	-2	0	2	3	5
y	-1	3	7	?	?

16.

x	-2	0	3	4	6
y	-7	-3	3	?	?

Write an equation of a line that satisfies each set of conditions in Exercises 17–24.

17. All points on the line are 5 units from the y-axis.

18. All points on the line are 2 units from the x-axis.

★ **19.** The distance of each point on the line from the x-axis is twice its distance from the y-axis.

20. Passes through $(-2, -5)$ and has an undefined slope

21. Passes through $(0, 0)$ and has a slope of 0

22. Passes through the origin and has the same slope as the line $3x - 4y = 6$

23. Is parallel to $2x + y = 5$ and has a y-intercept of 6

24. Forms a right triangle with the axes in Quadrant 1 and whose legs are 6 and 8

25. Do the points $A:(0, 0)$, $B:(5, 2)$, $C:(-10, -4)$ lie on a straight line? Why?

26. Do the points $P:(1, 3)$, $Q:(-2, 9)$, $R:(3, 11)$ lie on a straight line? Why?

27. For what value of k will the points $(5, 0)$, $(-2, 4)$, $(k, -4)$ lie on a straight line?

28. A test on a spring produced the following results where F is the force in pounds required to stretch the spring to the length L in inches. F and L are related by a linear equation, but L, being an observed measurement, was subject to experimental error.

L (inches)	5.05	6.95	9.00	13.01	15.06
F (pounds)	0	1	2	4	5

a. Graph the ordered pairs, using the horizontal axis as the L-axis.

b. Draw a straight line that comes closest to passing through all the points. (This is called the "line of best fit.")

c. Write an equation for this line.
d. The slope of the line is called the spring constant, k. Find k.

29. The prices listed below were quoted in a newspaper for a certain model car.

t *Age of Car in Years— from 1973*	*Year of Manufacture*	P *Quoted Price*
8	1965	\$988
7	1966	\$1199
6	1967	\$1395
5	1968	\$1598
4	1969	\$1795
3	1970	\$2199
2	1971	\$2698
1	1972	\$3399

a. Plot the set of ordered pairs (t, P). (Round off P to the nearest hundred dollars, and graph t versus P in hundreds of dollars.)
b. Join the points consecutively by straight line segments.
c. For what years does the data appear to be linear (lie on the same straight line)?
d. Find an equation of the line referred to in (c).
e. Assuming that the data continues to be linear for cars manufactured before 1965, what should be the quoted price of a 1963 model?

EXERCISES 5.4 B

1. The line L has slope $-\frac{1}{2}$ and passes through the point $(2, -3)$. Find an equation of the line.

2. Find an equation of the line that passes through the points $(3, -2)$ and $(2, -3)$.

For Exercises 3–8, find an equation of the form $Ax + By + C = 0$ *for the line satisfying the stated conditions.*

3. Parallel to $2x - y + 3 = 0$ and passing through the point $(-2, -1)$

4. $m = \frac{4}{3}$, y-intercept -2 (where m is the slope of the line)

5. $m = -\frac{4}{3}$, passing through $(-2, 0)$

6. Passing through the origin and the point $(-3, 5)$
7. Passing through the point $(-3, 5)$ and parallel to the x-axis
8. Passing through the point $(-3, 5)$ and parallel to the y-axis

In Exercises 9–13, find the slope, if it exists, and write an equation of the line containing the two points.

9. $(3, \frac{1}{2})$ and $(-\frac{1}{4}, 2)$
10. $(0, 0)$ and $(5, -2)$
11. $(2, 3)$ and $(-2, 3)$
12. $(-3, -4)$ and $(-4, -3)$
13. $(-2, 1)$ and $(-2, -1)$

14. Using the graph, select two points, find the slope of the line, and write its equation.

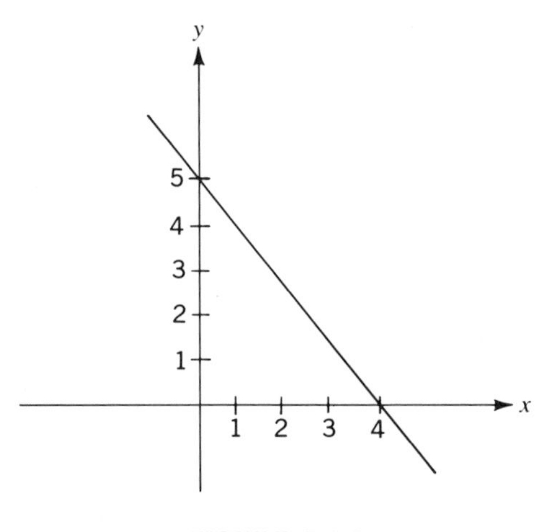

FIGURE 5.4.2

In Exercises 15–16, assuming that each given set of ordered pairs lies on a straight line, find an equation of the line and complete the table.

15.

x	-1	0	2	?	8
y	-1	1	5	9	?

16.

x	-3	1	2	4	?
y	10	2	0	?	8

Write an equation of a line that satisfies each set of conditions in Exercises 17–24.

17. All points on the line are 6 units from the x-axis.

18. All points on the line are 4 units from the y-axis.

★ **19.** The distance of each point on the line from the y-axis is 5 times its distance from the x-axis.

20. Passes through $(4, -6)$ and has a slope of 0

21. Passes through $(0, 0)$ and has an undefined slope

22. Passes through the origin and is parallel to $x + 3y = 6$

23. Is parallel to $x - 2y = 7$ and has a y-intercept of -3

24. Forms a right triangle with the axes in Quadrant 2 and whose legs are 5 and 12

25. Do the points $A:(1, -2)$, $B:(3, 0)$, $C:(0, 3)$ lie on a straight line? Why?

26. Do the points $P:(0, -1)$, $Q:(2, 0)$, $R:(4, 1)$ lie on a straight line? Why?

27. For what value of k will the points $(8, 2)$, $(5, 3)$, $(6, k)$ lie on a straight line?

★ **28.** The results of a study made by a physiologist for the purpose of trying to predict adult height from height as a child are shown in the table below, where C = height of two-year-old child in inches and A = adult height in inches.

C	30	31	33	34	34	35	37	38
A	60	61	65	67	66	70	73	74

a. Graph the ordered pairs, using the horizontal axis as the C-axis.
b. Draw a straight line that comes closest to passing through all the points plotted. (This is called the "line of best fit.")
c. Write the equation for this line.
d. Using the equation in (c), predict the adult height of a two-year-old child whose height is

1. 32 inches
2. 36 inches
3. 39 inches

29. The "Bromine number" test is a test used in chemistry to determine the number of double bonds of carbon in a compound. Measurements made at the end of 5 minutes, 10 minutes, and 15 minutes for the chemical reaction yielded the results shown in the table below.

t *Time in Minutes*	B "*Bromine Number*"
5	35
10	50
15	65

a. Graph the ordered pairs (t, B).
b. Join the ordered pairs with a straight line and extend this line so it intersects the vertical B-axis.
c. From the graph, find the value of B for $t = 0$. (This y-intercept is the desired number that indicates the number of double bonds of carbon in the compound.)
d. Check the result in (c) by writing the equation of the line in the graph and evaluating B for $t = 0$.

5.5 SYSTEMS OF LINEAR EQUATIONS: GRAPHICAL METHOD

In geometry it is learned that two straight lines on a plane either intersect in exactly one point or they are parallel. The algebraic counterpart of a set of two straight lines is a set, or system, of two linear equations in two variables. Whether or not the two linear equations have a common solution corresponds to whether their graphs have a common point of intersection or are parallel lines. "Intersection" is a key concept here, so first the intersection of sets will be reviewed.

DEFINITION

The intersection of two sets A and B is the set consisting of all elements that are in *both* A and B. In symbols,

$$A \cap B \quad \text{(read "}A\text{ intersection }B\text{")}$$

designates the intersection of A and B.

EXAMPLE 5.5.1 If $A = \{1, 2, 3, 4\}$ and $B = \{3, 4, 5\}$, list the elements in $A \cap B$.

Solution $A \cap B = \{3, 4\}$

EXAMPLE 5.5.2 List the elements in $A \cap B$ if $A = \{(2, 5), (3, 8), (4, 7)\}$ and $B = \{(3, 8), (-2, 5), (5, 2)\}$.

Solution $A \cap B = \{(3, 8)\}$

EXAMPLE 5.5.3 List the elements in $A \cap B$ if $A = \{(1, -1), (2, -2), (3, -3)\}$ and $B = \{(-1, 1), (-2, 2), (-3, 3)\}$.

Solution $A \cap B = \varnothing$, the empty set, and sets A and B are **disjoint.**

EXAMPLE 5.5.4 List the elements in $A \cap B$

if $$A = \{(x, y) \mid y = x\}$$

and $$B = \{(x, y) \mid y = x + 1\}$$

Solution If $y = x$ and $y = x + 1$, then by the transitive axiom of equality,

$$x = x + 1$$

and $$-x + x = -x + x + 1$$
$$0 = 1$$

Since this is impossible, there are no values of x such that $x = x + 1$.

Therefore, the sets do not have an element in common and $A \cap B = \varnothing$, the empty set.

EXAMPLE 5.5.5 List the elements in $A \cap B$

if $$A = \{(x, y \mid y = x + 1\}$$

and $$B = \{(x, y) \mid y = 9 - x\}$$

Solution If $y = x + 1$ and $y = 9 - x$, then

$$x + 1 = 9 - x$$
$$2x = 8 \text{ and } x = 4$$

For $y = x + 1$ and $x = 4$, $y = 5$ and $(4, 5)$ is a solution of $y = x + 1$.

For $y = 9 - x$ and $x = 4$, $y = 9 - 4 = 5$ and $(4, 5)$ is also a solution of $y = 9 - x$.

Therefore, $A \cap B = \{(4, 5)\}$.

It has been seen that the solution set of a linear equation in two variables is an infinite set of ordered pairs. It was also stated that the graph of a linear equation in two variables is a straight line.

When the graphs of two linear equations in two variables are drawn on the same set of axes, then there are three possibilities:

1. The lines intersect in exactly one point.
2. The lines are parallel and there is no point in common.
3. The lines coincide and have all their points in common.

These three possibilities are illustrated by the graphs in Figure 5.5.1.

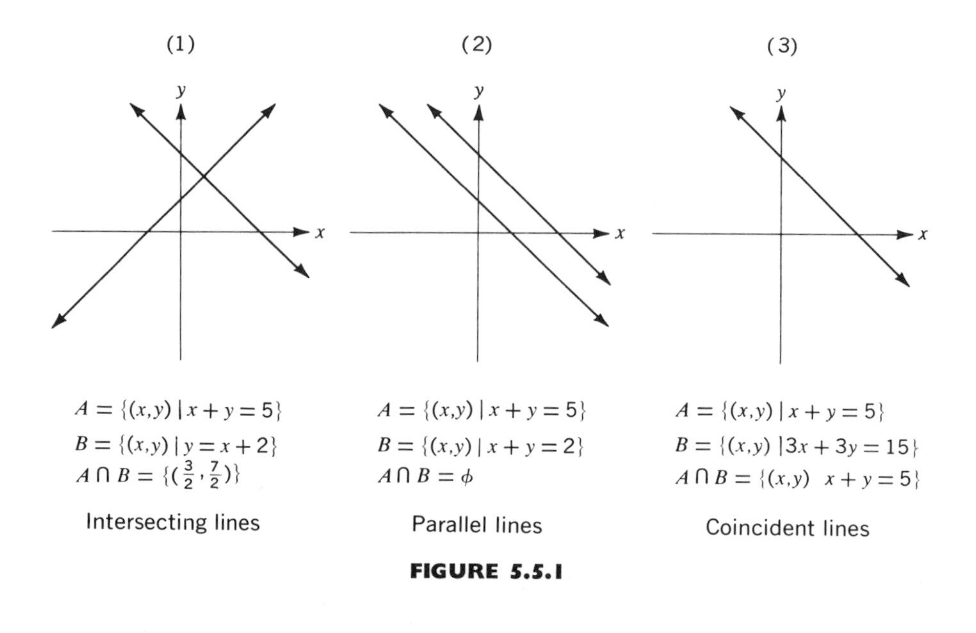

(1)	(2)	(3)
$A = \{(x,y) \mid x + y = 5\}$	$A = \{(x,y) \mid x + y = 5\}$	$A = \{(x,y) \mid x + y = 5\}$
$B = \{(x,y) \mid y = x + 2\}$	$B = \{(x,y) \mid x + y = 2\}$	$B = \{(x,y) \mid 3x + 3y = 15\}$
$A \cap B = \{(\frac{3}{2}, \frac{7}{2})\}$	$A \cap B = \phi$	$A \cap B = \{(x,y) \quad x + y = 5\}$
Intersecting lines	Parallel lines	Coincident lines

FIGURE 5.5.1

DEFINITION

A **system of equations** is a set of equations in two or more variables.

A system of equations is also referred to as a set of **simultaneous equations.**

DEFINITION

The **solution set of a system of equations** in two variables is the set of all ordered pairs that are common solutions to all the equations in the system.

Using the language of set theory, the solution set of a system of two equations in two variables is the intersection of the solution set of one of the equations with the solution set of the other. In symbols, if $A = \{(x, y) \mid ax + by = c\}$ and $B = \{(x, y) \mid dx + ey = f\}$, then $A \cap B$ is the solution of the system.

By examining the graphs in Figure 5.5.1, the conclusions shown in Table 5.1 can be made.

TABLE 5.1

The System	*Geometric Meaning*	*Algebraic Meaning*
1. $x + y = 5$ $y = x + 2$ $A \cap B$ $= \left\{\left(\frac{3}{2}, \frac{7}{2}\right)\right\}$	The lines have exactly one point in common. The lines intersect. The lines have different slopes.	The equations have exactly one solution in common.
2. $x + y = 5$ $x + y = 2$ $A \cap B = \varnothing$	The lines have no point in common. The lines are parallel. The lines have the same slope and different y-intercepts.	The equations have no common solution—that is, the intersection of their solution sets is the empty set, $\varnothing$.
3. $x + y = 5$ $3x + 3y = 15$ $A \cap B$ $= \{(x, y) \mid$ $x + y = 5\}$	The lines have all their points in common. The lines coincide. The lines have the same slope and the same y-intercept.	The equations have infinitely many solutions in common: the solution set of either equation.

It has been stated that two nonvertical lines are parallel if and only if their slopes are equal. It was also stated that the slope and y-intercept of a line completely determine the line. Thus two nonvertical lines coincide if and only if their slopes and their y-intercepts are equal. Two vertical lines are either parallel or coincide if their x-intercepts are equal. Vertical lines have the form $x = a$, so they can be identified immediately.

Thus it may be concluded that **two nonvertical lines intersect in exactly one point if and only if their slopes are unequal.**

EXAMPLE 5.5.6 Determine whether the lines represented by each of the following systems of equations intersect, are parallel, or are coincident:

a. $3x - 5y = 9$
   $3x + 2y = 6$

b. $2x - 2y = 7$
   $3x - 3y = 5$

c. $5x + y = 4$
   $x + 0.2y = 0.8$

Solution Solve each equation for y to find the slope of the line. If $y = mx + b$, then $m =$ the slope and $b =$ the y-intercept.

a. $3x - 5y = 9$

$$-5y = 9 - 3x$$

$$y = \frac{3}{5}x - \frac{9}{5}$$

Thus the slope is $\frac{3}{5}$.

$3x + 2y = 6$

$$2y = 6 - 3x$$

$$y = -\frac{3}{2}x + 3$$

Thus the slope is $-\frac{3}{2}$.

Since the slopes are not equal, the lines intersect in exactly one point.

b. $2x - 2y = 7$

$$-2y = -2x + 7$$

$$y = x - \frac{7}{2}$$

Thus the slope is 1, and the y-intercept is $-\frac{7}{2}$.

$3x - 3y = 5$

$$-3y = -3x + 5$$

$$y = x - \frac{5}{3}$$

Thus the slope is 1 and the y-intercept is $-\frac{5}{3}$.

Since the slopes are equal and the y-intercepts are not equal, the lines are parallel.

c. $5x + y = 4$

$$y = -5x + 4$$

Thus the slope is -5, and the y-intercept is 4.

$x + 0.2y = 0.8$

$$0.2y = -x + 0.8$$

$$y = -\frac{1}{0.2}x + 4$$

$$y = -5x + 4$$

Thus the slope is -5, and the y-intercept is 4.

Since the slopes are equal and the y-intercepts are equal, the lines coincide.

Two linear equations in two variables whose graphs intersect are solved by the **graphical method** by the following procedure:

1. Graph both equations on the same set of axes.
2. Determine the point of intersection. (It may be necessary to extend the lines to a considerable length to obtain the point of intersection.) The ordered pair that names the point of intersection is the solution set of the system.
3. Check the solution in both equations. It is necessary to check the solution because the point of intersection cannot always be read accurately from the graph.

EXAMPLE 5.5.7 Using the graphical method, solve the system

$$2x + 3y - 5 = 0$$
$$x - 2y + 8 = 0$$

Solution

$2x + 3y - 5 = 0$

x	y
1	1
4	−1
7	−3

$x - 2y + 8 = 0$

x	y
−8	0
−6	1
−4	2

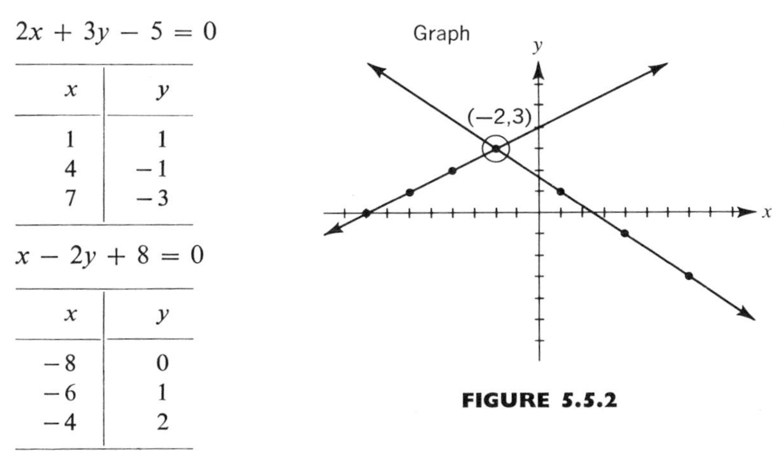

FIGURE 5.5.2

Solution set: $(-2, 3)$.

Check: $2x + 3y - 5 = 2(-2) + 3(3) - 5 = -4 + 9 - 5 = 0.$
$x - 2y + 8 = -2 - 2(3) + 8 = -2 - 6 + 8 = 0.$

EXERCISES 5.5 A

For each pair of sets in Exercises 1–10, find the intersection set, $A \cap B$.

1. $A = \{2, 4, 6, 8, 10\}$
$B = \{4, 8, 12, 16\}$

2. $A = \{-9, -4, -1, 0\}$
$B = \{0, 1, 4, 9\}$

3. $A = \{1, 3, 5, 7, 9, \ldots\}$
$B = \{2, 4, 6, 8, 10, \ldots\}$

4. $A = \{(2, 4), (3, 6), (4, 8)\}$
$B = \{(4, 2), (4, 8), (6, 3)\}$

5. $A = \{(0, 0), (1, -1), (2, -2), (3, -3), \ldots\}$
$B = \{(0, 0), (-1, 1), (-2, 2), (-3, 3), \ldots\}$

6. $A = \left\{x \mid \frac{36}{x} \text{ is a positive integer}\right\}$
$B = \left\{x \mid \frac{24}{x} \text{ is a positive integer}\right\}$

7. $A = \{(x, y) \mid x = 6\}$
$B = \{(x, y) \mid y = 4\}$

8. $A = \{(x, y) \mid y = 10 - x\}$
$B = \{(x, y) \mid x = 2\}$

9. $A = \{(x, y) \mid y = 10 - x\}$
$B = \{(x, y) \mid y = x - 4\}$

10. $A = \{(x, y) \mid y = x + 2\}$
$B = \{(x, y) \mid y = x - 2\}$

For each of the pairs of lines represented by the given equations in Exercises 11–15:

a. Find the slope of each line.
b. Find the y-intercept of each line.
c. State whether the lines intersect, are parallel, or coincide.
d. State whether each system of equations has exactly one solution, no solution, or a line of solutions.

11. $3x + 3y = 4$
$3x - 3y = 4$

12. $3x + 3y = 4$
$2x + 2y = 5$

13. $3x + 3y = 4$
$6x + 6y = 8$

14. $x + 3 = 0$
$y - 4 = 0$

15. $x + 3 = 0$
$3x + y = 0$

Using the graphical method, solve each of the systems in Exercises 16–20. Use 1 *unit* = 5 *squares.*

16. $x + y = 7$
$x - y = 1$

17. $4x - 3y = 0$
$2x - 3y = 6$

18. $2x + y = 7$
$x - 2y = 4$

19. $x + 3y = 5$
$2x + y = 5$

20. $x - 4y = 5$
$x + 2y = 10$

★ **21.** A car costing \$4800 will have a scrap value of \$800 at the end of 10 years. Using the straight-line method of depreciation, the book value y at the end of the xth year is given by the equation

$$y = 4800 - 400x$$

To offset the depreciation, a contribution is placed in a fund each year. The amount of money y in the fund at the end of the xth year is given by the equation

$$y = 400x$$

Solve this system of equations graphically, thereby finding the year when the amount in the depreciation fund is equal to the book value of the car.

★ **22.** Positions of airplanes and enemy positions can be determined by using the properties of sound and a mathematical curve called a hyperbola.

An explosion is heard at A, with coordinates (0, 0), $1\frac{1}{2}$ seconds after it is heard at B, with coordinates (10, 0), and $2\frac{1}{2}$ seconds after it is heard at C, with coordinates (0, 26). It can be shown that the place where the explosion occurred is (approximately) a point of intersection of two of the four hyperbolic asymptotes whose equations are

$$3y = 2(x - 5)$$
$$3y = -2(x - 5)$$
$$5x = 12(y - 13)$$
$$5x = -12(y - 13)$$

a. Graph these four equations on the same set of axes.
b. Find where the explosion occurred if 1 unit = 0.05 miles.

EXERCISES 5.5 B

For each pair of sets in Exercises 1–10, find the intersection set, $A \cap B$.

1. $A = \{-3, -1, 0, 1, 3\}$
$B = \{-2, -1, 0, 1, 2\}$

2. $A = \{1, 2, 3, 4\}$
$B = \{-1, -2, -3, -4\}$

3. $A = \left\{1, \frac{1}{2}, \frac{1}{3}, \frac{1}{4}, \frac{1}{5}, \ldots\right\}$
$B = \{1, 2, 3, 4, 5, \ldots\}$

4. $A = \{(-1, 2), (-2, 3), (-3, -4)\}$
$B = \{(-3, -4), (3, -2), (2, -1)\}$

5. $A = \{(x, x^2) \mid x \text{ is an integer}\}$
$B = \{(x, \sqrt{x}) \mid x \text{ is a nonnegative integer}\}$

6. $A = \{x \mid 10 - x \text{ is a natural number and } x \text{ is a natural number}\}$
$B = \{x \mid x + 5 \text{ is a natural number and } x \text{ is a natural number}\}$

7. $A = \{(x, y) \mid y = 2x\}$
$B = \{(x, y) \mid y = 3x\}$

8. $A = \{(t, v) \mid t = \sqrt{9}\}$
$B = \{(t, v) \mid v = 16t\}$

9. $A = \{(a, b) \mid b = a + 1\}$
$B = \{(a, b) \mid b = a - 1\}$

10. $A = \{(x, y) \mid x + y = y + x\}$
$B = \{(x, y) \mid x - y = y - x\}$

For each of the pairs of lines represented by the given equations in Exercises 11–15:

a. Find the slope of each line.
b. Find the y-intercept of each line.
c. State whether the lines intersect, are parallel, or coincide.
d. State whether each system of equations has exactly one solution, no solution, or a line of solutions.

11. $3x - 2y = 3$
$5x + 2y = 5$

12. $3x - 2y = 3$
$3x - 2 = 3y$

13. $2x - 1 = 0$
$3y + 4 = 0$

14. $3x - 6y = 3$
$2x - 4y = 2$

15. $x - 2y = 5$
$2y - x = 5$

Using the graphical method, solve each of the systems in Exercises 16–20. Use 1 *unit* = 5 *squares.*

16. $x + 2y = 4$
$2x - y = 3$

17. $x - y = 0$
$3x + 2y = 10$

18. $3x + y - 5 = 0$
$4x + y = 4$

19. $3x - 2y = 6$
$x - 2y = 10$

20. $3x + 2y = 7$
$2x + y = 6$

21. The heat h in calories needed to melt a grams of ice in b grams of water is related to the temperature t of the mixture by

$$h = (a + b)t + 80a$$

The heat h in calories given off by c grams of steam condensing is related to the temperature t by

$$h = 640c - ct$$

For 20 grams of steam introduced into a mixture of 15 grams of ice and 45 grams of water in equilibrium at 0 degrees centigrade, these equations become

$$h = 60t + 1200$$
$$h = 12{,}800 - 20t$$

By solving this system of equations graphically, find the final temperature of the resulting mixture—that is, the value of t for the point of intersection.

22. A dietitian wants to combine two foods so that the combination will supply 120 units of vitamins and 80 units of minerals. Food A supplies 2 units of vitamins and 2 units of minerals per ounce. Food B supplies 3 units of vitamins and 1 unit of minerals per ounce.
If x = the number of ounces of Food A needed and
y = the number of ounces of Food B needed
then $2x + 3y = 120$
and $2x + y = 80$
Solve this system graphically and state how many ounces of each food should be used.

5.6 SYSTEMS OF LINEAR EQUATIONS: ADDITION METHOD

While the graphical method affords a valuable visual representation of the solution of a system of equations, it has the disadvantage that the solution cannot always be obtained accurately. Thus it is desirable to investigate other methods of solution.

It should be obvious that the solution of the system

$$x = 2$$
$$y = 3$$

can be recognized immediately. The solution is (2, 3), since this is the only ordered pair that simultaneously satisfies both equations. This special case suggests the possibility of transforming a system whose solution is not obvious into the simpler equivalent form

$$x = p$$
$$y = q$$

DEFINITION

Two systems of equations are equivalent if and only if the systems have the same solution set.

Consider the system

$$7x + 3y - 13 = 0$$
$$2x + 3y + 2 = 0$$

If it is assumed that the ordered pair (p, q) is a solution of both equations, then $7p + 3q - 13 = 0$ and $2p + 3q + 2 = 0$. Moreover, by selecting *a* and b to be two real numbers that are not both zero, then it follows that (p, q) is also a solution of

$$a(7x + 3y - 13) + b(2x + 3y + 2) = 0$$

since

$$a(7p + 3q - 13) + b(2p + 3q + 2) = a(0) + b(0) = 0 + 0 = 0$$

The objective is to obtain an equivalent system having the form

$$\begin{aligned} x &= p \\ y &= q \end{aligned}$$

Either equation of this system can be obtained by choosing a and b in such a manner that the coefficients of one of the variables become additive inverses.

Choosing $a = 1$ and $b = -1$ and writing the system as

$$\begin{aligned} 7x + 3y &= 13 \\ 2x + 3y &= -2 \end{aligned}$$

then

$$\begin{array}{rcl} 1(7x + 3y) = 1(13) & \rightarrow & 7x + 3y = 13 \\ -1(2x + 3y) = -1(-2) & & -2x - 3y = 2 \\ \hline \text{Adding,} & & 5x \quad = 15 \\ & & x = 5 \end{array}$$

If $a = 2$ and $b = -7$, then the coefficients of x become additive inverses, and

$$\begin{array}{rcl} 2(7x + 3y) = 2(13) & \rightarrow & 14x + 6y = 26 \\ -7(2x + 3y) = -7(-2) & & -14x - 21y = 14 \\ \hline \text{Adding,} & & -15y = 40 \\ & & y = \dfrac{-8}{3} \end{array}$$

Thus the system

$$\begin{aligned} 7x + 3y - 13 &= 0 \\ 2x + 3y + 2 &= 0 \end{aligned} \quad \text{is equivalent to} \quad \begin{aligned} x &= 3 \\ y &= \frac{-8}{3} \end{aligned}$$

and the solution is $\left(3, \frac{-8}{3}\right)$.

Check:

For $7x + 3y - 13 = 0$, $7(3) + 3\left(\frac{-8}{3}\right) - 13 = 21 - 8 - 13 = 0$.

For $2x + 3y + 2 = 0$, $2(3) + 3\left(\frac{-8}{3}\right) + 2 = 6 - 8 + 2 = 0$.

EXAMPLE 5.6.1 Using the addition method, solve the system

$$2x - 3y = 18$$
$$7x - 2y = -5$$

Solution

$$\begin{array}{rl} 2(2x - 3y = 18) & \\ -3(7x - 2y = -5) & \end{array} \rightarrow \begin{array}{r} 4x - 6y = 36 \\ -21x + 6y = 15 \\ \hline -17x \quad = 51 \\ x = -3 \end{array}$$

$$\begin{array}{rl} 7(2x - 3y = 18) & \\ -2(7x - 2y = -5) & \end{array} \rightarrow \begin{array}{r} 14x - 21y = 126 \\ -14x + 4y = 10 \\ \hline -17y = 136 \\ y = -8 \end{array}$$

The solution is $(-3, -8)$.

Check: $2x - 3y = 2(-3) - 3(-8) = -6 + 24 = 18$ and $18 = 18$.
$7x - 2y = 7(-3) - 2(-8) = -21 + 16 = -5$ and $-5 = -5$.

EXAMPLE 5.6.2 Using the addition method, solve the system

$$4x - y = 3$$
$$5x + 2y = 20$$

Solution

$$\begin{array}{rl} 2(4x - y = 3) & \\ 1(5x + 2y = 20) & \end{array} \rightarrow \begin{array}{r} 8x - 2y = 6 \\ 5x + 2y = 20 \\ \hline 13x \quad = 26 \\ x = 2 \end{array}$$

Although this addition process can be used to find the value for y, it is also possible to replace x by 2 in either of the two equations and then solve this equation for y. Selecting the first,

$$\text{for } x = 2 \text{ and } 4x - y = 3$$
$$4(2) - y = 3$$
$$8 - y = 3$$
$$-y = 3 - 8 = -5$$
$$y = 5$$

The solution is $(2, 5)$.

Check: For $x = 2$ and $y = 5$,

$$4x - y = 4(2) - 5 = 8 - 5 = 3$$
$$5x + 2y = 5(2) + 2(5) = 10 + 10 = 20$$

EXERCISES 5.6 A

Using the addition method, solve each of the systems in Exercises 1–15 and check each solution set.

1. $x + y = 3$
$x - y = 7$

2. $2x - y = 3$
$3x + 2y = 8$

3. $x - 2y = 11$
$x - 3y = 18$

4. $3a + 8b = 14$
$a + 7b = 22$

5. $4c - 3d = 11$
$6c - 3d = 12$

6. $4x - 5y = 48$
$5x - 4y = 51$

7. $2p + 3q = 4$
$5p + 6q = 7$

8. $6r - 2t = 9$
$9r - 8t = 1$

9. $7x - 5y = 36$
$12x = 15y + 108$

10. $9a = 5b + 30$
$3b = 7a + 10$

11. $3r + s = 15$
$r - 2s = 12$

12. $x = 5y - 2$
$y = 3x - 1$

13. $8x + 20y = 6$
$4x + 10y - 3 = 0$

14. $y = 2x - 1$
$3y - 6x = 5$

15. $p + 3q = 4$
$2p - 4q = 10$

EXERCISES 5.6 B

Using the addition method, solve each of the systems in Exercises 1–15 and check each solution set.

1. $x - y = -2$
$x + y = -6$

2. $5x + y = 2$
$4x - 3y = 13$

3. $7p + 4q = 1$
$7p - 5q = -80$

4. $5a - 3b = 15$
$a - 5b = -63$

5. $8c + d = -3$
$12c - 5d = 28$

6. $3x = 17 + 4y$
$4x = 106 - 3y$

7. $5x = 8y - 2$
$10x = 7y - 3$

8. $4p - 5q = 3$
$8p + 10q = 78$

9. $8 = 2a - 5b$
$7 = 5a + 7b$

10. $6x = 51 - 7y$
$5y = 34 - 4x$

11. $4r - 2s = 10$
$3s + 5r = 4$

12. $m - n = 14$
$n = 4m + 1$

13. $x = 2y$
$3x - y + 1 = 0$

14. $y - 2x = 6$
$5y = 10x + 1$

15. $2a + 3b - 4 = 0$
$3a + 2b - 4 = 0$

5.7 SYSTEMS OF LINEAR EQUATIONS: SUBSTITUTION METHOD

Another method for solving a system of two linear equations in two variables is the substitution method. This method is sometimes more convenient to use than the addition method. Also, the same principle of substitution is used to solve systems of equations that are not linear. The procedure is as follows:

1. Solve one of the equations for one variable as a function of the other.
2. Substitute the expression that is obtained into the other equation.
3. Solve the resulting linear equation in one variable.
4. Substitute the solution obtained in step 3 into the equation obtained in step 1 and simplify.
5. Check the solution obtained in each of the original equations.

EXAMPLE 5.7.1 Using the substitution method, solve the system

$$\begin{aligned} 2x + y &= 8 \\ 8x - 5y &= 5 \end{aligned}$$

Solution

1. Solve $2x + y = 8$ for y:

$$y = 8 - 2x$$

2. Substitute $8 - 2x$ for y in $8x - 5y = 5$:

$$8x - 5(8 - 2x) = 5$$

3. Solve this equation for x:

$$\begin{aligned} 8x - 40 + 10x &= 5 \\ 18x - 40 &= 5 \\ 18x &= 45 \\ x &= \frac{45}{18} = \frac{5}{2} \end{aligned}$$

4. Substitute $\frac{5}{2}$ for x in $y = 8 - 2x$:

$$y = 8 - 2\left(\frac{5}{2}\right) = 8 - 5 = 3$$

5. Check the solution $\left(\frac{5}{2}, 3\right)$ in both equations:

$$2x + y = 2\left(\frac{5}{2}\right) + 3 = 8$$

$$\begin{aligned} 8x - 5y &= 8\left(\frac{5}{2}\right) - 5(3) \\ &= 20 - 15 = 5 \end{aligned}$$

The substitution axiom of the equal relation and the operational theorems of equivalence previously used in the solution of a linear equation in one variable guarantee that the system

$$x = \frac{5}{2}$$
$$y = 3$$

is equivalent to

$$2x + y = 8$$
$$8x - 5y = 5$$

In set notation, if

$$A = \{(x, y) \mid 2x + y = 8\}$$
$$B = \{(x, y) \mid 8x - 5y = 5\}$$
$$C = \left\{(x, y) \mid x = \frac{5}{2}\right\}$$
$$D = \{(x, y) \mid y = 3\}$$

then

$$A \cap B = C \cap D = \left\{\left(\frac{5}{2}, 3\right)\right\}$$

Sometimes it is more convenient to solve for one variable rather than the other. As a general rule, if the coefficient of a variable is 1 in one of the equations, then this variable and this equation are selected for the first step of the solution process.

EXAMPLE 5.7.2 Using the substitution method, solve the system

$$4x - 3y = 26$$
$$x - 5y = 15$$

Solution In the second equation, the coefficient of x is 1 and this equation is first solved for x:

1. $x - 5y = 15$ and $x = 5y + 15$
2. $4x - 3y = 26$

$$4(5y + 15) - 3y = 26$$
$$20y + 60 - 3y = 26$$
$$17y = -34$$
$$y = -2$$

3. $x = 5y + 15$

$$x = 5(-2) + 15 = 5$$

The solution is $(5, -2)$.

Check: For $4x - 3y = 26$, $4(5) - 3(-2) = 20 + 6 = 26$.
For $x - 5y = 15$, $5 - 5(-2) = 5 + 10 = 15$.

EXERCISES 5.7 A

Using the substitution method, solve each of the systems in Exercises 1–15 and check each solution set.

1. $2x + 3y = 16$
   $y = 2x$
2. $y = 3x - 4$
   $5x + 2y = 25$
3. $4a - 7b = 9$
   $b = 12 + 5a$
4. $2a - 3b = 1$
   $4a + b = 23$
5. $3x - 5y = 49$
   $4x - y = 3$
6. $5x - 4y = 22$
   $x - 2y = 5$
7. $2r - 3s = 17$
   $r + 4s = 3$
8. $3p + 4q = 10$
   $p + 6q = 1$
9. $3x - 3y = 1$
   $y = x + 5$
10. $3x - 9y = 3$
    $3y = x - 1$
11. $x = 3y - 1$
    $y = 2x + 1$
12. $5 = 3a + 2b$
    $a - b = 3$
13. $x + y - 2 = 0$
    $2x - y = 4$
14. $5x - 2y = 1$
    $5x = y + 7$
15. $2a + 6y = 10$
    $3y = 2 - 4a$

EXERCISES 5.7 B

Using the substitution method, solve each of the systems in Exercises 1–15 and check each solution set.

1. $3x + 2y = 36$
   $y = 3x$
2. $y = 2x - 3$
   $3y - 5x = 5$
3. $8a - 3b = 93$
   $b = 37 - 3a$
4. $8p - 5q = 1$
   $2p + q = 7$
5. $x + 2y = 4$
   $3x - y = 5$
6. $3x - 5y = 6$
   $x + 7y = -24$
7. $3r - 2s = 5$
   $r - 3s = 18$
8. $5r + 5t = 3$
   $r = t - 4$

9. $3x + y = 2$
$6x = 4 - 2y$

10. $2x = -2 - 3y$
$7y = x - 16$

11. $m = 2n - 3$
$n = m + 5$

12. $4 = 2x - 7y$
$x - y = 5$

13. $r + 2s - 5 = 0$
$2s + 3r = 1$

14. $3x - 4y = 2$
$3x = y - 5$

15. $9m + 4n = 6$
$3m = 5 - 2n$

5.8 VERBAL PROBLEMS

Verbal problems that involve two unknown numbers often can be solved more conveniently by using two variables to form two equations.

EXAMPLE 5.8.1 Find two numbers whose sum is 135 and whose difference is 61.

Solution

Let x = one number
y = the other number

Then

$$\begin{array}{r} x + y = 135 \\ x - y = 61 \\ \hline 2x \qquad = 196 \\ x = 98 \end{array}$$

(Solving by using the addition method)

Now if $x = 98$ and $x + y = 135$, then $98 + y = 135$ (by using the substitution axiom).

Thus $y = 135 - 98 = 37$, and the two numbers are 98 and 37.

Check:

$$x + y = 98 + 37 = 135$$
$$x - y = 98 - 37 = 61$$

EXAMPLE 5.8.2 A grocer has an imported tea that sells for \$1.50 a pound and a domestic tea that sells for \$1.20 a pound. How many pounds of each should he use to obtain a 200-pound mixture that will sell for \$1.32 a pound?

Solution

Let x = the number of pounds of imported tea
y = the number of pounds of domestic tea

	Price per Pound	*Number of Pounds*	*Value*
Imported	1.50	x	$1.50x$
Domestic	1.20	y	$1.20y$
Mixture	1.32	200	$1.32(200) = 264$

Equations:

$$x + y = 200$$
$$1.5x + 1.2y = 264$$

Using the substitution method,

$$y = 200 - x$$
$$1.5x + 1.2(200 - x) = 264$$
$$1.5x + 240 - 1.2x = 264$$
$$0.3x = 24$$
$$x = 80$$
$$y = 200 - 80 = 120$$

Thus 80 pounds of imported tea and 120 pounds of domestic tea should be used.

EXAMPLE 5.8.3 A jet plane traveling with the wind flies 2325 miles in 3 hours. Against the wind, it takes the jet 4 hours to go 2900 miles. Find the speed in still air of the plane and the speed of the wind.

Solution

Let x = the speed in still air of the plane
y = the speed of the wind

Formula	r ·	t =	d
With the wind	$x + y$	3	2325
Against the wind	$x - y$	4	2900

Equations:

$$3(x + y) = 2325 \rightarrow x + y = 775$$
$$4(x - y) = 2900 \rightarrow x - y = 725$$
$$2x = 1500$$
$$x = 750$$
$$750 + y = 775$$
$$y = 25$$

Thus the speed in still air of the jet plane is 750 mph, and the speed of the wind is 25 mph.

EXERCISES 5.8 A

Solve Exercises 1–10 by using two variables.

1. The sum of two numbers is 20 and their difference is 3. Find the numbers.
2. Find the number of quarters and dimes in a collection consisting of 22 coins worth a total of $3.55.
3. How many pounds of candy worth $1.20 a pound and candy worth 90 cents a pound must be mixed to obtain 150 pounds worth $1.00 a pound?
4. A man can walk uphill at 2 mph and downhill at 4 mph on a road no part of which is level. If he walked 24 miles in 9 hours, how much of the road was uphill?
5. Two sums of money are invested at 5 and $6\frac{1}{2}$ percent. The total investment is $10,000, and the total income is $552.50. How much is invested at each rate?
6. A plane, having a tail wind, flew 600 miles in 3 hours. The return trip, with no wind change, was made in 4 hours. Find the speed of the wind and the rate of the plane in still air.
7. Three years from now a father will be twice as old as his son. Seven years ago he was 3 times as old as his son. Find the present age of each.
8. A crew rows 30 miles downstream in 4 hours. Returning, they cover the same distance upstream in 6 hours and 40 minutes. Find the rate of the current and the rate of the crew in still water.
9. Three loaves of bread and 2 pounds of butter cost $2.80. Five loaves of bread and 3 pounds of butter cost $4.41. Find the cost of 1 loaf of bread and the price of 1 pound of butter if these were the same for both purchases.
10. ★ A mail order company charges a fixed fee for shipping and handling the first 10 pounds of merchandise and then an additional fee for each pound over 10 pounds. If the shipping and handling charge for 30 pounds of merchandise was $2.49 and if the charge for 42 pounds was $3.45, find the fixed fee and the additional fee.

EXERCISES 5.8 B

Solve Exercises 1–10 by using two variables.

1. Find two numbers whose sum is 4 and whose difference is 11.
2. The admission prices to a certain college dance were 75 cents for each person with a student association card and $2.00 for each person without

a student association card. A total of 350 persons attended the dance and a total of $450 was collected. How many of the persons who attended had student association cards?

3. How much butter worth 75 cents a pound should be mixed with an oleo worth 30 cents a pound to produce 500 pounds of a spread worth 40 cents a pound?
4. During part of its trip, a jet plane traveling 600 mph ran into a storm that reduced its speed by 80 mph. If the total trip of 2380 miles was made in 4 hours, how far did the plane travel during the storm?
5. A man received a total income of $480 from $5000 invested in bonds and $4000 invested in stocks. The same year his brother received a total income of $260 from $3000 invested in the same bonds and $2000 invested in the same stocks. Find the percentage they received from each investment.
6. A boat travels 120 miles up a river in 6 hours. Its return trip downstream takes 4 hours. Find the speed of the current and the speed of the boat in still water.
7. Twenty pounds of ore A combined with 30 pounds of ore B produce 26.5 pounds of silver. Ten pounds of ore A combined with 20 pounds of ore B produce 15.5 pounds of silver. Find the percentage of silver in each ore.
8. A plane with a head wind flies 300 miles from city A to city B in 2 hours. On the same day and under the same weather conditions, a plane with the same speed in still air flies from city B to city A in 1 hour and 30 minutes. Find the speed of the wind.
9. A manufacturer must use each of two different machines for the production of two different products. The number of hours required on each machine to produce the two products is given in the table, along with the maximum number of hours each machine can be run per week.

Machine	*Product*		*Maximum Hours for Machine*
	A	*B*	
I	3	2	48
II	1	4	46

How many items of product A and how many items of product B should be manufactured each week if the machines are used at their maximum capacity?

10. Five years ago, City College was 4 times as old as State College. Seventeen years from now, City College will be only twice as old as State College. Find the present ages of each. When will City College celebrate its centennial (100 years old)?

SUMMARY

- ☐ An **ordered pair** is an expression of the form (a, b) where a is called the **first component**, and b is called the **second component**.
- ☐ A **solution of an open equation in two variables**, x and y, is an ordered pair (a, b) such that the equation becomes true when x is replaced by a and y is replaced by b.
- ☐ Two perpendicular number lines in a plane and intersecting at their origins form a **rectangular** (or **Cartesian**) **coordinate system**.
- ☐ Every point in the plane is named by an ordered pair, called the **coordinates** of the point.
- ☐ If a point P has coordinates (x, y), then x, the first coordinate, is called the **abscissa** of P, and y, the second coordinate, is called the **ordinate** of P.
- ☐ The number lines are called the **vertical axis** and the **horizontal axis,** and the intersection of these axes is called the **origin**.
- ☐ An equation of the form $Ax + By + C = 0$ where A and B are not both zero is called a **linear equation in two variables**.
- ☐ The **graph of a linear equation in two variables** is a **straight line**, the graph of its solution set.
- ☐ The **slope** m of the line segment joining $P_1{:}(x_1, y_1)$ and $P_2{:}(x_2, y_2)$ where $x_1 \neq x_2$ is given by

$$m = \frac{y_2 - y_1}{x_2 - x_1}$$

- ☐ The slope of a horizontal line is 0.
- ☐ The slope of a vertical line is undefined.
- ☐ **Standard form of a line:**

$$Ax + By + C = 0$$

- ☐ **Slope-intercept form of a line:**

$$y = mx + b$$

- ☐ **Point-slope form of a line:**

$$y - y_1 = m(x - x_1)$$

Slope and Parallel Lines

- ☐ If two lines have the same slope, they are parallel.

☐ If two lines are parallel and nonvertical, they have the same slope.

☐ A **system of equations** (simultaneous equations) is a set of equations in two or more variables.

☐ The **solution set of a system of equations in two variables** is the set of ordered pairs that are common solutions to all the equations in the system.

☐ The coordinates (a, b) of the point of intersection of two intersecting lines is the solution of the system of two linear equations in two variables, having these lines for their graphs.

☐ A system of two linear equations in two variables may be solved algebraically by the **addition method** or the **substitution method**.

REVIEW EXERCISES

1. Define the following terms:
   a. Coordinate axes
   b. Ordinate
   c. Abscissa
   d. Origin
   e. Quadrants
   f. Slope of a line

2. Which of the following points lie on the graph of $x + 2y - 5 = 0$?
   a. $(1, 4)$ b. $(7, -1)$ c. $(0, 0)$ d. $(-4, -5)$ e. $(5, 0)$

3. Graph each of the following and find the slope and y-intercept from the graph:
   a. $3x + 2y = 12$ b. $\frac{x}{2} + \frac{y}{5} = 1$
   c. $\frac{1}{2}(x - 4) = \frac{1}{4}(y + 2)$ d. $x - 3y - 2 = 0$

4. Check the answers for Exercise 3 by changing each equation into an equivalent equation of the form $y = mx + b$.

5. Find an equation of the line through the given points:
   a. $(-3, 0)$ and $(6, 2)$ b. $(-3, 6)$ and $(3, -8)$
   c. $(a, 2a)$ and $(3a, 5a)$ d. $(p, 5)$ and $(-1, 5)$

6. Find an equation of the line through the given point and with slope m:
   a. $(0, 3), m = \frac{1}{2}$ b. $(1, 5), m = -3$
   c. $(-3, -1), m = 0$ d. $(-3, -1), m = -\frac{1}{3}$

7. Find an equation of the line which is parallel to the line whose equation is $2x - 3y = 5$ and which passes through the point $(-1, 4)$.

8. Which of the following lines are parallel to $2x - 5y = 10$?

a. $5x - 2y = 10$ b. $5y = 2x - 1$
c. $4x - 10y = 5$ d. $5y = 1 - 2x$

9. List the elements in each of the following sets:

a. $\{(x, y) \mid y = 2x - 1\} \cap \{(x, y) \mid y = 1 - 2x\}$
b. $\{(x, y) \mid x + 2y = 1\} \cap \{(x, y) \mid x - y + 5 = 0\}$
c. $\{(x, y) \mid x = 5\} \cap \{(x, y) \mid y = 2\}$
d. $\{(x, y) \mid x + 2y = 1\} \cap \{(x, y) \mid 2x + 4y = 1\}$

10. Determine an equation of the line satisfying the stated conditions:

a. Passing through $(3, -2)$ and $(4, 1)$
b. Parallel to the x-axis, with y-intercept 2
c. Passing through the origin with slope $-\frac{3}{5}$
d. Passing through $(5, -7)$ with slope undefined

11. Plot the following points on the same set of axes. Use 20 squares = 1 unit for the y scale.

TABLE OF RECIPROCALS

(correct to nearest hundredth)

Number, x	0.5	1	2	4	5	6
Reciprocal, y	2.00	1.00	0.50	0.25	0.20	0.17

a. Join the points by a smooth curve.
b. From the graph, estimate $\frac{1}{3.5}$ to the nearest hundredth (interpolate).
c. Extending the graph, estimate $\frac{1}{7}$ to the nearest hundredth (extrapolate).

12. The following table illustrates the distances required to stop an automobile traveling at different speeds.

Speed, mph	x	15	20	25	30	35
Distance, feet	y	21	37	58	83	114

a. Plot these points on the same set of axes.
b. Join these points by a smooth curve.
c. Using the graph, determine the speed of a car that requires 70 feet to stop.
d. Extending the graph, determine the speed of a car that requires 130 feet to stop.

13. Solve by the addition method and check by the graphical method:

a. $3x + 5y = 3$
$2x - y = -11$

b. $2x - y = 10$
$4x + 3y = 5$

14. Solve by the substitution method and check by the graphical method:

a. $2x - 3y = 1$
$x - 2y = 2$

b. $3x + y = 9$
$2x - 3y = 6$

15. Match column B with column A

Column A	*Column B*
a. $x + y = 50$, $y + 5x = 90$	1. Equations with no solution
b. $7x - 5y = 0$, $8x + 3y = 0$	2. Equations whose graphs are coincident lines
c. $2x + 6y = 5$, $3x + 9y = 1$	3. A system of equations that can be used to find the number of nickels and cents in a collection of 50 coins worth 90 cents
d. $y = 2 - x$, $x = 2 - y$	4. A system equivalent to $x = 15$, $y = 75$
e. $x + y = 90$, $y = 5x$	5. Equations whose linear graphs intersect at the origin

16. One week a plumber worked 20 hours and his helper worked 12 hours on a certain job. They sent the contractor of the job a bill of $228 for their combined wages. The next week the plumber worked 15 hours and his helper worked 10 hours. This time the bill for their combined wages was $175. Find the hourly wage of each.

17. Cruising at half-speed against a wind, a traffic helicopter flies 36 miles in 40 minutes. Flying with the same wind at full speed, the helicopter flies 93 miles in 30 minutes. Find the speed of the wind.

18. A certain rush order was received in an office on the first day. The regular typist worked 10 hours, an extra typist worked 5 hours, and together they completed two thirds of the order. The next day, both typists worked 4 hours each, and they finished the order. How long would it have taken the regular typist alone to do the work?

19. In competitive business, the price y for each unit of a commodity depends on the number of units x demanded by the consumers. The equation relating the price and the number of units demanded is called the demand law. A certain demand law is represented by

$$5x + 8y = 60$$

a. Graph this equation for $x \geq 0$ and $y \geq 0$.
b. Find the highest price that will be paid for this commodity—that is, the value of the y-intercept (where $x = 0$).
c. Find the greatest amount that will be demanded—that is, the value of the x-intercept (where $y = 0$).

20. Referring to Exercise 19, the price y for each unit of a commodity also depends on the number of units x that can be supplied. The equation relating the price and the number of units x that can be supplied is called the supply law. A supply law corresponding to the demand law in Exercise 19 is

$$8y = 6x + 16$$

a. Graph this equation on the same set of axes as used for Exercise 19. Again use $x \geq 0$ and $y \geq 0$.

b. Find the lowest price at which the supplier will sell the commodity—that is, the value of the y-intercept.

c. Find the point of intersection of the demand curve and the supply curve. (This point corresponds to "Market equilibrium"; the value of x and the value of y are called the equilibrium quantity and price, respectively. In this case, the quantity demanded equals the quantity supplied.)

CHAPTER 6

EXPONENTS AND RADICALS

Many problems in mathematics, physics, engineering, chemistry, biology, economics, and other areas involve equations. Solving equations is a major concern of algebra.

In Chapter 3 it was seen that a polynomial equation can readily be solved if the linear factors of the polynomial can be found easily. However, not all polynomials with integral coefficients can be factored over the set of integers; $x^2 - 2$ and $x^2 + x - 1$ are examples. Finding the factors in these cases is more difficult. It is desirable, then, to develop other techniques for solving equations.

Earlier it was noted that if $x^2 = 2$ and if x is positive, then $x = \sqrt{2}$. Similarly, if $x^3 = 2$ and x is a real number, then $x = \sqrt[3]{2}$. In general, if x and p are positive real numbers and if $x^n = p$, then $x = \sqrt[n]{p}$, where the natural number n used in x^n is called an exponent and where $\sqrt[n]{p}$ is called a radical.

Since a polynomial involves terms of the form cx^n, it is reasonable to conclude that the solutions of polynomial equations will involve exponents and radicals. This, indeed, is the case. Thus, to develop methods for solving equations, it is first necessary to study exponents and radicals.

This chapter is concerned with exponents and radicals, their meaning, and how to use them in algebraic calculations.

6.1 POSITIVE INTEGRAL EXPONENTS

The definition of x^n, introduced earlier, is restated below.

DEFINITION OF x^n

If x is a real number and if n is a natural number, then

$$x^n = \underbrace{x \cdot x \cdots x}_{n \text{ factors}} \text{ and } x^1 = x$$

The natural number n is called the **exponent**, the real number x is called the **base**, and the number x^n is called the **nth power** of x.

How to simplify and calculate with expressions involving nth powers is explained by the exponent theorems. These five theorems are introduced, one at a time, and then summarized at the end of this section. In the discussion that follows, x and y are real numbers and m and n are positive integers (that is, natural numbers).

THEOREM 1

$$x^m x^n = x^{m+n}$$

The product of two powers having the same base is a power having the same base and an exponent equal to the sum of the exponents.

Proof:

$$x^m x^n = \underbrace{(x \cdot x \cdots x)}_{m \text{ factors}} \underbrace{(x \cdot x \cdots x)}_{n \text{ factors}} = \underbrace{(x \cdot x \cdots x)}_{m + n \text{ factors}} = x^{m+n}$$

EXAMPLE 6.1.1

$$x^4 x^3 = (xxxx)(xxx) = x^7 = x^{4+3}$$
$$2^5 \cdot 2^2 = 2^{5+2} = 2^7 = 128$$
$$yy^7 = y^1 y^7 = y^{1+7} = y^8$$

THEOREM 2

$$(x^m)^n = x^{mn}$$

A power of a power is a power having the same base and an exponent equal to the product of the exponents.

Proof:

$$(x^m)^n = \underbrace{x^m \cdot x^m \cdots x^m}_{n \text{ factors}}$$

$$= \underbrace{(x \cdot x \cdots x)}_{m \text{ factors}}\underbrace{(x \cdot x \cdots x)}_{m \text{ factors}} \cdots \underbrace{(x \cdot x \cdots x)}_{m \text{ factors}}$$

$$= \underbrace{x \cdot x \cdots x}_{mn \text{ factors}} = x^{mn}$$

EXAMPLE 6.1.2

$$(x^2)^3 = (x^2)(x^2)(x^2)$$

$$= (x \cdot x)(x \cdot x)(x \cdot x)$$

$$= x^6 = x^{2 \cdot 3}$$

$$(x^3)^4 = x^{3 \cdot 4} = x^{12}$$

$$(2^3)^2 = 2^{3 \cdot 2} = 2^6$$

THEOREM 3

1. $\dfrac{x^m}{x^n} = x^{m-n}$ **if** $m > n$ **and** $x \neq 0$

2. $\dfrac{x^m}{x^n} = \dfrac{1}{x^{n-m}}$ **if** $n > m$ **and** $x \neq 0$

3. $\dfrac{x^m}{x^n} = \dfrac{x^n}{x^n} = 1$ **if** $m = n$

The quotient of two powers having the same base is:

1. A power having the same base and an exponent equal to the difference of the exponents (larger minus smaller), or

2. A reciprocal of a power having the same base and an exponent equal to the difference of the exponents (larger minus smaller), or

3. The number 1 if the exponents are equal.

EXAMPLE 6.1.3

$$\frac{x^5}{x^2} = x^{5-2} = x^3$$

$$\left(\frac{xxxxx}{xx} = xxx = x^3\right)$$

$$\frac{2^7}{2^3} = 2^{7-3} = 2^4 = 16$$

$$\frac{2^3}{2^7} = \frac{1}{2^{7-3}} = \frac{1}{2^4} = \frac{1}{16}$$

$$\frac{x^2}{x^5} = \frac{1}{x^{5-2}} = \frac{1}{x^3}$$

$$\frac{x^5}{x^5} = 1$$

THEOREM 4

$$(xy)^n = x^n y^n$$

A power of a product is a product of two powers whose bases are factors of the product and each of whose exponents is the same as the exponent on the product.

EXAMPLE 6.1.4

$$(xy)^3 = x^3y^3$$

$$(xy)^3 = (xy)(xy)(xy) = (xxx)(yyy) = x^3y^3$$

$$(-2x)^5 = (-2)^5x^5 = -32x^5$$

THEOREM 5

$$\left(\frac{x}{y}\right)^n = \frac{x^n}{y^n} \text{ if } y \neq 0$$

A power of a quotient is a quotient of two powers whose bases are the numerator and denominator of the quotient and each of whose exponents is the same as the exponent of the quotient.

EXAMPLE 6.1.5

$$\left(\frac{x}{y}\right)^4 = \frac{x^4}{y^4}$$

$$\left(\frac{x}{y}\right)^4 = \left(\frac{x}{y}\right)\left(\frac{x}{y}\right)\left(\frac{x}{y}\right)\left(\frac{x}{y}\right) = \frac{xxxx}{yyyy} = \frac{x^4}{y^4}$$

$$\left(\frac{-2}{x}\right)^6 = \frac{(-2)^6}{x^6} = \frac{64}{x^6}$$

The exponent theorems are summarized below. The examples that follow show how these theorems are used to simplify certain algebraic expressions.

THE EXPONENT THEOREMS

Theorem 1 $x^n x^m = x^{n+m}$

Theorem 2 $(x^n)^m = x^{nm}$

Theorem 3 $\dfrac{x^n}{x^m} = x^{n-m}$ if $n > m$ and $x \neq 0$

$$\frac{x^n}{x^m} = \frac{1}{x^{m-n}} \text{ if } m > n \text{ and } x \neq 0$$

$$\frac{x^n}{x^m} = 1 \text{ if } m = n \text{ and } x \neq 0$$

Theorem 4 $(xy)^n = x^n y^n$

Theorem 5 $\left(\dfrac{x}{y}\right)^n = \dfrac{x^n}{y^n}$ if $y \neq 0$

EXAMPLE 6.1.6 Simplify $(5x^2)^3$.

Solution
$$\begin{aligned} (5x^2)^3 &= 5^3(x^2)^3 && \text{(Theorem 4)} \\ &= 5^3x^6 && \text{(Theorem 2)} \\ &= 125x^6 && \text{(Definition)} \end{aligned}$$

EXAMPLE 6.1.7 Simplify

$$\frac{a}{b^2}\left(\frac{a^2}{b}\right)^3$$

Solution

$$\begin{aligned} \frac{a}{b^2}\left(\frac{a^2}{b}\right)^3 &= \frac{a}{b^2} \cdot \frac{(a^2)^3}{b^3} && \text{(Theorem 5)} \\ &= \frac{a(a^6)}{b^2(b^3)} && \text{(Theorem 2)} \\ &= \frac{a^7}{b^5} && \text{(Theorem 1)} \end{aligned}$$

EXAMPLE 6.1.8 Simplify

$$\frac{(abc)^5}{a^2b^5c^8}$$

Solution

$$\begin{aligned}\frac{(abc)^5}{a^2b^5c^8} &= \frac{a^5b^5c^5}{a^2b^5c^8} && \text{(Theorem 4)}\\ &= \frac{a^5}{a^2}\cdot\frac{b^5}{b^5}\cdot\frac{c^5}{c^8}\\ &= a^{5-2}\cdot 1\cdot\frac{1}{c^{8-5}} && \text{(Theorem 3)}\\ &= a^3\cdot\frac{1}{c^3}\\ &= \frac{a^3}{c^3}\end{aligned}$$

EXAMPLE 6.1.9 Simplify

$$\frac{(9^2x^2)^3}{(3^3x^3)^2}$$

Solution

$$\begin{aligned}\frac{(9^2x^2)^3}{(3^3x^3)^2} &= \frac{9^6x^6}{3^6x^6} && \text{(Theorems 4 and 2)}\\ &= \left(\frac{9}{3}\right)^6\frac{x^6}{x^6} && \text{(Theorem 5)}\\ &= 3^6\cdot 1\\ &= 729\end{aligned}$$

Note that $(x^2)^3 = (x^3)^2$ since $2\cdot 3 = 3\cdot 2 = 6$.
In general, $(x^m)^n = (x^n)^m$ since $mn = nm$.

EXERCISES 6.1 A

Simplify in Exercises 1–34 by using one or more exponent theorems. In Exercises 1–10, state which exponent theorem was used.

1. $(x^2)^4$

2. x^3x^4

3. $\dfrac{x^8}{x^3}$

4. $(-2y)^3$

5. $\left(\frac{-3}{y}\right)^4$

6. $\frac{x^4}{x^8}$

7. $2^4 5^4$

8. $-5(2^3)^2$

9. $\frac{(100)^6}{(50)^6}$

10. $(-a)^4(-a)^6$

11. $(-5x^2)^3$

12. $-(5x^3)^2$

13. $(a^2b^3)(ab^5)$

14. $\frac{3^8x^7}{3^5x^{10}}$

15. $\frac{12y^{12}}{4y^4}$

16. $\frac{(8^2)^3}{(4^3)^2}$

17. $(-x^2)(-x)^2$

18. $\frac{(10a^2)^3}{(10a^3)^2}$

19. $(5xy^2)^3$

20. $(x^4y^2)(x^3y^4)$

21. $\left(\frac{12a^2bc}{10ab^3}\right)^2$

22. $(-2a^2b^5)^5$

23. $\frac{(-y^4)^4}{-(y^4)^4}$

24. $\frac{10^5 10^7}{10^4}$

25. $\frac{(10^5)^2}{10^2 10^5}$

26. $\frac{2^6 \cdot 5^4}{2^3 \cdot 5^6}$

27. $(2^2)^3(5^2)^2$

28. $\frac{3^5 \cdot 3^{12}}{3^7 \cdot 3^{10}}$

29. $\frac{3^5 \cdot 4^5}{6^5}$

30. $\frac{x^n}{x}$

31. $(y^n)^4$

32. $\left(\frac{x^2}{5^2}\right)^n$

33. $\frac{x^n x}{x^{n+1}}$

34. $\frac{(x^{n+1})^2}{x^2x^n}$

35. $x^{n+1}x^n$

EXERCISES 6.1 B

Simplify in Exercises 1–34 by using one or more exponent theorems. In Exercises 1–10, state which exponent theorem was used.

1. $\frac{x^7}{x^3}$

2. $\frac{c^3}{c^6}$

3. $(t^4)^3$

4. $\frac{x^2}{x^5}$

5. $(-5t)^4$

6. a^5a^7

7. $\left(\frac{-n}{2}\right)^5$

8. $5^6 2^6$

9. $-7(x^2)^5$

10. $\dfrac{(24)^7}{(48)^7}$

11. $(-2y^3)^4$

12. $-2(y^4)^3$

13. $(ab^2)^3(a^2b)^4$

14. $\dfrac{6y^6}{2y^2}$

15. $\dfrac{a^3b^5}{a^4b^2}$

16. $\left(\dfrac{a^3b^5}{ab^3}\right)^4$

17. $(-x^3)(-x)^3$

18. $\dfrac{(10^4)^3}{(10^2)^5}$

19. $(-3x^2y^3z^4)^4$

20. $-3(a^2b^3c^4)^2$

21. $(c^3d^5)(cd^3)$

22. $\dfrac{5x^5}{2y^4}\cdot\dfrac{3y^6}{10x^3}$

23. $\dfrac{-(x^3)^2}{(-x^3)^2}$

24. $\dfrac{10^6}{10^2 10^8}$

25. $\dfrac{(10^6)^3}{10^3 10^6}$

26. $\dfrac{(2^3\cdot 3^4)^3}{(2^2\cdot 3^2)^5}$

27. $\dfrac{(2^4)^3(5^6)^2}{(10^3)^2}$

28. $\dfrac{9^6\cdot 4^6}{6^6}$

29. $\dfrac{x}{x^n}$

30. $\dfrac{3^{2n}}{3^n}$

31. $\left(\dfrac{x^3}{y^2}\right)^n$

32. $(x^2y)^n$

33. $\dfrac{(x^n)^2}{(x^2)^{n+1}}$

34. $x(x^2)^n$

35. $\left(\dfrac{x^n}{y^n}\right)^n$

6.2 INTEGRAL EXPONENTS

It is desirable to establish a meaning for the expressions x^0 and x^{-n} where x is any nonzero real number and n is a natural number. If the first theorem of exponents is to remain valid when zero is used as an exponent, then

$$x^n x^0 = x^{n+0} = x^n$$

But

$$x^n \cdot 1 = x^n,$$

and there is only one multiplicative identity. Therefore, if x^0 is to be defined, it must be defined as the number 1.

DEFINITION OF x^0

$$x^0 = 1 \text{ if } x \text{ is a real number and } x \neq 0$$

As examples, $5^0 = 1$, $(-3)^0 = 1$, and $(x + 2)^0 = 1$ if $x \neq -2$.

It can be shown that all the exponent theorems remain valid when x^0 is defined as 1.

Now if the first theorem of exponents is to remain valid when an exponent is a negative integer, then

$$x^n x^{-n} = x^{n+(-n)} = x^0 = 1$$

However,

$$x^n \cdot \frac{1}{x^n} = 1$$

Since each nonzero real number has exactly one reciprocal, if x^{-n} is to be defined, x^{-n} must be defined as $\frac{1}{x^n}$, the reciprocal of x^n.

DEFINITION OF x^{-n}

$$x^{-n} = \frac{1}{x^n} \text{ if } x \text{ is any real number and } x \neq 0$$

Noting that

$$\frac{1}{x^n} = \frac{1}{\underbrace{x \cdot x \cdots x}_{n \text{ factors}}} = \underbrace{\frac{1}{x} \cdot \frac{1}{x} \cdots \frac{1}{x}}_{n \text{ factors}} = \left(\frac{1}{x}\right)^n$$

the following theorem can be stated.

THEOREM

$$\frac{1}{x^n} = \left(\frac{1}{x}\right)^n \text{ if } x \neq 0$$

As examples, $5^{-1} = \frac{1}{5}$

$$2^{-3} = \frac{1}{2^3} = \frac{1}{8}$$

$$\left(\frac{1}{3}\right)^{-2} = \left(\frac{1}{\frac{1}{3}}\right)^2 = 3^2 = 9$$

$$x^{-4} = \frac{1}{x^4}$$

$$\left(-\frac{2}{5}\right)^{-3} = \left(-\frac{5}{2}\right)^3 = \frac{-125}{8} \text{ since } \frac{1}{\frac{-2}{5}} = \frac{-5}{2}$$

$$\frac{1}{3^{-4}} = \left(\frac{1}{3}\right)^{-4} = 3^4 = 81$$

It can be shown that the five theorems of exponents remain valid when the exponent is a negative integer. These theorems are illustrated in the following examples.

EXAMPLE 6.2.1

(Theorem 1) $$2^{-3}2^{-2} = 2^{(-3)+(-2)} = 2^{-5} = \frac{1}{2^5} = \frac{1}{32}$$

$$2^4 2^{-7} = 2^{4+(-7)} = 2^{-3} = \frac{1}{2^3} = \frac{1}{8}$$

EXAMPLE 6.2.2

(Theorem 2) $$(2^{-3})^2 = 2^{-6} = \frac{1}{2^6} = \frac{1}{64}$$

$$(2^{-2})^{-4} = 2^{(-2)(-4)} = 2^8 = 256$$

EXAMPLE 6.2.3

(Theorem 3) $$\frac{3^4}{3^7} = 3^{4-7} = 3^{-3} = \frac{1}{3^3} = \frac{1}{27}$$

$$\frac{x^4}{x^6} = x^{4-6} = x^{-2} = \frac{1}{x^2}$$

$$\frac{x^{-2}}{x^{-5}} = x^{-2-(-5)} = x^{-2+5} = x^3$$

EXAMPLE 6.2.4

(Theorem 4) $$(2x^{-2})^{-3} = 2^{-3}(x^{-2})^{-3} = 2^{-3}x^6 = \frac{x^6}{8}$$

EXAMPLE 6.2.5

(Theorem 5) $$\left(\frac{x^{-1}}{5}\right)^{-4} = \frac{(x^{-1})^{-4}}{5^{-4}} = \frac{x^4}{5^{-4}} = 5^4x^4 = 625x^4$$

$$\left(\frac{2^{-3}}{y^{-3}}\right)^{-2} = \frac{(2^{-3})^{-2}}{(y^{-3})^{-2}} = \frac{2^6}{y^6}$$

EXAMPLE 6.2.6 Simplify $(x^{-2}y^{-2})^{-2}$.

Solution $(x^{-2}y^{-2})^{-2} = (x^{-2})^{-2}(y^{-2})^{-2}$ (Theorem 4)

$= x^4y^4$ (Theorem 2)

EXAMPLE 6.2.7 Simplify $(2^{-3} \cdot 2^{-5})^{-1}$.

Solution $(2^{-3}\cdot 2^{-5})^{-1} = (2^{-3-5})^{-1} = (2^{-8})^{-1}$ (Theorem 1)

$= 2^{8}$ (Theorem 2)

$= 256$ (Definition)

EXAMPLE 6.2.8 Simplify $(2^{-3} + 2^{-5})^{-1}$.

Solution There is no exponent theorem that applies to a power of a sum (or difference), so the definition must be used:

$$(2^{-3} + 2^{-5})^{-1} = \left(\frac{1}{2^3} + \frac{1}{2^5}\right)^{-1}$$

$$= \left(\frac{2^2}{2^5} + \frac{1}{2^5}\right)^{-1}$$ (2^5 is the L.C.D. for the two fractions)

$$= \left(\frac{4}{32} + \frac{1}{32}\right)^{-1}$$

$$= \left(\frac{5}{32}\right)^{-1}$$

$$= \frac{32}{5}$$ $\left(\text{Definition of } x^{-1} = \frac{1}{x}\right)$

EXAMPLE 6.2.9 Simplify

$$\frac{2^{-3} + 2^{-5}}{2^{-7}}$$

Solution

$$\frac{2^{-3} + 2^{-5}}{2^{-7}} = \frac{2^{-3} + 2^{-5}}{2^{-7}} \cdot \frac{2^7}{2^7}$$

$$= \frac{2^7(2^{-3} + 2^{-5})}{2^7 \cdot 2^{-7}}$$

$$= \frac{(2^7 2^{-3}) + (2^7 2^{-5})}{2^7 2^{-7}}$$ (Distributive axiom)

$$= \frac{2^4 + 2^2}{2^0}$$ (Theorem 1)

$$= \frac{16 + 4}{1} = 20$$ (Definition)

The preceding example illustrates the use of the distributive axiom to distribute the operation of multiplication over a sum. Theorem 1 for exponents is also used in the simplification. It is important to note the differences illustrated by the last 4 examples. It is especially important to remember that the exponent theorems apply to products and quotients and *not* to sums and differences.

EXAMPLE 6.2.10 Does $(x + 2)^2 = x^2 + 4$? Why?

Solution By the definition of squaring,

$$(x + 2)^2 = (x + 2)(x + 2) = x^2 + 4x + 4$$

If
$$(x + 2)^2 = x^2 + 4$$
then
$$x^2 + 4x + 4 = x^2 + 4$$
and
$$4x = 0 \quad \text{and } x = 0$$
Thus $(x + 2)^2 = x^2 + 4$ only if $x = 0$.

EXERCISES 6.2 A

Use the definitions and theorems to express Exercises 1–55 in simplest form without zero or negative exponents. Assume all variables to be nonzero.

1. 2^{-1}

2. 5^{-3}

3. $\left(\frac{1}{2}\right)^{-4}$

4. $\left(\frac{1}{10}\right)^{-5}$

5. $\frac{1}{3^{-2}}$

6. $\frac{1}{5^{-4}}$

7. 10^{-3}

8. $\left(\frac{3}{5}\right)^{-1}$

9. $\left(\frac{-10}{3}\right)^{-4}$

10. $3^5 3^{-2}$

11. $2^5 2^{-9}$

12. $10^6 10^{-8}$

13. $10^{-2} 10^{-3}$

14. $5^4 5^{-4}$

15. $\frac{10^2}{10^{-4}}$

16. $\frac{10^{-3}}{10^2}$

17. $\frac{10^{-4}}{10^{-4}}$

18. $\frac{10^{-5}}{10^{-2}}$

19. $(2^{-3})^2$

20. $(3^{-2}5)^{-2}$

21. $(-2^{-2})^{-1}$

22. $\left(\frac{5^{-2}}{2^{-2}}\right)^{-1}$

23. $\frac{2^{-5}3^4}{2^{-4}3^{-1}}$

24. $\left(\frac{2^{-3}}{5}\right)^{-2}$

25. $10^{-3}(10^4 10^{-1})$

26. $5^{-2}(5^3 + 5^2)$

27. $4^7(4^{-7} - 4^{-5})$

28. $\left(\frac{2^{-2}}{5^0}\right)^{-3}$

29. $(2^{-1}5^{-1})^{-1}$

30. $(2^{-1} + 5^{-1})^{-1}$

31. $10^{-2}(2^{-1}5^{-1})$

32. $10^{-2}(2^{-1} + 5^{-1})$

33. $(2x)^{-3}$

34. $2x^{-3}$

35. $(4^{-1}x)^{-2}$

36. $\left(\frac{5}{x}\right)^{-3}$

37. $(2^{-1}x^{-1})^{-2}$

38. $(2^{-1} + x^{-1})^{-2}$

39. $\frac{y}{y^{-1}}$

40. $\frac{1}{2x^{-2}}$

41. $(xx^{-4}x^3)^{10}$

42. $(x^{-2}y)^{-1}$

43. $(x^{-3} + x^{-5})^0$

44. $(x^{-1} + y^{-1})^{-1}$

45. $x^0 + y^0$

46. $\left(\frac{2x^3}{y^{-4}}\right)\left(\frac{x^{-2}y^5}{8}\right)$

47. $\left(\frac{-1}{x^{-1}}\right)^{-1}$

48. $\frac{(a^{-3}b)(a^4b^2)}{a^{-4}b^4}$

49. $(3y^{-3})^{-2}$

50. $(3y^{-3})(y^{-2})$

★ 51. $\frac{xx^{-n}}{(x^{-1})^n}$

★ 52. $x^{-n}(x^n - x^{n-1})$

★ 53. $(x^{-n} + y^{-n})^0(x^{-n}y^n)^{-1}$

★ 54. $\frac{(x^{n+1}x^{2n-1})^{-2}}{x^{4n}}$

★ 55. $\frac{x^{-n}y^{2n}}{x^{2n}y^{-n}}$

EXERCISES 6.2 B

Use the definitions and theorems to express Exercises 1–55 in simplest form without zero or negative exponents. Assume all variables to be nonzero.

1. 10^{-1}

2. 6^{-2}

3. $\left(\frac{1}{5}\right)^{-3}$

4. $\left(\frac{1}{10}\right)^{-6}$

5. $\frac{1}{2^{-3}}$

6. $\frac{1}{10^{-5}}$

7. 10^{-4}

8. $\left(\frac{4}{3}\right)^{-2}$

9. $\left(-\frac{3}{16}\right)^{-1}$

10. $5^4 5^{-7}$

11. $10^{-4}10^{2}$

12. $10^{-5}10^{-4}$

13. $10^{6}10^{-6}$

14. $6^{-5}6^{5}$

15. $\dfrac{10^{6}}{10^{-3}}$

16. $\dfrac{10^{-5}}{10^{-7}}$

17. $(10^{-2})^{-1}$

18. $\dfrac{10^{-6}}{10^{-6}}$

19. $(2^{-1}3^{-2})^{2}$

20. $(-3^{-1})^{2}$

21. $(-5^{-1})^{-1}$

22. $\left(\dfrac{5}{8^{-1}}\right)^{-2}$

23. $\dfrac{3^{-2}5^{-1}}{3^{2}5^{-2}}$

24. $\left(\dfrac{5^{-3}}{3^{-4}}\right)^{-1}$

25. $5^{4}(5^{-3}5^{-1})$

26. $10^{8}(10^{-6} + 10^{-8})$

27. $3^{-5}(6^{5} - 3^{6})$

28. $\left(\dfrac{4^{0}}{4^{-3}}\right)^{-1}$

29. $(2^{-2}5^{-2})^{-2}$

30. $(2^{-2} + 5^{-2})^{-2}$

31. $\dfrac{2^{-2}5^{-2}}{10^{-2}}$

32. $\dfrac{2^{-2} + 5^{-2}}{10^{-2}}$

33. $-3(x^{2})^{-4}$

34. $(-3x^{2})^{-4}$

35. $(10^{-3}x^{3})^{-2}$

36. $\left(\dfrac{x}{4}\right)^{-2}$

37. $(x^{-2} + 5^{-2})^{-1}$

38. $(x^{-2}\cdot 5^{-2})^{-1}$

39. $\dfrac{x^{-3}}{y^{-2}}$

40. $\dfrac{1}{4x^{-3}}$

41. $\left(\dfrac{x^{-3}}{y^{-2}}\right)\left(\dfrac{x^{-5}}{x^{7}}\right)^{0}$

42. $(y^{-5} - y^{2})^{-3}(y^{-5} - y^{2})^{3}$

43. $\left(\dfrac{x}{y}\right)^{-4}\left(\dfrac{y^{-4}}{x^{-2}}\right)$

44. $\dfrac{x^{-1} + y^{-1}}{x^{-1} - y^{-1}}$

45. $(x^{-1} - y^{-1})(x - y)^{-1}$

46. $\dfrac{a + b^{-1}}{a^{-1} + b}$

47. $\dfrac{a^{4}b - a^{-1}b^{-4}}{a^{5}b^{-5}}$

48. $(-2x^{-2}y^{-2})^{-3}$

49. $(-2x^{-2}y^{-2})y^{-3}$

50. $\dfrac{2x^{3}y^{-2}}{3x^{-2}y^{3}}$

★ **51.** $(x^{-2} + y^{-2})^{n}(x^{-2} + y^{-2})^{-n}$

★ **52.** $x^{-n-1}(x^{2n+1} + x^{n+2})$

★ **53.** $(2^{n} + 2^{-n})(2^{n} - 2^{-n})$

★ **54.** $(2^{n} + 2^{n})^{-n}$

★ **55.** $\dfrac{2^{-3n}3^{-2n}}{2^{2n}3^{-n}}$

6.3 SQUARE ROOTS

6.3.1 Symbolic Representation of Square Roots

DEFINITION

The number a is a **square root** of b if and only if $a^2 = b$.

This definition implies that the operations of squaring and extracting square roots are inverses of each other, just as addition and subtraction are inverse operations and multiplication and division are inverse operations. For example,

$$3 \text{ is a square root of } 9 \text{ since } 3^2 = 9$$

$$-3 \text{ is a square root of } 9 \text{ since } (-3)^2 = 9$$

$$\frac{5}{8} \text{ is a square root of } \frac{25}{64} \text{ since } \left(\frac{5}{8}\right)^2 = \frac{25}{64}$$

$$\frac{-5}{8} \text{ is a square root of } \frac{25}{64} \text{ since } \left(\frac{-5}{8}\right)^2 = \frac{25}{64}$$

and

$$0 \text{ is a square root of } 0 \text{ since } 0^2 = 0$$

Examination of these examples and similar ones reveals the following properties:

1. If x is a real number, then $x^2 \geq 0$. (In other words, the square of a real number is never negative.)

2. Every positive real number has two square roots, a positive real number and its negative. Thus if a is a positive real number ($a > 0$), then the square roots of a^2 are a and $-a$.

Now it is useful to denote a square root by the symbol $\sqrt{x}$. However, the symbolic expression $\sqrt{x}$ must represent exactly one number. Suppose, for example, that $\sqrt{9} = 3$ and $\sqrt{9} = -3$. Then, by the transitive axiom of the equal relation, it would follow that $3 = -3$, and this is impossible. Therefore, $\sqrt{9}$ and, in general, $\sqrt{x}$ must represent exactly one number in order to avoid contradictions. Mathematicians agree that $\sqrt{9}$ shall be used to designate $+3$, the positive square root of 9. The negative square root, -3, is indicated by $-\sqrt{9}$.

In general, the following definition is made.

DEFINITION OF $\sqrt{x}$

If $x > 0$, then $\sqrt{x}$ is the unique positive real number such that $(\sqrt{x})^2 = x$.

Note: If $x < 0$, then $\sqrt{x}$ is not a real number.

The following useful theorem aids in the solution of problems involving square roots.

THEOREM

$\sqrt{x^2} = x$ if and only if $x \geq 0$.

Using the absolute value concept introduced in an earlier chapter, the following theorem can be stated:

THEOREM

For all real numbers x, $\sqrt{x^2} = |x|$.

Note that since $|x|$ is always positive or zero, $\sqrt{x^2}$ is never negative.

For example, $\sqrt{5^2} = 5$ but $\sqrt{(-5)^2} = \sqrt{25} = \sqrt{5^2} = 5$.

The positive real number $\sqrt{x}$ is called the principal square root of x. Its negative is also a square root and is designated by the symbol $-\sqrt{x}$.

Thus $\sqrt{25} = 5$, and 5 is the principal square root of 25.

Also, $-\sqrt{25} = -5$, and -5 is the other square root of 25.

Thus there is exactly one symbolic representation for each of the square roots of a positive real number.

6.3.2 Irrational Square Roots

Although every positive real number x has two real square roots—namely, $\sqrt{x}$ and $-\sqrt{x}$—it does not follow that every positive rational number has two rational roots.

Rational numbers, such as 49, 81, and $\frac{9}{16}$, are called perfect squares because their square roots are also rational.

On the other hand, the number 2 is not a perfect square and its square roots, $\sqrt{2}$ and $-\sqrt{2}$, are not rational but irrational.

Q, the set of rational numbers, was defined as the set of quotients of integers, $\frac{p}{q}$, where $q \neq 0$. It can be shown that $\sqrt{2}$ cannot be expressed as the quotient of two integers, and thus $\sqrt{2}$ is irrational.

The terminating decimals and the nonterminating repeating decimals represent the rational numbers. Therefore, it is convenient to think of the irrational numbers as the set of nonterminating, nonrepeating decimals and the real numbers as the union of these two sets—that is, the set of all decimals.

The table of squares and square roots inside the book cover is useful in providing a first approximation to the square root of a number.

The square root values in the table are not exact but are approximations to the nearest thousandth.

6.3.3 Geometric Interpretation of Square Roots

The set of real numbers has a property called the **axiom of completeness.**

THE AXIOM OF COMPLETENESS

Each point on the number line corresponds to exactly one real number, and each real number corresponds to exactly one point on the number line.

A point on the number line that corresponds to an irrational square root can be found by using the theorem of Pythagoras.

THE THEOREM OF PYTHAGORAS

The square of the length of the hypotenuse of a right triangle is equal to the sum of the squares of the lengths of the legs of the right triangle.

The **hypotenuse** of a right triangle, which is opposite the right angle, is the longest side. The **legs** of a right triangle are the other two sides—that is, the sides that form the right angle.

Thus if c represents the length of the hypotenuse in Figure 6.3.1 and if a and b represent the lengths of the legs, then

$$c^2 = a^2 + b^2$$

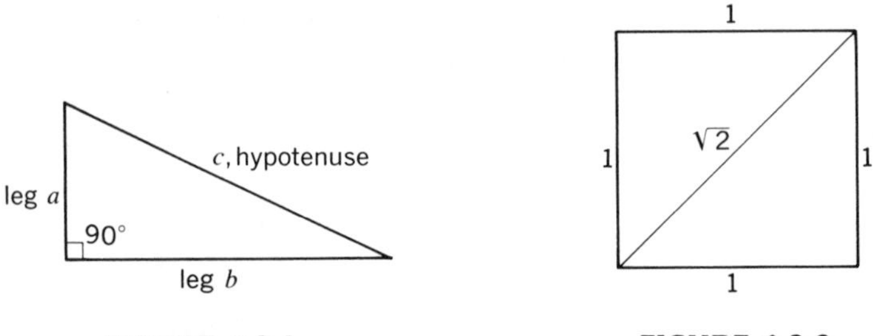

FIGURE 6.3.1 **FIGURE 6.3.2**

Therefore, a length of $\sqrt{2}$ can be represented by the diagonal of a square (see Figure 6.3.2).

Since $$c^2 = 1^2 + 1^2$$

$$c^2 = 2$$

Thus $$c = \sqrt{2}$$

A length of $\sqrt{3}$ can be represented as the diagonal of a rectangle whose sides are 1 and $\sqrt{2}$ (see Figure 6.3.3).

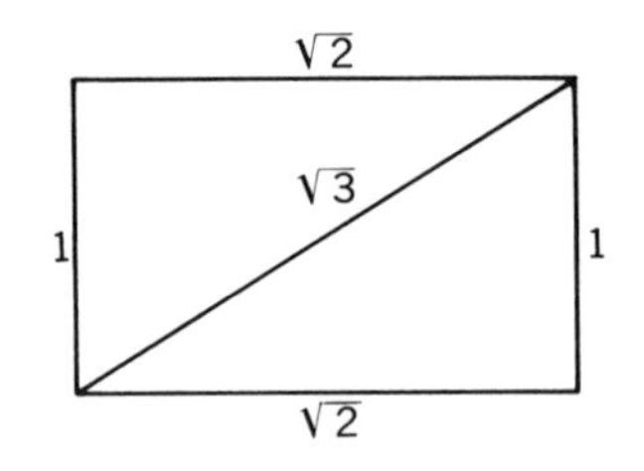

FIGURE 6.3.3

Since $$c^2 = (\sqrt{2})^2 + 1^2$$

$$c^2 = 2 + 1 = 3$$

Thus $$c = \sqrt{3}$$

In a similar way, the numbers $\sqrt{4} = 2$, $\sqrt{5}$, $\sqrt{6}$, and so on can be represented as the lengths of the diagonals of rectangles. Now using circles with centers at the origin and with radii equal successively to these diagonals, the square roots can be located on the number line. This is illustrated in Figure 6.3.4.

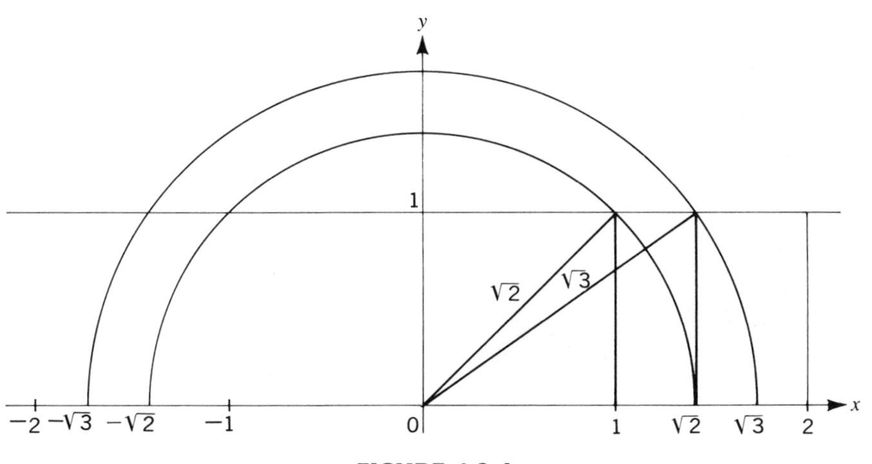

FIGURE 6.3.4

EXAMPLE 6.3.1 Simplify $\sqrt{64} - \sqrt{(-4)^2}$.

Solution $\sqrt{64} - \sqrt{(-4)^2} = \sqrt{64} - \sqrt{16}$

$$= 8 - 4 = 4$$

EXAMPLE 6.3.2 Approximate $\sqrt{12} - \sqrt{20}$ to the nearest hundredth by using the table inside the book cover.

Solution From the table,

$$\sqrt{12} = 3.464$$

and

$$\sqrt{20} = 4.472$$

$$\sqrt{12} - \sqrt{20} = 3.464 - 4.472$$

$$= -1.008$$

$$= -1.01 \text{ to the nearest hundredth}$$

EXAMPLE 6.3.3 Find the hypotenuse c of a right triangle if the lengths of its legs a and b are such that $a = 12$ and $b = 35$.

Solution Using the theorem of Pythagoras,

$$c^2 = a^2 + b^2$$

$$c^2 = (12)^2 + (35)^2$$

$$c^2 = 144 + 1225 = 1369$$

$$c = \sqrt{1369} = 37 \text{ (See the table)}$$

HISTORICAL NOTE

(ca. 580–501 B.C.)

Although there are many stories about the life of Pythagoras, little is known for certain. He was born about 580 B.C., probably on the island of Samos, probably traveled to Egypt and Babylonia, and is known to have settled in Crotona on the Italian coast. In Crotona he founded a brotherhood composed of some 300 wealthy young aristocrats. This group, known as the Pythagoreans, became the prototype of all the secret societies of Europe and America. Their motto, "Number rules the universe," expressed the combination of mathematics and mysticism in which they believed. Shakespeare refers to the Pythagorean belief in immortality and transmigration of the soul in *The Merchant of Venice:*

> Thou almost mak'st me waver in
> my faith,
> To hold opinion with Pythagoras,
> That souls of animals infuse
> themselves
> Into the trunks of men.

The name of Pythagoras is most famous in connection with the relationship of the squares of the sides of a right triangle. While Pythagoras did not discover this property (it was already known to the Babylonians), he may have offered the first proof of this statement.

EXAMPLE 6.3.4 If a, b, and c are the sides of a right triangle whose hypotenuse is c, find b if $a = 16$ and $c = 18$.

Solution Using

$$a^2 + b^2 = c^2$$

$$(16)^2 + b^2 = (18)^2$$

$$b^2 = (18)^2 - (16)^2 = (18 + 16)(18 - 16)$$

$$b^2 = 68$$

$$b = 8.25 \text{ to the nearest hundredth}$$

EXAMPLE 6.3.5 Solve:

a. $x = \sqrt{9}$ b. $x^2 = 9$

Solution

a. Since $\sqrt{9} = 3$, the positive square root of 9, $x = \sqrt{9}$ means $x = 3$. The solution set is $\{3\}$.

b. Since $(+3)^2 = 9$ and $(-3)^2 = 9$, the solution set of $x^2 = 9$ is $\{3, -3\}$. An alternate solution to (b) is as follows:

$$x^2 = 9$$

$$x^2 - 9 = 0$$

$$(x + 3)(x - 3) = 0$$

$$x + 3 = 0 \quad \text{or} \quad x - 3 = 0$$

$$x = -3 \quad \text{or} \quad x = 3$$

and the solution set is $\{3, -3\}$.

EXERCISES 6.3 A

Simplify in Exercises 1–10. (If necessary, use the table of squares and square roots inside the book cover.)

1. $\sqrt{16}$

2. $-\sqrt{36}$

3. $\sqrt{(-3)^2}$

4. $\sqrt{\dfrac{4}{25}}$

5. $\sqrt{400}$

6. $\sqrt{5^2 - 4^2}$

7. $\sqrt{5^2} - \sqrt{4^2}$

8. $\sqrt{(-5)^2 + (12)^2}$

9. $\sqrt{(-5)^2} + \sqrt{(12)^2}$

10. $\dfrac{3\sqrt{64 + 36}}{3(\sqrt{64} + \sqrt{36})}$

Use the table of squares and square roots inside the book cover to approximate Exercises 11–20 to the nearest hundredth.

11. $\sqrt{45}$

12. $\sqrt{10}$

13. $\sqrt{6500}$

14. $\sqrt{0.05}$

15. $2\sqrt{5}$

16. $5 - \sqrt{3}$

17. $\sqrt{7} + \sqrt{2}$

18. $\dfrac{1 + \sqrt{6}}{2}$

19. $\dfrac{-5 + \sqrt{12}}{6}$

20. $\sqrt{2}\sqrt{6}$

If a and b represent the lengths of the legs of a right triangle and c represents the length of the hypotenuse of that triangle, evaluate the length of the missing side in Exercises 21–30.

21. $a = 12, b = 5$
22. $b = 5, c = 13$
23. $a = 1, b = 1$
24. $a = 15, c = 17$
25. $a = \sqrt{3}, b = \sqrt{5}$
26. $c = \sqrt{3}, a = 1$
27. $a = 2, c = 6$
★ **28.** $a = 2t, c = t^2 + 1$
★ **29.** $a = 4u^2 - v^2, b = 4uv$
★ **30.** $a = x^2, b = \sqrt{6x^2 + 9}$

Solve Exercises 31–35 over the set of real numbers.

31. $x = \sqrt{25}$
32. $x^2 = 25$
33. $x^2 = 36$
34. $x = -\sqrt{36}$
35. $x^2 = -36$

EXERCISES 6.3 B

Simplify in Exercises 1–10. (If necessary, use the table of squares and square roots inside the book cover.)

1. $\sqrt{169}$
2. $-\sqrt{1225}$
3. $\sqrt{\frac{25}{64}}$
4. $\sqrt{(-7)^2}$
5. $\sqrt{0.01}$
6. $\sqrt{(25)^2 - 7^2}$
7. $\sqrt{(25)^2} - \sqrt{7^2}$
8. $\sqrt{(-8)^2 + (-15)^2}$
9. $\sqrt{(-8)^2} + \sqrt{(-15)^2}$
10. $\dfrac{4\sqrt{225 - 144}}{4\sqrt{225} - \sqrt{144}}$

Use the table of squares and square roots inside the book cover to approximate each of the following to the nearest hundredth.

11. $\sqrt{86}$
12. $\sqrt{2300}$
13. $\sqrt{18}$
14. $\sqrt{0.57}$
15. $2 + \sqrt{2}$
16. $\sqrt{5} - \sqrt{3}$
17. $5\sqrt{8}$
18. $\dfrac{4 - \sqrt{10}}{2}$

19. $\dfrac{-6 - \sqrt{15}}{3}$

20. $\sqrt{5}\sqrt{10}$

If a and b represent the lengths of the legs of a right triangle and c represents the length of the hypotenuse of that triangle, evaluate the length of the missing side in Exercises 21–30.

21. $a = 12, b = 35$

22. $b = 5, c = 15$

23. $c = 10, a = b$

24. $a = 2, c = \sqrt{3}$

25. $b = \sqrt{10}, c = \sqrt{30}$

26. $a = 1, b = 2$

★ **27.** $a = p^2 - q^2, b = 2pq$

★ **28.** $b = t^2 - 9, c = t^2 + 9$

★ **29.** $a = x^2, b = \sqrt{2x^2 + 1}$

★ **30.** $a = 4u^2, b = \sqrt{25 + 40u^2}$

Solve Exercises 31–35 over the set of real numbers.

31. $x = \sqrt{16}$

32. $x^2 = 16$

33. $x^2 = 49$

34. $x = -\sqrt{49}$

35. $x^2 = -49$

6.4 CUBE ROOTS AND *n*th ROOTS

6.4.1 Cube Roots

DEFINITION

The number a is a **cube root** of b if and only if $a^3 = b$.

For example, 2 is a cube root of 8 because $2^3 = 8$ and -3 is a cube root of -27 because $(-3)^3 = -27$.

Since the product of three positive numbers is positive, and since the product of three negative numbers is negative, then a positive real number has exactly one real cube root and it is positive, and a negative real number has exactly one real cube root and it is negative. Therefore, the radical sign can be used to designate this unique real cube root of a real number.

DEFINITION OF $\sqrt[3]{x}$

$\sqrt[3]{x}$ is the unique real number such that $(\sqrt[3]{x})^3 = x$.

Again, a useful theorem follows the definition.

THEOREM

$\sqrt[3]{x^3} = x$ for all real numbers x.

For example, $\sqrt[3]{8} = 2$ and $\sqrt[3]{-27} = -3$

Also $\sqrt[3]{\frac{8}{27}} = \frac{2}{3}$ because $\left(\frac{2}{3}\right)^3 = \frac{2}{3} \cdot \frac{2}{3} \cdot \frac{2}{3} = \frac{8}{27}$

and $\sqrt[3]{\frac{-64}{125}} = \frac{-4}{5}$ because $\left(\frac{-4}{5}\right)^3 = \left(\frac{-4}{5}\right)\left(\frac{-4}{5}\right)\left(\frac{-4}{5}\right) = \frac{-64}{125}$

$\sqrt[3]{0} = 0$ because $0^3 = 0 \cdot 0 \cdot 0 = 0$

Rational numbers, such as 8, -27, and $\frac{-64}{125}$, are called perfect cubes because their cube roots are rational numbers.

As is the case with square roots, some cube roots are irrational and can be named only by using the radical sign.

For example, $\sqrt[3]{2}$, $\sqrt[3]{-3}$, and $\sqrt[3]{\frac{2}{7}}$ are irrational real numbers. Their nonterminating, nonrepeating decimal representation can be approximated by arithmetical calculations. A table of cubes and cube roots is provided inside the book cover.

The values of the cube roots in the table are approximations to the nearest thousandth.

6.4.2 *n*th Roots

DEFINITION

The number a is an nth root of b if and only if $a^n = b$.

For example,

$$5 \text{ is a 4th root of } 625 \text{ because } 5^4 = 625$$

$$-2 \text{ is a 5th root of } -32 \text{ because } (-2)^5 = -32$$

Roots such as the square roots, the 4th roots, and the 6th roots have the following common properties:

1. Every positive real number has exactly two real nth roots for n even—one positive and one negative. For example, 5 and -5 are 4th roots of 625.

2. Negative real numbers do *not* have real nth roots for n even. For example, there is no number x such that $x^4 = -625$.

Roots such as the cube root, the 5th root, and the 7th root have the following common properties:

1. Every real number has exactly one real nth root for n odd.
2. The real nth root of a positive number is positive for n odd.
3. The real nth root of a negative number is negative for n odd. For example, 2 is the real 5th root of 32, and -2 is the real 5th root of -32.

The symbol $\sqrt[n]{x}$ (read "the principal nth root of x") is used to indicate exactly one nth root of the real number x.

DEFINITION OF $\sqrt[n]{x}$, x POSITIVE

$\sqrt[n]{x}$ is the unique positive real number such that $(\sqrt[n]{x})^n = x$.

DEFINITION OF $\sqrt[n]{-x}$, x POSITIVE, n ODD

$$\sqrt[n]{-x} = -\sqrt[n]{x}$$

Note that $\sqrt[n]{-x}$ is not defined for n even and $-x$ negative.

The symbol $\sqrt[n]{x}$ is called a **radical**, the natural number n is called the **index**, and the real number x is called the **radicand**.

THEOREM

If x is a positive real number and n is a natural number, then

$$\sqrt[n]{x^n} = x$$

EXAMPLE 6.4.1 Simplify, if possible:

a. $\sqrt[3]{8}$ b. $\sqrt[3]{-8}$ c. $\sqrt[4]{81}$ d. $\sqrt[4]{-81}$ e. $\sqrt[5]{243}$ f. $\sqrt[5]{-243}$

Solution

a. $\sqrt[3]{8} = \sqrt[3]{2^3} = 2$

b. $\sqrt[3]{-8} = \sqrt[3]{(-2)^3} = -2$

c. $\sqrt[4]{81} = \sqrt[4]{3^4} = 3$ since 3 is positive.

d. $\sqrt[4]{-81}$ is not a real number.

e. $\sqrt[5]{243} = \sqrt[5]{3^5} = 3$

f. $\sqrt[5]{-243} = \sqrt[5]{(-3)^5} = -3$

EXAMPLE 6.4.2 Simplify

$$\frac{(\sqrt[3]{25})^3}{\sqrt[3]{64}}$$

Solution

$$\frac{(\sqrt[3]{25})^3}{\sqrt[3]{64}} = \frac{25}{\sqrt[3]{4^3}} = \frac{25}{4}$$

EXAMPLE 6.4.3 Approximate $2\sqrt[3]{15} - \sqrt[3]{6}$ correct to the nearest hundredth.

Solution Using the table inside the book cover,

$$\begin{aligned} \sqrt[3]{15} &= 2.466 \\ 2\sqrt[3]{15} &= 4.932 \\ \sqrt[3]{6} &= 1.817 \\ \hline 2\sqrt[3]{15} - \sqrt[3]{6} &= 3.115 \\ &= 3.12 \text{ correct to the nearest hundredth} \end{aligned}$$

EXERCISES 6.4 A

Simplify in Exercises 1–15. If necessary, use the table inside the book cover.

1. $\sqrt[3]{64}$
2. $\sqrt[3]{-27}$
3. $\sqrt[4]{16}$
4. $\sqrt[3]{(-8)^3}$
5. $\sqrt[4]{(-7)^4}$
6. $\sqrt[5]{-1}$
7. $\sqrt[6]{10^6}$
8. $(\sqrt[3]{15})^3$
9. $\sqrt[4]{5}\,\sqrt[4]{5}\,\sqrt[4]{5}\,\sqrt[4]{5}$
10. $\sqrt[5]{\frac{1}{32}}$
11. $\sqrt[3]{\frac{27}{125}}$
12. $\sqrt[3]{1000}$

13. $\sqrt[3]{1,000,000}$

14. $\sqrt[4]{0.0001}$

15. $(\sqrt[4]{25x^2})^4$

Use the table inside the book cover to approximate Exercises 16–25 correct to the nearest hundredth.

16. $\sqrt[3]{90}$

17. $\sqrt[3]{31}$

18. $\sqrt[3]{57}$

19. $\sqrt[3]{12}$

20. $2\sqrt[3]{5}$

21. $\dfrac{\sqrt[3]{4}}{2}$

22. $\sqrt[3]{20} - \sqrt[3]{10}$

23. $\dfrac{\sqrt[3]{316 - 100}}{8}$

24. $\sqrt[3]{\dfrac{316 - 100}{8}}$

25. $\sqrt[3]{8}\,\sqrt[3]{2}$

List the elements in the sets in Exercises 26–31, where x is a real number.

26. $\{x \mid x^3 = 64\}$

27. $\{x \mid x^3 = -64\}$

28. $\{x \mid x = \sqrt[3]{64}\}$

29. $\{x \mid x = \sqrt[3]{-64}\}$

30. $\{x \mid x = \sqrt[4]{16}\}$

31. $\{x \mid x^4 = 16\}$

32. $\{x \mid x^4 = -16\}$

EXERCISES 6.4 B

Simplify in Exercises 1–15. If necessary, use the table inside the book cover.

1. $\sqrt[3]{-125}$

2. $\sqrt[3]{729}$

3. $\sqrt[4]{625}$

4. $\sqrt[4]{(-6)^4}$

5. $\sqrt[3]{(-12)^3}$

6. $\sqrt[5]{-32}$

7. $\sqrt[8]{10^8}$

8. $(\sqrt[4]{24})^4$

9. $\sqrt[3]{7}\,\sqrt[3]{7}\,\sqrt[3]{7}$

10. $\sqrt[5]{\dfrac{1}{5^3 \cdot 5^2}}$

11. $\sqrt[3]{\dfrac{343}{8}}$

12. $\sqrt[5]{100{,}000}$

13. $\sqrt[3]{0.001}$

14. $\sqrt[4]{0.0016}$

15. $(\sqrt[3]{-3x^3y^6})^3$

Use the table inside the book cover to approximate Exercises 16–25 correct to the nearest hundredth.

16. $\sqrt[3]{80}$

17. $\sqrt[3]{96}$

18. $\sqrt[3]{65}$

19. $\sqrt[3]{29}$

20. $5\sqrt[3]{10}$

21. $\dfrac{\sqrt[3]{100}}{10}$

22. $\sqrt[3]{6} + \sqrt[3]{2}$

23. $\sqrt[3]{\dfrac{100 - 612}{8}}$

24. $\dfrac{\sqrt[3]{100 - 612}}{8}$

25. $\sqrt[3]{125}\,\sqrt[3]{4}$

List the elements in the sets in Exercises 26–31, where x is a real number.

26. $\{x \mid x^5 = 10^5\}$

27. $\{x \mid x^5 = -10^5\}$

28. $\{x \mid x = \sqrt[5]{10^5}\}$

29. $\{x \mid x = \sqrt[5]{-10^5}\}$

30. $\{x \mid x^6 = 64\}$

31. $\{x \mid x = \sqrt[6]{64}\}$

32. $\{x \mid x^6 = -64\}$

6.5 FRACTIONAL EXPONENTS

It is useful to assign a meaning to expressions such as $9^{1/2}$, $8^{1/3}$, $x^{3/4}$, and $x^{-2/3}$. To see if it is possible to define a power having a fractional exponent and still have the five exponent theorems remain valid, the second theorem of exponents, $(x^m)^n = x^{mn}$, is examined first. If the second theorem is to be valid, then

$$(9^{1/2})^2 = 9^{1/2 \cdot 2} = 9^1 = 9$$

But

$$(9^{1/2})^2 = 9^{1/2} . 9^{1/2} = 9$$

and

$$\sqrt{9}\,\sqrt{9} = 9$$

Thus a possibility is $9^{1/2} = \sqrt{9} = 3$.

Similarly

$$(8^{1/3})^3 = 8^{1/3 \cdot 3} = 8^1 = 8$$

$$(8^{1/3})(8^{1/3})(8^{1/3}) = 8$$

$$\sqrt[3]{8}\,\sqrt[3]{8}\,\sqrt[3]{8} = 8$$

This suggests the possibility $8^{1/3} = \sqrt[3]{8} = 2$.

Accordingly, if x is a nonnegative real number and n is a natural number, then $x^{1/n}$ is defined as $\sqrt[n]{x}$, the principal real root of x.

DEFINITIONS OF $x^{1/n}$ AND $(-x)^{1/n}$

If x is a positive real number, and n is a natural number,

$$x^{1/n} = \sqrt[n]{x}$$

$(-x)^{1/n} = -\sqrt[n]{x}$ **if n is odd (1, 3, 5, 7, . . .)**

$(-x)^{1/n}$ **is undefined if n is even (2, 4, 6, 8, . . .)**

$$0^{1/n} = 0$$

For example,

$$(64)^{1/3} = \sqrt[3]{64} = 4$$

$$(-64)^{1/3} = -\sqrt[3]{64} = -4$$

$$(64)^{1/2} = \sqrt{64} = 8$$

$$(-64)^{1/2} = \sqrt{-64}$$

The last equation is undefined since the index, 2, is even and the radicand, -64, is negative.

The following definitions are similar.

DEFINITIONS

$x^{m/n}$ **is defined as** $(x^m)^{1/n} = \sqrt[n]{x^m}$

and

$x^{-m/n}$ **is defined as** $\dfrac{1}{x^{m/n}}$

Note that $x^{m/n} = (x^{1/n})^m = (\sqrt[n]{x})^m$.

It can be shown that the five exponent theorems remain valid with these definitions. The results on exponents developed so far are summarized in the following table.

TABLE 6.1 EXPONENT DEFINITIONS AND THEOREMS

Assumptions: x and y are real numbers, $x \neq 0$ and $y \neq 0$
n and m are natural numbers
a and b are rational numbers

Definitions	*Theorems*
1. $x^n = xx\ldots x$ (n factors)	1. $x^a x^b = x^{a+b}$
2. $x^1 = x$	2. $(x^a)^b = x^{ab}$
3. $x^0 = 1$	3. $\dfrac{x^a}{x^b} = x^{a-b}$
4. $x^{-n} = \dfrac{1}{x^n}$	4. $(xy)^a = x^a y^a$
5. $x^{1/n} = \sqrt[n]{x}$ ($x > 0$ if n even)	5. $\left(\dfrac{x}{y}\right)^a = \dfrac{x^a}{y^a}$
6. $x^{m/n} = \sqrt[n]{x^m}$ ($x > 0$ if n even)	
7. $x^{-m/n} = \dfrac{1}{\sqrt[n]{x^m}}$ ($x > 0$ if n even)	
8. $0^a = 0$ if $a \neq 0$	

EXAMPLE 6.5.1 Simplify $(16)^{3/4}$.

Solution Factoring the base, $16 = 2^4$, then

$$(16)^{3/4} = (2^4)^{3/4} = 2^{4\cdot 3/4} = 2^3 = 8$$

EXAMPLE 6.5.2 Simplify $(64)^{-2/3}$.

Solution Factoring the base, $64 = 2^6$, then

$$(64)^{-2/3} = (2^6)^{-2/3} = 2^{6(-2/3)} = 2^{-4} = \frac{1}{2^4} = \frac{1}{16}$$

EXAMPLE 6.5.3 Simplify

$$\left(\frac{1}{16}\right)^{-1/2}$$

Solution

$$\left(\frac{1}{16}\right)^{-1/2} = (16)^{1/2} = \sqrt{16} = 4$$

EXAMPLE 6.5.4 Simplify $[(-6)^2]^{1/2}$.

Solution $[(-6)^2]^{1/2} = \sqrt{(-6)^2} = \sqrt{36} = 6$

It is important to note that the theorems do not apply whenever the base is negative and an exponent indicates a root whose index is even. Thus

$$[(-6)^2]^{1/2} \neq (-6)^1$$

EXAMPLE 6.5.5 Simplify $(10^{-2/3}10^{5/6})^3$.

Solution

$$\begin{aligned}(10^{-2/3}10^{5/6})^3 &= (10^{-4/6}10^{5/6})^3 \\ &= (10^{1/6})^3 = 10^{3/6} = 10^{1/2} = \sqrt{10}\end{aligned}$$

EXAMPLE 6.5.6 Simplify

$$\left(\frac{2^{3/4}x^{-3/4}}{2^{1/2}}\right)^{-4}$$

Solution

$$\begin{aligned}\left(\frac{2^{3/4}x^{-3/4}}{2^{1/2}}\right)^{-4} &= (2^{3/4-2/4}x^{-3/4})^{-4} \\ &= (2^{1/4})^{-4}(x^{-3/4})^{-4} \\ &= 2^{-1}x^3 = \frac{x^3}{2}\end{aligned}$$

EXERCISES 6.5 A

In Exercises 1–5, verify each given example of an exponent theorem by simplifying each side. State which theorem is illustrated.

1. $16^{1/2}16^{1/4} = 16^{1/2+1/4}$

2. $[(64)^{1/2}]^{1/3} = 64^{1/2 \cdot 1/3}$

3. $\dfrac{16^{3/4}}{16^{1/4}} = 16^{3/4-1/4}$

4. $(16 \cdot 625)^{1/4} = (16)^{1/4}(625)^{1/4}$

5. $\left(\dfrac{10{,}000}{16}\right)^{1/4} = \dfrac{(10{,}000)^{1/4}}{(16)^{1/4}}$

In Exercises 6–50, express each as a simplified rational number or as a simplified quotient. Assume that all variables are positive. No exponents should appear in the answer.

6. $4^{1/2}$

7. $(16)^{1/2}$

8. $(16)^{1/4}$

9. $(27)^{1/3}$

10. $(32)^{1/5}$

11. $\left(\frac{27}{125}\right)^{1/3}$

12. $\left(\frac{81}{256}\right)^{1/4}$

13. $(64)^{2/3}$

14. $(64)^{3/2}$

15. $8^{-2/3}$

16. $(25)^{-3/2}$

17. $(100{,}000)^{-3/5}$

18. $\left(\frac{1}{27}\right)^{-2/3}$

19. $\left(-\frac{1}{125}\right)^{-2/3}$

20. $\left(\frac{216}{343}\right)^{2/3}$

21. $\left(\frac{1}{16}\right)^{3/4}$

22. $\left(\frac{1}{16}\right)^{-3/4}$

23. $(-8)^{2/3}$

24. $-8^{2/3}$

25. $(-8)^{-2/3}$

26. $-8^{-2/3}$

27. $(5^{3/2})^4$

28. $(8^{3/2})^{-2/3}$

29. $(10^{-3/5})^{-5/6}$

30. $(4^2)^{1/6}$

31. $\left(\frac{1}{14}\right)^{-1/4}$

32. $8^{1/2}8^{1/6}$

33. $5^{1/2}5^{-1/6}$

34. $10^{1.2}10^{3.8}$

35. $10^{-1.4}10^{1.9}$

36. $10^{-0.4}10^{-0.6}$

37. $(10^{-2}10^{-1/2})^{-1/2}$

38. $(5 \cdot 5^{1/2})^{1/3}$

39. $\frac{16^{1/3}}{4^{1/6}}$

40. $\frac{10^{2/3}}{10^{1/6}}$

★ **41.** $2^{1/2} \cdot 4^{1/3} \cdot (32)^{1/6}$

42. $(x^2y^4)^{1/4}$, $x > 0$, $y > 0$

43. $(4x^{2/3})^{-3}$

44. $4(x^{2/3})^{-3}$

45. $(9x^{-4}y^2)^{1/2}$

46. $(10x^{1/6}y^{5/6})^{-6}$

★ **47.** $(x^{1/2} + y^{1/2})^2$

★ **48.** $(x^{1/2} + 3^{1/2})(x^{1/2} - 3^{1/2})$

★ **49.** $x^{1/2}(2x^{1/2} + 5x^{-1/2})$

50. $(3^2 + 4^2)^{1/2}$

EXERCISES 6.5 B

In Exercises 1–5, verify each given example of an exponent theorem by simplifying each side. State which theorem is illustrated.

1. $(64)^{1/2}(64)^{1/6} = (64)^{1/2+1/6}$

2. $[(81)^{1/2}]^{1/2} = (81)^{1/2 \cdot 1/2}$

3. $\frac{(64)^{1/2}}{(64)^{1/6}} = 64^{1/2-1/6}$

4. $(8 \cdot 125)^{1/3} = 8^{1/3}(125)^{1/3}$

5. $\left(\frac{1000}{125}\right)^{1/3} = \frac{(1000)^{1/3}}{(125)^{1/3}}$

In Exercises 6–50, express each as a simplified rational number or as a simplified quotient. Assume that all variables are positive. No exponents should appear in the answer.

6. $(25)^{1/2}$

7. $(81)^{1/4}$

8. $(81)^{1/2}$

9. $(216)^{1/3}$

10. $(243)^{1/5}$

11. $\left(\frac{64}{81}\right)^{1/2}$

12. $\left(\frac{1000}{27}\right)^{1/3}$

13. $(32)^{3/5}$

14. $(100)^{3/2}$

15. $(16)^{-3/2}$

16. $(125)^{-2/3}$

17. $(32)^{-2/5}$

18. $\left(\frac{1}{125}\right)^{-2/3}$

19. $\left(-\frac{729}{512}\right)^{-2/3}$

20. $\left(\frac{1}{16}\right)^{3/2}$

21. $\left(\frac{1}{16}\right)^{-3/2}$

22. $\left(\frac{25}{16}\right)^{-3/2}$

23. $-125^{2/3}$

24. $(-125)^{2/3}$

25. $-125^{-2/3}$

26. $(-125)^{-2/3}$

27. $(6^{3/4})^{-4/3}$

28. $(7^{-2/5})^{-5/4}$

29. $(5^{6})^{-1/2}$

30. $(25^{3})^{1/6}$

31. $(25^{3})^{-1/6}$

32. $2^{-2/3}2^{-1/3}$

33. $4^{3/4}4^{-1/4}$

34. $10^{3.5}10^{-1.5}$

35. $10^{-1.47}10^{1.72}$

36. $10^{-0.08}10^{-0.17}$

37. $(2 \cdot 2^{1/3})^{1/2}$

38. $(2^{3} \cdot 2^{1/2})^{1/2}$

39. $\frac{7^{1/2}}{7^{1/4}}$

40. $\frac{(216)^{1/4}}{6^{1/12}}$

41. $7^{1/2} \cdot 7^{1/3} \cdot 7^{1/6}$

42. $8x^{-2/3}$

43. $(8x)^{-2/3}$

44. $(x^{-3/4})^{-4/3}$

★ **45.** $\{[(81)(81)^{1/2}]^{1/3}\}^{1/2}$

46. $(x^{1/2} - x^{-1/2})^{2}$

★ **47.** $(5 - x^{1/2})(5 + x^{1/2})$

48. $(x^{1/2} - y^{1/2})(x^{1/2} + y^{1/2})$

★ **49.** $x^{-1/2}(x^{-1/2} + x^{1/2})$

50. $x^{1/4}(x^{3/4} + x^{-3/4} + x^{-1/4})$

6.6 RADICALS: SIMPLIFICATION AND PRODUCTS

A radical of the form $\sqrt{M}$, where M is a monomial with an integer for its coefficient, is said to be **simplified** *if no perfect squares are factors of the radicand, M.*

Similarly, a radical of the form $\sqrt[3]{M}$ is said to be **simplified** *if no perfect cubes are factors of the radicand.*

Renaming a number denoted by a radical so that any resulting radical is simplified is called **reducing the radicand**.

To reduce the radicand of a square root radical, the product of square roots theorem is used.

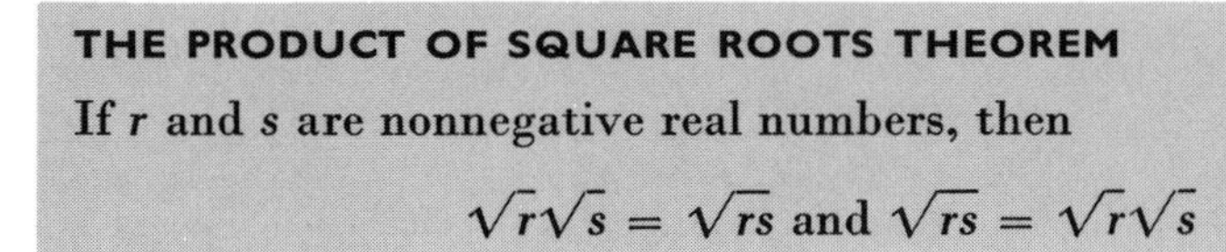

THE PRODUCT OF SQUARE ROOTS THEOREM

If r and s are nonnegative real numbers, then

$$\sqrt{r}\sqrt{s} = \sqrt{rs} \text{ and } \sqrt{rs} = \sqrt{r}\sqrt{s}$$

Since $r \geq 0$ and $s \geq 0$, by using exponents,

$$\sqrt{r}\,\sqrt{s} = r^{1/2}s^{1/2} = (rs)^{1/2} = \sqrt{rs}$$

EXAMPLE 6.6.1 Simplify $\sqrt{3}\,\sqrt{12}$.

Solution $\sqrt{3}\,\sqrt{12} = \sqrt{3 \cdot 12} = \sqrt{36} = 6$

The following theorem is useful for simplifying square root radicals.

THEOREM

If x and y are any nonnegative real numbers,

$$\sqrt{x^2y} = \sqrt{x^2}\sqrt{y} = x\sqrt{y}$$

EXAMPLE 6.6.2 Simplify $\sqrt{75}$.

Solution The basic idea is to factor the radicand to find the perfect square factors:

$$75 = 3 \cdot 5^2$$

$$\sqrt{75} = \sqrt{5^2 \cdot 3} = \sqrt{5^2}\,\sqrt{3} = 5\sqrt{3}$$

Alternate Solution $\sqrt{75} = (5^2 \cdot 3)^{1/2} = (5^2)^{1/2}(3^{1/2}) = 5\sqrt{3}$

EXAMPLE 6.6.3 Simplify $\sqrt{6x}\,\sqrt{12xy^3}$, where $x \geq 0$ and $y \geq 0$.

Solution

$$\begin{aligned}\sqrt{6x}\,\sqrt{12xy^3} &= \sqrt{6x(12xy^3)}\\ &= \sqrt{(2\cdot 3)(2\cdot 2\cdot 3)x^2y^2y}\\ &= \sqrt{2^2 3^2 x^2 y^2}\,\sqrt{2y}\\ &= \sqrt{(6xy)^2}\,\sqrt{2y}\\ &= 6xy\,\sqrt{2y}\end{aligned}$$

Alternate Solution

$$\begin{aligned}(6x)^{1/2}(12xy^3)^{1/2} &= (72x^2y^3)^{1/2}\\ &= (2^2\cdot 3^2\cdot 2x^2y^2y)^{1/2}\\ &= (6^2x^2y^2)^{1/2}(2y)^{1/2}\\ &= 6xy\,\sqrt{2y}\end{aligned}$$

Reducing the radicand of a cube root involves the product of cube roots theorem.

THE PRODUCT OF CUBE ROOTS THEOREM

If a and b are any real numbers, then

$$\sqrt[3]{a}\sqrt[3]{b} = \sqrt[3]{ab} \text{ and } \sqrt[3]{ab} = \sqrt[3]{a}\sqrt[3]{b}$$

$$\sqrt[3]{a}\sqrt[3]{b} = a^{1/3}b^{1/3} = (ab)^{1/3} = \sqrt[3]{ab}$$

EXAMPLE 6.6.4 Simplify $\sqrt[3]{20}\,\sqrt[3]{50}$.

Solution $\sqrt[3]{20}\,\sqrt[3]{50} = \sqrt[3]{20(50)} = \sqrt[3]{1000} = \sqrt[3]{10^3} = 10$

The following theorem is useful for simplifying cube root radicals.

THEOREM

If x and y are any real numbers, then

$$\sqrt[3]{x^3y} = \sqrt[3]{x^3}\sqrt[3]{y} = x\sqrt[3]{y}$$

EXAMPLE 6.6.5 Simplify $\sqrt[3]{500}$.

Solution $500 = 2 \cdot 2 \cdot 125 = 2^2(125) = 2^2 5^3$

$$\sqrt[3]{500} = \sqrt[3]{5^3 2^2} = \sqrt[3]{5^3}\sqrt[3]{2^2}$$
$$= 5\sqrt[3]{4}$$

Alternate Solution $\sqrt[3]{500} = (5^3 \cdot 2^2)^{1/3} = (5^{3 \cdot 1/3})(2^{2/3}) = 5\sqrt[3]{4}$

EXAMPLE 6.6.6 Simplify $\sqrt[3]{-81x^6}$.

Solution
$$\sqrt[3]{-81x^6} = -\sqrt[3]{81x^6}$$
$$= -\sqrt[3]{3^4x^6}$$
$$= -\sqrt[3]{3^3 \cdot 3 \cdot (x^2)^3}$$
$$= -\sqrt[3]{3^3(x^2)^3}\sqrt[3]{3}$$
$$= -3x^2\sqrt[3]{3}$$

Alternate Solution
$$\sqrt[3]{-81x^6} = -(3^4x^6)^{1/3}$$
$$= -(3^{4/3})(x^{6/3})$$
$$= -(3^{1+1/3})(x^2)$$
$$= -(3 \cdot 3^{1/3})(x^2)$$
$$= -3x^2\sqrt[3]{3}$$

EXERCISES 6.6 A

Simplify in Exercises 1–40. Assume all variables to be positive.

1. $\sqrt{5}\sqrt{20}$

2. $\sqrt{50}\sqrt{2}$

3. $\sqrt{3x}\sqrt{12x}$

4. $\sqrt{y}\sqrt{y^3}$

5. $\sqrt{12}$

6. $\sqrt{125}$

7. $\sqrt{405}$

8. $\sqrt{9x}$

9. $\sqrt{24x^2}$

10. $\sqrt{40y^3}$

11. $\sqrt{x}\sqrt{x^3}$

12. $\sqrt{y^3}\sqrt{y^4}$

13. $\sqrt{6x}\sqrt{6x^2}$

14. $\sqrt{40x^5}$

15. $\sqrt{32x}$

16. $\sqrt{1452}$

17. $\sqrt{343x^3}$

18. $5\sqrt{72}$

19. $2x\sqrt{54x^4}$

20. $10xy\sqrt{800x^2y}$

21. $(\sqrt{5} + \sqrt{2})(\sqrt{5} - \sqrt{2})$

22. $(4 - \sqrt{3})(4 + \sqrt{3})$

23. $(\sqrt{6} - 2)(\sqrt{6} + 2)$

24. $(1 - \sqrt{2})(1 + \sqrt{2})$

25. $(\sqrt{8} - \sqrt{2})(\sqrt{8} + \sqrt{2})$

26. $\sqrt[3]{54}$

27. $\sqrt[3]{-54}$

28. $\sqrt[3]{72}$

29. $\sqrt[3]{16}$

30. $\sqrt[3]{2}\,\sqrt[3]{4}$

31. $\sqrt[3]{9}\,\sqrt[3]{81}$

32. $\sqrt[3]{20x}\,\sqrt[3]{50x^2}$

33. $\sqrt[3]{32x}$

34. $\sqrt[3]{108y^2}$

35. $\sqrt[3]{40y^4}$

36. $\sqrt[3]{x}\,\sqrt[3]{x^2}$

37. $\sqrt[3]{4y^2}\,\sqrt[3]{4y^4}$

38. $2y\sqrt[3]{54y^2}$

39. $\sqrt[3]{6x^2}\,\sqrt[3]{36x}$

40. $-2\sqrt[3]{-81}$

EXERCISES 6.6 B

Simplify in Exercises 1–40. Assume all variables to be positive.

1. $\sqrt{2}\,\sqrt{18}$

2. $\sqrt{27}\,\sqrt{3}$

3. $\sqrt{5x}\,\sqrt{45x}$

4. $\sqrt{xy^3}\,\sqrt{x^3y}$

5. $\sqrt{24}$

6. $\sqrt{128}$

7. $\sqrt{500}$

8. $\sqrt{49y}$

9. $\sqrt{18y^3}$

10. $\sqrt{28y^4}$

11. $\sqrt{x^3}\,\sqrt{x^5}$

12. $\sqrt{2x}\,\sqrt{8x^3}$

13. $\sqrt{x^3y^5}$

14. $\sqrt{54y^7}$

15. $\sqrt{108y}$

16. $\sqrt{1849}$

17. $\sqrt{112x^3}$

18. $4\sqrt{363}$

19. $3x\sqrt{200x^6}$

20. $10xy\sqrt{81xy^2z^3}$

21. $(\sqrt{6} - \sqrt{2})(\sqrt{6} + \sqrt{2})$

22. $\dfrac{\sqrt{3} + 1}{2} \cdot \dfrac{\sqrt{3} - 1}{2}$

23. $(\sqrt{x} - \sqrt{3})(\sqrt{x} + \sqrt{3})$

24. $(2\sqrt{x} + y)(2\sqrt{x} - y)$

25. $(\sqrt{3x} - \sqrt{2x})(\sqrt{3x} + \sqrt{2x})$

26. $\sqrt[3]{128}$

27. $\sqrt[3]{-128}$

28. $\sqrt[3]{500}$

29. $\sqrt[3]{625}$

30. $\sqrt[3]{5}\,\sqrt[3]{25}$

31. $\sqrt[3]{12}\,\sqrt[3]{18}$

32. $\sqrt[3]{98x^2}\,\sqrt[3]{28x}$

33. $\sqrt[3]{24x^2}$

34. $\sqrt[3]{256y}$

35. $\sqrt[3]{56y^5}$

36. $\sqrt[3]{x^2}\,\sqrt[3]{x^4}$

37. $\sqrt[3]{25y^3}\,\sqrt[3]{25y^6}$

38. $5y^2\sqrt[3]{80y^4}$

39. $\sqrt[3]{49x}\,\sqrt[3]{7x^2}$

40. $-5\sqrt[3]{-625}$

6.7 RADICALS: RATIONALIZING

In order to rename an expression that has a radical in a denominator or a fraction in a radicand, the quotient of square roots theorem is used for expressions involving square roots. The process is called **rationalizing the denominator**, or **simplification.**

THE QUOTIENT OF SQUARE ROOTS THEOREM

If r and s are positive real numbers, then

$$\sqrt{\frac{r}{s}} = \frac{\sqrt{r}}{\sqrt{s}} \text{ and } \frac{\sqrt{r}}{\sqrt{s}} = \sqrt{\frac{r}{s}}$$

Since r and s are positive,

$$\sqrt{\frac{r}{s}} = \left(\frac{r}{s}\right)^{1/2} = \frac{r^{1/2}}{s^{1/2}} = \frac{\sqrt{r}}{\sqrt{s}}$$

EXAMPLE 6.7.1 Simplify

$$\sqrt{\frac{2}{3}}$$

Solution

$$\sqrt{\frac{2}{3}} = \frac{\sqrt{2}}{\sqrt{3}}$$ (By the quotient of square roots theorem)

$$= \frac{\sqrt{2}\,\sqrt{3}}{\sqrt{3}\,\sqrt{3}}$$ (By the Fundamental Theorem of Fractions—the numerator and denominator are multiplied by that number that causes the radicand of the denominator to become a perfect square)

Thus $$\sqrt{\frac{2}{3}} = \frac{\sqrt{6}}{3}$$

EXAMPLE 6.7.2 Rationalize

$$\frac{2}{\sqrt{27}}$$

Solution

$$\frac{2}{\sqrt{27}} = \frac{2\sqrt{3}}{\sqrt{27}\,\sqrt{3}}$$ (Multiplying numerator and denominator by the smallest number possible so that the resulting radicand in the denominator is a perfect square)

$$= \frac{2\sqrt{3}}{\sqrt{81}}$$

Thus

$$\frac{2}{\sqrt{27}} = \frac{2\sqrt{3}}{9}$$

If the denominator is a sum or difference of terms involving a radical, then the difference of squares theorem provides a technique for rationalizing denominators involving a square root. In other words, a fraction having an irrational denominator may be renamed as a fraction with a rational denominator. The basic idea involved is as follows:

$$X^2 - Y^2 = (X + Y)(X - Y)$$

$$a - b = (\sqrt{a} + \sqrt{b})(\sqrt{a} - \sqrt{b})$$

If the denominator has the form $\sqrt{a} + \sqrt{b}$, then multiplication by $\sqrt{a} - \sqrt{b}$ (called its conjugate) will produce the rational number $a - b$. This assumes, of course, that a and b are positive rational numbers.

Similarly, multiplying $\sqrt{a} - \sqrt{b}$ by its conjugate $\sqrt{a} + \sqrt{b}$ will produce the rational number $a - b$.

EXAMPLE 6.7.3 Rationalize

$$\frac{1}{\sqrt{5} - 1}$$

Solution

$$\frac{1}{\sqrt{5} - 1} \cdot \frac{\sqrt{5} + 1}{\sqrt{5} + 1} = \frac{\sqrt{5} + 1}{5 - 1} = \frac{\sqrt{5} + 1}{4}$$

EXAMPLE 6.7.4 Simplify

$$\frac{2}{6+\sqrt{2}}$$

Solution

$$\frac{2}{6+\sqrt{2}}\cdot\frac{6-\sqrt{2}}{6-\sqrt{2}} = \frac{2(6-\sqrt{2})}{36-2} = \frac{2(6-\sqrt{2})}{34} = \frac{6-\sqrt{2}}{17}$$

EXAMPLE 6.7.5 Express with a rational denominator

$$\frac{3}{\sqrt{5}-\sqrt{2}}$$

Solution

$$\frac{3}{\sqrt{5}-\sqrt{2}}\cdot\frac{\sqrt{5}+\sqrt{2}}{\sqrt{5}+\sqrt{2}} = \frac{3(\sqrt{5}+\sqrt{2})}{5-2} = \frac{3(\sqrt{5}+\sqrt{2})}{3}$$

$$= \sqrt{5}+\sqrt{2}$$

EXERCISES 6.7 A

In Exercises 1–25, rationalize the denominator and write in simplest radical form. Assume all variables to be positive.

1. $\sqrt{\frac{1}{2}}$

2. $\sqrt{\frac{1}{6}}$

3. $\sqrt{\frac{5}{12}}$

4. $\frac{1}{\sqrt{5}}$

5. $\frac{1}{\sqrt{20}}$

6. $\sqrt{\frac{2}{25}}$

7. $\sqrt{\frac{3}{50}}$

8. $\frac{8}{\sqrt{12}}$

9. $\frac{14\sqrt{9}}{3\sqrt{7}}$

10. $\frac{\sqrt{64x^4y^6}}{\sqrt{128x^6y^6}}$

11. $\sqrt{\frac{2}{x}}$

12. $\sqrt{\frac{8}{y}}$

13. $\frac{6}{\sqrt{72x^3}}$

14. $3\sqrt{\frac{1}{3}}\cdot 5\sqrt{\frac{3}{5}}$

15. $\dfrac{\sqrt{3}}{\sqrt{6x}}$

16. $\dfrac{1}{\sqrt{2} - 1}$

17. $\dfrac{4}{\sqrt{7} + \sqrt{3}}$

18. $\dfrac{\sqrt{5}}{2 - \sqrt{5}}$

19. $\dfrac{5}{\sqrt{11} + 1}$

20. $\dfrac{2}{\sqrt{x} - 1}$

21. $\dfrac{\sqrt{6}}{\sqrt{3} + \sqrt{27}}$

22. $\dfrac{\sqrt{25}}{\sqrt{16} + \sqrt{9}}$

23. $\dfrac{\sqrt{25}}{\sqrt{16 + 9}}$

★ 24. $\dfrac{x - y}{\sqrt{x} - \sqrt{y}}$

★ 25. $\dfrac{x - 4y}{\sqrt{x} + \sqrt{2y}}$

EXERCISES 6.7 B

In Exercises 1–25, rationalize the denominator and write in simplest radical form. Assume all variables to be positive.

1. $\sqrt{\dfrac{1}{3}}$

2. $\sqrt{\dfrac{3}{5}}$

3. $\sqrt{\dfrac{3}{8}}$

4. $\dfrac{1}{\sqrt{10}}$

5. $\dfrac{1}{\sqrt{18}}$

6. $\sqrt{\dfrac{3}{12}}$

7. $\sqrt{\dfrac{3}{32}}$

8. $\dfrac{9}{\sqrt{45}}$

9. $\dfrac{4}{2\sqrt{200}}$

10. $\dfrac{21\sqrt{8y^3}}{\sqrt{49y}}$

11. $\sqrt{\dfrac{1}{3x}}$

12. $\sqrt{\dfrac{125}{y^3}}$

13. $\dfrac{4}{\sqrt{32x}}$

14. $\sqrt{\dfrac{7}{8x}}\sqrt{\dfrac{2x^3}{49}}$

15. $\dfrac{\sqrt{2x}}{\sqrt{10x}}$

16. $\dfrac{1}{1 + \sqrt{3}}$

17. $\dfrac{\sqrt{3}}{\sqrt{7} - \sqrt{3}}$

18. $\dfrac{4}{\sqrt{7} + \sqrt{5}}$

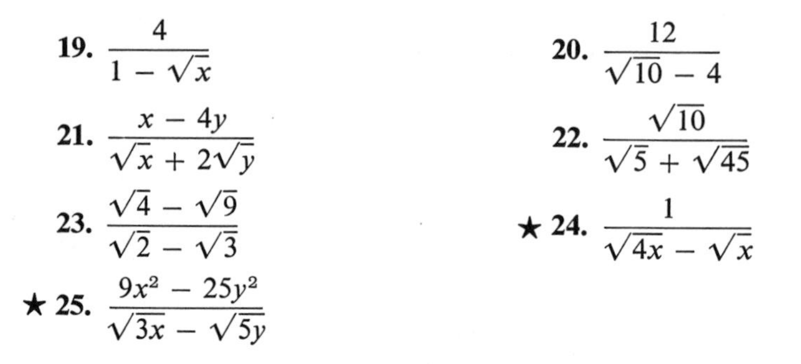

19. $\dfrac{4}{1 - \sqrt{x}}$

20. $\dfrac{12}{\sqrt{10} - 4}$

21. $\dfrac{x - 4y}{\sqrt{x} + 2\sqrt{y}}$

22. $\dfrac{\sqrt{10}}{\sqrt{5} + \sqrt{45}}$

23. $\dfrac{\sqrt{4} - \sqrt{9}}{\sqrt{2} - \sqrt{3}}$

★ 24. $\dfrac{1}{\sqrt{4x} - \sqrt{x}}$

★ 25. $\dfrac{9x^2 - 25y^2}{\sqrt{3x} - \sqrt{5y}}$

6.8 RADICALS: SUMS AND DIFFERENCES

It is often possible to express a sum or difference of two radical terms as a single term. For example, since $\sqrt{5}$ is a real number, so are $3\sqrt{5}$ and $4\sqrt{5}$ and $3\sqrt{5} + 4\sqrt{5}$ by the closure axioms for real numbers. Now using the distributive axiom,

$$3\sqrt{5} + 4\sqrt{5} = (3 + 4)\sqrt{5} = 7\sqrt{5}$$

Radicals that have the same radicands *and* the same indices are called **like radicals**. Radicals that are not like are called unlike. For example,

$3\sqrt{5x}$ and $4\sqrt{5x}$ contain like radicals

$\sqrt{5x}$ and $\sqrt[3]{5x}$ are unlike radicals

$\sqrt{5x}$ and $\sqrt{3x}$ are unlike radicals

Sometimes it is necessary to reduce each radical term to simplified radical form in order to identify like radicals.

EXAMPLE 6.8.1 Simplify $\sqrt{50x} + \sqrt{18x}$ if possible.

Solution

$$\sqrt{50x} + \sqrt{18x} = 5\sqrt{2x} + 3\sqrt{2x} \quad \text{(Simplifying each radical)}$$

$$= (5 + 3)\sqrt{2x} \quad \text{(Using the distributive axiom)}$$

$$= 8\sqrt{2x}$$

EXAMPLE 6.8.2 Simplify $\sqrt{48} + \sqrt{18} - \sqrt{12}$.

Solution

$$\begin{aligned}\sqrt{48} + \sqrt{18} - \sqrt{12} &= 4\sqrt{3} + 3\sqrt{2} - 2\sqrt{3} && \text{(Simplifying)}\\ &= (4\sqrt{3} - 2\sqrt{3}) + 3\sqrt{2} && \text{(Collecting like radicals)}\\ &= (4 - 2)\sqrt{3} + 3\sqrt{2} && \text{(Using the distributive axiom)}\\ &= 2\sqrt{3} + 3\sqrt{2} && \text{(Simplified form)}\end{aligned}$$

EXAMPLE 6.8.3 Simplify

$$\sqrt{40} - \sqrt{\frac{2}{5}}$$

Solution

$$\begin{aligned}\sqrt{40} - \sqrt{\frac{2}{5}} &= \sqrt{4 \cdot 10} - \sqrt{\frac{2}{5} \cdot \frac{5}{5}}\\ &= 2\sqrt{10} - \frac{\sqrt{10}}{5}\\ &= \left(2 - \frac{1}{5}\right)\sqrt{10}\\ &= \frac{9\sqrt{10}}{5}\end{aligned}$$

EXERCISES 6.8 A

Write the expressions in Exercises 1–28 in simplest radical form. Assume all variables to be positive.

1. $6\sqrt{5} + 2\sqrt{5}$

2. $7\sqrt{2} - 4\sqrt{2}$

3. $5\sqrt{6} + \sqrt{6}$

4. $9\sqrt{10} - 8\sqrt{10}$

5. $3\sqrt{5} + 5\sqrt{3}$

6. $5\sqrt{3} - 4\sqrt{3} + 2\sqrt{3}$

7. $6\sqrt{14x} - 4\sqrt{14x}$

8. $2\sqrt{2x} + 3\sqrt{3x}$

9. $\sqrt{27} + \sqrt{3}$

10. $\sqrt{40} - \sqrt{10}$

11. $3\sqrt{24} + 2\sqrt{54}$

12. $\sqrt{32x} + \sqrt{98x}$

13. $\sqrt{52x} - \sqrt{13x}$

14. $\sqrt{63x^2} + \sqrt{28x^2}$

15. $4\sqrt{9x^2} - 3\sqrt{4x^2}$

16. $\sqrt{4y^2} + 4\sqrt{y^2}$

17. $\sqrt{\frac{2}{5}} + \sqrt{\frac{1}{10}}$

18. $\sqrt{18} - \sqrt{\frac{1}{18}}$

19. $\sqrt{20} + \sqrt{12} + \sqrt{45}$

20. $\sqrt{16x} - \sqrt{4y} - \sqrt{9y}$

21. $\sqrt{\frac{42}{25}} + \sqrt{2\frac{5}{8}}$

22. $\sqrt{3x} - \sqrt{18x} + \sqrt{12x}$

23. $\sqrt{224} + 4\sqrt{\frac{1}{2}} - \frac{1}{2}\sqrt{50}$

24. $(\sqrt{3} + \sqrt{2})^2$

25. $(2\sqrt{3} - 3\sqrt{2})^2$

26. $(\sqrt{6x} + \sqrt{3x})^2$

27. $\sqrt{3}(\sqrt{6} + \sqrt{12} + \sqrt{24})$

28. $\sqrt{5}(\sqrt{10} - \sqrt{15} - \sqrt{40})$

29. Find the value of $x^2 - 10x + 23$ for $x = 5 + \sqrt{2}$.

30. Find the value of $x^2 + x - 1$ for $x = \sqrt{5} - 1$.

EXERCISES 6.8 B

Write the expressions in Exercises 1–28 in simplest radical form. Assume all variables to be positive.

1. $7\sqrt{3} - 5\sqrt{3}$

2. $5\sqrt{7} + 4\sqrt{7}$

3. $8\sqrt{6} - 7\sqrt{6}$

4. $2\sqrt{15} + \sqrt{15}$

5. $2\sqrt{30} + 5\sqrt{30} - 4\sqrt{30}$

6. $7\sqrt{5} - 5\sqrt{7}$

7. $8\sqrt{3x} - 5\sqrt{3y}$

8. $6\sqrt{7x} + 2\sqrt{7x}$

9. $\sqrt{20} + \sqrt{5}$

10. $\sqrt{72} - \sqrt{50}$

11. $5\sqrt{18} - 4\sqrt{8}$

12. $\sqrt{75x} + \sqrt{108x}$

13. $\sqrt{20x} - \sqrt{45x}$

14. $\sqrt{150x^2} - \sqrt{24x^2}$

15. $3\sqrt{25x^2} - 5\sqrt{9x^2}$

16. $\sqrt{9y^2} + 9\sqrt{y^2}$

17. $\sqrt{24} + \sqrt{\frac{3}{8}}$

18. $\sqrt{\frac{2}{5}} - \sqrt{\frac{5}{2}}$

19. $\sqrt{54} + \sqrt{32} - \sqrt{24}$

20. $\sqrt{25x} - \sqrt{36y} - \sqrt{x}$

21. $\sqrt{\frac{7}{8}} + \sqrt{\frac{5}{6}} - \sqrt{\frac{7}{2}}$

22. $\sqrt{\frac{2}{x}} + \sqrt{\frac{1}{2x}} + \sqrt{2x}$

23. $\frac{2}{\sqrt{12}} + 2\sqrt{81} + \frac{\sqrt{27}}{\sqrt{3}}$

24. $(\sqrt{5} - \sqrt{3})^2$

25. $(2\sqrt{5} + 1)^2$

26. $(\sqrt{10x} - \sqrt{5x})^2$

27. $\sqrt{2}(\sqrt{8} - \sqrt{18} - \sqrt{12})$

28. $\sqrt{6}(\sqrt{12} + \sqrt{3} - \sqrt{2})$

29. Find the value of $x^2 - x + 2$ for $x = \sqrt{2} + \sqrt{3}$.

30. Find the value of $x^2 + 2x - 2$ for $x = \sqrt{3} - 1$.

6.9 SCIENTIFIC NOTATION

Many scientific measurements require the use of extremely large or extremely small numbers. For example, the speed of light is about 30,000,000,000 centimeters per second; the number of molecules in 1 cubic centimeter of gas at 0 degrees centigrade is about 30,000,000,000,000,000,000; the time for an electronic computer to do a certain arithmetical operation is 0.000 000 0024 second; and the mass of the hydrogen atom is about 0.000 000 000 000 000 000 000 001 672 gram.

A convenient and effective system for expressing such numbers is called scientific notation.

A number is expressed in scientific notation when it is written as a product of a decimal fraction between 1 and 10 and an integral power of 10.

DEFINITION

In symbols, a number written in scientific notation has the form

$$N \times 10^k$$

where N is a number between 1 and 10 in decimal form and k is an integer.

EXAMPLE 6.9.1

Ordinary Notation	*Scientific Notation*
3.14	3.14×10^0
20.5	2.05×10^1
608	6.08×10^2
5,000,000	5×10^6
0.14	1.4×10^{-1}
0.025	2.5×10^{-2}
0.000 000 167	1.67×10^{-7}

It may be noted that the exponent on 10 for the scientific notation indicates the number of places to move the decimal point to obtain the ordinary notation (to the right if the exponent is positive, to the left if the exponent is negative, and no change if the exponent is 0).

EXAMPLE 6.9.2 Express 30,000,000,000 centimeters per second, the speed of light, in scientific notation.

Solution 3.00×10^{10}

EXAMPLE 6.9.3 Express 3.1×10^{-5} inches, the diameter of an average red blood corpuscle, in ordinary notation.

Solution 0.000 031

EXAMPLE 6.9.4 Calculate and express the answer in ordinary notation:

$$\frac{(2.5 \times 10^{-3})(4.2 \times 10^{5})}{1.4 \times 10^{-1}}$$

Solution First multiply each number in scientific notation by

$$10^{k} \times 10^{-k}$$

to eliminate the decimal point:

$$\begin{aligned}\frac{(2.5 \times 10^{-3})(4.2 \times 10^{5})}{1.4 \times 10^{-1}} &= \frac{(25 \times 10^{-4})(42 \times 10^{4})}{14 \times 10^{-2}} \\ &= \frac{(25)(42)}{14} \cdot \frac{(10^{-4})(10^{4})}{10^{-2}} \\ &= (25)(3) \times 10^{-4+4-(-2)} \\ &= 75 \times 10^{2} = 7500\end{aligned}$$

EXAMPLE 6.9.5 Evaluate

$$\sqrt{(3.0 \times 10^{-3})^2 - 4(4.8 \times 10^{-4})(0.3 \times 10^{-2})}$$

Solution $\sqrt{(3.0 \times 10^{-3})^2 - 4(4.8 \times 10^{-4})(0.3 \times 10^{-2})}$

$$\begin{aligned}&= \sqrt{(3)^2(10^{-3})^2 - 4(4.8)(0.3)(10^{-4})(10^{-2})} \\ &= \sqrt{(9 \times 10^{-6}) - (5.76)(10^{-6})} \\ &= \sqrt{(9.00 - 5.76)10^{-6}} \\ &= \sqrt{3.24 \times 10^{-6}} \\ &= \sqrt{(3.24 \times 10^{2})(10^{-2}10^{-6})} \\ &= \sqrt{(324)(10^{-8})} \\ &= \sqrt{324}\,\sqrt{10^{-8}} \\ &= \sqrt{(4)(81)} \times (10^{-8})^{1/2} \\ &= (2)(9) \times 10^{-4} = 18 \times 10^{-4} = 0.0018\end{aligned}$$

EXERCISES 6.9 A

Write each number in Exercises 1–10 in scientific notation.

1. 92,900,000 miles; distance from the earth to the sun
2. 6,600,000,000,000,000,000,000 tons; weight of the earth
3. 11,400,000; population of Tokyo
4. 0.000 000 095 centimeter; wave length of certain X rays
5. 0.000 000 0024 second; time for an electronic computer to perform an addition
6. 2,210,000,000; heartbeats per normal lifetime
7. 0.00061 atmosphere; a gas pressure
8. 120,000; seating capacity of a football stadium
9. 0.002 205 pound; weight of one gram
10. 0.000 000 0667; constant of gravitation

Write each number in Exercises 11–20 in ordinary notation.

11. 2.3×10^3; pounds of pollution per car per year
12. -4.60×10^2 degrees Fahrenheit; absolute zero
13. 1.80×10^{-5}; ionization constant of acetic acid
14. 8.64×10^5 miles; diameter of the sun
15. 3.03×10^{-8} centimeters; grating space in calcite crystals
16. 1.745×10^{-2}; number of radians in 1 degree
17. 1.6667×10^{-1} inches; width of an em space (printing industry)
18. 4.80×10^{-10} absolute electrostatic units; electronic charge
19. 1.87×10^9 dollars; a congressional appropriation
20. 6.3×10^{18} electrons per second; for one ampere of current

Calculate, using scientific notation, in Exercises 21–25. Express the answer in ordinary notation.

21. $(5.4 \times 10^{-3})(2.0 \times 10^5)$
22. $\sqrt{(1.25 \times 10^{-2})(8.0 \times 10^{-3})}$
23. $\dfrac{6.9 \times 10^{-8}}{2.3 \times 10^{-6}}$
24. $\dfrac{(3.75 \times 10^{-6})(2.00 \times 10^9)}{2.5 \times 10^{-2}}$
25. $\dfrac{(1.2 \times 10^{-2})^3}{3.6 \times 10^{-5}}$

26. Find the number of radians in 60 degrees if 1 degree $= 1.745 \times 10^{-2}$ radians.

27. Find the number of em spaces across the width of a page 7 inches wide if 1 em space $= 1.6667 \times 10^{-1}$ inches.

28. How long does it take a spaceship traveling 2.8×10^4 mph to travel the 2.48×10^5 miles from the earth to the moon?

29. (Nuclear physics) Find the force F with which a helium nucleus and a neon nucleus repel each other when separated by a distance of 4×10^{-9} meters using

$$F = \frac{khn}{d^2}$$

where

$$k = 9 \times 10^9$$

$$h = 3.2 \times 10^{-19}$$

$$n = 1.6 \times 10^{-18}$$

$$d = 4 \times 10^{-9}$$

30. (Chemistry) Use

$$[H^+] = \frac{K}{ac}$$

to calculate the hydrogen ion concentration of a certain 0.1-molar solution given that

$$K = 1.0 \times 10^{-14}$$

$$a = 7.5 \times 10^{-5}$$

$$c = 0.1$$

EXERCISES 6.9 B

Write each number in Exercises 1–10 in scientific notation.

1. 2,000,000,000 light years; probable diameter of the universe

2. 5,870,000,000,000 miles; the distance light travels in a year, called a light-year

3. 300,000,000,000 dollars; national debt

4. 0.000 000 015 centimeter; radius of an atom

5. 0.000 011 foot per degree centigrade; expansion of steel pipe

6. 0.000 005 centimeter; size of a certain virus

7. 0.03 millimeter per second; rate of certain plant growth
8. 0.001 5625 square mile; area of one acre
9. 603,000,000,000,000,000,000,000; Avogadro's number
10. 3,500,000,000; approximate world population

Write each number in Exercises 11–20 in ordinary notation.

11. 4.9×10^{10} dollars; federal investment in public water supplies
12. 9.11×10^{-28} grams; mass of electron
13. 8.31×10^{7} ergs per degree-mole; molar gas constant
14. 3.4×10^{4} centimeters per second; velocity of sound
15. 1×10^{-8} centimeter; equals one angstrom, unit used to measure wave lengths
16. 3.937×10^{-1}; number of inches in 1 centimeter
17. 2.5×10^{-2} seconds; shutter speed of a motion picture camera
18. 7.1×10^{8} years; half-life of uranium 235
19. 5×10^{-7} centimeters; thickness of an oil film
20. 1.256×10^{8} cubic yards; of earth in Fort Peck Dam, largest dam in the world

Calculate, using scientific notation, in Exercises 21–25. Express the answer in ordinary notation.

21. $(3.6 \times 10^{-5})(1.1 \times 10^{4})$

22. $(4.3 \times 10^{-1})(4.7 \times 10^{-3})$

23. $\sqrt[3]{\dfrac{1.5 \times 10^{-12}}{1.2 \times 10^{-17}}}$

24. $\dfrac{(2.4 \times 10^{-5})(1.5 \times 10^{4})}{1.8 \times 10^{3}}$

25. $\dfrac{(1.4 \times 10^{-3})^{2}}{2.8 \times 10^{-4}}$

26. Find the number of inches in 100 centimeters if 1 centimeter $= 3.937 \times 10^{-1}$ inches.
27. Find how many seconds it takes a sound to travel 1.70×10^{6} centimeters if sound travels 3.4×10^{4} centimeters per second.
28. Find I (electric current flowing in diode) if

$$I = KE^{3/2} \text{ amperes}$$

where

$$K = 16 \times 10^{-6}$$

$$E = 225 \text{ volts}$$

29. (Chemistry) Find the pH of a solution if the electric potential E is measured as 4.57×10^{-1} volts where

$$\text{pH} = \frac{E - (2.80 \times 10^{-1})}{5.9 \times 10^{-2}}$$

30. (Television: vacuum-tubes) Find f if

$$f = \frac{1}{RC + rc}$$

where

$$R = 10^4$$

$$C = 0.05 \times 10^{-6}$$

$$r = 5 \times 10^4$$

$$c = 0.03 \times 10^{-6}$$

SUMMARY

☐ **The axiom of completeness.** Each point on the number line corresponds to exactly one real number, and each real number corresponds to exactly one point on the number line.

☐ **The theorem of Pythagoras.** The square of the length of the hypotenuse of a right triangle is equal to the sum of the squares of the lengths of the legs of the right triangle.

Radicals

☐ **Definition of radicals.** If x is a real number and n is an odd natural number, then $\sqrt[n]{x^n} = x$. If x is a nonnegative real number and n is an even natural number, then $\sqrt[n]{x^n} = x$.

☐ **The product of square roots theorem.** If r and s are nonnegative real numbers, then $\sqrt{rs} = \sqrt{r}\sqrt{s}$ and $\sqrt{r}\sqrt{s} = \sqrt{rs}$.

☐ **The product of cube roots theorem.** If r and s are any real numbers, then $\sqrt[3]{rs} = \sqrt[3]{r}\sqrt[3]{s}$ and $\sqrt[3]{r}\sqrt[3]{s} = \sqrt[3]{rs}$.

☐ **The quotient of square roots theorem.** If r and s are positive real numbers, then $\dfrac{\sqrt{r}}{\sqrt{s}} = \sqrt{\dfrac{r}{s}}$ and $\sqrt{\dfrac{r}{s}} = \dfrac{\sqrt{r}}{\sqrt{s}}$.

Exponents

x and y are real numbers; a and b are rational numbers; m and n are natural numbers.

☐ **Definition.** $x^n = \underbrace{x \cdot x \cdots x}_{n \text{ factors}}$ and $x^1 = x$

☐ **Definition.** If $x \neq 0$, then $x^0 = 1$.

☐ **Definition.** If $x \neq 0$, then $x^{-a} = \dfrac{1}{x^a}$.

☐ **Definition.** If $x \geq 0$, then $x^{1/n} = \sqrt[n]{x}$; if $x < 0$ and n is odd, $x^{1/n} = \sqrt[n]{x}$.

☐ **Definition.** If $x \geq 0$, then $x^{m/n} = \sqrt[n]{x^m} = (\sqrt[n]{x})^m$.

☐ **Definition.** If $x > 0$, then $x^{-m/n} = \dfrac{1}{\sqrt[n]{x^m}}$.

Theorems

☐ **Theorem 1.** $x^a x^b = x^{a+b}$

☐ **Theorem 2.** $(x^a)^b = x^{ab}$

☐ **Theorem 3.** $\dfrac{x^a}{x^b} = x^{a-b}$ if $x \neq 0$.

☐ **Theorem 4.** $(xy)^a = x^a y^a$

☐ **Theorem 5.** $\left(\dfrac{x}{y}\right)^a = \dfrac{x^a}{y^a}$ if $y \neq 0$.

REVIEW EXERCISES

1. If a and b are the legs of a right triangle and c is the hypotenuse, find the missing side:

a. $a = \sqrt{7}$, $b = \sqrt{5}$

b. $a = 24$, $c = 26$

2. Approximate the following to the nearest hundredth:

a. $\sqrt{73}$

b. $\sqrt{5 \times 10^{-4}}$

c. $\sqrt[3]{25}$

d. $\sqrt[3]{4 \times 10^{-5}}$

3. Simplify:

a. $\sqrt[3]{-81}$

b. $\sqrt{(-7)^2}$

c. $\sqrt[3]{\dfrac{8}{27}}$

d. $\sqrt[7]{(0.0013)^7}$

4. Write in simplest radical form:

a. $\sqrt{192}$ b. $2\sqrt{45}$

c. $\sqrt[3]{144}$ d. $\sqrt{60} + \sqrt{1500}$

5. Rationalize the denominator and write in simplest radical form:

a. $\dfrac{3}{\sqrt{5}}$ b. $\dfrac{\sqrt{3}}{\sqrt{5}}$

c. $\dfrac{24\sqrt{60}}{8\sqrt{5}}$ d. $\dfrac{\sqrt{24} - \sqrt{75}}{\sqrt{3}}$

e. $\dfrac{\sqrt{3}}{1 - \sqrt{2}}$

6. Simplify:

a. $\sqrt{10}\sqrt{15}$ b. $\sqrt{5}(\sqrt{10} - \sqrt{5} - \sqrt{45})$

c. $\sqrt{252} + 4\sqrt{\dfrac{1}{7}} - \dfrac{1}{2}\sqrt{72}$ d. $(2\sqrt{8} + 8\sqrt{2})^2$

e. $\sqrt{6} - \dfrac{1}{2}\sqrt{54}$ f. $\dfrac{2}{\sqrt{120}} + \sqrt{27} + \sqrt{30}$

7. Simplify:

a. $(a^2b^3)(a^3b^5)^2$ b. $\dfrac{a^3b^4}{5m^2n^2} \cdot \dfrac{-36a^5b^3}{10m^6n}$

c. $\left(\dfrac{1}{3}a^2b\right)\left(\dfrac{1}{4}a^2b\right)^3$ d. $-(y)^2(-y)^2$

8. Simplify by writing equivalent statements containing only positive exponents (assume all variables $\neq 0$):

a. m^{-4} b. $\dfrac{1}{p^{-2}}$

c. $\dfrac{x^0y^{-1}z^2}{x^{-2}y^0z}$ d. $\dfrac{a^{-2} - b^{-2}}{a^{-1} - b^{-1}}$

9. Write each of the following in radical form:

a. $x^{1/4}$ b. $2a^{1/2}$

c. $x^{3/4}$ d. $(xy)^{-2/3}$

10. Write each of the following in exponential form:

a. $\sqrt[3]{y}$ b. $\sqrt[6]{x^3y^5}$

c. $x\sqrt{y}$ d. $\sqrt[3]{(xy)^2}$

11. Simplify:

a. $2(4)^{-1/2}$ b. $2^3 \cdot 2^{-1} + 2^0$

c. $16^{-3/4}$ d. $5^0 - 36^{-1/2} + 32^{2/5}$

12. Find the value of $x^2 - 8x + 10$ for:

a. $x = 4 + \sqrt{6}$ b. $x = 4 - \sqrt{6}$

c. $x = 4 + 2\sqrt{6}$

13. Find the value of $\sqrt{b^2 - 4ac}$ for each of the following:

a. $a = 3, b = -5, c = -2$ b. $a = 1, b = -7, c = 1$

c. $a = 4, b = 4, c = -5$

14. Find the value of $100(1 - R^{-2/5})$ for:

a. $R = \dfrac{3125}{243}$ b. $R = 32 \times 10^5$

15. Evaluate using scientific notation; express the answer in ordinary notation:

a. $\sqrt{(2.4 \times 10^9)(1.5 \times 10^{-4})^3}$

b. $\dfrac{(1.60 \times 10^{-5})(2.50 \times 10^{-3})^2}{1.25 \times 10^{-15}}$

CHAPTER 7

QUADRATIC EQUATIONS

The solution of linear equations in one variable and the solution of a system of two linear equations in two variables have been treated in previous chapters. In the section on factoring, a method was shown for the solution of polynomial equations. This method was restricted to the case where the polynomials had linear factors with integers for coefficients. This is not always the case. For example, $x^2 - 2$ and $x^2 + 1$ cannot be factored over the integers. On the other hand, using the factoring theorem,

$$x^2 - a^2 = (x - a)(x + a)$$

one could write

$$x^2 - 2 = x^2 - (\sqrt{2})^2 = (x - \sqrt{2})(x + \sqrt{2})$$

and

$$x^2 + 1 = x^2 - (-1) = x^2 - (\sqrt{-1})^2 = (x - \sqrt{-1})(x + \sqrt{-1})$$

The radical $\sqrt{2}$ has been defined, and it designates a positive real irrational number. The expression $\sqrt{-1}$ has not been defined so far. If $\sqrt{-1}$ has the property $(\sqrt{-1})^2 = -1$, then $\sqrt{-1}$ cannot be a real number since the square of a real number is either positive or zero. By defining $\sqrt{-1}$ to be a new kind of number, called an imaginary number, factorization can be extended and solutions can be provided for the equation $x^2 + 1 = 0$.

$$x^2 + 1 = (x - \sqrt{-1})(x + \sqrt{-1}) = 0$$

$$x - \sqrt{-1} = 0 \text{ or } x + \sqrt{-1} = 0$$

$$x = \sqrt{-1} \text{ or } x = -\sqrt{-1}$$

Numbers having the form $a + b\sqrt{-1}$ where a and b are real numbers with $b \neq 0$ are called imaginary numbers.

These numbers are needed if every polynomial equation is to have a solution, and they will be discussed in this chapter. Having done this, it is then possible to provide a more general solution method for quadratic polynomial equations having the form

$$ax^2 + bx + c = 0$$

where $a \neq 0$. This equation is called a quadratic equation. Its solution and some of its many applications are the major themes of this chapter.

7.1 COMPLEX NUMBERS: DEFINITION, ADDITION, AND SUBTRACTION

7.1.1 Square Roots of Negative Numbers

Up to this point, a meaning has not been assigned to an even root of a negative real number, for example, $\sqrt{-1}$, $\sqrt{-9}$, and $\sqrt[4]{-16}$. In the discussions of the properties of the set of real numbers, it was observed that some real numbers are not rational but irrational, such as $\sqrt{2}$, $\sqrt{3}$, $\sqrt{5}$, $\sqrt[3]{5}$, and $\sqrt[3]{-5}$. The set of rational numbers was established as being closed under the operations of addition, subtraction, multiplication, division, and raising to a power. However, it is not closed with respect to the root-extraction operation. For example, $\sqrt{2}$ is not a rational number. To have a solution for $x^2 = 2$, it is necessary to include the irrational numbers $\sqrt{2}$ and $-\sqrt{2}$ in the number system, since $(\sqrt{2})^2 = \sqrt{2} \cdot \sqrt{2} = 2$ and $(-\sqrt{2})^2 = (-\sqrt{2})(-\sqrt{2}) = 2$.

Thus the set of rationals was extended to the set of real numbers for two basic reasons:

1. To help close the number system with respect to the root extraction operation.
2. To establish a one-to-one correspondence between the real numbers and the points of the number line (the completeness axiom).

Although the real number system has the property that exactly one real number can be assigned to each point on the number line, still the set of real numbers is not closed with respect to the operation of root extraction. There is no real number whose square is a negative real number.

In particular, there is no real number x such that $x^2 = -1$. To obtain closure, a number is invented with the property that its square is -1. This number is assigned the symbolic name $\sqrt{-1}$. Thus

$$(\sqrt{-1})^2 = \sqrt{-1}\,\sqrt{-1} = -1$$

It is convenient to designate this new number by the letter i.

DEFINITION

$\boldsymbol{i = \sqrt{-1}}$ **and** $\boldsymbol{i^2 = -1}$**.**

Now it is still necessary to include numbers whose squares are the other negative real numbers. In other words, a meaning must be established for expressions such as $\sqrt{-4}$, $\sqrt{-5}$, and $\sqrt{\frac{-4}{9}}$. Therefore, the following definition is stated.

DEFINITION

$\boldsymbol{\sqrt{-a} = i\sqrt{a}}$ **if** $\boldsymbol{a}$ **is a positive real number.**

The number named by the symbol $\sqrt{-a}$ is called the principal square root of $-a$. There is another square root, $-\sqrt{-a}$, because

$$(-\sqrt{-a})(-\sqrt{-a}) = +(\sqrt{-a})^2 = -a$$

The definition states that the principal square root of a negative real number can be expressed as the product of the new number i and a positive real number.

For example,

$$\sqrt{-4} = i\sqrt{4} = i \cdot 2 = 2i$$

$$\sqrt{-5} = i\sqrt{5} = \sqrt{5}i$$

Numbers of the form bi, where b is a real number, are called **pure imaginary numbers.** Numbers of the form $a + bi$, where a and b are real numbers and $b \neq 0$, are called **imaginary numbers.**

For example, $5i$, $-\frac{2}{3}i$, $\frac{1}{2} + 6i$, and $-7 - i\sqrt{2}$ are imaginary numbers, but of these, only $5i$ and $-\frac{2}{3}i$ are pure imaginary numbers.

Note that $i\sqrt{5} = \sqrt{5}\,i$ and that $i\sqrt{2} = \sqrt{2}\,i$. When the multiplier of i is a radical, it is conventional to write $i\sqrt{5}$ rather than $\sqrt{5}\,i$ to avoid the accidental error of writing $\sqrt{5i}$, since $\sqrt{5i} \neq \sqrt{5}\,i$.

The set of **complex numbers,** C (Figure 7.1.1), is the union of the set of real numbers, R, with the set of imaginary numbers, Im.

The complex numbers, $\{a + bi\}$

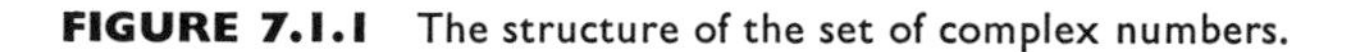

The reals, $b = 0$ — R

The imaginaries, $b \neq 0$ — Im

$$R \cup Im = C$$
$$R \cap Im = \phi$$

FIGURE 7.1.1 The structure of the set of complex numbers.

Every complex number can be expressed in the **standard form** $a + bi$, where a and b are real numbers and i is the imaginary unit such that $i^2 = -1$.

The real number a is called the **real part** of $a + bi$.

The real number b is called the **imaginary part** of $a + bi$.

Two complex numbers are said to be equal if and only if their real parts are equal and their imaginary parts are equal.

DEFINITION

$a + bi = c + di$ **if and only if** $a = c$ **and** $b = d$.

7.1.2 Complex Numbers: Sums and Differences

The sum and difference of two complex numbers are defined as follows.

DEFINITION OF SUM

$$(a + bi) + (c + di) = (a + c) + (b + d)i$$

DEFINITION OF DIFFERENCE

$$(a + bi) - (c + di) = (a - c) + (b - d)i$$

The above definitions state that two complex numbers are added or subtracted in the usual way where the imaginary unit i is treated as a literal constant. The addition properties for the set of complex numbers are similar to those for the set of real numbers, as can be seen in the following summary.

ADDITION PROPERTIES OF COMPLEX NUMBERS

1. *Closure* $(a + bi) + (c + di)$ is a complex number.
   $(a + bi) - (c + di)$ is a complex number.
2. *Commutativity for Addition*
   $$(a + bi) + (c + di) = (c + di) + (a + bi)$$
3. *Associativity (Addition)*
   $$[(a + bi) + (c + di)] + (e + fi) =$$
   $$(a + bi) + [(c + di) + (e + fi)]$$
4. *Identity (Addition)*. There is exactly one complex number, $0 = 0 + 0 \cdot i$, so that $(a + bi) + 0 = a + bi$ for any $a + bi$.
5. *Inverse (Addition)*. For each $a + bi$, there is exactly one complex number $-(a + bi)$, called the additive inverse of $a + bi$, so that
   $$(a + bi) + [-(a + bi)] = 0$$

Moreover, $$-(a + bi) = -a + -bi$$

EXAMPLE 7.1.1 Express each of the following in terms of the imaginary unit i:

a. $\sqrt{-36}$ b. $\sqrt{-5}$ c. $\sqrt{-4x^2}$ where $x > 0$

Solution

a. $\sqrt{-36} = \sqrt{+36}\,\sqrt{-1} = 6i$

b. $\sqrt{-5} = \sqrt{5}\,\sqrt{-1} = \sqrt{5}\,i = i\sqrt{5}$ (Preferred form)

c. $\sqrt{-4x^2} = \sqrt{4x^2}\,\sqrt{-1} = 2xi$

EXAMPLE 7.1.2 Write each of the following in standard form:

a. $3 + \sqrt{-4}$ b. $5 - \sqrt{-9}$ c. $\sqrt{-12}$ d. 6

Solution

a. $3 + \sqrt{-4} = 3 + 2i$

b. $5 - \sqrt{-9} = 5 - 3i$

c. $\sqrt{-12} = \sqrt{4}\,\sqrt{3}\,\sqrt{-1} = 2\sqrt{3}\,i = 0 + 2i\sqrt{3}$ (Preferred form)

d. $6 = 6 + 0i$

EXAMPLE 7.1.3 Determine the real numbers x and y for which

$$x + yi = 3 - 5i$$

is true.

Solution Since

$$a + bi = c + di \text{ if and only if } a = c \text{ and } b = d$$
$$x + yi = 3 - 5i \text{ if and only if } x = 3 \text{ and } y = -5$$

EXAMPLE 7.1.4 Determine the real numbers x and y for which

$$(x - 2y) + 2xi = 8i$$

Solution

$$a + bi = c + di \text{ if and only if } a = c \text{ and } b = d$$
$$(x - 2y) + 2xi = 0 + 8i \text{ if and only if}$$
$$x - 2y = 0 \text{ and } 2x = 8$$
$$x = 2y \text{ and } x = 4$$
$$2y = 4 \text{ and } y = 2$$

EXAMPLE 7.1.5 Add $(3 - 4i) + (5 + 2i)$.

Solution

$$(3 - 4i) + (5 + 2i) = (3 + 5) + (-4 + 2)i = 8 - 2i$$

EXAMPLE 7.1.6 Subtract $(-2 - i) - (7 + 6i)$.

Solution $(-2 - i) - (7 + 6i) = (-2 - 7) + (-1 - 6)i = -9 - 7i$

EXAMPLE 7.1.7 Simplify $(5 + 2\sqrt{-9}) - (4 - \sqrt{-25})$.

Solution First, express each complex number in the standard form, $a + bi$:

$$\begin{aligned}(5 + 2\sqrt{-9}) - (4 - \sqrt{-25}) &= (5 + 2\cdot 3i) - (4 - 5i)\\ &= (5 + 6i) - (4 - 5i)\\ &= (5 + 6i) + (-4 + 5i)\\ &= (5 - 4) + (6 + 5)i\\ &= 1 + 11i\end{aligned}$$

HISTORICAL NOTE

The Italian mathematician Girolamo Cardano (1501–1576) was the first to use the square root of a negative number in a computation, in his *Ars Magna* of 1545. In 1572 the Italian mathematician Raffael Bombelli (ca. 1530–after 1572) introduced a consistent theory of imaginary numbers in his *Algebra*. The French mathematician René Descartes (1596–1650) classified numbers as "real" and "imaginary" and discussed complex numbers as the solutions of equations in his *La Géométrie* of 1637. The letter i to designate $\sqrt{-1}$ was introduced in 1748 by the Swiss mathematician Leonard Euler (1707–1783). The name "complex number" was introduced in 1832 by the great German mathematician Karl Friedrich Gauss (1777–1855).

EXERCISES 7.1 A

In Exercises 1–10, express each in terms of the imaginary unit i.

1. $\sqrt{-4}$

2. $\sqrt{-81}$

3. $\sqrt{-x^2}$ where $x \geq 0$

4. $\sqrt{-\frac{1}{4}}$

5. $\sqrt{-2}$

6. $\sqrt{-8}$

7. $5\sqrt{-64}$

8. $-5\sqrt{-64}$

9. $2\sqrt{-18}$

10. $-2\sqrt{-18}$

Write the numbers in Exercises 11–18 in standard form ($a + bi$).

11. $4 + \sqrt{-25}$

12. $6 - \sqrt{-49}$

13. $7 + 2\sqrt{-9}$

14. $8 - \sqrt{-12}$

15. 5

16. 0

17. $\sqrt{-4}$

18. $\sqrt{-49} + \sqrt{-36}$

Determine the real numbers x and y for which each of the equations in Exercises 19–24 is true.

19. $x + yi = 4 + 2i$

20. $3x + yi = 12 - 5i$

21. $x - 2yi = 10i$

22. $4x + 3yi = 16$

23. $(x + y) + (x - y)i = 4 + 6i$

24. $(3x - y - 1) + (2x + y - 4)i = 0$

In Exercises 25–35, perform the indicated operations and express the result in standard form.

25. $(2 + 3i) + (5 - i)$

26. $(3 + 4i) + (2 + 3i)$

27. $(4 - 5i) - (6 - 2i)$

28. $(5 - 2i) - (4 + 3i)$

29. $(2 + \sqrt{-1}) + (3 + \sqrt{-4})$

30. $(6 - \sqrt{-9}) - (9 - 2\sqrt{-16})$

31. $(8 + \sqrt{-12}) + (1 - \sqrt{-27})$

32. $(\sqrt{2} + \sqrt{-6}) - (\sqrt{50} + \sqrt{-24})$

33. $(1 + i) + (3 - 2i) - (5 + 3i)$

34. $(-1 + \sqrt{-3}) - (1 + \sqrt{-3}) + 2$

35. $\sqrt{-4} + \sqrt{-9} + \sqrt{-16}$

★ *What restriction should be placed on the value of x so that the expressions in Exercises 36–40 denote real numbers?*

36. $\sqrt{x - 4}$

37. $\sqrt{9 - x}$

38. $\sqrt{x^2 - 16}$

39. $\sqrt{9 - x^2}$

40. $\dfrac{1}{\sqrt{x^2 - 25}}$

41. Under what conditions will $a + bi = a - bi$?

EXERCISES 7.1 B

In Exercises 1–10, express each in terms of the imaginary unit i.

1. $\sqrt{-16}$

2. $\sqrt{-25y^2}$ where $y > 0$

3. $\sqrt{-\dfrac{36}{49}}$

4. $5\sqrt{-9}$

5. $-2\sqrt{-100}$

6. $\sqrt{-98}$

7. $4\sqrt{-27}$

8. $-4\sqrt{-27}$

9. $3\sqrt{-49x^2}$ where $x > 0$

10. $-5\sqrt{-121y^4}$ where $y > 0$

Write the numbers in Exercises 11–18 in standard form ($a + bi$).

11. $1 - \sqrt{-36}$

12. $9 + \sqrt{-50}$

13. $4 - 2\sqrt{-72}$

14. -6

15. $-2\sqrt{-25}$

16. i^2

17. $\sqrt{-16} + \sqrt{-4}$

18. $\sqrt{-25} - \sqrt{-9}$

Determine the real numbers x and y for which each of the equations in Exercises 19–24 is true.

19. $x + yi = 3 - 4i$

20. $2x - 3yi = 6 + 9i$

21. $2x + 3yi = 6i$

22. $5x - 2yi = 10$

23. $(x - 2y) + (x + y)i = 12i$

24. $(3x + 2y + 1) + (x + 2y)i = 0$

In Exercises 25–35, perform the indicated operations and express the result in standard form.

25. $(6 - 4i) + (-2 + 2i)$

26. $(-6 - i) + (3 - 4i)$

27. $(2 + 5i) - (4 - i)$

28. $(3 - 2i) - (7 + 3i)$

29. $(5 + \sqrt{-9}) + (2 - \sqrt{-25})$

30. $(5 - \sqrt{-36}) - (8 + 2\sqrt{-49})$

31. $(8 - \sqrt{-8}) - (9 - \sqrt{-18})$

32. $(\sqrt{45} + \sqrt{-24}) + (\sqrt{20} - \sqrt{-54})$

33. $(4 - 2i) - (5 + 6i) - (3 - 8i)$

34. $(4 - \sqrt{-9}) + (1 + \sqrt{-16}) + (3 - \sqrt{-1})$

35. $\sqrt{-25} + \sqrt{-36} - \sqrt{-49}$

★ *What restriction should be placed on the values of x so that the expressions in Exercises 36–40 denote real numbers?*

36. $\sqrt{32 - 2x}$

37. $\sqrt{15 - 3x}$

38. $\sqrt{50 - 2x^2}$

39. $\sqrt{4x^2 - 36}$

40. $\dfrac{1}{\sqrt{100 - x^2}}$

★ **41.** If z is any complex number, under what conditions does

a. $z + (a + bi) = 2a$?

b. $z - (a + bi) = 2bi$?

7.2 COMPLEX NUMBERS: PRODUCTS

The product of two complex numbers is defined to be the result obtained by applying the distributive, associative, and commutative properties, treating i as a literal constant, and then finally replacing i^2 by -1.

$$\begin{aligned}(a + bi)(c + di) &= (a + bi)c + (a + bi)di \\ &= (ac + bci) + (adi + bdi^2) \\ &= ac + (bc + ad)i + bd(-1) \\ &= (ac - bd) + (bc + ad)i\end{aligned}$$

DEFINITION

$$(a + bi)(c + di) = (ac - bd) + (bc + ad)i$$

With this definition, the following properties can now be established for the multiplication of two complex numbers.

MULTIPLICATION PROPERTIES OF COMPLEX NUMBERS

1. *Closure* $(a + bi)(c + di)$ is a complex number.
2. *Commutativity* $(a + bi)(c + di) = (c + di)(a + bi)$
3. *Associativity* $[(a + bi)(c + di)](e + fi) = (a + bi)[(c + di)(e + fi)]$
4. *Identity* For any complex number $a + bi$, there is exactly one number, $1 = 1 + 0i$, so that $(a + bi)\cdot 1 = a + bi$.
5. *Inverse* For each $a + bi \neq 0$, there is exactly one complex number, $\dfrac{1}{a + bi}$, called the reciprocal of $a + bi$, so that $(a + bi)\left(\dfrac{1}{a + bi}\right) = 1$.
6. *Distributive Property*
   $(a + bi)[(c + di) + (e + fi)] = (a + bi)(c + di) + (a + bi)(e + fi)$

While the definition is necessary to show how these properties are logically derived, it is too difficult to remember to use it directly in computations. Instead, the product is obtained by multiplying in the usual way, as for real numbers, treating i as a literal constant *but* replacing i^2 by -1.

EXAMPLE 7.2.1 Multiply $(3 + 2i)(2 - 5i)$.

Solution

$(3 + 2i)(2 - 5i) = 6 - 15i + 4i - 10i^2$ (Using the distributive property twice)

$= 6 - 11i - 10i^2$ (Collecting like terms)

$= 6 - 11i - 10(-1)$ (Replacing i^2 by -1)

$= 16 - 11i$

EXAMPLE 7.2.2 Simplify $(4 + \sqrt{-9})(3 - \sqrt{-25})$.

Solution Before any calculations are performed, the expression $\sqrt{-k^2}$ must be replaced by ki with $k > 0$, since all the properties are established for expressions having the form $a + bi$. Thus

$$\begin{aligned}(4 + \sqrt{-9})(3 - \sqrt{-25}) &= (4 + \sqrt{9}\sqrt{-1})(3 - \sqrt{25}\sqrt{-1})\\ &= (4 + 3i)(3 - 5i)\\ &= 12 - 20i + 9i - 15i^2\\ &= 12 - 11i - 15(-1)\\ &= 27 - 11i\end{aligned}$$

EXERCISE 7.2 A

In Exercises 1–25, multiply and express the result in simplified form.

1. $\sqrt{-5}\sqrt{-20}$

2. $\sqrt{-8}\sqrt{-9}$

3. $\sqrt{-12}\sqrt{3}$

4. $\sqrt{2}\sqrt{-3}$

5. $\sqrt{-2}\sqrt{-3}\sqrt{-6}$

6. $2i(4 + 3i)$

7. $3i(5 - 2i)$

8. $\sqrt{-2}(\sqrt{18} - \sqrt{-18})$

9. $(3 + \sqrt{-4})(2 - \sqrt{-9})$

10. $(4 + 5i)(3 - 2i)$

11. $(4 + 2i)(4 - 2i)$

12. $(\sqrt{-2} + \sqrt{-3})(\sqrt{-2} - \sqrt{-3})$

13. $(3 - i)(1 + 2i)$

14. $5i(-3i)(1 - 6i)$

15. $(3 + 2i)^2$

16. $2i(3 + 4i)(5 - 6i)$

17. $(2 + i)[(3 + 2i) + (4 - 5i)]$

18. $4(3 + 2i) - 5i(3 + 2i)$

19. $i(i - 1)(i - 2)$

20. $(1 + 2i)(1 - 2i)(1 - 3i)$

21. $(1 - 3i)(1 - 2i)(1 + 2i)$

22. $(5 - 4i)^2$

23. $(2 + i)^2 - 4(2 + i) + 5$

24. $(2 - i)^2 - 4(2 - i) + 5$

25. $\left(\dfrac{-1 + \sqrt{-3}}{2}\right)^3$

26. Show that $3 + i$ and $3 - i$ are solutions of $x^2 - 6x + 10 = 0$.

27. Show that $2i$ is a solution of $x^2 + ix + 6 = 0$.

★ *Find real values for x and y that satisfy the equations in Exercises 28–30.*

28. $x + yi = i^2 + 1$

29. $(ix)^2 + 9 = 0$

30. $x^2 + (1 + i)x + i = 0$

Using $i^1 = i$, $i^2 = -1$, $i^3 = i^2 \cdot i = -i$, *and* $i^4 = i^2 \cdot i^2 = (-1)(-1) = 1$, *simplify the expressions in Exercises 31–38.*

31. i^5

32. i^6

33. i^7

34. i^8

35. i^{15}

36. i^{22}

37. $i^3 + i^9$

38. $i^3 + i^5 + i^7$

EXERCISES 7.2 B

In Exercises 1–25, multiply and express the result in simplified form.

1. $\sqrt{20}\sqrt{-5}$

2. $\sqrt{-20}\sqrt{-5}$

3. $\sqrt{-27}\sqrt{-4}$

4. $\sqrt{-27}\sqrt{4}$

5. $\sqrt{-5}\sqrt{-10}\sqrt{-20}$

6. $5i(6 - i)$

7. $(6 + 7i)(3 + 2i)$

8. $(3 - 7i)(2 - 5i)$

9. $(4 + i)(3 - 2i)$

10. $(3 + 2i)(2 + 3i)(5 - 4i)$

11. $(4 + i\sqrt{2})(3 - i\sqrt{5})$

12. $(4 + 2i)^2$

13. $(5 + \sqrt{-36})(5 - \sqrt{-36})$

14. $(3 - 2i)[(2 - 4i) + (5 + 2i)]$

15. $\sqrt{-5}(\sqrt{80} + \sqrt{-80})$

16. $6(5 - 4i) - 2i(5 - 4i)$

17. $(2i + 7)(2i - 7)$

18. $(\sqrt{-2} + \sqrt{-8})^2$

19. $(i + 1)(4 + 3i)(4 - 3i)$

20. $(4 - 3i)(i + 1)(4 + 3i)$

21. $(1 - \sqrt{-3})^2$

22. $(1 - \sqrt{-3})^2 - 2(1 - \sqrt{-3}) + 4$

23. $(1 - \sqrt{-3})^3$

24. $(3 + 4i)(1 + 2i)^2$

25. $2i[(25 - 24i)(25 + 24i)]$

26. Show that $2 + i\sqrt{5}$ and $2 - i\sqrt{5}$ are solutions of $x^2 - 4x + 9 = 0$.

27. Show that $1 + i$ is a solution of $x^2 - 2ix - 2 = 0$.

★ *Find real values for x and y that satisfy the equations in Exercises 28–30.*

28. $(x + i)^2 = y$

29. $(x + yi)^2 = 8i$

30. $(x + yi)^2 = i$

Using $i^1 = i$, $i^2 = -1$, $i^3 = i^2 \cdot i = -i$, *and* $i^4 = i^2 \cdot i^2 = (-1)(-1) = 1$, *simplify the expressions in Exercises 31–38.*

31. $i^4 + i^7$

32. i^{-2}

33. i^{13}

34. i^{120}

35. i^{54}

36. $i^2 + i^3 + i^4$

37. $(i^3)^4$

38. $(i^2)^{-3}$

7.3 COMPLEX NUMBERS: QUOTIENTS

The quotient of two complex numbers is defined so that division retains its meaning as the inverse operation of multiplication. Thus

$$\frac{a + bi}{c + di} = (a + bi)\left(\frac{1}{c + di}\right)$$

Therefore, it is necessary to investigate the reciprocal of a complex number. To do this, the concept of the **conjugate** of a complex number is introduced.

DEFINITION

The conjugate of $a + bi$ is $a - bi$.

The conjugate of $a - bi$ is $a + bi$.

THE PRODUCT OF CONJUGATES THEOREM

$$(a + bi)(a - bi) = a^2 + b^2, \text{ a real number}$$

For example, $(3 + 4i)(3 - 4i) = 9 + 16 = 25$

Similarly, $(-2 + 5i)(-2 - 5i) = 4 + 25 = 29$

EXAMPLE 7.3.1 Express

$$\frac{1}{3 + 5i}$$

in the $a + bi$ form.

Solution Multiplying numerator and denominator by $3 - 5i$, the conjugate of $3 + 5i$,

$$\frac{1}{3 + 5i} = \frac{3 - 5i}{(3 + 5i)(3 - 5i)}$$

$$= \frac{3 - 5i}{9 + 25} = \frac{3}{34} + \frac{-5}{34}i$$

EXAMPLE 7.3.2 Divide

$$\frac{2 + \sqrt{-9}}{3 - \sqrt{-4}}$$

Solution First rewriting in terms of i,

$$\frac{2 + \sqrt{-9}}{3 - \sqrt{-4}} = \frac{2 + 3i}{3 - 2i}$$

$$= \frac{(2 + 3i)(3 + 2i)}{(3 - 2i)(3 + 2i)} \quad \text{(Multiplying numerator and denominator by } 3 + 2i\text{, the conjugate of } 3 - 2i\text{)}$$

$$= \frac{6 + 4i + 9i + 6i^2}{9 + 4}$$

$$= \frac{6 + 13i + 6(-1)}{13}$$

$$= \frac{13i}{13} = i$$

Using more advanced methods, it may be shown that the set of complex numbers is closed with respect to the operation of root extraction. Thus the set of complex numbers is closed with respect to all six operations of elementary algebra. It may also be shown that every polynomial in one variable with coefficients from the set of real or complex numbers has a solution in the set of complex numbers. This statement is called the *Fundamental Theorem of Algebra*. The German mathematician Karl Friedrich Gauss (1777–1855) gave the first satisfactory proof of this theorem when he was only twenty-one years old.

EXERCISES 7.3 A

Find the conjugate of each of the complex numbers in Exercises 1–8.

1. $5 + 7i$

2. $6 - 4i$

3. $2 + i$

4. $4 - 3i$

5. i

6. $-i$

7. 3

8. $i - 1$

Express each of the quotients in Exercises 9–18 in standard form.

9. $\dfrac{1}{3 + 2i}$

10. $\dfrac{1}{3i - 5}$

11. $\dfrac{1 + i}{1 - i}$

12. $\dfrac{3 + 4i}{2i}$

13. $\dfrac{3i}{2 + 4i}$

14. $\dfrac{1 - \sqrt{-3}}{1 + \sqrt{-3}}$

15. $\dfrac{\sqrt{-3} + 3\sqrt{-1}}{\sqrt{-3} - 3\sqrt{-1}}$

16. $\dfrac{\sqrt{2} + 3i}{1 + \sqrt{2}i}$

17. $\dfrac{a + bi}{a - bi}$ (a and b are real numbers)

18. $\dfrac{i + 2}{i + 1}$

In Exercises 19–20, simplify each, expressing the result in standard form.

19. $\dfrac{2 + i}{1 + 2i} + \dfrac{2 - i}{1 - 2i}$

20. $\dfrac{2 + 3i}{3 + i} - \dfrac{1 + i}{1 + 2i}$

EXERCISES 7.3 B

Find the reciprocal of each of the complex numbers in Exercises 1–8 and write the answer in standard form.

1. $3 - 4i$

2. $8 + 5i$

3. $1 - i$

4. $1 + i$

5. 1

6. -1

7. i

8. $-i$

Express each of the quotients in Exercises 9–18 in standard form.

9. $\dfrac{2}{3 + 2i}$

10. $\dfrac{4}{2i - 3}$

11. $\dfrac{2i}{2i + 3}$

12. $\dfrac{2i + 3}{2i}$

13. $\dfrac{2i + 3}{3 + 2i}$

14. $\dfrac{1 + 2i}{1 - 2i}$

15. $\dfrac{\sqrt{2} + i}{\sqrt{3} - 2i}$

16. $\dfrac{2 + 3\sqrt{-3}}{3 + 2\sqrt{-2}}$

17. $\dfrac{\sqrt{-2} + 2\sqrt{-1}}{\sqrt{-3} - 3\sqrt{-1}}$

18. $\dfrac{1 + 2i}{3 + 4i} \div \dfrac{2i}{2 - i}$

★ *In Exercises 19–20, simplify each, expressing the result in standard form.*

19. $(\sqrt{1 + \sqrt{-3}} + \sqrt{1 - \sqrt{-3}})^2$

20. $\dfrac{1}{a + bi} + \dfrac{1}{a - bi}$ (a and b are real numbers)

7.4 A GEOMETRIC MODEL FOR COMPLEX NUMBERS (OPTIONAL)

A complex number, $a + bi$, may be represented graphically by interpreting the real part, a, as the distance along a horizontal axis, called the real axis, and by interpreting the imaginary part, b, as the distance along a vertical axis, called the imaginary axis.

For example, the complex numbers $3 + 2i$, $-4 + i$, $-2 - 3i$, and $3 - 3i$ are represented in Figure 7.4.1 as the points A, B, C, and D, respectively.

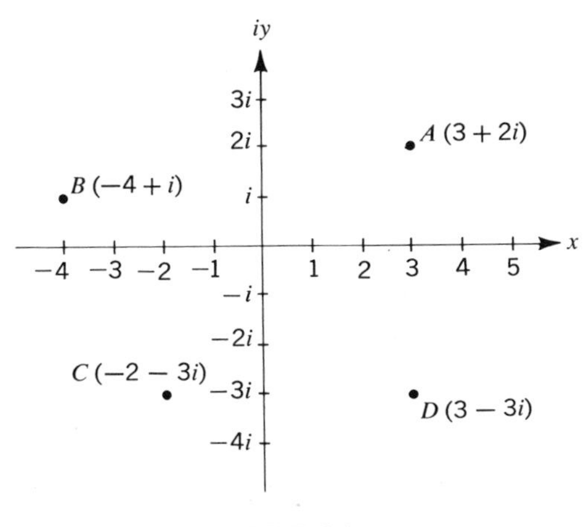

FIGURE 7.4.1

Vectors

Many physical concepts, such as force, velocity, and acceleration, can be described by vectors. A **vector** is a quantity that has both magnitude and direction. Since a vector can be considered as the sum of a horizontal component and a vertical component, which, when expressed symbolically, must be identified in some way, a complex number can be interpreted as a vector. The real part of the complex number is the horizontal component of the vector, and the imaginary part is the vertical component. A geometric model of a vector is the directed line segment from the origin to the point that is the graph of the complex number representing the vector.

In Figure 7.4.2, the sum of two complex numbers (or vectors) is represented graphically:

$$(1 + 3i) + (2 + i) = (1 + 2) + (3 + 1)i = 3 + 4i$$

The methods of geometry can be used to show that the sum of two vectors is the directed diagonal of the parallelogram formed by using the two vectors as two adjacent sides.

The complex number $1 + 3i$, illustrated in Figure 7.4.2, can be interpreted as a force vector whose horizontal x-component is 1 pound and whose vertical y-component is 3 pounds.

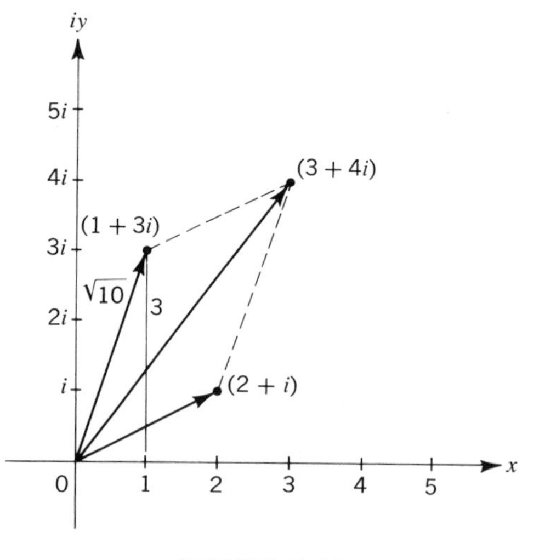

FIGURE 7.4.2

The magnitude of the force is the length of the directed line segment that represents the vector. For the vector $1 + 3i$, the magnitude is

$$\sqrt{1^2 + 3^2} = \sqrt{10}$$

using the theorem of Pythagoras.

The direction of the force is often given by stating the slope of the line.

Thus the direction of the vector $1 + 3i$ is $\frac{3}{1}$. In general, if $a + bi$ represents a vector,

$$\text{the } \textbf{magnitude} = \sqrt{a^2 + b^2}$$

and

$$\text{the } \textbf{direction} = \frac{b}{a} \text{ (the slope)}$$

When two forces act on the same object, the resulting force is that represented by the sum of the complex numbers that represent the two forces.

Thus the resulting force of the forces $1 + 3i$ and $2 + i$ is

$$(1 + 3i) + (2 + i) = (1 + 2) + (3 + 1)i = 3 + 4i$$

The resulting magnitude is $\sqrt{3^2 + 4^2} = \sqrt{25} = 5$, and the resulting direction is $\frac{4}{3}$ (the slope of the line from 0 to $3 + 4i$).

EXAMPLE 7.4.1 A plane directed toward the east flies with a speed in still air of 240 mph. A wind from the south blowing at 70 mph forces the plane off its course. What is the resultant speed and direction?

Solution

The resulting velocity is $a + bi = 240 + 70i$.

The resulting speed is $\sqrt{a^2 + b^2} = \sqrt{(240)^2 + (70)^2} = 250$ mph.

The resulting direction is $\frac{b}{a} = \frac{70}{240} = \frac{7}{24}$.

As a result, the plane flies 250 mph in the direction of a line from east to north with slope $\frac{7}{24}$.

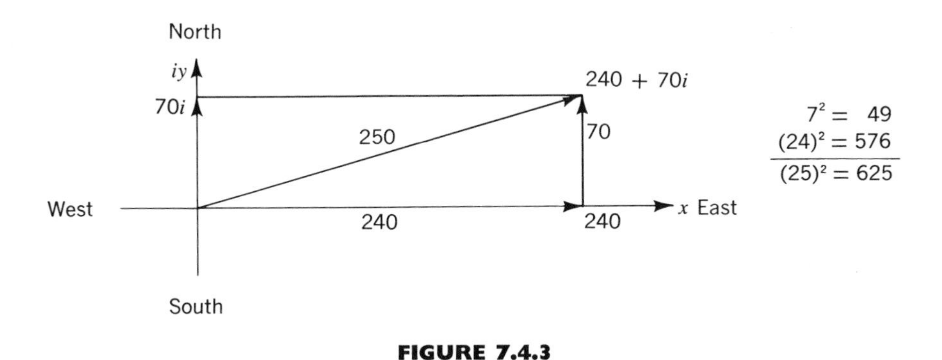

FIGURE 7.4.3

HISTORICAL NOTE

The Irish mathematician William Rowan Hamilton (1805–1865) presented the modern rigorous treatment of complex numbers as number pairs in 1835. Later he extended these numbers to a space of three dimensions in his *Lectures on Quaternions* in 1853. Quaternions can be used to describe the three-dimensional rotations of an object in space, just as multiplication of a complex number by the imaginary unit i can be interpreted as a rotation of 90 degrees about the origin (see Figure 7.4.4).

In 1752, Jean Le Rond d'-Alembert of France (1717–1783) used complex numbers in his study of hydrodynamics, which led to the modern theory of aerodynamics.

In 1772, Johann Heinrich Lambert of Germany (1728–1777) used complex numbers to construct maps by a technique called "conformal conic projection."

In the twentieth century, the American electrician and mathematician Charles Proteus Steinmetz (born in Breslau, Germany, in 1865 and died in the United States in 1923) used complex numbers to develop the theory of electrical circuits.

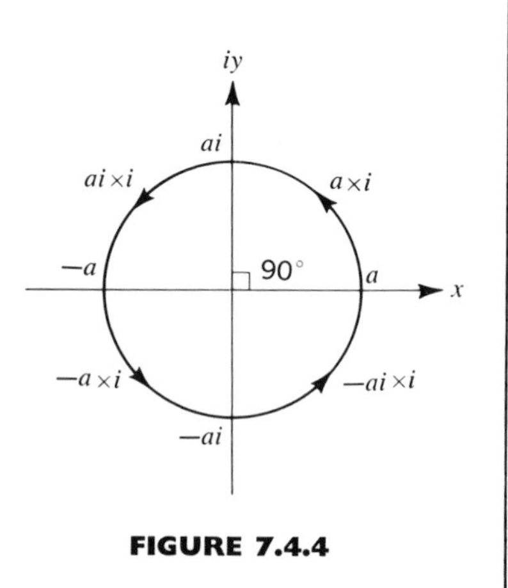

FIGURE 7.4.4

EXERCISES 7.4 A

Plot each of the points in Exercises 1–10 in a complex plane. Be sure to label the axes and to label each point you plot.

1. 2

2. $-4 + i$

3. $2i$

4. $3 + 2i$

5. $2 - 3i$

6. $-2 - 3i$

7. -5

8. $-i$

9. $2 + i^2$

10. 0

11. A boy who wants to row his boat across a river heads for a point directly opposite on the other shore. His rowing rate in still water is 8 mph, and the rate of the current is 6 mph.

a. Represent his resultant velocity vector as a complex number, $a + bi$.

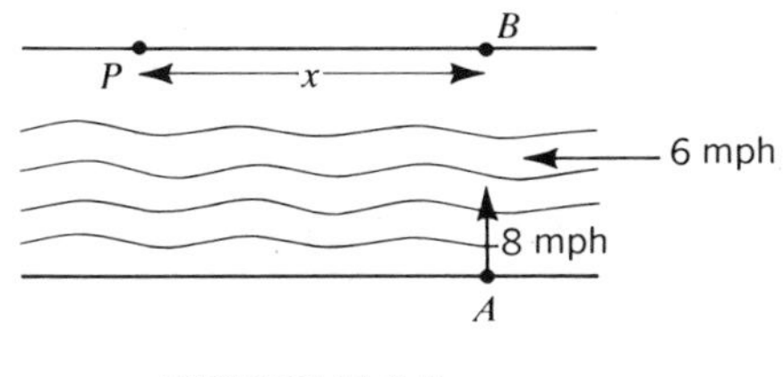

FIGURE 7.4.5

b. Find his resultant speed (magnitude of the velocity vector).

c. How far downstream from his intended destination does he land if it takes him one half hour to cross the river?

12. Two forces represented by the vectors $-4 + 20i$ and $-4 - 4i$ act on an object. Find the resultant force vector, its magnitude, and its direction. Illustrate geometrically.

EXERCISES 7.4 B

Plot each of the points in Exercises 1–10 in a complex plane.

1. $1 + 2i$

2. -3

3. $-i^2$

4. $-3 - 4i$

5. $-2i$

6. $1 - 3i$

7. $i^2 + 1$

8. $3 + 4i$

9. $3(2 - i)$

10. $\dfrac{2i + 1}{i}$

11. A plane flying 175 mph south suddenly encounters a 60 mph wind blowing from the east.

a. Represent the resultant velocity vector as a complex number, $a + bi$.

b. Find the resultant speed.

c. Find the resultant direction, as a slope.

d. How far west from his intended destination will the plane be at the end of 45 minutes?

12. Two men pull on a heavy object at right angles to each other, each with a force of 100 pounds.

a. Represent the resultant force vector as a complex number, $a + bi$.

b. Find the direction of the resultant force.

c. Find the magnitude of the resultant force.

7.5 QUADRATIC EQUATIONS: SOLUTION BY FACTORING

A **quadratic equation** is an equation that can be expressed in the form

$$ax^2 + bx + c = 0 \text{ where } a \neq 0$$

In preceding chapters, the coefficients a, b, and c of the quadratic polynomial

$$ax^2 + bx + c \quad (a \neq 0)$$

have been restricted to belong to the set of integers. The factorization of such polynomials was also restricted to polynomial factors with integral coefficients. These polynomials will now be extended so that a, b, and c may be any complex numbers, and polynomial factors may also have complex coefficients. Now expressions such as $x^2 + 4$, which were not factorable over the set of integers, can be factored over the set of complex numbers.

$$x^2 + 4 = x^2 - (-4) = (x + \sqrt{-4})(x - \sqrt{-4}) = (x + 2i)(x - 2i)$$

where $i^2 = -1$.

The solution of quadratic equations by factoring was introduced earlier and involved four basic steps:

1. Find an equivalent quadratic equation whose right side is zero.
2. Factor the quadratic polynomial.
3. Set each factor equal to zero.
4. Solve for the variable.

This method is based on the zero-product theorem, which is restated below for convenience.

THE ZERO-PRODUCT THEOREM

Let r and s be any real numbers. If $rs = 0$, then $r = 0$ or $s = 0$.

In solving an equation by the factoring method, it is important to remember that the right side of the equation must be 0 before the factors on the left side can be equated to 0.

EXAMPLE 7.5.1 Solve for x: $x^2 - x = 6$.

Solution

$$x^2 - x = 6$$

1. $x^2 - x - 6 = 0$ (Equivalent equation whose right side is zero)

2. $(x - 3)(x + 2) = 0$ (Factor the quadratic polynomial)

3. $x - 3 = 0$ or $x + 2 = 0$ (Zero-product theorem: If $rs = 0$, then $r = 0$ or $s = 0$)

4. $x = 3$ or $x = -2$

Check:

$$\begin{array}{rr} x^2 - x = 6 & x^2 - x = 6 \\ (3)^2 - 3 = 6 & (-2)^2 - (-2) = 6 \\ 9 - 3 = 6 & 4 + 2 = 6 \\ 6 = 6 & 6 = 6 \end{array}$$

Therefore, the solution set is $\{3, -2\}$.

EXAMPLE 7.5.2 Solve $(x - 2)(x - 5) = 40$.

Solution

$$\begin{aligned} (x - 2)(x - 5) &= 40 \\ x^2 - 7x + 10 &= 40 \\ x^2 - 7x - 30 &= 0 \\ (x + 3)(x - 10) &= 0 \\ x + 3 = 0 \quad &\text{or } x - 10 = 0 \\ x = -3 \text{ or} \quad &x = 10 \end{aligned}$$

The solution set is $\{-3, 10\}$.

Check:

For $x = -3$, $(x - 2)(x - 5) = (-3 - 2)(-3 - 5) = (-5)(-8) = 40$.
For $x = 10$, $(x - 2)(x - 5) = (10 - 2)(10 - 5) = (8)(5) = 40$.

A special case of the quadratic equation $ax^2 + bx + c = 0$ occurs when $b = 0$. Then

$$ax^2 + c = 0$$

Since $a \neq 0$, we can divide by a.

$$x^2 + \frac{c}{a} = 0$$

$$x^2 - \left(-\frac{c}{a}\right) = 0$$

$$\left(x + \sqrt{-\frac{c}{a}}\right)\left(x - \sqrt{-\frac{c}{a}}\right) = 0$$

$$x = -\sqrt{-\frac{c}{a}} \text{ or } x = +\sqrt{-\frac{c}{a}}$$

The term $\sqrt{-\frac{c}{a}}$ designates a real number if either c or a (but not both) is a negative number and an imaginary number if a and c both are positive or both negative numbers.

A useful shortcut to this problem is:

$$ax^2 + c = 0$$

$$ax^2 = -c$$

$$x^2 = -\frac{c}{a}$$

$$x = \pm\sqrt{-\frac{c}{a}} \left(\text{read } x = +\sqrt{-\frac{c}{a}} \text{ or } x = -\sqrt{-\frac{c}{a}}\right)$$

EXAMPLE 7.5.3 Solve for x: $3x^2 - 2 = 0$.

Solution In this problem $b = 0$.

$$3x^2 - 2 = 0$$

$$3x^2 = 2$$

$$x^2 = \frac{2}{3}$$

$$x = \pm\sqrt{\frac{2}{3}} = \pm\frac{\sqrt{6}}{3}$$

The solution set is $\left\{-\frac{\sqrt{6}}{3}, \frac{\sqrt{6}}{3}\right\}$.

Check: For $x = +\sqrt{\frac{2}{3}}$,

$$3x^2 - 2 = 3\left(\sqrt{\frac{2}{3}}\right)^2 - 2 = 3\left(\frac{2}{3}\right) - 2 = 2 - 2 = 0$$

For $x = -\sqrt{\frac{2}{3}}$,

$$3x^2 - 2 = 3\left(-\sqrt{\frac{2}{3}}\right)^2 - 2 = 3\left(\frac{2}{3}\right) - 2 = 2 - 2 = 0$$

It is important to remember that the expression $\sqrt{a^2}$ designates exactly one number, the principal square root of a^2. However, the quadratic equation $x^2 = a^2$ has two solutions, as most quadratic equations do, and both these solutions must be found—that is,

$$x = +\sqrt{a^2} \text{ and } x = -\sqrt{a^2}$$

Another special case occurs when $c = 0$. The equation reads

$$ax^2 + bx = 0$$

Now there is a common factor, x, for the left member of the equation.

$$ax^2 + bx = 0$$
$$x(ax + b) = 0$$
$$x = 0 \text{ or } ax + b = 0 \text{ (Zero-product theorem)}$$
$$ax = -b$$
$$x = -\frac{b}{a}$$

The solution set is $\left\{0, -\frac{b}{a}\right\}$.

EXAMPLE 7.5.4 Solve by factoring $3x^2 + 2x = 0$.

Solution $3x^2 + 2x = 0$

$$x(3x + 2) = 0$$
$$x = 0 \text{ or } 3x + 2 = 0$$
$$3x = -2$$
$$x = -\frac{2}{3}$$

Therefore, the solution set is $\left\{0, -\frac{2}{3}\right\}$.

Check: If $x = 0$, $3(0)^2 + 2(0) = 0 + 0 = 0$.

If $x = -\frac{2}{3}$, $3x^2 + 2x = 3\left(-\frac{2}{3}\right)^2 + 2\left(-\frac{2}{3}\right)$

$$= 3\left(\frac{4}{9}\right) - \frac{4}{3} = \frac{4}{3} - \frac{4}{3} = 0$$

EXAMPLE 7.5.5 Solve $4x^2 = 12x$.

Solution $4x^2 = 12x$

$$4x^2 - 12x = 0$$
$$4x(x - 3) = 0$$
$$4x = 0 \text{ or } x - 3 = 0$$

Since $4 \neq 0$, $x = 0$ or $x = 3$.

Check: For $x = 0$, $4x^2 = 4(0)^2 = 0$ and $12x = 12(0) = 0$.

For $x = 3$, $4(3)^2 = 4(9) = 36$ and $12x = 12(3) = 36$.

Note in the preceding examples that each quadratic equation had *two* solutions. If both sides of the equation $4x^2 = 12x$ had been divided by x, this would have yielded $4x = 12$ and $x = 3$, thus losing the solution $x = 0$. Since the equation $4x^2 = 12x$ is true for $x = 0$, dividing both sides by x is the same as dividing both sides by 0, and this is not permitted. In general, division by an expression containing a variable may lose a solution, and multiplication by an expression containing a variable may introduce an extraneous solution (a number that is not a solution of the original equation).

A quadratic equation may have only one number in its solution set. Such a solution is called a **double root**.

In solving a quadratic equation, it is always necessary to account for *two* solutions. By considering a double root as a solution counted twice, it can then be stated that a quadratic equation, an equation of degree 2, always has two solutions in the set of complex numbers.

EXAMPLE 7.5.6 Solve $x^2 - 6x + 9 = 0$.

Solution

$$x^2 - 6x + 9 = (x - 3)^2$$
$$x^2 - 6x + 9 = 0$$
$$(x - 3)^2 = 0$$
$$(x - 3)(x - 3) = 0$$
$$x - 3 = 0 \text{ or } x - 3 = 0$$
$$x = 3 \text{ or } \quad x = 3$$

The solution set is $\{3\}$, where 3 is called a double root.

EXERCISES 7.5 A

Find all the values of x that are solutions of each of Exercises 1–30. Check each solution.

1. $(x + 5)(x - 1) = 0$

2. $5x(x - 4) = 0$

3. $(x + c)(x - 2c) = 0$

4. $x^2 + 5x + 6 = 0$

5. $4x^2 = 25$

6. $x^2 + 3x = 0$

7. $x^2 + 3x = -2$

8. $3x^2 + x = 2$

9. $x^2 + 4 = 0$

10. $(x + 3)(x - 2) = 0$

11. $(x + 3)(x - 2) = 14$

12. $5x^2 = 20x$

13. $x^2 + 10x + 25 = 0$

14. $x^2 = kx$

15. $3x^2 - 15 = 0$

16. $4x^2 = a^2$ where $a > 0$

17. $(x - 2)^2 = 1$

18. $(x - 2)^2 = n^2$ where $n > 0$

19. $(x - a)^2 = n^2$ where $n > 0$

20. $(ax + b)^2 = 25$ where $a \neq 0$

21. $(cx - d)^2 = 0$ where $c \neq 0$

22. $x^2 - 6ax = 16a^2$

23. $x^3 - 5x^2 + 4x = 0$

24. $x^2 + 4bx + 4b^2 = 0$

25. $\frac{x - 4}{3} = \frac{3}{x + 4}$

26. $\frac{3}{x - 1} - \frac{3}{x + 1} = \frac{2}{5}$

27. $2 - \frac{1}{x^2} + \frac{1}{x} = 0$

28. $x^2 - (c + d)x + cd = 0$

29. $\frac{x}{x + 5} + \frac{2}{x - 5} = \frac{4x}{x^2 - 25}$

30. $\frac{x}{x + 1} + \frac{3}{x + 3} + \frac{2}{(x + 1)(x + 3)} = 0$

★ *Solve each of the formulas in Exercises 31–35 for the letter indicated.*

31. $s = \frac{1}{2}gt^2;\ t$

32. $A = \pi(R^2 - r^2);\ r$

33. $k = \frac{n_1^2}{n_1^2 + n_2^2};\ n_1$

34. $cd^2 = c^2d;\ c$

35. $4wd^2 - 5wdL + wL^2 = 0;\ d$

EXERCISES 7.5 B

Find all the values of x that are solutions of each of Exercises 1–30. Check each solution.

1. $3x(x + 1) = 0$

2. $(ax + b)(cx + d) = 0;\ a \neq 0,\ c \neq 0$

3. $cx(dx + 1) = 0;\ d \neq 0$

4. $x^2 + x - 6 = 0$

5. $x^2 = 3x$

6. $x^2 + 9 = 0$

7. $2x^2 - 4x = 0$

8. $-2x^2 + 4x - 2 = 0$

9. $10x = 4x^2$

10. $(x - 2)(x - 1) = 6$

11. $2x^2 - 1 = x$

12. $2x^2 - 1 = 0$

13. $(x - 4)(x - 3) = 42$

14. $4x^2 - 4x - 8 = 0$

15. $4x^2 - 12x + 9 = 0$

16. $(ax + b)^2 = 0;\ a \neq 0$

17. $(ax + b)^2 = c^2;\ a \neq 0,\ c > 0$

18. $(5x - 2)^2 = 100$

19. $x^2 - 2ax - 63a^2 = 0$

20. $ax^2 - x^2 - ax = 0;\ a \neq 1$

21. $8mx^2 + 28mx - 60m = 0;\ m \neq 0$

22. $x^3 + 5x = 6x^2$

23. $x^4 - 13x^2 + 40 = 4$

24. $\frac{x}{a} - \frac{a}{x} = 0; a > 0$

25. $x + 3 = \frac{15}{2x - 1}$

26. $\frac{4}{3x + 12} + \frac{x}{9} = \frac{1}{x + 4}$

27. $x + \frac{12}{x} = 7$

28. $\frac{3}{x + 2} + \frac{x + 8}{x(x + 2)} = 1$

29. $1 + \frac{a - b}{x} = \frac{ab}{x^2}$

30. $\frac{3}{x - 2} - \frac{x + 1}{x} = \frac{2}{x(x - 2)}$

★ *Solve each of the formulas in Exercises 31–35 for the letter indicated.*

31. $K = \frac{mv^2}{2}; v$

32. $c^2 = a^2 + b^2; a$

33. $5wd^2 - 59wd + 90w = 0; d$

34. $F = \frac{GMm}{R^2}; R$

35. $\frac{c_1}{r_1^2} = \frac{c_2}{r_2^2}; r_2$

7.6 QUADRATIC EQUATIONS: COMPLETING THE SQUARE

Not all quadratic polynomials can be factored easily, and the solution of quadratic equations could be very time-consuming and cumbersome if the factors had to be determined by the trial-and-error method. In order to find the solution more rapidly, other methods are available. One such method is called **completing the square.** The aim of this method is to make the left side of the equation a perfect square trinomial which is equal to a constant—that is, $(x + a)^2 = k$.

The equation $(x + a)^2 = k$ is equivalent to the equation

$$(x + a)^2 - k = 0$$

whose left side can now be factored by the difference of two squares theorem:

$$(x + a)^2 - (\sqrt{k})^2 = 0$$

$$(x + a - \sqrt{k})(x + a + \sqrt{k}) = 0$$

$$x + a - \sqrt{k} = 0 \text{ or } x + a + \sqrt{k} = 0$$

$$x = -a + \sqrt{k} \quad \text{or } x = -a - \sqrt{k}$$

In practice, it is convenient to write the solution in the shorter form indicated below:

$$(x + a)^2 = k$$

$$x + a = \pm \sqrt{k}$$

$$x = -a + \sqrt{k} \text{ or } x = -a - \sqrt{k}$$

EXAMPLE 7.6.1 Solve $x^2 + 6x + 4 = 0$ by completing the square.

Solution

1. Subtract the constant from both sides of the equation: $x^2 + 6x = -4$
2. Add to both sides the number that makes $x^2 + 6x$ a perfect square trinomial: $x^2 + 6x + 9 = -4 + 9$
3. Write the left side as the square of a binomial and simplify the right side: $(x + 3)^2 = 5$
4. Take the square root of both sides: $x + 3 = \pm \sqrt{5}$
5. Solve for x: $x = -3 + \sqrt{5}$ or $x = -3 - \sqrt{5}$

Check:

If $x = -3 + \sqrt{5}$,

$$x^2 + 6x + 4 = 0$$

$$(-3 + \sqrt{5})^2 + 6(-3 + \sqrt{5}) + 4 = 0$$

$$9 - 6\sqrt{5} + 5 - 18 + 6\sqrt{5} + 4 = 0$$

$$0 = 0$$

If $x = -3 - \sqrt{5}$,

$$x^2 + 6x + 4 = 0$$

$$(-3 - \sqrt{5})^2 + 6(-3 - \sqrt{5}) + 4 = 0$$

$$9 + 6\sqrt{5} + 5 - 18 - 6\sqrt{5} + 4 = 0$$

$$0 = 0$$

Therefore, the solution set is $\{-3 + \sqrt{5}, -3 - \sqrt{5}\}$.

It is useful to recall that the constant term of a perfect square trinomial of the form $x^2 + bx + c$ is found by taking half of the coefficient of the linear term and squaring this number.

If in the quadratic trinomial $ax^2 + bx + c$, $a \neq 1$, it is convenient to divide by a first.

EXAMPLE 7.6.2 Solve $3x^2 + 6x + 1 = 0$ by completing the square.

Solution

1. Divide by 3: $$3x^2 + 6x + 1 = 0$$ $$x^2 + 2x + \frac{1}{3} = 0$$
2. Subtract $\frac{1}{3}$ from both sides of the equation: $$x^2 + 2x = -\frac{1}{3}$$
3. Complete the square and add this term to both sides $$x^2 + 2x + 1 = -\frac{1}{3} + 1$$
4. Write left side as square of a binomial and simplify right side: $$(x + 1)^2 = \frac{2}{3}$$
5. Take the square root of both sides of the equation: $$x + 1 = \pm\sqrt{\frac{2}{3}} = \pm\frac{\sqrt{6}}{3}$$
6. Solve for x: $$x = -1 + \frac{\sqrt{6}}{3} \text{ or } x = -1 - \frac{\sqrt{6}}{3}$$

Check: For $x = -1 + \dfrac{\sqrt{6}}{3}$,

$$\begin{aligned} 3x^2 + 6x + 1 &= 3\left(-1 + \frac{\sqrt{6}}{3}\right)^2 + 6\left(-1 + \frac{\sqrt{6}}{3}\right) + 1 \\ &= 3\left(1 - \frac{2\sqrt{6}}{3} + \frac{6}{9}\right) - 6 + 2\sqrt{6} + 1 \\ &= 3 - 2\sqrt{6} + 2 - 5 + 2\sqrt{6} \\ &= 5 - 5 - 2\sqrt{6} + 2\sqrt{6} \\ &= 0 \end{aligned}$$

For $x = -1 - \dfrac{\sqrt{6}}{3}$,

$$\begin{aligned} 3x^2 + 6x + 1 &= 3\left(-1 - \frac{\sqrt{6}}{3}\right)^2 + 6\left(-1 - \frac{\sqrt{6}}{3}\right) + 1 \\ &= 3\left(1 + \frac{2\sqrt{6}}{3} + \frac{6}{9}\right) - 6 - 2\sqrt{6} + 1 \\ &= 3 + 2\sqrt{6} + 2 - 5 - 2\sqrt{6} \\ &= 0 \end{aligned}$$

EXAMPLE 7.6.3 Solve the equation $x^2 + x + 1 = 0$ by completing the square.

Solution

$$x^2 + x + 1 = 0$$
$$x^2 + x = -1$$
$$x^2 + x + \frac{1}{4} = -1 + \frac{1}{4}$$
$$\left(x + \frac{1}{2}\right)^2 = -\frac{3}{4}$$
$$x + \frac{1}{2} = \pm\frac{i\sqrt{3}}{2}$$
$$x = \frac{-1 \pm i\sqrt{3}}{2}$$

Note that both roots are imaginary.

Check: For $x = \dfrac{-1 + i\sqrt{3}}{2}$,

$$\begin{aligned} x^2 + x + 1 &= \left(\frac{-1 + i\sqrt{3}}{2}\right)^2 + \left(\frac{-1 + i\sqrt{3}}{2}\right) + 1 \\ &= \frac{1 - 2i\sqrt{3} - 3}{4} + \frac{-1 + i\sqrt{3}}{2} + 1 \\ &= \frac{-2 - 2i\sqrt{3}}{4} + \frac{-1 + i\sqrt{3}}{2} + 1 \\ &= \frac{-1 - i\sqrt{3}}{2} + \frac{-1 + i\sqrt{3}}{2} + 1 \\ &= \frac{-2}{2} + 1 = 0 \end{aligned}$$

The check is similar for $x = \dfrac{-1 - i\sqrt{3}}{2}$.

EXERCISES 7.6 A

Solve each quadratic equation in Exercises 1–22 by completing the square. Check each solution.

1. $(x - 5)^2 = 2$

2. $(x + 3)^2 = -4$

3. $x^2 - 4x + 1 = 0$

4. $x^2 - 2x + 3 = 0$

5. $x^2 - 2x - 3 = 0$

6. $y^2 - 8y + 15 = 0$

7. $y^2 - 8y + 17 = 0$
8. $x^2 = 2x + 19$
9. $x(x - 1) = 1$
10. $t + 2 = 3t^2$
11. $t^2 + 2 = 2t$
12. $4x - x^2 = 3$
13. $y^2 + 4y = 0$
14. $4x(3 - x) = 7$
15. $14z = z^2 + 53$
16. $6z^2 = 5z + 4$
17. $u(u + 6) + 2 = 0$
18. $(u + 3)^2 + 16 = 0$
19. $x - 10 + \frac{95}{x} = 0$
20. $\frac{x}{2} + \frac{2}{3} = \frac{x^2}{6}$
21. $\frac{y + 3}{y - 2} = \frac{13}{y}$
22. $\frac{t}{2} = \frac{1 - t}{2 - t}$

★ *Solve each equation in Exercises 23–30 for the letter indicated.*

23. $x^2 - 6xy + 5y^2 = 0; x$
24. $x^2 - 4y^2 = 4; x$
25. $4x^2 + 25y^2 = 100; y$
26. $x^2 + xy + y^2 = 0; y$
27. $s = vt - 16t^2; t$
28. $P = EI + RI^2; I$
29. $x^2 - 2bx + c = 0; x$
30. $ax^2 + bx + c = 0; x$

EXERCISES 7.6 B

Solve each quadratic equation in Exercises 1–22 by completing the square. Check each solution.

1. $(x - 2)^2 = 7$
2. $(x + 4)^2 = -5$
3. $x^2 - 10x + 20 = 0$
4. $x^2 - 6x - 3 = 0$
5. $x^2 - 6x + 3 = 0$
6. $x^2 + 6x - 3 = 0$
7. $x^2 + 6x + 3 = 0$
8. $y^2 = 2(5y - 4)$
9. $y(2y - 1) = 1$
10. $u^2 + 3 = u$
11. $4z = z^2 + 9$
12. $z + 10 = 2z^2$
13. $4u = 15 - 4u^2$
14. $t^2 + 8t + 20 = 0$
15. $2(6 - t^2) = 5t$
16. $3y^2 = 7y + 6$
17. $x(3 - x) = 4$
18. $x + 2 = \frac{3}{x}$
19. $y + \frac{5}{y} = 2$
20. $\frac{10}{x^2} = \frac{20}{(10 - x)^2}$
21. $\frac{1}{t} + \frac{3 - t}{1 - t} = 0$
22. $x^2 - 2\pi x + 1 = 0$

★ *Solve each equation in Exercises 23–30 for the letter indicated.*

23. $x^2 - xy - y^2 = 0; x$

24. $4x^2 + 9y^2 = 36; y$

25. $16x^2 - y^2 = 16; y$

26. $x^2 - xy - 2y^2 = 0; x$

27. $K = \frac{x^2}{1 - x}; x$

28. $\frac{a}{x^2} = \frac{b}{(c - x)^2}; x$

29. $Q = a + bt + ct^2; t$

30. $Ax^2 + 2Bx + C = 0; x$

7.7 THE QUADRATIC FORMULA

In the preceding section it was seen that any quadratic equation of the form

$$ax^2 + bx + c = 0 \qquad (a \neq 0)$$

can be solved by completing the square. The process is sometimes lengthy, so the method is applied to the general equation and a formula is developed:

$$ax^2 + bx + c = 0$$

1. Divide each side by a: $\quad x^2 + \frac{b}{a}x + \frac{c}{a} = 0$

2. Subtract $\frac{c}{a}$ from each side: $\quad x^2 + \frac{b}{a}x = -\frac{c}{a}$

3. Complete the square: $\quad x^2 + \frac{b}{a}x + \left(\frac{b}{2a}\right)^2 = -\frac{c}{a} + \left(\frac{b}{2a}\right)^2$

4. Write the left side as the square of a binomial and simplify the right side:

$$\left(x + \frac{b}{2a}\right)^2 = -\frac{c}{a} + \frac{b^2}{4a^2}$$

$$= \frac{b^2 - 4ac}{4a^2}$$

5. Take the square root of each side:

$$x + \frac{b}{2a} = \pm \frac{\sqrt{b^2 - 4ac}}{2a}$$

6. Solve for x:

$$x = \frac{-b + \sqrt{b^2 - 4ac}}{2a}$$

or

$$x = \frac{-b - \sqrt{b^2 - 4ac}}{2a}$$

If $b^2 - 4ac \geq 0$, then $\sqrt{b^2 - 4ac}$ is a real number, and x designates a real number. If $b^2 - 4ac < 0$, then $\sqrt{b^2 - 4ac}$ represents an imaginary number, and the solutions of the equations are imaginary.

If $b^2 - 4ac = 0$, then the equation has a double root—namely, $\frac{-b}{2a}$. The roots are also said to be equal in this case.

Because the expression $b^2 - 4ac$ determines whether the roots are real, imaginary, or equal, $b^2 - 4ac$ is called **the discriminant** of the quadratic equation.

Since a, b, and c were chosen completely arbitrarily except $a \neq 0$, the set of equations

$$\left\{x = \frac{-b + \sqrt{b^2 - 4ac}}{2a}, x = \frac{-b - \sqrt{b^2 - 4ac}}{2a}\right\}$$

can be used as a formula for the solution of a quadratic equation.

THE QUADRATIC FORMULA

The quadratic equation $ax^2 + bx + c = 0$, $a \neq 0$, has the solutions

$$x = \frac{-b + \sqrt{b^2 - 4ac}}{2a}, x = \frac{-b - \sqrt{b^2 - 4ac}}{2a}$$

DEFINITION

$b^2 - 4ac$ is the discriminant of the quadratic equation $ax^2 + bx + c = 0$, $a \neq 0$.

If $b^2 - 4ac$ is positive, the solutions are real.
If $b^2 - 4ac$ is negative, the solutions are imaginary.
If $b^2 - 4ac = 0$, the solutions are real and equal—that is, the equation has a double root which is a real number.

EXAMPLE 7.7.1 Solve $2x^2 + 3x + 1 = 0$ by using the quadratic formula.

Solution Comparing with $ax^2 + bx + c = 0$,

$$2x^2 + 3x + 1 = 0$$

it is seen that $a = 2$, $b = 3$, and $c = 1$. Therefore

$$x = \frac{-b \pm \sqrt{b^2 - 4ac}}{2a}$$

yields

$$x = \frac{-3 + \sqrt{9 - 4(2)(1)}}{4} = \frac{-3 + 1}{4} = \frac{-2}{4} = -\frac{1}{2}$$

or

$$x = \frac{-3 - \sqrt{9 - 4(2)(1)}}{4} = \frac{-3 - 1}{4} = \frac{-4}{4} = -1$$

Check:

If $x = -\frac{1}{2}$,

$$2x^2 + 3x + 1 = 0$$

$$2\left(-\frac{1}{2}\right)^2 + 3\left(-\frac{1}{2}\right) + 1 = 0$$

$$\frac{1}{2} - \frac{3}{2} + 1 = 0$$

$$0 = 0$$

If $x = -1$,

$$2x^2 + 3x + 1 = 0$$

$$2(-1)^2 + 3(-1) + 1 = 0$$

$$2 - 3 + 1 = 0$$

$$0 = 0$$

Therefore, $\left\{-\frac{1}{2}, -1\right\}$ is the solution set.

Note in this example that the discriminant

$$b^2 - 4ac = 9 - 4(2)(1) = 9 - 8 = 1$$

a positive number, indicating that the roots are real numbers. Moreover, the discriminant is a perfect square since $1^2 = 1$. Whenever the discriminant is a perfect square, the roots are rational numbers.

EXAMPLE 7.7.2 Solve $2x^2 - 3x = x^2 - 1$ by using the quadratic formula.

Solution First an equivalent equation of the form $ax^2 + bx + c = 0$ must be found:

$$2x^2 - 3x = x^2 - 1$$

$$x^2 - 3x + 1 = 0$$

$$a = 1, b = -3, c = 1$$

$$x = \frac{-(-3) \pm \sqrt{(-3)^2 - 4(1)(1)}}{2}$$

$$x = \frac{3 + \sqrt{5}}{2} \text{ or } x = \frac{3 - \sqrt{5}}{2}$$

The solution set is $\left\{\frac{3 + \sqrt{5}}{2}, \frac{3 - \sqrt{5}}{2}\right\}$.

Check:

$$\text{For } x = \frac{3 + \sqrt{5}}{2},$$

$$\begin{aligned}
2x^2 - 3x &= 2\left(\frac{3 + \sqrt{5}}{2}\right)^2 - 3\left(\frac{3 + \sqrt{5}}{2}\right) \\
&= \frac{2(14 + 6\sqrt{5})}{4} - \frac{3(3 + \sqrt{5})}{2} \\
&= \frac{14 + 6\sqrt{5} - 9 - 3\sqrt{5}}{2} = \frac{5 + 3\sqrt{5}}{2}
\end{aligned}$$

$$\begin{aligned}
x^2 - 1 = \left(\frac{3 + \sqrt{5}}{2}\right)^2 - 1 &= \frac{14 + 6\sqrt{5}}{4} - \frac{4}{4} \\
&= \frac{10 + 6\sqrt{5}}{4} \\
&= \frac{5 + 3\sqrt{5}}{2}
\end{aligned}$$

The check for $x = \dfrac{3 - \sqrt{5}}{2}$ is similar.

Note that the discriminant $b^2 - 4ac = (-3)^2 - 4(1)(1) = 9 - 4 = 5$, a positive number, indicating that the roots are real. Also, since 5 is not a perfect square, the roots are irrational.

EXAMPLE 7.7.3 Solve for x:

$$\frac{x^2}{4} - \frac{x}{2} + 1 = 0$$

Solution Although this problem can be worked with $a = \frac{1}{4}$, $b = -\frac{1}{2}$, and $c = 1$, it is easier to solve by first finding an equivalent equation with integers for coefficients. Multiplying each term by 4, the desired equivalent equation is obtained:

$$x^2 - 2x + 4 = 0$$

Then $a = 1$, $b = -2$, and $c = 4$.

$$\begin{aligned}
x &= \frac{-(-2) \pm \sqrt{(-2)^2 - 4(1)(4)}}{2(1)} \\
&= \frac{2 \pm \sqrt{4 - 16}}{2} = \frac{2 \pm \sqrt{-12}}{2} \\
&= \frac{2 \pm i2\sqrt{3}}{2} = 1 \pm i\sqrt{3}
\end{aligned}$$

The solution set is $\{1 + i\sqrt{3}, 1 - i\sqrt{3}\}$.

Check: For $x = 1 + i\sqrt{3}$,

$$\begin{aligned}\frac{x^2}{4} - \frac{x}{2} + 1 &= \frac{(1 + i\sqrt{3})^2}{4} - \frac{1 + i\sqrt{3}}{2} + 1\\ &= \frac{-2 + 2i\sqrt{3}}{4} - \frac{1 + i\sqrt{3}}{2} + 1\\ &= \frac{-1 + i\sqrt{3} - 1 - i\sqrt{3}}{2} + 1\\ &= -1 + 1 = 0\end{aligned}$$

The check is similar for $x = 1 - i\sqrt{3}$.

Note that the discriminant

$$b^2 - 4ac = (-2)^2 - 4(1)(4) = 4 - 16 = -12,$$

a negative number, indicating that the roots are imaginary.

EXAMPLE 7.7.4 Solve $9x^2 + 30x + 25 = 0$.

Solution $a = 9$, $b = 30$, $c = 25$.

$$\begin{aligned}x &= \frac{-30 \pm \sqrt{(30)^2 - 4(9)(25)}}{2(9)}\\ &= \frac{-30 \pm \sqrt{900 - 900}}{2(9)}\\ &= \frac{-30}{6(3)}\\ &= \frac{-5}{3}\end{aligned}$$

The solution set is $\left\{-\frac{5}{3}\right\}$.

Check: For $x = -\frac{5}{3}$,

$$\begin{aligned}9x^2 + 30x + 25 &= 9\left(\frac{-5}{3}\right)^2 + 30\left(\frac{-5}{3}\right) + 25\\ &= 9\left(\frac{25}{9}\right) + 10(-5) + 25\\ &= 25 - 50 + 25\\ &= 0\end{aligned}$$

In this case, $-\frac{5}{3}$ is a double root. Note that the discriminant $b^2 - 4ac = 0$:

$$b^2 - 4ac = (30)^2 - 4(9)(25) = 900 - 900 = 0$$

EXERCISES 7.7 A

In Exercises 1–5: (a) Write the equations as equivalent equations in the form $ax^2 + bx + c = 0$ *and state the value of a, b, and c in each equation; (b) find the value of the discriminant,* $b^2 - 4ac$, *and state whether the roots of the equation are real and unequal, imaginary, or real and equal.*

1. $2x^2 = 3x + 1$

2. $4x^2 = 12x - 9$

3. $\frac{5}{x} = x$

4. $(x + 2)(x + 3) = 6$

5. $4x - 8 = x^2$

Solve each of the equations in Exercises 6–20 by using the quadratic formula. Check each solution.

6. $3x^2 + 5x + 2 = 0$

7. $3x^2 - 5x - 2 = 0$

8. $x^2 - 3x + 1 = 0$

9. $x^2 + 3x - 1 = 0$

10. $x^2 - 3x + 3 = 0$

11. $x^2 + x + 1 = 0$

12. $y^2 - 2y - \frac{3}{4} = 0$

13. $3y^2 = 2(y - 1)$

14. $y + 2 = 3y(y + 1)$

15. $t^2 + 16 = 0$

16. $5t = 3t^2$

17. $4x - 36x^2 = \frac{1}{9}$

18. $80t - 16t^2 = 80$

19. $\frac{1}{2}d^2 - \frac{30}{8}d + \frac{36}{8} = 0$

20. $\frac{w}{w + 10} = \frac{w - 10}{10}$

Solve Exercises 21–28 for the variable indicated.

21. $x^2 + 2px + q = 0;\ x$

22. $y^2 - ny - n^2 = 0;\ y$

23. $3x^2 + 8xy - 3y^2 = 0;\ x$

24. $y^2 - 4y + 2x = 6;\ y$

25. $s = vt + \frac{1}{2}gt^2;\ t$

26. $s = (k - s)^2;\ s$

27. $\frac{1}{R} + \frac{1}{a - R} = \frac{1}{b};\ R$

28. $A = \pi r(r + s);\ r$

EXERCISES 7.7 B

In Exercises 1–5: (a) Write the equations as equivalent equations in the form $ax^2 + bx + c = 0$ and state the value of a, b, and c in each equation; (b) find the value of the discriminant, $b^2 - 4ac$, and state whether the roots of the equation are real and unequal, imaginary, or real and equal.

1. $4x^2 - 2 = 2x$

2. $(x + 1)^2 = x + 3$

3. $9 = 6x + x^2$

4. $(5 - x)(4 + x) = 40$

5. $2x + \frac{1}{8x} = 1$

Solve each of the equations in Exercises 6–20 by using the quadratic formula. Check each solution.

6. $4x^2 + 7x + 3 = 0$

7. $4x^2 + 7x + 4 = 0$

8. $4x^2 + 7x - 2 = 0$

9. $4x^2 - 4x + 1 = 0$

10. $x^2 - 4x - 4 = 0$

11. $x^2 + 2x + 4 = 0$

12. $y^2 - y + \frac{1}{5} = 0$

13. $\frac{y^2 + 1}{6} = \frac{y}{2}$

14. $3(y + 1) = 5 - 2y^2$

15. $9 = 6t + t^2$

16. $\frac{t - 1}{t} - \frac{t - 1}{6} = \frac{1}{6}$

17. $25x - 100x^2 = 0$

18. $25u^2 = 40u - 16$

19. $\frac{1}{2}d^2 = 6d + 9$

20. $\frac{c^2}{0.01 - c} = 1.7 \times 10^{-5}$

Solve Exercises 21–28 for the variable indicated.

21. $cx^2 + 2bx + a = 0; x$

22. $y^2 - 4ny + 2n^2 = 0; y$

23. $4x^2 - 6xy + y^2 = 0; x$

24. $\frac{W}{L} = \frac{L - W}{W}; W$

25. $P = EI + RI^2; I$

26. $F^2 = \frac{1}{LC} - \frac{R}{4L^2}; L$

27. $y = mx - \frac{16(m^2 + 1)}{v^2}x^2; x$

28. $V = \frac{h}{3}(a^2 + ab + b^2); a$

7.8 VERBAL PROBLEMS

There are many practical applications that involve solving a quadratic equation. When finding the solutions to verbal problems that involve the application of quadratic equations, it is especially important to check each root of the equation in the statement of the problem to see if the necessary conditions are met. Often the equation will have two roots, but only one may apply to a given problem. For example, lengths of sides of rectangles, triangles, etc. are always positive numbers; ages of individuals are positive numbers; digits in a numeral cannot be fractions; and the number of people present at a certain gathering cannot be fractional or negative.

EXAMPLE 7.8.1 If the legs of a right triangle measure 5 inches and 12 inches, respectively, what is the length of the hypotenuse?

Solution The theorem of Pythagoras states that if a and b are the measures of the legs of a right triangle and c represents the measure of the hypotenuse, then

$$a^2 + b^2 = c^2$$

Let $a = 5$, $b = 12$; then

$$\begin{aligned} 5^2 + 12^2 &= c^2 \\ 25 + 144 &= c^2 \\ 169 &= c^2 \\ \pm 13 &= c \end{aligned}$$

Since length is a positive number, the condition $c > 0$ is implied. Thus the common solution of $c = 13$ or $c = -13$, and $c > 0$ is $c = 13$.

EXAMPLE 7.8.2 If the hypotenuse of a right triangle is 25 inches long and one leg measures 24 inches, how long is the other leg?

Solution Applying the theorem of Pythagoras as in the preceding example, let $c = 25$, $a = 24$, and b designate the other leg:

$$\begin{aligned} a^2 + b^2 &= c^2 \\ (24)^2 + b^2 &= (25)^2 \\ b^2 &= (25)^2 - (24)^2 \\ b^2 &= (25 - 24)(25 + 24) \\ b^2 &= 49 \\ b &= \pm 7 \end{aligned}$$

Again we disregard the solution $b = -7$. Therefore, the other leg is 7 inches long.

EXAMPLE 7.8.3 One leg of a right triangle is 1 foot longer than the other leg and 8 feet shorter than the hypotenuse. Find the lengths of the sides of the right triangle.

Solution If the lengths of the legs of the triangle are designated by a and b, where b is the shorter leg, and the length of the hypotenuse by c, the following relationships are given:

(1) $a^2 + b^2 = c^2$ (Theorem of Pythagoras)

(2) $a = c - 8$ (Given)

(3) $b = a - 1$ (Given, and the assumption that b is the shorter leg)

From (2) we obtain the equivalent equation

$$c = a + 8$$

Substituting $(a + 8)$ for c and $(a - 1)$ for b in equation (1) yields the equivalent equation

$$a^2 + (a - 1)^2 = (a + 8)^2$$

$$a^2 + a^2 - 2a + 1 = a^2 + 16a + 64$$

(4) $a^2 - 18a - 63 = 0$

Equation (4) is a quadratic equation that can be solved by factoring:

$$(a - 21)(a + 3) = 0$$

$$a - 21 = 0 \quad \text{or } a + 3 = 0$$

$$a = 21 \text{ or } \qquad a = -3$$

Since the verbal problem contains the assumption $a > 0$, we disregard the solution $a = -3$, and accept $a = 21$. From (2) we know that $a = c - 8$; therefore, $c = 29$. From (3) we know that $b = a - 1$; therefore, $b = 20$. The sides of the triangle are 20 feet, 21 feet, and 29 feet.

EXAMPLE 7.8.4 A plane flies 300 miles with a tail wind of 10 mph and returns against a wind of 20 mph. What is the speed of the plane in still air if the total flying time is $4\frac{1}{2}$ hours?

Solution Let x represent the speed of the plane in still air.

Then $\quad x + 10 =$ speed with the tail wind

and $\quad x - 20 =$ speed against the wind

Formula	r ·	t =	d
With wind	$x + 10$	$\frac{300}{x+10}$	300
Against wind	$x - 20$	$\frac{300}{x-20}$	300

Equation:

Time going + time returning = total time $= 4\frac{1}{2}$ hours $= \frac{9}{2}$ hours:

$$\frac{300}{x+10} + \frac{300}{x-20} = \frac{9}{2}$$

Multiplying both sides of this equation by the L.C.D.,

$$2(x + 10)(x - 20)$$

yields

$$600(x - 20) + 600(x + 10) = 9(x + 10)(x - 20)$$

Dividing both sides by 3,

$$200(x - 20) + 200(x + 10) = 3(x + 10)(x - 20)$$
$$200x - 4000 + 200x + 2000 = 3x^2 - 30x - 600$$
$$3x^2 - 430x + 1400 = 0$$
$$(3x - 10)(x - 140) = 0$$
$$3x - 10 = 0 \quad \text{or } x - 140 = 0$$
$$x = 3\frac{1}{3} \text{ or } x = 140$$

Since speed is always a positive number, $x \neq 3\frac{1}{3}$, because

$$x - 20 = 3\frac{1}{3} - 20$$

which is a negative number.

Therefore, the speed of the plane in still air is 140 mph.

EXAMPLE 7.8.5 A square flower bed has a 3-foot walk surrounding it. If the walk were to be replaced and planted with flowers, the new flower bed would have 4 times the area of the original bed. What is the length of one side of the original bed?

Solution

Let x = length of a side of the original flower bed
Then x^2 = area of original bed
$x + 6$ = length of a side of the new flower bed
and $(x + 6)^2$ = area of the new flower bed

$$(x + 6)^2 = 4x^2$$
$$x^2 + 12x + 36 = 4x^2$$
$$3x^2 - 12x - 36 = 0$$
$$3(x^2 - 4x - 12) = 0$$
$$x^2 - 4x - 12 = 0$$
$$(x + 2)(x - 6) = 0$$
$$x + 2 = 0 \quad \text{or } x - 6 = 0$$
$$x = -2 \text{ or } \quad x = 6$$

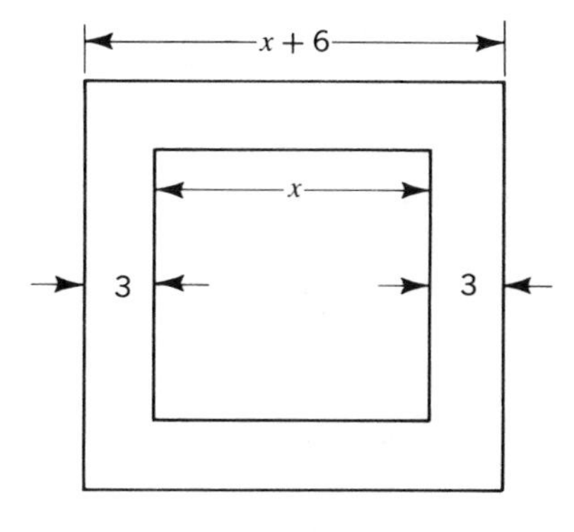

FIGURE 7.8.1

Again we discard the negative answer, and the length of the original side is 6 feet.

Check:

Area of original bed $= x^2 = 6^2 = 36$ square feet
Area of new bed $= (x + 6)^2 = (6 + 6)^2 = 12^2 = 144$ square feet
Area of new bed = 4 times area of old bed

$$144 = 4 \times 36$$
$$144 = 144$$

EXAMPLE 7.8.6 The sum of a number and its reciprocal is 4. Find the number.

Solution

Let $x =$ the number (Clearly $x \neq 0$. Why?)

Then $\frac{1}{x} =$ its reciprocal by the definition of reciprocal.

$$x + \frac{1}{x} = 4$$
$$x^2 + 1 = 4x$$
$$x^2 - 4x + 1 = 0$$

Since this quadratic polynomial cannot be factored easily, the method of completing the square is used. (This is a good technique when $\frac{b}{a}$ is an even integer.)

$$x^2 - 4x + 1 = 0$$
$$x^2 - 4x = -1$$
$$x^2 - 4x + (-2)^2 = -1 + (-2)^2$$
$$(x - 2)^2 = -1 + 4 = 3$$
$$x - 2 = \pm\sqrt{3}$$
$$x = 2 \pm \sqrt{3}$$

Check: If $x = 2 + \sqrt{3}$, then

$$\frac{1}{x} = \frac{1}{2 + \sqrt{3}}$$

and
$$x + \frac{1}{x} = 2 + \sqrt{3} + \frac{1}{2 + \sqrt{3}} = \frac{8 + 4\sqrt{3}}{2 + \sqrt{3}} = \frac{4(2 + \sqrt{3})}{(2 + \sqrt{3})} = 4$$

If $x = 2 - \sqrt{3}$, then

$$\frac{1}{x} = \frac{1}{2 - \sqrt{3}}$$

and
$$x + \frac{1}{x} = 2 - \sqrt{3} + \frac{1}{2 - \sqrt{3}} = \frac{8 - 4\sqrt{3}}{2 - \sqrt{3}} = \frac{4(2 - \sqrt{3})}{(2 - \sqrt{3})} = 4$$

Therefore, the solution set is $\{2 + \sqrt{3}, 2 - \sqrt{3}\}$.

EXAMPLE 7.8.7 Working alone, a carpenter can make a set of cabinets in 3 hours less time than his helper can. Working together, they can make the set of cabinets in 6 hours. Find the time (correct to the nearest minute) that each requires to make the set alone.

Solution

Let $x =$ time it takes helper alone

Then $x - 3 =$ time it takes carpenter alone

Formulas:

$tr = w$, and w of first $+$ w of second $=$ w of both $= 1$,

where $t =$ time, $r =$ rate, and $w =$ amount of work done.

	Working Alone			*Working Together (Whole Job)*		
Formula	t	$\cdot\ r$	$= w$	t	$\cdot\ r$	$= w$
Carpenter	$x - 3$	$\frac{1}{x-3}$	1	6	$\frac{1}{x-3}$	$\frac{6}{x-3}$
Helper	x	$\frac{1}{x}$	1	6	$\frac{1}{x}$	$\frac{6}{x}$

Equation: Work of helper + work of carpenter = whole job

$$\frac{6}{x} + \frac{6}{x-3} = 1$$

with restrictions $x > 0$ and $x - 3 > 0$

$$6(x-3) + 6x = x(x-3)$$
$$12x - 18 = x^2 - 3x$$
$$x^2 - 15x + 18 = 0$$

Since $x^2 - 15x + 18$ is not readily factorable, and since $\frac{b}{a} = -15$ is not an even integer, the method using the quadratic formula is recommended for solving the equation.

Solution of Equation:

$ax^2 + bx + c = 0$ if and only if

$$x = \frac{-b \pm \sqrt{b^2 - 4ac}}{2a}$$

If $x^2 - 15x + 18 = 0$, then $a = 1$, $b = -15$, $c = 18$. Thus

$$x = \frac{-(-15) \pm \sqrt{(-15)^2 - 4(18)}}{2} = \frac{15 \pm \sqrt{225 - 72}}{2}$$

$$x = \frac{15 + \sqrt{153}}{2} \text{ or } x = \frac{15 - \sqrt{153}}{2}$$

Approximating (by using the tables), $\sqrt{153} = 12.37$. Thus, approximately to the nearest hundredth,

$$x = \frac{15 + 12.37}{2} \text{ or } x = \frac{15 - 12.37}{2}$$

$$x = \frac{27.37}{2} \qquad \text{or } x = \frac{2.63}{2}$$

$$x = 13.68 \qquad \text{or } x = 1.32$$

and

$x - 3 = 10.68$ or $x - 3 = -1.68$ (This solution rejected, since $x - 3 > 0$)

Thus $x = 13.68$ and $x - 3 = 10.68$ are correct to the nearest hundredth. Since there are 60 minutes in an hour,

$$0.68 \text{ hours} = 0.68(60) \text{ minutes} = 40.8 \text{ minutes}$$

Therefore, correct to the nearest minute,

$$x = 13 \text{ hours, 41 minutes, time of helper alone}$$
$$x - 3 = 10 \text{ hours, 41 minutes, time of carpenter alone}$$

EXERCISES 7.8 A

Solve Exercises 1–16.

1. The hypotenuse of a right triangle is 3 units and the legs are equal in length. Find the length of a leg of the triangle.
2. The length of a rectangle exceeds 3 times its width by 1 inch. The area is 52 square inches. Find the dimensions of the rectangle.
3. A certain number exceeds its reciprocal by $\frac{15}{4}$. Find the number.
4. It takes John 3 hours longer to do a certain job than it does his brother Bob. For 3 hours they worked together; then John left and Bob finished the job in 1 hour. How many hours would it have taken Bob to do the whole job by himself?
5. A rectangular piece of sheet metal is twice as long as it is wide. From each of its four corners a square piece 2 inches on a side is cut out. The flaps are then turned up to form an uncovered metal box. If the volume of this box is 320 cubic inches, find the dimensions of the original piece of sheet metal.

6. The sum of two numbers is 7 and the difference of their reciprocals is $\frac{1}{12}$. Find the numbers.

7. The roof line of a certain house has a pitch of 3 to 12. This means that for each vertical rise of 3 feet, there is a horizontal run of 12 feet. To find the length of lumber he must cut for the roof, the carpenter has to calculate the length of the hypotenuse of a right triangle whose legs are the height of the roof and the half-span of the roof. Find the length of this hypotenuse, correct to the nearest inch, for this house having a half-span of 13 feet.

8. A fisherman trolled upstream in a motorboat to a spot 6 miles from his campsite and then returned to camp. If the rate of the current was $1\frac{1}{2}$ mph and if the round trip took 3 hours, find the rate of the motorboat in still water.

9. The span s of a circular arch is related to its height h and its radius r by the formula

$$s^2 = 8rh - 4h^2$$

Find the height of a circular arch whose span is 80 feet and whose radius is 50 feet. (Assume that the circular arch is less than a semicircle.)

10. A circular hole has a radius of 5 inches. How much larger should the radius be so that a new circular hole will have a cross-sectional area twice as large as the original area? (Area of circle, $A = \pi r^2$.)

11. The ionization constant K of a weak acid is related to the hydrogen ion concentration x and the molarity M of the acid solution by the equation

$$\frac{x^2}{M - x} = K$$

Find the hydrogen ion concentration for a certain solution of acetic acid having $M = 0.0002$ and $K = 1.8 \times 10^{-5}$.

12. A private plane flew from San Francisco to Lake Tahoe, a distance of 180 miles, with a tail wind and then returned flying against the same wind. If the total flying time was $2\frac{1}{2}$ hours and if the speed of the plane in still air was 150 mph, find the speed of the wind.

13. One man can do a job in 8 days less time than another man, but he charges $50 a day, whereas the slower man charges $20 a day. When the two men work together, they take 3 days to complete the job. Which would cost less—to have the faster man do the job alone, to have the slower man do the job alone, or to have both work together? State the cost for each case.

14. A fisherman in a boat on a small lake sees some plants growing in the water with their roots at the bottom of the lake. To find the depth of the lake, he pushes a plant extending 8 inches out of the water so that the plant is completely submerged with its tip just touching the water. He measures the distance from the point where the plant first emerged from the water to the new position of its tip and finds this to be 32 inches. Find the depth of the lake.

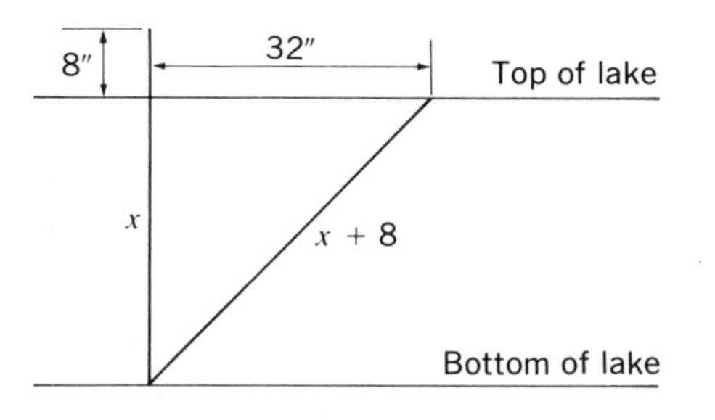

FIGURE 7.8.2

15. A dive bomber traveling 480 feet per second vertically downward releases a bomb 1024 feet above the earth. If s is the distance of the bomb below the point where it was dropped and t is the time in seconds after the bomb was dropped, then

$$s = 480t + 16t^2$$

How long does the bomb take to reach the earth? With what velocity v does the bomb hit the earth if

$$v = 480 + 32t$$

16. For a certain electric motor having a mechanical output of 25,000 watts, a resistance in the armature of 0.04 ohm, and a line voltage of 110 volts, the armature current I in amperes is given by

$$25{,}000 = 110I - 0.04I^2$$

Find the current I.

EXERCISES 7.8 B

Solve Exercises 1–14.

1. A concrete walk of uniform width extends around a rectangular lawn having dimensions 20 feet by 80 feet. Find the width of the walk if the area of the walk is 864 square feet.

2. The hypotenuse of a right triangle is 13 inches. If one leg is 7 inches longer than the other, how long are the legs of the triangle?

3. If 4 times a number is added to 3 times its square, the sum is 95. Find the number.

4. Find two consecutive odd integers the sum of whose squares is 514.

5. A 36-inch length of copper tubing is bent to form a right triangle having a 15-inch hypotenuse. Find the lengths of the other two sides of the triangle.

6. A wire is stretched from the top of a 4-foot fence to the top of a 20-foot vertical pole. If the fence and the pole are 30 feet apart, find the length of the wire.

7. One of two outlets can fill a swimming pool in 6 hours. The time for the other outlet to fill the pool is 2 hours longer than the two outlets together. Find the time it takes for the two outlets together to fill the tank, correct to the nearest minute.

8. One inlet pipe takes 12 minutes longer than another pipe inlet to fill a certain tank. An outlet pipe can empty the tank in 45 minutes. When all three pipes are open, it takes 15 minutes to fill the tank. Find the time it takes to fill the tank if only the larger inlet pipe is open.

9. A jet plane flying against a head wind of 20 mph takes 20 minutes longer to fly a distance of 2610 miles than a plane with the same still air speed flying in the opposite direction. Find the still air speed of the plane.

10. A boat that travels 12 mph in still water takes 2 hours less time to go 45 miles downstream than to return the same distance upstream. Find the rate of the current.

11. A baseball diamond has the shape of a square with each side 90 feet long. The pitcher's mound is 60.5 feet from home plate on the line joining home plate to second base. Find the distance from the pitcher's mound to second base.

12. The bending moment M of a beam fixed at one end and simply supported at the other is related to its length L, its uniform load distribution w, and the distance x from the fixed end by the relation

$$M = \frac{-w}{8}(4x^2 - 5Lx + L^2)$$

A certain beam has a length L of 12 feet and a uniform load distribution w of 100 pounds per foot.
a. For what position x on the beam is the bending moment zero?
b. For what position x on the beam is the bending moment 100 foot-pounds?

13. If I is the intensity of illumination in lumens, c is the candlepower of the source of light, and s is the distance in feet from the source, then

$$I = \frac{c}{s^2}$$

A 20-candlepower light is 3 feet to the left of a 45-candlepower light. Find the distance x from the 20-candlepower source on the line joining the two sources so that the illumination is the same from each source. Use

$$\frac{20}{x^2} = \frac{45}{(3 - x)^2}$$

★ **14.** A mechanized division 2 miles long is moving at the rate of 30 mph. A messenger from the rear rushes to the head of the division and immediately returns to the rear, arriving $7\frac{1}{2}$ minutes after he left. Find the speed of the messenger.

SUMMARY

Complex Numbers

☐ **Definition.** $i = \sqrt{-1}$ and $i^2 = -1$.

☐ **Definition.** $\sqrt{-a} = i\sqrt{a}$ if a is a positive real number.

☐ **Standard Form.** Every complex number can be expressed in the form $a + bi$, called the standard form, where a and b are real numbers and i is the imaginary unit such that $i^2 = -1$.

☐ **Definition. Two complex numbers are equal** if and only if their real parts are equal and their imaginary parts are equal—that is,

$$a + bi = c + di \text{ if and only if } a = c \text{ and } b = d$$

☐ **Definition.** The **conjugate** of $a + bi$ is $a - bi$.
The **conjugate** of $a - bi$ is $a + bi$.

☐ **Definition. Sum:** $(a + bi) + (c + di) = (a + c) + (b + d)i$
Product: $(a + bi)(c + di) = (ac - bd) + (ad + bc)i$

☐ **The Quadratic Formula.** The quadratic equation $ax^2 + bx + c = 0$, $a \neq 0$, has the solutions

$$x = \frac{-b \pm \sqrt{b^2 - 4ac}}{2a}$$

☐ The **discriminant** of a quadratic equation is $b^2 - 4ac$.

If $b^2 - 4ac > 0$, then the roots are real and unequal.

If $b^2 - 4ac = 0$, then the roots are real and equal.

If $b^2 - 4ac < 0$, then the roots are imaginary.

REVIEW EXERCISES

1. Write each of the following complex numbers in the standard form:

a. $2 + \sqrt{-16}$
b. $(3 + 5i) - (2 - 3i)$
c. $\dfrac{4 + \sqrt{-12}}{2}$
d. $(2 + 3i)(3 - 4i)$
e. $2i(3 + 4i)^2$
f. $\dfrac{1 - i}{2 + 3i}$

2. Determine the real numbers x and y for which the equations are true:

a. $x + yi = 3 + 2i$
b. $(x + 2y + 2) + (2x + y)i = 0$

3. Two forces represented by the vectors $7 - 8i$ and $5 + 3i$ act on an object. Find the resultant force vector, its magnitude, and its direction. Illustrate geometrically.

4. Show that

$$\sqrt{\frac{9}{2} + \frac{9i\sqrt{3}}{2}} = \frac{3\sqrt{3}}{2} + \frac{3i}{2}$$

by showing that their squares are equal.

5. Determine the solution set of each of the following equations over the complex numbers by the factoring method and check each solution:

a. $x^2 - 6x - 40 = 0$
b. $12x^2 + 32x + 5 = 0$
c. $3x^2 - 4x = 0$
d. $x^2 + 9 = 0$

6. Solve the following equations by completing the square and check each solution:

a. $x^2 + 8x + 15 = 0$
b. $2x - x^2 - 3 = 0$
c. $3x^2 = 12x + 3$
d. $(x + 4)(x + 6) = 7$

7. Solve the following equations by the quadratic formula and check each solution:

a. $2x^2 + 5x - 1 = 0$
b. $2 = 2x^2 - x$
c. $6x^2 - 7x - 20 = 0$
d. $16x = 3x^2 + 8$

8. Solve the following equations by any method you wish and check each solution:

a. $\frac{3}{4}x^2 = \frac{7}{8}$ b. $x^2 + 4x + 8 = 0$

c. $(2x + 3)(x - 2) = 4$ d. $3x^2 + 5x = -1$

9. Find the value of k for which the roots of the given equation are equal:

a. $x^2 + 6x + k = 0$ b. $x^2 + 2kx + 4 = 0$

c. $kx^2 + 5x + 5 = 0$ d. $kx^2 + 8x + k = 0$

10. Solve each of the following for y:

a. $10x^2 = y^2 - 3xy$ b. $y^2 + 2x^2y = x^4$

11. John can mow his lawn in 20 minutes less time with his power mower than with his hand mower. One day his power mower broke down 15 minutes after he started mowing, and he had to complete the job with his hand mower. It took him 25 minutes to finish mowing by hand. How long does it take John to do the complete job with the power mower?

12. Find a number whose additive inverse is equal to its multiplicative inverse.

13. The sum of the reciprocals of two consecutive numbers is $\frac{11}{30}$. Find the numbers.

14. The perimeter of a rectangle is 36 feet and its area is 45 square feet. Find the dimensions of the rectangle.

15. A pilot left a Chicago airport and flew 200 miles south to a town T with a tail wind of 20 mph. From T he flew back to Chicago against a head wind of 30 mph. If his total flying time was $2\frac{1}{3}$ hours, what was the average speed of the plane in still air?

16. One leg of a right triangle is 9 inches longer than the other leg. The hypotenuse is 45 inches long. Find the lengths of the legs of the triangle.

17. The height H of a projectile at the end of t seconds is given by

$$H = cvt - \frac{1}{2}gt^2$$

Solve for t.

18. The total surface area T of a right circular cylinder of radius r and height h is given by

$$T = 2\pi r(r + h)$$

Solve for r.

ANSWERS

This section includes the answers to the Exercise Sets A and the Review Exercises.

Exercises 1.1 A, page 6

1. {4, 8, 12, 16, 20, . . .}, infinite

2. {3, 4, 5, 6, 7}, finite

3. {1, 2, 3, 4, 5}, finite

4. {11, 12, 13, 14, 15, . . .}, infinite

5. $\varnothing$ (or { }), finite

6. {0, 2, 4, 6, 8}, finite

7. {2, 4, 6, 8, 10, . . .}, infinite

8. {1, 3, 5, 7, 9, . . .}, infinite

9. {0}, finite

10. {19, 20, 21, 22, 23, 24}, finite

11. {0, 4, 8}, finite

12. {4, 8, 12, 16, 20, . . .}, infinite

13. {10, 11, 12, 13, 14, 15}, finite

14. {0, 1, 2, 3, 4}, finite

15. $\varnothing$ (or { }), finite

16. {26, 27, 28, 29, 30, 31, 32, 33, 34, 35}, finite

17. $\varnothing$ (or { }), finite

18. {12, 24, 36, 48, 60, . . .}, infinite

19. {1, 2, 3, 4, 6, 12}, finite

20. {1, 2, 3, 6}, finite

Exercises 1.2 A, page 11

1. 15

2. 10

3. 15

4. 8

5. 25

6. 64

7. 63

8. 49

9. 3

10. 4

11. 729

12. 200

13. 6

14. 1000

15. 11

16. 100

17. 20

18. 64

19. 5

20. 216

21. $x + 5$

22. $5x$

23. n^2

24. t^3

25. $y - 4$

26. $\frac{x}{6}$

27. $n - m$

28. st

29. c^2

30. z^3

31. $x + y$

32. xy

33. $y - x$

34. $\frac{x}{y}$

35. (25)(45)

36. $\sqrt{15 + y}$

37. $\sqrt[3]{7 - p}$

38. $m\sqrt{t}$

39. $t^3 \sqrt[3]{s}$

40. $\frac{x}{y^2}$

41. {1, 2, 3, 4, 5, 6, 7, 8, 9}, finite

42. {3, 6, 9, 12, 15, . . .}, infinite

43. {1, 4, 9, 16, 25, . . .}, infinite

44. {5, 10, 15, 20, 25, . . .}, infinite

45. {1, 2, 3, 4, 6, 12}, finite

46. {1, 8, 27, 64, 125, . . .}, infinite

47. {1, 2, 3, 4, 5, 6, 7}, finite

48. {9, 10, 11, 12, 13, . . .}, infinite

49. {2, 3, 4, 5}, finite

50. {2, 4, 6, 10}, finite

Exercises 1.3 A, page 19

1. {1, 2, 3, 4, 5, 6, 7}

2. {1, 2, 3, 5, 6, 10, 15, 30}

3. {30, 60, 90, 120, 150, . . .}

4. {16, 17, 18, 19, 20, . . .}

5. {1, 4, 9, 16, 25, . . .}

6. {1, 8, 27, 64, 125, . . .}

7. {12, 20, 30}

8. {2, 3, 5}

9. {31, 37, 41, 43, 47}

10. {31, 33, 35, 37, 39}

11. {18, 20, 22, 24}

12. {20, 21, 22, 24, 25, 26}

13. {1, 2, 4, 7, 14, 28}

14. {13, 26, 39, 52, 65, . . .}

15. {1, 2, 3, 6, 9, 18}

16. {25, 50, 75, 100, 125, . . .}

17. a. {1, 5, 12, 15, 20, 25} b. {5, 20}

18. a. {1, 2, 3, 4, 5, . . .} = N b. ∅

19. a. {2, 3, 4, 5, 6, 7, 8, 9} b. ∅

20. a. {31, 33, 35, 36, 37, 39} b. {33, 39}

21. a. {1, 2, 3, 5, 6, 7, 10, 11, 13, 15, 17, 19, 23, 29, 30} b. {2, 3, 5}

22. { }, {2}, {3}, {5}, {2, 3}, {2, 5}, {3, 5}, {2, 3, 5}

23. { }, {1}, {3}, {5}, {7}, {1, 3}, {1, 5}, {1, 7}, {3, 5}, {3, 7}, {5, 7}, {1, 3, 5}, {1, 3, 7}, {1, 5, 7}, {3, 5, 7}, {1, 3, 5, 7}

24. True

25. False

26. False

27. True

28. False

29. False, because 1 is not prime or composite

30. True

31. True

32. True

33. True

Exercises 1.4 A, page 26

1. 24

2. 24

3. 8

4. 2

5. 60

6. 60

7. 30

8. 26

9. 12

10. 3

11. 2

12. 8

13. 64

14. 512

15. 144

16. 6

17. 50

18. 100

19. 27

20. 49

21. 2

22. 8

23. 10

24. 2560

25. 11

26. 3

27. 7

28. 13

29. 6

30. 6

31. 2

32. 0

33. 4

34. 12

35. 73

36. 6

37. 16

38. 720

39. 2

40. 6

41. $x + 2y$

42. $2(x + y)$

43. $2xy$

44. $5x^2$

45. $y - (x + 3)$

46. $\frac{x}{x + 3}$

47. $(y + 2)^2 - 7$

48. $\frac{1}{2} \cdot 3(a + b)$, or $\frac{3(a + b)}{2}$

49. $6[9 - (x + 2)]$

50. $6 + 3[(x + 9) - 2]$

Exercises 1.5 A, page 32

1. 4576

2. 48

3. 12

4. 2

5. 180

6. 75

7. 45
8. 7
9. 120
10. 98
11. 64
12. 64
13. 34
14. 6
15. 25
16. 5
17. 196
18. 108
19. 49
20. 150
21. 14
22. 169
23. 60
24. 405
25. 38
26. 128
27. 9
28. 10,100
29. 8000
30. 9
31. $A = \frac{1}{2}bh$
32. $A = \frac{1}{2}h(a + b)$
33. $V = s^3$
34. $V = \frac{1}{3}\pi hr^2$
35. $E = \frac{I}{r^2}$
36. $C = \frac{5}{9}(F - 32)$
37. $T = 2\pi\sqrt{\frac{L}{g}}$
38. $R = \frac{ST}{S + T}$
39. $R = \frac{C - S}{n}$
40. $C = \frac{100W}{L}$
41. 180
42. 63
43. 1728
44. $733\frac{1}{3}$
45. 10
46. 100
47. 11
48. 12
49. 200
50. 75

Exercises 1.6 A, page 44

1. Positive
2. Signed
3. Signed
4. Signed
5. Signed
6. Signed
7. Positive
8. Positive
9. Signed
10. Signed
11. 80 degrees west longitude
12. $50 loss
13. 70 feet below sea level
14. A loss of 5 yards
15. 15 mph north wind
16. 2800 mph downward
17. 6 steps downward
18. 8 hours earlier
19. 35 pounds underweight
20. decrease of 20 cubic centimeters
21. $+3$
22. -2

23. -4 **24.** -1

25. $+3$ **26.** $+4$

27. -2 **28.** -5

29. $+2$ **30.** -1

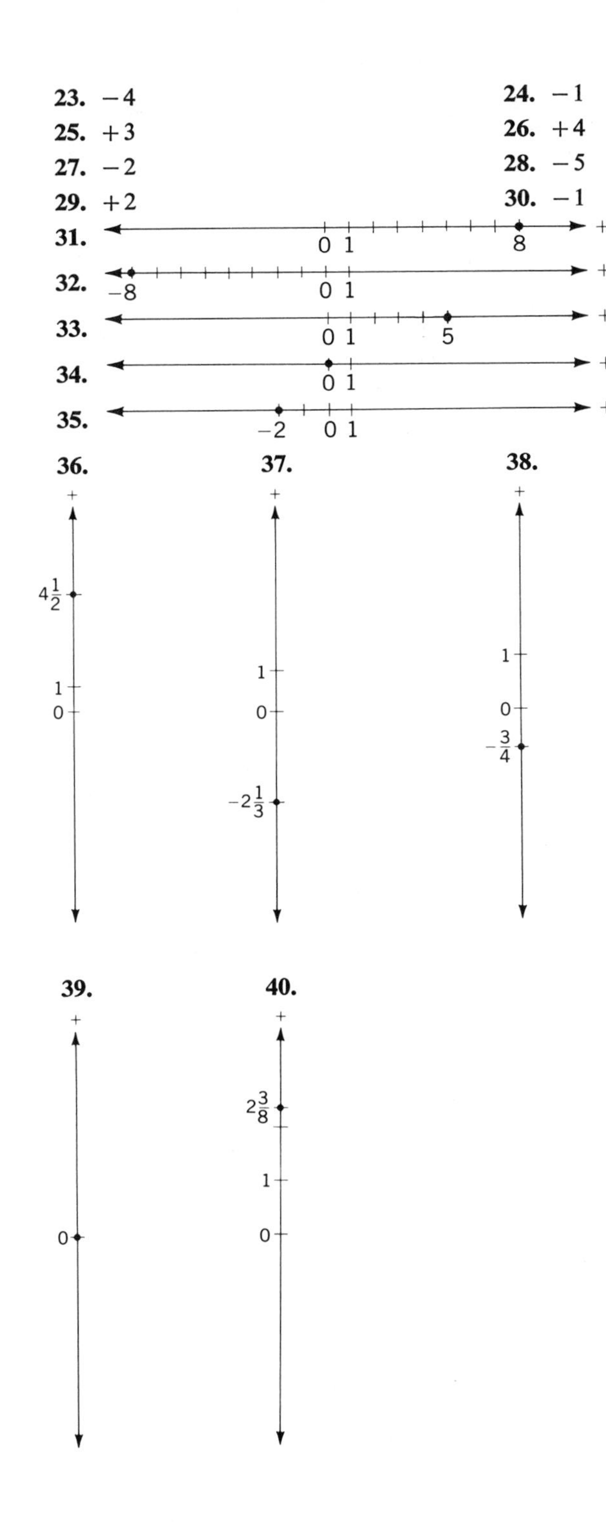

41. 5
42. 4
43. 0
44. 1
45. $1\frac{3}{4}$
46. $7\frac{1}{2}$
47. 12
48. 9
49. 11
50. 15
51. 24
52. 7
53. 7
54. 16
55. 0
56. 4
57. 0
58. -6
59. 40
60. -5
61. $\{-15, -1, 0, 1, 14\}$
62. $\left\{-15, -4\frac{1}{2}, -1, -\frac{2}{3}, 0, \frac{3}{4}, 1, 5\frac{7}{8}, 14\right\}$
63. $\left\{-4\frac{1}{2}, -\frac{2}{3}, \frac{3}{4}, 5\frac{7}{8}\right\}$
64. $\left\{-15, -4\frac{1}{2}, -\sqrt{2}, -1, -\frac{2}{3}, 0, \frac{3}{4}, 1, \sqrt{2}, 5\frac{7}{8}, 14\right\} = U$
65. $\{-\sqrt{2}, \sqrt{2}\}$

Exercises 1.7 A, page 53

1. 3 is less than 7; 3 is to the left of 7.
2. 8 is greater than 5; 8 is to the right of 5.
3. -7 is less than 0; -7 is to the left of 0.
4. 4 is greater than -7; 4 is to the right of -7.
5. 7 is greater than 0; 7 is to the right of 0.
6. -5 is less than 1; -5 is to the left of 1.
7. -6 is less than -2; -6 is to the left of -2.
8. -9 is greater than -15; -9 is to the right of -15.
9. -1 is greater than -4; -1 is to the right of -4.
10. -8 is less than -6; -8 is to the left of -6.
11. The absolute value of -8 is greater than the absolute value of -6.
12. The absolute value of -5 is greater than zero.
13. The absolute value of -4 is less than the absolute value of -7.
14. 0 is less than the absolute value of -9.
15. 7 is greater than 4 and less than 10; or 7 is between 4 and 10.
16. $12 > 5$
17. $4 < 9$
18. $-4 < 2$
19. $5 > -6$
20. $0 > -3$
21. $0 < 8$
22. $-5 > -7$
23. $-9 < -7$
24. $-1 > -2$
25. $1 < 2$
26. $|-3| < |-6|$

27. $|-2| = |+2|$
28. $|-2| > |+1|$
29. $|-7| > |-4|$
30. $|-5| > 0$
31. $x + 8 < 20$
32. $5y > 6$
33. $\frac{x}{2} \leq 7$
34. $x - 4 \geq 12$
35. $3x > 0$
36. $x + y < 0$
37. $1 < x + 2 < 8$
38. $2(n + 3) \geq 25$
39. $x^2 - 9 > 0$
40. $|x| \geq 0$
41. x is less than 7.
42. x is greater than or equal to 4.
43. x is less than or equal to -3.
44. x is greater than -5.
45. The sum of x and y is greater than 10.
46. The sum of x and y is less than or equal to 15.
47. x is between -3 and 4.
48. x is between 1 and 6 or x equals 1.
49. The product of 2 and x is between 0 and 5 or equals 5.
50. The absolute value of the sum of x and 2 is less than or equal to 7.

51. a. 0 4 b. $\{\ldots, -3, -2, -1, 0, 1, 2, 3\}$

52. a. 0 4 b. $\{4, 5, 6, 7, 8, \ldots\}$

53. a. -5 0 b. $\{-5, -6, -7, -8, \ldots\}$

54. a. -5 0 b. $\{-4, -3, -2, -1, 0, 1, 2, \ldots\}$

55. a. -2 0 b. $\{-1, 0, 1, 2, \ldots\}$

56. a. 0 6 b. $\{\ldots, -2, -1, 0, 1, 2, 3, 4, 5, 6\}$

57. a. -3 0 b. $\{\ldots, -6, -5, -4\}$

58. a. -1 0 b. $\{-1, 0, 1, 2, \ldots\}$

59. a. 0 b. $\{\ldots, -3, -2, -1, 0\}$

60. a. 0 b. $\{0, 1, 2, 3, \ldots\}$

61. a. 0 3 5 b. $\{4\}$

62. a. -1 0 3 b. $\{0, 1, 2, 3,\}$

63. a. -4 -1 0 b. $\{-4, -3, -2\}$

64. a. 0 2 b. $\{0, 1, 2\}$

65. a. -1 0 1 b. $\{0, 1\}$

Exercises 1.8 A, page 59

1. $+\$600$
2. $+\$350$
3. $-\$175$
4. $-\$125$
5. $-\$55$
6. 217 pounds
7. 180 pounds
8. 227 pounds
9. 179 pounds
10. 208 pounds
11. $+10$
12. -10
13. $+4$
14. -4
15. -5
16. -8
17. -30
18. $+6$
19. $+5$
20. -10
21. -10
22. -15
23. -40
24. -50
25. -7
26. 0
27. -12
28. $+14$
29. -15
30. $+12$
31. -20
32. -8
33. -4
34. -40
35. -27
36. $+20$
37. -1
38. 0
39. 0
40. 10
41. \$404
42. -9
43. 149 pounds
44. \$55.25
45. $+10$ pound-feet, counterclockwise

Exercises 1.9 A, page 65

1. $+4$
2. -4
3. $+10$
4. -10
5. -10
6. -2
7. 0
8. 0
9. -14
10. $+16$
11. -28
12. -5
13. $+10$
14. -2
15. -2
16. $+5$
17. $+7$
18. -7
19. -23
20. -19
21. -9
22. -2
23. -10
24. 6
25. 59
26. -26

27. 46
28. -9
29. 50
30. 3
31. 11 degrees
32. 25 degrees
33. 16 degrees
34. 25 degrees
35. 8 degrees
36. $22.625
37. $30.75
38. $43.00
39. $18.75
40. $41.75
41. 51 degrees
42. 13 degrees
43. 6 degrees
44. 8 degrees
45. 66 degrees
46. False
47. True
48. True
49. False
50. True
51. True
52. True
53. False
54. True
55. True
56. n
57. n
58. $-n$
59. 0
60. 0
61. n
62. n
63. m
64. 0
65. $n - m$
66. Yes

Exercises 1.10 A, page 72

1. 35
2. -35
3. 35
4. -16
5. 16
6. -100
7. -1000
8. 100
9. -1000
10. 105
11. -48
12. 460
13. -54
14. -64
15. -125
16. 36
17. -32
18. 36
19. -686
20. 6
21. -15
22. 4
23. 0
24. 10
25. -36
26. 0
27. 24
28. 24
29. 32
30. -24
31. 1100 miles east of airport
32. 1100 miles west of airport
33. 1860 miles east of airport
34. 1860 miles west of airport

35. 750 miles east of airport

36. $\dfrac{-88}{10{,}000{,}000} = \dfrac{-11}{1{,}250{,}000}$

37. $\dfrac{3}{100{,}000}$

38. $\dfrac{-12}{1{,}000{,}000{,}000} = \dfrac{-3}{250{,}000{,}000}$

39. $\dfrac{564}{1{,}000{,}000{,}000} = \dfrac{141}{250{,}000{,}000}$

40. $\dfrac{-99}{100{,}000{,}000}$

Exercises 1.11 A, page 76

1. -3
2. -3
3. 3
4. 3
5. 0
6. Undefined
7. -7
8. -9
9. 5
10. -6
11. -7
12. 9
13. -11
14. -13
15. 25
16. -125
17. 5
18. Undefined
19. 0
20. Undefined
21. -1
22. 2
23. -60
24. 0
25. -10
26. 4
27. -1
28. 1
29. -4
30. -1
31. -12
32. $-\dfrac{4}{3}$
33. -20
34. -1
35. -1
36. -8
37. -2
38. $-\dfrac{1}{2}$
39. -2
40. 19

Exercises 1.12 A, page 80

1. -6
2. -2
3. 30
4. -35
5. 0
6. -6
7. -4
8. 0
9. -40
10. 0
11. -7
12. 17
13. 9
14. -17
15. 7
16. 127

17. 35
18. 3
19. 28
20. -23
21. -20
22. 5
23. 25
24. $-4\frac{1}{2}$
25. 60
26. -6
27. -5
28. 500
29. -2.5
30. $3\frac{5}{7}$

Chapter 1 Review Exercises, page 88

1. $N + 5$
2. $5N$
3. $N - 5$
4. N^2
5. $\frac{5}{N}$
6. N^3
7. $\sqrt{5}$
8. $\sqrt[3]{N}$
9. $5(x + 8)$
10. $6x - 4$
11. $\left(\frac{y}{5}\right)^2$
12. $\frac{1}{2}(x^2 - \sqrt{y})$
13. $n^3 + (n - 1)^2$
14. $\frac{2xy}{x + y}$
15. $10 - 3x = 7$
16. $(x + 4)^2 < \frac{x}{4}$
17. $9 + 6x \geq 8$
18. $4(3x + 6) - 12x = 24$
19. $\frac{4x}{5} - 5 = 3(x + 2)$
20. a. 4 b. -6 c. -10
21. a. 14 b. -6 c. 24
22. a. -16 b. 0 c. -1
23. a. 6 b. 4 c. -1
24. a. 9 b. -7 c. -15
25. a. 5 b. -2 c. -3
26. a. 12 b. -4 c. 36
27. a. -56 b. 0 c. 20
28. a. 13 b. -29
29. a. 9 b. 1
30. a. -25 b. -10
31. a. 700 b. -24
32. a. -4 b. $\frac{-4}{3}$
33. True
34. False
35. True
36. True
37. True
38. False
39. False
40. True
41. True
42. True
43.
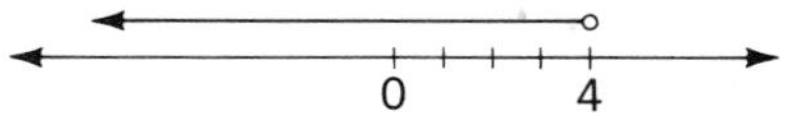

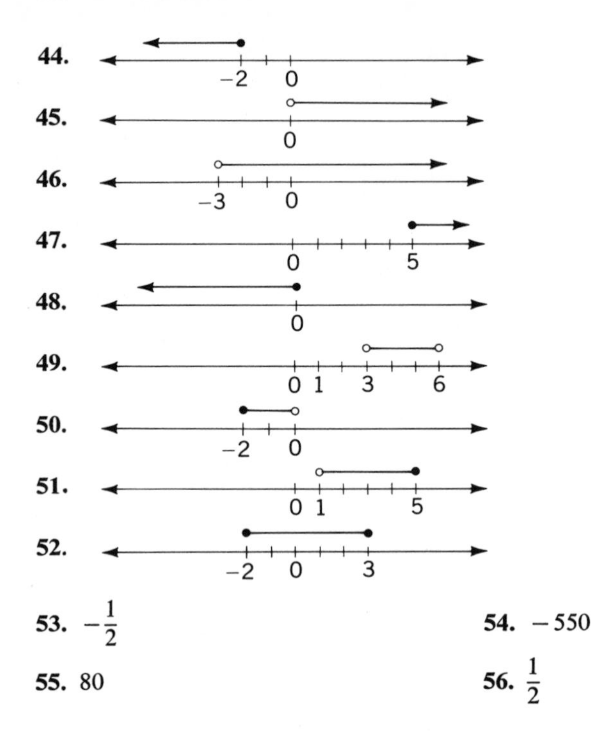

53. $-\frac{1}{2}$

54. -550

55. 80

56. $\frac{1}{2}$

Exercises 2.1 A, page 97

1. Reflexive

2. Symmetric

3. Symmetric

4. Transitive

5. False

6. False

7. Reflexive

8. Transitive

9. Transitive

10. False

11. Commutative, addition

12. Associative, multiplication

13. Commutative, multiplication

14. Associative, addition

15. Commutative, addition

16. Commutative and associative, addition

17. Commutative, addition and multiplication

18. Commutative, addition

19. Commutative and associative, multiplication

20. Commutative and associative, multiplication

21. $x + y + 4$

22. $-15x^2$

23. $-x$

24. $x + y - 9$

25. $36x^2y^2$

26. 3

27. $2ab - 4$

28. $24abc$

29. $m + n - 7$

30. $-abcn^3$

31. $(35 + 65) + 87 = 187$

32. $69\,(4 \cdot 25) = 6900$

33. $\left(7\frac{3}{8} + 2\frac{5}{8}\right) + 9 = 19$

34. $(0.5 \cdot 20)45 = 450$

35. $(5 + 195) + (6 + 194) = 400$

Exercises 2.2 A, page 105

1. Closed

2. $\frac{1}{2}$ is not a natural number, for example.

3. Closed, excluding division by 0

4. Closed

5. Closed

6. $1 + 1 = 2$ and $2 \notin \{0, 1\}$

7. $0 - 1 = -1$ and $-1 \notin \{0, 1\}$

8. Closed

9. $\frac{1}{0}$ is undefined.

10. a. Closed b. $2 - 4 = -2$ and $-2 \notin N$

c. Closed d. $\frac{2}{4} = \frac{1}{2}$ and $\frac{1}{2} \notin N$

11. -4

12. 3

13. $-\frac{1}{5}$

14. $\frac{2}{7}$

15. -1256

16. $\frac{1}{12}$

17. $-\frac{1}{5}$

18. 5

19. $-\frac{7}{2}$

20. $\frac{10}{3}$

21. 8

22. $-\frac{2}{5}$ or -0.4

23. $\frac{3}{13}$

24. 10,000

25. $-\frac{250}{3}$

26. x

27. y

28. 25

29. x

30. b

31. x

32. 0

33. -1

34. x

35. x

36. x

37. 0

38. x

39. $-\frac{1}{9}$

40. $8y$

41. 0

42. 0

43. x

44. x

45. -1

46. 17

47. $\frac{3}{2}$

48. -3

49. -4

50. 1

Exercises 2.3 A, page 112

1. $2x + 10$
2. $-3y - 6$
3. $5n - 35$
4. $-12x + 18$
5. $-x + 5$
6. $-3x - 4$
7. $-6 + y = y - 6$
8. $3x + 3y - 6$
9. $-x + 3y$
10. $ax + bx$
11. $6ay - 3by$
12. $a - b + c$
13. $x - y - 5$
14. $8x - 12y + 16$
15. $a - 2b + 2c - 6$
16. $2x + y - 4z + 10$
17. $3a - 6b + 4x + 8y$
18. $-c + d$
19. $4x^3 - 4x^2 - x + 1$
20. $ax + 5a - 2x - 10$
21. $5x$
22. $10x$
23. x
24. 0
25. $-2xy$
26. z
27. $2x + 5$
28. $8k$
29. $-7t + 6$
30. $5x^2 - 3x - 2$
31. $3z^2 + 5z + 4$
32. $p - q$
33. $6x^2 + 7xy - 10y^2$
34. $x^2 - y^2$
35. $-3ab + ac$
36. $3x + 1$
37. $y + 3$
38. $3x^2 - 3x + 3$
39. $3y^2 - 4y + 2$
40. $8x + 8y$
41. $-8c - 8d$
42. $a - b$
43. $-7x$
44. $2xy - 12x + 15y$
45. $-x + 5$
46. $k - 4$
47. $ab - 16a + 6b$
48. $4n + 21$
49. $5x - 8$
50. $-3c - 12d$
51. $45(37 + 63) = 4500$
52. $24\left(\frac{1}{2}\right) + 24\left(\frac{1}{3}\right) = 12 + 8 = 20$
53. $\left(\frac{1}{7} + \frac{6}{7}\right)65 = 65$
54. $58(99 + 1) = 5800$
55. $\frac{3}{2}(12) + \frac{2}{3}(12) = 18 + 8 = 26$
56. $(87 - 86)450 = 450$
57. $246\left(\frac{7}{8} - \frac{3}{8}\right) = 123$

58. $60\left(\frac{5}{12} + \frac{1}{12}\right) - 60\left(\frac{1}{15}\right) = 30 - 4 = 26$

59. $18\left(\frac{9}{7} - \frac{2}{7}\right) - 18\left(\frac{5}{6}\right) = 18 - 15 = 3$

60. $497(8 + 7 - 5) = 4970$

Exercises 2.4 A, page 124

1. Yes
2. Yes
3. No
4. Yes
5. Yes
6. No
7. No
8. Yes
9. Yes
10. No
11. 1
12. 7
13. -7
14. -6
15. -1
16. 5
17. 3
18. $\frac{12}{5}$
19. $\frac{3}{2}$
20. 4
21. $-\frac{2}{3}$
22. -8
23. $-\frac{1}{2}$
24. 7
25. -2
26. 2
27. $\varnothing$
28. 1
29. 5
30. -8
31. 2
32. 14
33. 20
34. 0
35. -8
36. 13
37. R (Identity)
38. 6
39. 1
40. 1
41. -3
42. -6
43. $\varnothing$
44. R (Identity)
45. 0

Exercises 2.5 A, page 129

1. -4
2. -1
3. -3
4. 0
5. 10
6. -20
7. $\varnothing$
8. -1

9. -30

10. $\frac{1}{3}$

11. R

12. 0

13. -6

14. 11

15. R

16. -5

17. 4

18. -5

19. 4

20. -4

21. -2

22. -8

23. 2

24. $\varnothing$

25. 7

26. 3

27. 1

28. 4

29. 24

30. 3

Exercises 2.6 A, page 133

1. $L = \frac{A}{W}$

2. $W = \frac{V}{LH}$

3. $h = \frac{V}{\pi r^2}$

4. $L = \frac{P - 2W}{2}$

5. $t = \frac{D}{r}$

6. $x = \frac{y - b}{m}$

7. $h = \frac{2A}{b}$

8. $a = \frac{2A}{h} - b$

9. $r = \frac{A - P}{Pt}$

10. $b = \frac{ad}{c}$

11. $B = 180 - A - C$

12. $a = \frac{2S}{n} - l$

13. $s = \frac{x - \bar{x}}{y}$

14. $b = P - a - c$

15. $n = \frac{D}{A} + 1$

16. $H = \frac{f + Sh}{S}$

17. $a = \frac{bf}{b - f}$

18. $C = \frac{100M}{Q}$

19. $0 = I - EI$

20. $n = \frac{C - S}{R}$

21. $a^2 = c^2 - b^2$

22. $C = \frac{2D}{dAV^2}$

23. $W = \frac{150C}{A}$

24. $A = 2(17 - H)$

25. $G = \frac{E(R - r)}{R}$

26. $y = 8 - x$

27. $y = 5x + 10$

28. $y = 3x - 6$

29. $x = 12 - 4y$

30. $x = 2y + 9$

31. $x = -y - 1$

32. $x = 2y + 4$

33. $y = \dfrac{6 - 3x}{2}$

34. $x = \dfrac{7y + 8}{2}$

35. $y = \dfrac{-ax - c}{b}$

Exercises 2.7 A, page 140

1. $3 + 2x = 3x - 7$
$x = 10$

2. $\dfrac{x + 6}{2} = 10 - (x - 5)$
$x = 8$

3. $(x + 6) + (3 + 2x) + \dfrac{x}{2} = 100$
$x = 26$
$x + 6 = 32$ singles
$3 + 2x = 55$ doubles
$\dfrac{x}{2} = 13$ twins

4. $x + 25x + 43x + 18x + 36x = 984$
$123x = 984$
$x = 8$

5. $x + (3x - 4) + \dfrac{x}{3} = 165$
$x = 39$ bicycles
$3x - 4 = 113$ scooters
$\dfrac{x}{3} = 13$ buggies

6. $x + (2x - 6) + (x + 9) + (x - 3) = 460$
$x = 92$ French
$2x - 6 = 178$ Spanish
$x + 9 = 101$ Swahili
$x - 3 = 89$ Japanese

7. $x + (x + 1) + (x + 2) = 42$
13, 14, 15

8. $4(x + 1) - (x + 3) = x + 53$
26, 27, 28, 29

9. $2(x + 4) + 3x = 73$
13, 15, 17

10. $2(x + 4) + 3x = 8$
0, 2, 4

11. $\dfrac{x + (x + 2) + (x + 4) + (x + 6)}{4} = 34$
31, 33, 35, 37

12. $\dfrac{x + (x + 2) + (x + 4) + (x + 6) + (x + 8) + (x + 10)}{6} = -9$
$-14, -12, -10, -8, -6, -4$

13. $x - 14 = \dfrac{x + 7}{4}$
21

14. $x + 12 = 2x \qquad y + 12 = 3(12)$
$x = 12$, age now $\quad y = 24$

15. $(x + 25) + (x + 3 + 25) = 97$
22, 25

16. $2x - 7 = 3(x - 7)$
14, 28

17. $44 - x - 4 = 5(x - 4)$
10, 34

Exercises 2.8 A, page 143

1. $2x + 2\left(\dfrac{x}{3} + 2\right) = 52$
18 feet by 8 feet

2. $10x + 10(x + 1) + 10(x + 2) = 180$
50 degrees, 60 degrees, 70 degrees

3. $2w + 2(2w - 6) = 78$
$0.80(24)(15) = \$288$

4. $60(70) + 60x + 70x = 5240$
$x = 8$, $x^2 = 64$ (Answer)

5. $4s = 40$
$s = 10$
$s^2 = 100$

6. $8(x + 3) + 10x = 234$
13 feet, 16 feet

7. $2(12 + 2x) + 2(18 + 2x) = 84$
3 inches

8. $x + (3 + 2x) + 1.5x = 48$
10 inches, 23 inches, 15 inches

9. a. $(8)(10)x = 112(110)$
$x = 154$
b. $(11)(14)x = 112(110)$
$x = 80$

10. $x(x + 2) - x(x - 2) = 12$
Original, 3 feet by 1 foot
New, 3 feet by 5 feet

Exercises 2.9 A, page 148

1. $40x + 35(300) = 37(x + 300)$
200 gallons

2. $12x + 20(50 - x) = 15(50)$
$31\frac{1}{4}$ pounds at 12 cents and $18\frac{3}{4}$ pounds at 20 cents

3. $4a + 3(2a) + 2(a + 3) = 54$
4 units of A, 8 units of B, 7 units of C
$\text{GPA} = \dfrac{54}{4 + 8 + 7} = \dfrac{54}{19} \approx 2.84$

4. $4a + 3(a + 4) + 2(a + 3) = 45$
3 units of A, 7 units of B, 6 units of C
$\text{GPA} = \dfrac{45}{16} \approx 2.81$

5. $100d + 10(50d) + 25(10d + 60) = 10{,}000$
10 dollars, 500 dimes, 160 quarters

6. $0.06x + 0.0525(2x) = 825$
   \$5000 at 6 percent, \$10,000 at $5\frac{1}{4}$ percent
7. $0.05x + 0.035(7500 - x) = 0.04(7500)$
   \$2500 in stocks
8. $250x + 175(420 - x) = 81{,}750$
   110 loges, 310 general
9. $0.75(100 - x) = 0.30(100)$
   60 liters
10. $0.88x = 40\left(5\frac{1}{2}\right)$
    250 grams

Exercises 2.10 A, page 152

1. $60x + 45x = 210$
   2 hours
2. $60\left(\frac{40}{60}\right) + 45\left(\frac{40}{60}\right) = 70$ miles
3. $3x + 3(x + 15) = 465$
   70 mph
4. $45x = 60(x - 2)$
   4 p.m.
5. $3x + 3(x + 80) = 2190$
   325 mph, 405 mph
6. $6x + 6(x - 12) = 648$
   60 mph, 48 mph
7. $60x = 0.5 + 40x$
   $1\frac{1}{2}$ minutes
8. $50x + 60\left(2\frac{5}{6} - x\right) = 150$
   100 miles
9. $95x = 35(x + 6)$
   332.5 miles
10. $4(x + 10) + x = 340$
    70 mph

Chapter 2 Review Exercises, page 157

1. Commutative, addition
2. Inverse, addition
3. Inverse, multiplication
4. Distributive
5. Associative, addition
6. Commutative, multiplication
7. Identity, multiplication
8. Associative, multiplication
9. Closure, addition and multiplication
10. Identity, addition
11. Addition, division
12. Multiplication, subtraction
13. Subtraction, multiplication
14. Subtraction, division
15. Subtraction, division
16. Addition, subtraction, division
17. Addition (twice), division
18. Subtraction (twice)
19. Multiplication, subtraction, division
20. Addition, division
21. -12
22. 1

23. -5

24. $\frac{3}{2}$

25. 8

26. 6

27. -7

28. 4

29. 1

30. $\varnothing$

31. R

32. 4

33. $\frac{1}{2}$

34. -2

35. -5

36. 5

37. -3

38. 7

39. $\varnothing$

40. 3

41. 0

42. -9

43. 2

44. -3

45. R

46. -5

47. 3

48. -5

49. 2

50. 4

51. $6x - 4$

52. $\frac{x}{2} + 9$

53. $10x + 5(2x) = 20x$

54. $5000 - x$

55. $5x + 5(3x) = 20x$

56. $24 - x$

57. $x + 40$

58. $x + \frac{1}{2}$

59. $0.05x + 0.065(2000)$

60. $x + (x + 1) + (x + 2) = 3x + 3$

61. $6 + 2x = 5(x - 3)$
7

62. $3(2x - 15) = 90 + 3x$
Bus, 45 mph; train, 75 mph

63. $x + 0.4(18) = 0.5(x + 18)$
3.6 ounces

64. $2x + 2(2x - 2) = 50$
$x = 9$
$\frac{3(9)(16)}{400} = 1.08$ gallons

65. $29x + 25(2x) + 10(2x - 1) = 221$
$2\frac{1}{3}$ pounds peaches, $4\frac{2}{3}$ pounds plums, $3\frac{2}{3}$ pounds bananas

66. $\frac{1}{2}x(x + 11) - \frac{1}{2}x(x - 10) = 2310$
$x + 11 = 231$ feet

67. $x + (x + 3) + \frac{x}{2} + 17 = 70$
20 years

68. $3.2(2x + 5 - 88) = 67.2$
52 pounds, 57 pounds

69. $65x + 50(x + 1) = 280$
$x = 2$ hours; 2 hours after 11 a.m. = 1 p.m.

70. $2(x - 9) = x + 8$
26

Exercises 3.1 A, page 163

1. x^5

2. $3x^5$

3. $6y^6$

4. $35z^{11}$

5. $-6x^6$

6. $5x^{13}$

7. $2y^8$

8. $-3x^6$

9. $-z^6$

10. $3x^2y^3$

11. $15x^3yz$

12. $a^3b^3c^3$

13. x^8

14. x^8

15. y^{12}

16. x^8

17. xy^{11}

18. x^2

19. $-x^6$

20. $-2x^2y^2z^3$

21. x^{m+2}

22. $2x^{p+q}$

23. x^{3n}

24. x^{3n}

25. $x^{n+1}y^{m+2n}$

26. y^{2n}

27. $r^{k+1}t^{2k}$

28. $r^n t^{2n-2}$

29. $x^{2n}y^{2n}$

30. t^{abc}

31. $2^5 = 32$

32. $3^6 = 729$

33. $2^6 = 64$

34. $3^8 = 6561$

35. $2^7 3^6 = (128)(729) = 93{,}312$

36. $(-2)^5 = -32$

37. $(-10)^6 = 1{,}000{,}000$

38. $(-2^7)(5^6) = -(2\cdot 5)^6 2 = -2(10)^6 = -2{,}000{,}000$

39. $2^8 5^8 = 10^8 = 100{,}000{,}000$

40. $(-2)^9(-5)^8 = -200{,}000{,}000$

Exercises 3.2 A, page 168

1. $7x^2 + 5x + 1$

2. $-y^3 + y^2 + 4$

3. $x^3 + x^2 + 5x + 1$

4. $7z^2 + 7z + 15$

5. $2x^2 + 3x + z$

6. $8x^2 - 4x - 4$

7. $5a$

8. $-6y - 8$

9. $-6a + 13$

10. $y^2 - y + 1$

11. $a - 2b - 2c$

12. $-2r - 2s + 2t$

13. -1

14. $-x^4 + x - 1$

15. $14a - 7b$

16. $10y + 5$

17. $25x^4 - 19x^2 - 6x$

18. $x^3 + 1$

19. $-3x^2 + 3y^2$

20. $a - 3b + c$

21. $4x^2 + 1;\ 145 = 145$

22. $-a^2 + 4b^2 + 5;\ 6 - (-6) = -9 + 21 = 12$

23. $2t - 9;\ (-45) - (-46) = 10 - 9 = 1$

24. $-5c^3 - 6c^2 + 3c - 1;\ (-10) + (1) = -9$

25. $x^3 - 2x^2 - 2x - 1;\ (-7) + 2 = -5$

26. $2x^2 + x + 18$
27. $-a^2 - 3ab + 4b^2$
28. $-x^4 - 4x^2 - 2x + 1$
29. $3x^3 - 10x^2y - 2xy^2 + y^3 + 4$
30. $3a^3 + 8a^2b + 18ab^2 - 10b^3$

Exercises 3.3 A, page 172

1. $x^2 + 3x$
2. $2x^3 + x^2$
3. $3y^4 + y^2 + 2y$
4. $2x^7 + 3x^4 + x^3$
5. $3x^3y^3 + 3x^2y^3 + 3xy^4$
6. $-2x^5 + 3x^4 + 14x^3$
7. $3x^4 + 2x^2$
8. $2a^3b + 6a^2b^2 + 2ab^3$
9. $-6z^5 + 12z^4 + 3z^3$
10. $5x^3z + 10x^2z^2 - 20xz^3$
11. $x^2 + 5x + 6$
12. $x^2 + 2x - 8$
13. $x^2 - 5x + 4$
14. $x^2 + 2x - 15$
15. $6x^2 + 13x + 5$
16. $15x^2 - 4x - 4$
17. $20x^2 - 17x + 3$
18. $2x^2 - x - 15$
19. $2x^2 + xy - y^2$
20. $6x^2 - xy - 2y^2$
21. $x^3 + 5x^2 + 6x$
22. $x^3 + 5x^2 + 7x + 2$
23. $6x^3 - 5x^2 + 3x - 1$
24. $x^4 - x^3 - 2x^2$
25. $-x^4 + x^3 + 3x^2 - 2x$
26. $x^2 + 4x + 4$
27. $x^2 - 6x + 9$
28. $x^3 + 2x^2 + x$
29. $x^3 - 4x^2 + 4x$
30. $x^4 - x^3 - 3x^2 + x + 2$
31. $3x^4 + 2x^3 - 10x^2 - 11x - 2$
32. $x^2 - 9$
33. $x^2 - 4$
34. $4x^2 - 1$
35. $x^4 - 9$
36. $2x^5 - 98x$
37. $-5y^3 + 405y$
38. $-100x^3y + 9xy^3$
39. $x^3 + 6x^2 + 12x + 8$
40. $y^3 - 15y^2 + 75y - 125$
41. $64x^3 + 144x^2y + 108xy^2 + 27y^3$
42. $x^3 - 216$
43. $8x^3 + 125y^3$
44. $x^4 - y^4$
45. $-x^8 + 16 = 16 - x^8$
46. $a^3b^3 + 3a^2b^2 + 3ab + 1$
47. $x^2 + 2x + 1 - y^2$
48. $a^2 - 2ab + b^2 - 16$
49. $x^2 - y^2 + 10y - 25$
50. $x^2 + 2xy + y^2 - z^2 - 6z - 9$
51. $x^2 + y^2 + z^2 + 2xy + 2xz + 2yz$
52. $a^2 + b^2 + c^2 - 2ab + 2ac - 2bc$
53. $x^{2n} - 1$
54. $y^{3n} + y^{2n} - y^n - 1$
55. $a^{n+2} - 2a^{n+1}b - a^nb^2$
56. $x^{2a+1} + x^{a+1} + x$
57. $x^{3n+1} - x^{2n}$
58. $x^{3n} - 4x^n - 3$
59. $x^{2n} - 4$
60. $x^{2a} + 6x^a + 9$

Exercises 3.4 A, page 177

1. $x^2 + 2x + 1$
2. $x^2 - 2x + 1$
3. $y^2 + 4y + 4$
4. $y^2 - 4y + 4$

5. $4y^2 + 4y + 1$
6. $4y^2 - 4y + 1$
7. $9x^2 + 24x + 16$
8. $16x^2 - 24x + 9$
9. $4x^2 + 12x + 9$
10. $x^2 - 8x + 16$
11. $x^3 + 3x^2 + 3x + 1$
12. $x^3 - 3x^2 + 3x - 1$
13. $y^3 + 6y^2 + 12y + 8$
14. $y^3 - 6y^2 + 12y - 8$
15. $125y^3 + 75y^2 + 15y + 1$
16. $8y^3 - 12y^2 + 6y - 1$
17. $27x^3 + 108x^2 + 144x + 64$
18. $64x^3 - 144x^2 + 108x - 27$
19. $8x^3 + 36x^2 + 54x + 27$
20. $-x^3 + 30x^2 - 300x + 1000$
21. $x^2 - 4$
22. $y^2 - 9$
23. $25x^2 - 1$
24. $9x^2 - 4$
25. $x^2 - y^2$
26. $36x^2 - 49y^2$
27. $a^2 - 64b^2$
28. $x^2 + 2xy + y^2$
29. $x^3 + 3x^2y + 3xy^2 + y^3$
30. $x^2 - 2xy + y^2$
31. $x^3 - 3x^2y + 3xy^2 - y^3$
32. $100x^2 + 180xy + 81y^2$
33. $1000x^3 + 2700x^2y + 2430xy^2 + 729y^3$
34. $x^4 - 36$
35. $x^4 - y^4$
36. $49 - x^2$
37. $-9x^2 + 4 = 4 - 9x^2$
38. $x^{2n} + 2x^n + 1$
39. $x^{3n} + 3x^{2n} + 3x^n + 1$
40. $x^{2n} - 1$

Exercises 3.5 A, page 182

1. $4(x + 3)$
2. $3(2x - 1)$
3. $a(y - 1)$
4. $5x(x + 5)$
5. $4y^2(y - 25)$
6. $2x(2x + 1)$
7. $x^4(1 - x)$
8. $x^4(x - 1)$
9. $cn^2(c + n)$
10. $6x^2(4x - 5)$
11. $6(p - 2q + 1)$
12. $9(4x^2 + 4x + y)$
13. $a(x + y - z)$
14. $7(2x^2 + 3xy + 1)$
15. $4an(4an + 1)$
16. $3an(2x + b)$
17. $ax(a - x)$
18. $ax(x^2 + ax - a^2)$
19. $6(4x^2 + 2x + 1)$
20. $a(y^2 + by + b)$
21. $c(c^3 - c^2 + c - 2)$
22. $x(x + y - 1)$
23. $-5x(5x^2 + 3)$
24. $a^2b^3(a + b - 1)$
25. $-xy(y^2 + y + 1)$
26. $-y(xz + x + z)$
27. $rs(r - s - 4)$
28. $6axy(x - 2y + 1)$
29. $9(3a^2b^2 + 2ab - 7)$
30. $-a(b + c)$
31. $-x^2(x^2 - 2x + 6)$
32. $8mn(9mn - 7)$
33. $3p^2q(1 + 3q - 4q^2)$
34. $-5x^2(3x^2 + 2x - 3) = 5x^2(3 - 2x - 3x^2)$

35. $x^{50}(x + 1)$
36. $3x$
37. $x(3y + x - 3) = x(x + 3y - 3)$
38. ad
39. $c(c - x + y)$
40. $-bct$
41. $x^n(x + 1)$
42. $x^n(x^n - 1)$
43. $x^n(x^{2n} + 1)$
44. $y^n(y^{3n} - z)$
45. $x^n(x^2 + x + 1)$
46. $(a + 3)(x + y)$
47. $(x^2 - 5)(x - 4)$
48. $(x^2 - 2y^2)(x + 5)$
49. $(y + 1)(y^2 - 3)$
50. $(5a^2 - 1)(t^2 + 4)$

Exercises 3.6 A, page 187

1. $x + 2$
2. $x - 12$
3. $x - 13$
4. $x + 3$
5. $x + 4$
6. $(x + 6)(x + 5)$
7. $(x - 1)(x + 6)$
8. $(x + 1)(x - 6)$
9. $(x - 1)(x - 6)$
10. $(x + 10)(x + 1)$
11. $(x - 4)(x - 5)$
12. $(x - 3)(x - 7)$
13. $(r - 4)(r + 8)$
14. $(p + 4)(p - 10)$
15. $(x + 2)(x + 3)$
16. $(p - 2)(p - 3)$
17. $(y - 3)(y - 4)$
18. $(x + 1)(x - 3)$
19. Not factorable
20. $(y + 1)(y - 2)$
21. $(x + y)^2$
22. $(x + 3y)(x - 5y)$
23. $(r - s)(r + 2s)$
24. $(x - 7z)(x + 10z)$
25. $(a + 2b)(a + 5b)$
26. Not factorable
27. $(x - 2a)(x + 5a)$
28. $(x - 8)^2$
29. $(x + 2)(x - 5)$
30. $-(y - 2)(y + 12)$
31. $4(x + 3)(x + 5)$
32. $5(x - 1)(x + 7)$
33. $100x(x + 2)(x - 4)$
34. $6y^2(y + 2)(y - 6)$
35. $15(y^2 + 2y + 3)$
36. $-7(a + 3)(a - 4)$
37. $-cd(d - 4)(d + 5)$
38. $30(u^2 + 5u - 4)$
39. $2x(x - 6y)(x + 10y)$
40. $-uv(u^2 + v^2 - 5)$

Exercises 3.7 A, page 192

1. $(4x + 3)(2x - 5)$
2. $(4x - 3)(2x + 5)$
3. $(4x - 3)(2x - 5)$
4. $(4x + 3)(2x + 5)$
5. $(4x - 5)(2x - 3)$
6. $(8x - 1)(x + 15)$
7. $(8x + 3)(x - 5)$
8. $(8x - 5)(x - 3)$
9. $(8x + 5)(x - 3)$
10. $(8x + 3)(x + 5)$
11. $2x + 11$
12. $2x - 5$
13. $3x - 2$
14. $3y - 5$

15. $3y - 5$

16. $4a - 7$

17. $5x - 1$

18. $5x + 1$

19. $5x + 1$

20. $5x - 1$

21. $(x - 1)(2x - 1)$

22. $(2x - 5)^2$

23. $(4x - 1)(3x - 5)$

24. $(4x + 1)(3x - 5)$

25. $(2x - 1)(6x - 5)$

26. $(3a - 7)(a + 1)$

27. $(3a + 7)(a - 1)$

28. Not factorable

29. $(3a + 1)(a - 7)$

30. $(3a - 1)(a + 7)$

31. Not factorable

32. $(3x - 2)(5x + 1)$

33. $(3y + 4)^2$

34. $(5x - 3)^2$

35. $(2x + 7)(5x - 1)$

36. $(5x - 7)(2x + 1)$

37. $(2x + 1)(x + 1)$

38. $(2a - 1)^2$

39. $(2x + 5)(2x + 3)$

40. $-(3a - 1)(3a + 8)$

41. $5, -5, 4, -4$

42. $k = n(18 - n)$ where n is any integer

43. $4, -4, 12, -12, 44, -44$

44. a. $(x - a)(x - b)$ b. $(x + a)(x - b)$ c. $(x - a)(x + b)$ d. $(x + a)(x + b)$

45. a. Factorable b. Not factorable c. Factorable d. Not factorable

Exercises 3.8 A, page 197

1. 9

2. 100

3. 49

4. 400

5. 1

6. $18y$

7. $20y$

8. $4xy$

9. $22uv$

10. $9x^2$

11. $(x + 4)^2$

12. $(y + 1)^2$

13. Not a square

14. Not a square

15. $(x - 7y)^2$

16. $(x - 6)^2$

17. Not a square

18. Not a square

19. Not a square

20. Not a square

21. $(x + 2)^2$

22. $(x - 1)^2$

23. $(x + 9)^2$

24. $(x - 5)^2$

25. $(y + 8)^2$

26. $(3a + 1)^2$

27. $(4p - 1)^2$

28. $(x - 6)^2$

29. $(8y - 3)^2$

30. $(x + 7)^2$

31. $(2x + 7)^2$

32. $(3p - 5)^2$

33. $(a + 6)^2$

34. $(2x + y)^2$

35. Not factorable

36. $(x^2 + 10)^2$

37. $(11u - 1)^2$

38. Not factorable

39. Not factorable

40. $(u - 21)^2$

41. $(x + y + 1)^2$
42. $(8a + 8b - 1)^2$
43. $(10a + b - c)^2$
44. $(x - y - z)^2$
45. $(x + y - 2a - 2b)^2$

Exercises 3.9 A, page 201

1. $(x - 6)(x + 6)$
2. $(2x - 1)(2x + 1)$
3. Not factorable
4. $(4 - 3a)(4 + 3a)$
5. $(3y - 10z)(3y + 10z)$
6. $(n^2 - 2x)(n^2 + 2x)$
7. $(x - y)(x + y)$
8. $(x - y)(x^2 + xy + y^2)$
9. Not factorable
10. Not factorable
11. $(x + y)(x^2 - xy + y^2)$
12. $(x - 1)(x^2 + x + 1)$
13. $(x + 1)(x^2 - x + 1)$
14. $(10a - 1)(10a + 1)$
15. $(2a - 1)(4a^2 + 2a + 1)$
16. $(2a + 1)(4a^2 - 2a + 1)$
17. $(2a - 11)(2a + 11)$
18. $(7r - 8s)(7r + 8s)$
19. $(9ab - 5c^2)(9ab + 5c^2)$
20. $(4x + 5y^2)(16x^2 - 20xy^2 + 25y^4)$
21. $3(x^2 - 49) = 3(x - 7)(x + 7)$
22. $a(a^2 - 16) = a(a - 4)(a + 4)$
23. $y(y^3 - 1) = y(y - 1)(y^2 + y + 1)$
24. $2(x^3 + 343) = 2(x + 7)(x^2 - 7x + 49)$
25. $2y(y^2 - 8y - 12)$
26. $4(25x^2 - y^2) = 4(5x - y)(5x + y)$
27. $4x(25x - 1)$
28. $4x^2(25 - x^2) = 4x^2(5 - x)(5 + x)$
29. $x(x^3 + 216) = x(x + 6)(x^2 - 6x + 36)$
30. $2yz(y^3 - 27z^3) = 2yz(y - 3z)(y^2 + 3yz + 9z^2)$
31. $(x^n - y^n)(x^n + y^n)$
32. $(x^n - 5)(x^{2n} - 5x^n + 25)$
33. $(a^n + 6b^n)(a^{2n} - 6a^nb^n + 36b^{2n})$
34. $(a^{2n} + 1)(a^{4n} - a^{2n} + 1)$
35. $([x + y] - [x - y])([x + y]^2 + [x + y][x - y] + [x - y]^2)$
$= (x + y - x + y)(x^2 + 2xy + y^2 + x^2 - y^2 + x^2 - 2xy + y^2)$
$= (2y)(3x^2 + y^2) = 2y(3x^2 + y^2)$
36. $x + 5$
37. $x^2 + 7x + 49$
38. $64x^2 - 8xy + y^2$
39. $a^n - b^n$
40. $a^{2n} - a^nb^n + b^{2n}$
41. $(x - 1)(x^4 + x^3 + x^2 + x + 1)$
42. $(x - 1)(x^6 + x^5 + x^4 + x^3 + x^2 + x + 1)$

Exercises 3.10 A, page 206

1. $3(y + 2)(y - 2)$
2. $2(x + 9)(x - 5)$
3. $p(p + 6)(p - 6)$
4. $100(1 + x^2)(1 + x)(1 - x)$
5. $2(2x - 1)(x - 1)$
6. $9(2a - 1)^2$
7. $2(2x - 5)(x + 1)$
8. $2k(2 - x)(1 + x)$
9. $x(2x - 5)^2$
10. $3x(4x - 1)(3x - 5)$
11. $p^2(1 - 7p)(1 + 7p)$
12. $x^2(x^2 + 4)$

13. $x(4 + 9x)(4 - 9x)$

14. $a(9a^2 + 4)(3a + 2)(3a - 2)$

15. $a(2a + 3)(3a - 4)$

16. $(x^2 + 4)(x + 1)(x - 1)$

17. $-x(x - 4)(x + 6)$

18. $2(5x^2 + 8y^2)(5x^2 - 8y^2)$

19. $2x(2x - 1)(3x + 5)$

20. $(y + 4)(y - 4)(y + 5)(y - 5)$

21. $4(u^2 + 4)(u + 5)(u - 5)$

22. $-2x^2(2x + 1)^2$

23. $4x^2(x^2 + 4x + 16)$

24. $3x(x + 7)(x - 6)$

25. $m(m^2 + 1)(m + 1)(m - 1)$

26. $5(x - 5)(x^2 + 5x + 25)$

27. $x(8x + 1)(8x - 1)$

28. $x(4x + 1)(16x^2 - 4x + 1)$

29. $(x^2 + y^2)(x + y)(x - y)$

30. $(x - y)(x + y)(x^2 + xy + y^2)(x^2 - xy + y^2)$

31. $m^2(m^2 - n^2 - 1)$

32. $(x - 2)(x + 2)(x^2 + 2x + 4)(x^2 - 2x + 4)$

33. $3x(x + 3)(x^2 - 3x + 9)$

34. $(t^2 + 1)(t + 7)(t - 7)$

35. $(t + 1)(t - 10)(t^2 - t + 1)(t^2 + 10t + 100)$

36. $5(x^n + 1)(x^n - 1)$

37. $6(x^n - 2)^2$

38. $y^n(y + 5)(y - 3)$

39. $a^2(b + c + d)(b + c - d)$

40. $(x + 2y)(1 - a)$

Exercises 3.11 A, page 210

1. $-2, 3$

2. $1, -4$

3. $\frac{3}{2}, -2$

4. $-\frac{1}{3}, \frac{2}{5}$

5. $1, 2$

6. $0, \frac{3}{2}$

7. $0, 5$

8. $4, -4$

9. $0, -2$

10. $2, 6$

11. $-2, -5$

12. $3, -2$

13. $15, -3$

14. $2, -12$

15. $-3, -4$

16. $-1, -6$

17. $9, -1$

18. 3

19. 1

20. $4, -4$

21. $5, -5$

22. $6, -6$

23. $7, -7$

24. $\varnothing$

25. $\varnothing$

26. $\varnothing$

27. 0

28. All real numbers

29. $0, -1$

30. $\varnothing$

31. $x^2 = 6x + 91$; 13 or -7

32. $x + \frac{1}{x} = 2$; 1

33. $x(x + 2) = 255$; 15, 17 or $-17, -15$

34. $x(x + 2) = 528$; 22, 24 or $-24, -22$

35. $10x - 18 = \frac{x^2}{2}$; 2 or 18

36. a. $A = \pi(R - r)(R + r)$ b. $A = \frac{22}{7}(2)(84) = 528$

37. a. $M = \frac{W(4x - L)(x - L)}{8}$ b. -200

38. $-16(t - 10)(t - 115) = 0$; $t = 10$ or $t = 115$

39. a. $s^2 = 4h(2r - h)$ b. $s^2 = 80(80)$, $s = 80$

40. $360 = 110I - 5I^2$; $I = 4$ or $I = 18$

41. 1591

42. 7224

43. 396

44. 484

45. 1156

46. 1209

47. 81

48. 7225

49. 621

50. 172

Chapter 3 Review Exercises, page 214

1. x^5

2. r^{13}

3. x^4

4. p^9

5. $-12x^4$

6. $-2a^8$

7. $-10n^3x^4$

8. $3a^6$

9. $12x^3$

10. x^{2n}

11. $-2a^2 + 2ab$

12. $2a^2b^3 - 2a^2b^2 + 2a^3b^2$

13. $-2x^2y + xy^2 - 3xyz$

14. $-3p^2x - 3p^4$

15. $x^2 - 5x + 6$

16. $y^2 - 3y - 4$

17. $6a^2 + 7a - 3$

18. $a^2 + 10a + 25$

19. $25x^2 - 1$

20. $4y^2 - 12y + 9$

21. $2x^3 - 3x^2 - 14x$

22. $2x^5 + x^4 - 2x^3$

23. $-3x^7 - 2x^5 + 4x^4$

24. $6x^2 - 29xy - 16y^2$

25. $2x^3 + 3x^2 - 20x$

26. $18x^3 + 24x^2 + 8x$

27. $x^3 + 8x^2 + 21x + 18$

28. $x^3 - x^2 - 10x - 8$

29. $6x^3 - x^2 + 8x + 3$

30. $x^4 + 2x^3 - 8x^2 - 13x + 6$

31. $3x(x - 2 + 8y)$

32. $(t + 3)(t + 8)$

33. $(x + 5)^2$

34. $(7y^2 - 1)(7y^2 + 1)$

35. $(x - 1)(x^2 + x + 1)$

36. $(a + 1)(a^2 - a + 1)$

37. $-3x(x^2 - 9) = -3x(x - 3)(x + 3)$

38. $(3a + 1)(2a - 3)$

39. $3x(12 + x^2 - x^4) = -3x(x^4 - x^2 - 12)$
$= -3x(x^2 - 4)(x^2 + 3)$
$= -3x(x - 2)(x + 2)(x^2 + 3)$

40. $25(4p^2 - q^2) = 25(2p - q)(2p + q)$

41. $4y(16y^2 - 9) = 4y(4y - 3)(4y + 3)$

42. $(x^3 - y^3)(x^3 + y^3) = (x - y)(x + y)(x^2 + xy + y^2)(x^2 - xy + y^2)$

43. $2y(6x^2 - 11x - 10) = 2y(2x - 5)(3x + 2)$

44. $(a + b)(8a - 3b)$

45. $2y(y - 3)(y + 3)(y^2 + 9)$

46. $3ax^4(5x^2 + 14x - 3) = 3ax^4(x + 3)(5x - 1)$

47. $(5r + 1)^2$

48. $(6x - 1)^2$

49. $(2p - 5q)^2$

50. $(x + 3)(y + 2)$

51. $(p + 5)(x - 3)$

52. $-x^3 - 4x^2 + 3x = -x(x^2 + 4x - 3)$

53. $yz(y[a - 1] + 1) = yz(ay - y + 1)$

54. $(a + 2)[x(a + 2) + y] = (a + 2)(ax + 2x + y)$

55. $(x + y)(3 - 2[x + y]) = (x + y)(3 - 2x - 2y)$

56. 9

57. 64

58. 16

59. $12ab$

60. $14yz$

61. $\left\{-2, \frac{7}{2}\right\}$

62. $\{0, -7\}$

63. $\{5, -12\}$

64. $\{0, 3\}$

65. $\{2, 9\}$

66. $\left\{\frac{1}{6}, -\frac{1}{2}\right\}$

67. $\{-1\}$

68. $\{1, 6\}$

69. $\{0, 3\}$

70. $\{8, -8\}$

Exercises 4.1 A, page 223

1. $\frac{1}{3}$

2. $-\frac{1}{3}$

3. $-\frac{3}{5}$

4. $\frac{2}{5}$

5. $-\frac{3}{4}$

6. $\frac{2}{3}$

7. 1

8. a

9. $\frac{1}{3x^2}$

10. $3x^2$

11. $\frac{-1}{m^2n}$

12. $-\frac{x}{2}$

13. $4x^2y$

14. $5ab$

15. -1

16. $3xy$

17. $\frac{1}{3xy}$

18. $\frac{-1}{xz}$

19. $\frac{1}{ac}$

20. $\frac{a + b}{4x}$

21. $\frac{x+3}{x}$

22. $\frac{1}{y-2}$

23. $\frac{x-4}{x-2}$

24. $\frac{x+2}{x}$

25. $\frac{y}{3}$

26. $\frac{3a}{4b}$

27. $\frac{c-d}{6}$

28. $\frac{1}{3}$

29. $\frac{y}{3}$

30. $3(5n-1)$

31. $\frac{x-y}{x+y}$

32. -1

33. $\frac{a^2+1}{a+1}$

34. $\frac{1}{4(1-3t)}$

35. $x+2$

36. $a-b$

37. -1

38. $\frac{3}{x-2}$

39. $\frac{x+1}{x+4}$

40. $\frac{x+4}{x+2}$

41. $\frac{2(t+2)}{3(t-3)}$

42. $\frac{z-5}{z+4}$

43. $\frac{m-n+1}{m+n-1}$

44. $\frac{x}{z}$

45. $\frac{x+1}{x+7}$

46. $\frac{-1}{x+5}$

47. $\frac{a-4}{2(a+4)}$

48. $\frac{5n-2}{5n+2}$

49. $-(3x+2y)$

50. $\frac{x^2+2x+4}{x+4}$

51. $\frac{x-5}{x^2-5x+25}$

52. $\frac{s-r}{t-s}$

Exercises 4.2 A, page 229

1. $\frac{12}{21}$

2. $\frac{15}{25}$

3. $\frac{-28}{42}$

4. $\frac{-175}{35}$

5. $\frac{3x}{36}$

6. $\frac{6x}{15x}$

7. $\frac{36xy}{48xy}$

8. $\frac{6x}{21x^2}$

9. $\frac{-15x^3}{40x^2}$

10. $\frac{-12x^2y}{30xy^3}$

11. $\frac{-18x^2y^2}{42xyz}$

12. $\frac{5a^2b}{5ab^3}$

13. $\frac{10}{2x+2}$

14. $\frac{3x}{6x-15}$

15. $\frac{5x-15}{5x+20}$

16. $\frac{2x^2}{x^2-2x}$

17. $\frac{2x^2}{2x^2-10x}$

18. $\frac{2y+4}{y^2+2y}$

19. $\frac{3y^2-9y}{y^2-9}$

20. $\frac{t^2+6t+9}{t^2+t-6}$

21. $\frac{x^2+3x+2}{x+1}$

22. $\frac{-1}{3-4x}$

23. $\frac{x}{x-7}$

24. $\frac{-x}{2x^2-5x}$

25. $\frac{-y^2-4y}{y^2-16}$

26. $\frac{x^2+5x+6}{x^2+6x+8}$

27. $\frac{a^2+2a+1}{a^2-2a-3}$

28. $\frac{6x^3-6x^2}{6x^2-6x}$

29. $\frac{t^2+5t-14}{14+9t+t^2}$

30. $\frac{a^2-b^2}{(a+b)^2}$

31. $\frac{12a^3+18a^2}{10a^2+13a-3}$

32. $\frac{7r^4-7r^3-14r^2}{7r^4-28r^2}$

33. $\frac{-2a^3+5a^2}{4a^2-10a}$

34. $\frac{9x+27}{(x-3)(x+3)^2}$

35. $\frac{15n^3+30n^2+15n}{3n^3-3n}$

36. $\frac{ab}{(b-a)^2}$

37. $\frac{x^3+x^2}{xy(x+1)^2}$

38. $\frac{3x^2y+6xy^2+3y^3}{y^2(x+y)^3}$

Exercises 4.3 A, page 237

1. $\frac{5}{3}$

2. 1

3. $\frac{4-5y}{x}$

4. $\frac{x+7}{x+3}$

5. $\frac{2y-1}{z-1}$

6. $\frac{17}{12}$

7. $\frac{31}{36}$

8. $\frac{5}{12}$

9. $\frac{47}{60}$

10. $\frac{41}{36y}$

11. $\frac{1}{3x}$

12. $-\frac{2x+3}{x^2}$

13. $\dfrac{3x^2 + 8x - 1}{6x^3}$

14. $\dfrac{5x + 3}{x(x + 1)}$

15. $\dfrac{1}{b(b - 1)}$

16. $\dfrac{x^3 + 4x^2 + 2x + 4}{2x(x + 2)}$

17. $\dfrac{y + 5}{(y + 1)(y + 2)}$

18. $\dfrac{-a^2 + 18a + 45}{3a(a + 3)}$

19. $\dfrac{yz + xz + xy}{xyz}$

20. $\dfrac{5x + 4}{(x + 2)^2}$

21. $\dfrac{10y^2 + 39y - 72}{(2y - 3)^2(2y + 3)}$

22. $\dfrac{2x^2 + 3}{x}$

23. $\dfrac{x^2 + 2x}{x + 1}$

24. $\dfrac{x + 3}{x + 1}$

25. $\dfrac{x^2 + 6x + 12}{(x + 1)(x + 3)(x + 4)}$

26. $\dfrac{-8x - 15}{(3x + 2)(x + 1)(x - 6)}$

27. $\dfrac{4y^2 - y + 9}{(y - 2)(y - 3)(2y + 1)}$

28. $\dfrac{1}{2 - x}$ or $\dfrac{-1}{x - 2}$

29. $\dfrac{5a - 13}{6(a + 1)}$

30. $\dfrac{3x + 10}{15(x - 1)}$

31. $\dfrac{13x}{2(x - 4)(x + 6)}$

32. $\dfrac{-c^2 + 5c + 20}{c^2 - 25}$

33. $\dfrac{1}{y(y - 5)}$

34. $\dfrac{x^2 - 28x - 49}{(x + 7)^2(x - 7)^2}$

35. $\dfrac{x + 1}{x}$

36. $\dfrac{4a^2}{2a + 5}$

37. $\dfrac{x - y}{x + y}$

38. $\dfrac{y}{(2y - 1)(3y - 1)}$

39. $\dfrac{a^2}{(a^3 + 1)(a - 1)}$

40. $\dfrac{2}{(3x + 1)(3x - 2)}$

Exercises 4.4 A, page 243

1. $\dfrac{1}{12}$

2. $\dfrac{4}{5}$

3. $\dfrac{2}{21}$

4. $\dfrac{x^2}{21}$

5. $\dfrac{x^2 - y^2}{16}$

6. $\dfrac{x}{2}$

7. $\dfrac{5y}{x^3}$

8. $\dfrac{1}{15}$

9. $\dfrac{-b}{az}$

10. $\dfrac{rs}{r + s}$

11. $\dfrac{2a(a - 5)}{a - 1}$

12. $\dfrac{x - 2y}{2x^2}$

13. $\frac{x+3}{3x}$

14. $\frac{3(x-2)}{x+2}$

15. $\frac{(a-b)^2}{a}$

16. $\frac{-5y^4(y+1)}{24}$

17. $\frac{2(n-2)}{n(n+3)}$

18. $\frac{x^2}{5(x-2)}$

19. 1

20. $\frac{4a}{b}$

21. 1

22. $x - y$

23. $\frac{(a^2+2a+4)(b+1)}{a(a+2)}$

24. -1

25. a

26. $2x$

27. $x^2 - 4$

28. $30x + 180$

29. $6x + 18$

30. $x^2 + 2x - 8$

Exercises 4.5 A, page 248

1. $\frac{1}{2}$

2. $\frac{8}{15}$

3. 3

4. $\frac{xz}{9y}$

5. $\frac{8}{5}$

6. $\frac{-3bc}{2ad}$

7. $\frac{7x(x+7)}{(x-7)^3}$

8. $\frac{2b+1}{b+2}$

9. $\frac{12y}{y+4}$

10. $\frac{a-5}{a+5}$

11. $\frac{-3(x+8)^2}{2(x-8)^2}$

12. $\frac{a^2+2a+4}{(a+2)^2}$

13. $\frac{b-5}{3b^3}$

14. $\frac{2x-1}{5}$

15. $\frac{y}{y-1}$

16. $2t$

17. $\frac{3-x}{3}$

18. 1

19. $\frac{(a-5)^2}{(a+5)^2}$

20. $\frac{x+1}{x+3}$

Exercises 4.6 A, page 252

1. $\frac{22}{31}$

2. $\frac{6}{5}$

3. $\frac{20}{13}$

4. $\frac{12}{17}$

5. 1

6. $\frac{1}{x-2}$

7. $\frac{10x+2y^2}{xy^2}$

8. $\frac{a}{a^2-3}$

9. $\frac{y+3}{y-4}$

10. $\frac{n+3}{3n-1}$

11. $\frac{xy}{x+y}$

12. $\frac{75}{2}$

13. x

14. $\frac{x-1}{x+1}$

15. $\frac{-3}{5}$

16. $\frac{a+1}{a(a+3)}$

17. $1-x$

18. $\frac{y^2+1}{y^3}$

19. $\frac{4}{x^2-3}$

20. $\frac{8}{13}$

Exercises 4.7 A, page 257

1. $2x^2 - 3x + 9 - \frac{29}{x+4}$

2. $y^3 + 2y + \frac{1}{y^2+2}$

3. $2x^2 - 3x - 2$

4. $9y^2 + 6y + 4$

5. $x^2 - 2x + 2$

6. $2x^2 - 2x - 1 - \frac{7}{2x+7}$

7. $x^2 + 5 - \frac{8}{x+4}$

8. $5x^2 + \frac{2x-1}{x^2-6}$

9. $a^2 - 4a + 16$

10. $2x^2 - x - 1 + \frac{4}{x-1}$

11. $2x^2 + 3 + \frac{1}{x-5}$
   $\left(\frac{32}{3} = \frac{32}{3}\right)$

12. $t^2 - 3t + 2$
   $(0 = 0)$

13. $x^3 + x^2 + x + 1$
   $(40 = 40)$

14. $2x + 1 + \frac{8}{2x-1}$
   $(11 = 11)$

15. $2x^3 + 4 + \frac{7}{x^2-3}$
   $(27 = 27)$

16. $y^2 + 4y + 8$
   $(20 = 20)$

17. $5a^2 - 2ab + b^2 - \frac{b^3}{5a+2b}$
   $(21 = 21)$

18. $x^4 + 1$
   $(17 = 17)$

19. $x + y + 3$
   $(9 = 9)$

20. $x - y - 1$
   $(0 = 0)$

Exercises 4.8 A, page 263

1. $x \neq 0; \{5\}; \frac{1}{12} = \frac{1}{12}$
2. $x \neq 0; \{-3\}; -\frac{1}{6} = -\frac{1}{6}$
3. $x \neq 0, x \neq -5; \{-11\}; -\frac{1}{6} = -\frac{1}{6}$
4. $y \neq 0, y \neq 2; \{3\}; 1 = 1$
5. $y \neq 0, y \neq 1, y \neq -1; \left\{-\frac{3}{2}\right\}; 0 = 0$
6. $\{90\}; 3 = 3$
7. $x \neq 0; \{\frac{1}{4}\}; 20 = 20$
8. $x \neq \frac{1}{2}, x \neq -\frac{1}{3}; \{2\}; 1 = 1$
9. $y \neq -\frac{1}{2}; \{6\}; \frac{2}{5} = \frac{2}{5}$
10. $\left\{\frac{4}{3}\right\}; \frac{1}{10} = \frac{1}{10}$
11. $x \neq 0; \varnothing$
12. $x \neq 5, x \neq -2; \{-3\}; \frac{1}{4} = \frac{1}{4}$
13. $x \neq -1, x \neq -6; \varnothing$
14. $t \neq 2, t \neq -2; \{-10\}; \frac{7}{9} = \frac{7}{9}$
15. All real numbers except $y = 7$ and $y = -7$
16. $x \neq 2, x \neq -\frac{1}{3}; \{0\}; -6 = -6$
17. $t \neq 0, t \neq 1, \{-2\}; \frac{1}{3} = \frac{1}{3}$
18. $x \neq 3; \varnothing$
19. $x \neq -5, x = -3; \{-7\}; 2 = 2$
20. $x \neq 0, x \neq 3, x \neq -2; \left\{\frac{1}{2}\right\}; \frac{8}{5} = \frac{8}{5}$
21. $x \neq -\frac{3}{5}, x \neq \frac{1}{2}; \left\{\frac{5}{2}\right\}; \frac{-11}{62} = \frac{-11}{62}$
22. $x \neq b; x = 2a - 3b$
23. $x = \frac{a^2c + ab^2}{bc}$
24. $A = \frac{BR}{B - R}; A \neq 0$
25. $V = \frac{Fv}{F - f}; V \neq 0, V \neq v$

Exercises 4.9 A, page 269

1. 15 by 24 inches
2. $88\frac{8}{9}$ grams oxygen, $11\frac{1}{9}$ grams hydrogen
3. 20 cups
4. \$1125 and \$675
5. 13
6. $208\frac{1}{3}$ miles
7. 44 amperes
8. \$0.54
9. $10\frac{1}{2}$ hours
10. $1066\frac{2}{3}$ pounds

11. $14\frac{6}{11}$ pounds
12. 300 degrees
13. 12
14. 24 and 32
15. \$36
16. \$24,000; \$14,400; \$9600
17. \$7000
18. 75 pounds
19. 55 inches
20. 2.88 lumens

Exercises 4.10 A, page 274

1. $\frac{x-2}{2x-2} = \frac{3}{7}$; $x = 8$; original fraction $= \frac{8}{16}$

2. $\frac{x+4}{(x+12)-3} = \frac{3}{4}$; $x = 11$; original fraction $= \frac{11}{23}$

3. $\frac{x}{8} + \frac{x}{12} = 1$; $4\frac{4}{5}$ days

4. $\frac{x}{18} + \frac{x}{30} + \frac{x}{45} = 1$; 9 minutes

5. $\frac{6}{15} + \frac{6}{x} = 1$; 10 days

6. $\frac{x+10}{40} + \frac{x}{60} = 1$; 18 minutes

7. $\frac{50}{x+8} = \frac{30}{x-8}$; 32 mph

8. $\frac{800}{175+x} = \frac{600}{175-x}$; 25 mph

9. $\frac{9}{x} + \frac{9}{3x} = 1$; 12 hours and 36 hours

10. $\frac{500}{x+5} = \frac{400}{x}$; 20 hours and 25 hours

11. $\frac{30}{x+3} = \frac{3}{5}\frac{30}{x-3}$; 12 mph

12. $\frac{\text{total distance}}{\text{total time}} = \frac{2d}{\frac{d}{60} + \frac{d}{40}} = \frac{120(2d)}{2d + 3d} = \frac{240}{5} = 48$ mph

13. $\frac{1}{x} + \frac{1}{3x} = \frac{1}{15}$; 20 ohms and 60 ohms

14. 5 volts

Chapter 4 Review Exercises, page 279

1. $\frac{2}{5}$

2. $\frac{5x^2}{9yz}$

3. -1

4. $\frac{9(x+2y)}{x-2y}$

5. $\frac{2x-15}{2x-3}$

6. $\frac{-4x^3(x-1)}{3(x+1)}$

7. 1

8. $\frac{x-1}{3}$

9. $\frac{-1}{3(x+3)}$

10. $\frac{xy+3}{xy-3}$

11. $\frac{5-x}{x+3}$

12. $\frac{6x^2-3x}{x^2-9}$

13. $\dfrac{4x - 15}{x^2 - 6x + 5}$

14. $\dfrac{4}{x(x + 1)(x - 1)}$

15. $\dfrac{x^2 + 14x + 4}{2x(x + 2)}$

16. $\dfrac{1}{(x + 3)(x + 4)}$

17. $\dfrac{16}{(x - 2)^2(x + 2)^2}$

18. $\dfrac{5(5x + 2)}{12(3x + 1)}$

19. $\dfrac{9x - 2}{3}$

20. Undefined

21. $x^2 - x + \dfrac{-1}{x - 3}$

22. $x^2 - 2 + \dfrac{4}{x^2 + 3}$

23. $3x^2 - 2x + 4 - \dfrac{1}{5x + 10}$

24. $a^2 + a + 1$

25. $5x^2 - 2x - 3 + \dfrac{-7}{2x + 1}$

26. a. -3 b. 3 c. 2

27. $x \neq 6, x \neq -6; \left\{-\frac{1}{2}\right\}; -\dfrac{32}{143} = -\dfrac{32}{143}$

28. $x \neq -2; \{0\}; \dfrac{3}{2} = \dfrac{3}{2}$

29. $x \neq 3; x \neq -3; \{9\}; \dfrac{-1}{6} = \dfrac{-1}{6}$

30. $x \neq 1, x \neq -3; \left\{-\frac{1}{4}\right\}; -2 = -2$

31. $x \neq 0, x \neq -\frac{1}{3}; \left\{\frac{1}{7}\right\}; 0 = 0$

32. $x \neq 4, x \neq -4; \varnothing$

33. All real numbers except $x = 1$

34. $x = \dfrac{a - b}{2}; x \neq a, x \neq -b, a \neq -b$

35. $x = \dfrac{p - 1}{c + 1}; c \neq -1, x \neq 0$

36. $\dfrac{\frac{2}{7}}{2} + \dfrac{2}{x} = 1; 4\frac{2}{3}$ hours

37. $\dfrac{860}{x - 35} = \dfrac{1140}{x + 35}$; 250 mph

38. $k \cdot \dfrac{5}{2} = 18; \dfrac{36}{5}$

39. $\dfrac{x - 3 - 2}{x + 1} = \dfrac{1}{4}; \dfrac{4}{7}$

40. $F(6^2) = 6(2^2)$

$F = \dfrac{2}{3}$

41. $\dfrac{x}{10} + \dfrac{x}{15} = 1$; 6 hours

42. $\dfrac{A}{bh} = k; \dfrac{A}{12(7)} = \dfrac{6}{3(4)}; A = 42$

43. 3

44. $\dfrac{720}{16} = \dfrac{450}{x}$; 10 gallons

Exercises 5.1 A, page 288

1. $(-3, -1)$

2. $(2, -2)$

3. $(2, 4)$

4. $(0, 0)$

5. $(3, -5)$

6.–13.

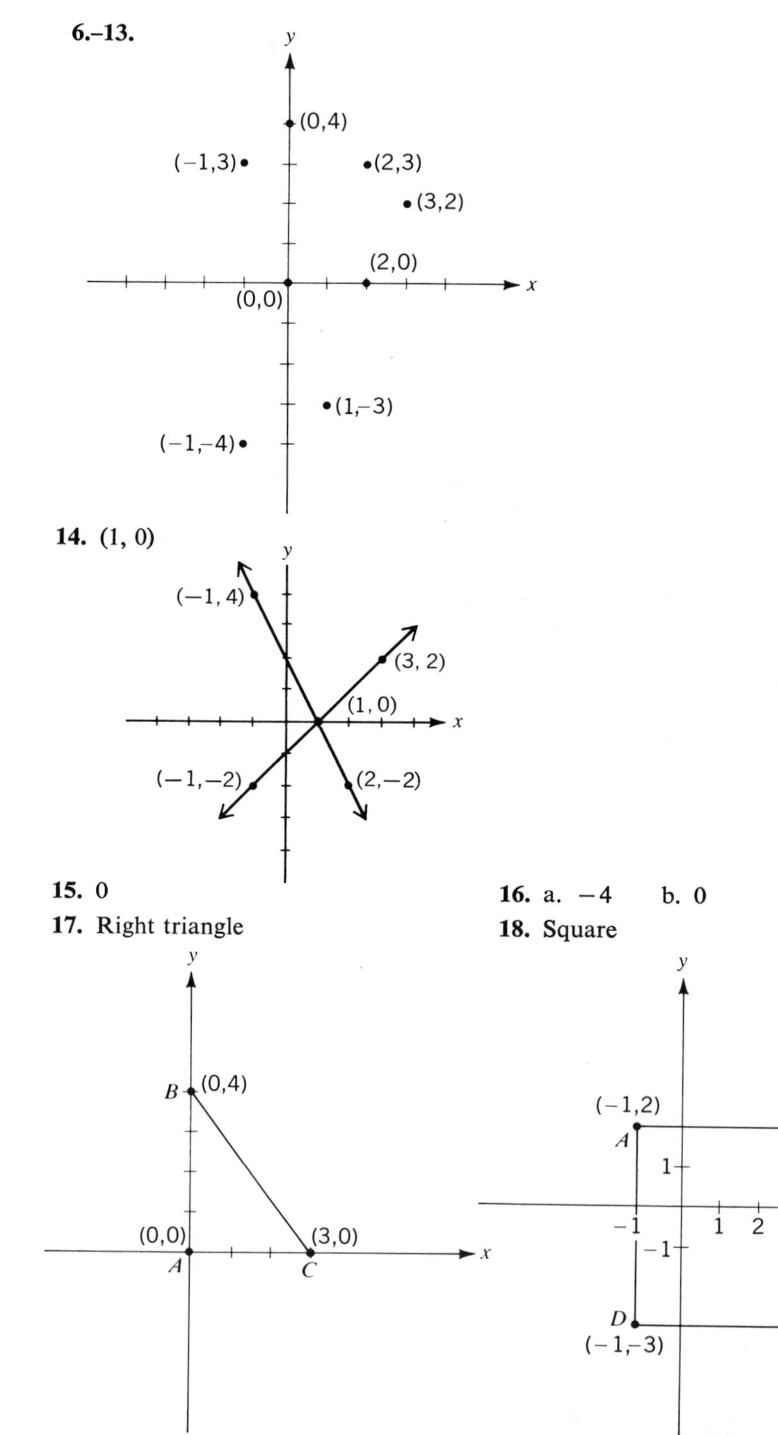

14. (1, 0)

15. 0

16. a. -4 b. 0

17. Right triangle

18. Square

19. Rectangle

20. Solution

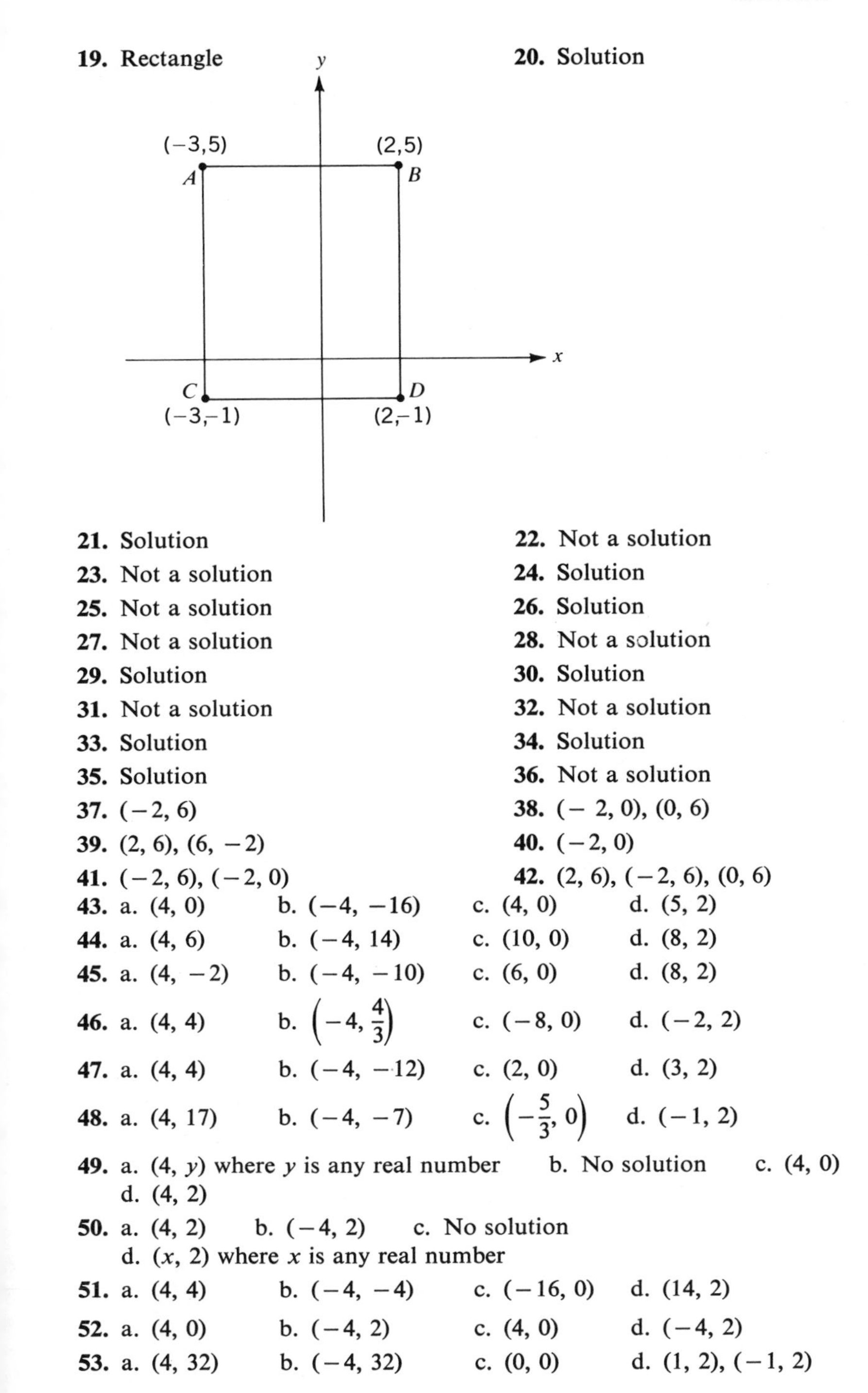

21. Solution

22. Not a solution

23. Not a solution

24. Solution

25. Not a solution

26. Solution

27. Not a solution

28. Not a solution

29. Solution

30. Solution

31. Not a solution

32. Not a solution

33. Solution

34. Solution

35. Solution

36. Not a solution

37. $(-2, 6)$

38. $(-2, 0), (0, 6)$

39. $(2, 6), (6, -2)$

40. $(-2, 0)$

41. $(-2, 6), (-2, 0)$

42. $(2, 6), (-2, 6), (0, 6)$

43. a. $(4, 0)$ b. $(-4, -16)$ c. $(4, 0)$ d. $(5, 2)$

44. a. $(4, 6)$ b. $(-4, 14)$ c. $(10, 0)$ d. $(8, 2)$

45. a. $(4, -2)$ b. $(-4, -10)$ c. $(6, 0)$ d. $(8, 2)$

46. a. $(4, 4)$ b. $\left(-4, \frac{4}{3}\right)$ c. $(-8, 0)$ d. $(-2, 2)$

47. a. $(4, 4)$ b. $(-4, -12)$ c. $(2, 0)$ d. $(3, 2)$

48. a. $(4, 17)$ b. $(-4, -7)$ c. $\left(-\frac{5}{3}, 0\right)$ d. $(-1, 2)$

49. a. $(4, y)$ where y is any real number b. No solution c. $(4, 0)$
d. $(4, 2)$

50. a. $(4, 2)$ b. $(-4, 2)$ c. No solution
d. $(x, 2)$ where x is any real number

51. a. $(4, 4)$ b. $(-4, -4)$ c. $(-16, 0)$ d. $(14, 2)$

52. a. $(4, 0)$ b. $(-4, 2)$ c. $(4, 0)$ d. $(-4, 2)$

53. a. $(4, 32)$ b. $(-4, 32)$ c. $(0, 0)$ d. $(1, 2), (-1, 2)$

54. a. $(4, 3), (4, -3)$ b. $(-4, 1), (-4, -1)$ c. $(-5, 0)$ d. $(-1, 2)$

55. a. $(4, 0)$ b. $(-4, 0)$ c. $(4, 0), (-4, 0)$ d. $(0, 2)$

56. a. $(4, 0)$ b. $(-4, 0)$ c. $(4, 0), (-4, 0)$ d. $(0, 2)$

57.

x	y
0	-6
2	0
1	-3
-2	-12
1	-3

Straight line

58.

x	y
0	3
5	0
-5	6
15	-6
10	-3

Straight line

Exercises 5.2 A, page 297

1.–4.

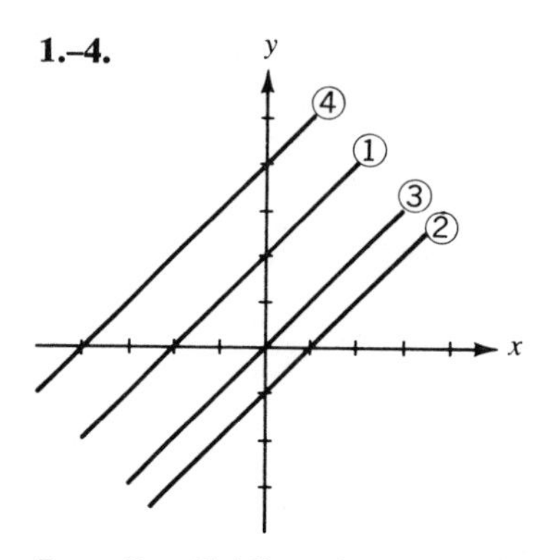

5. a. Parallel lines (same direction) b. Intersect axes at different points

6. a. $x = 5$, vertical line 5 units right of y-axis
b. $y = 6$, horizontal line 6 units above x-axis

7.

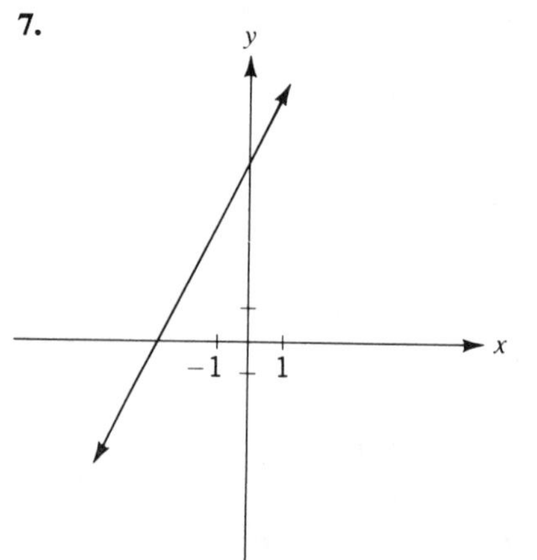

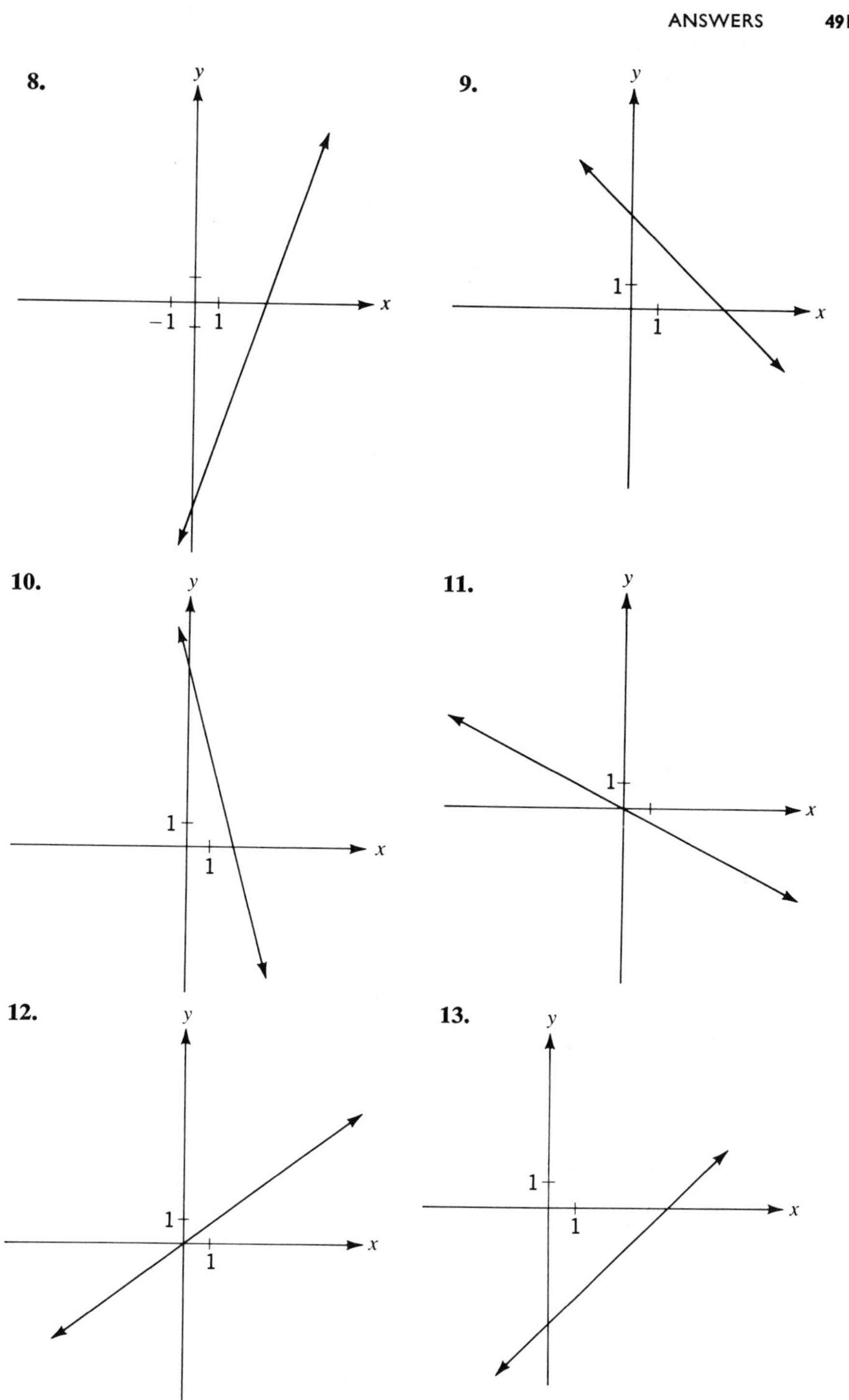
8.
y
x
−1
1
9.
y
x
1
1
10.
y
x
1
1
11.
y
x
1
12.
y
x
1
1
13.
y
x
1
1

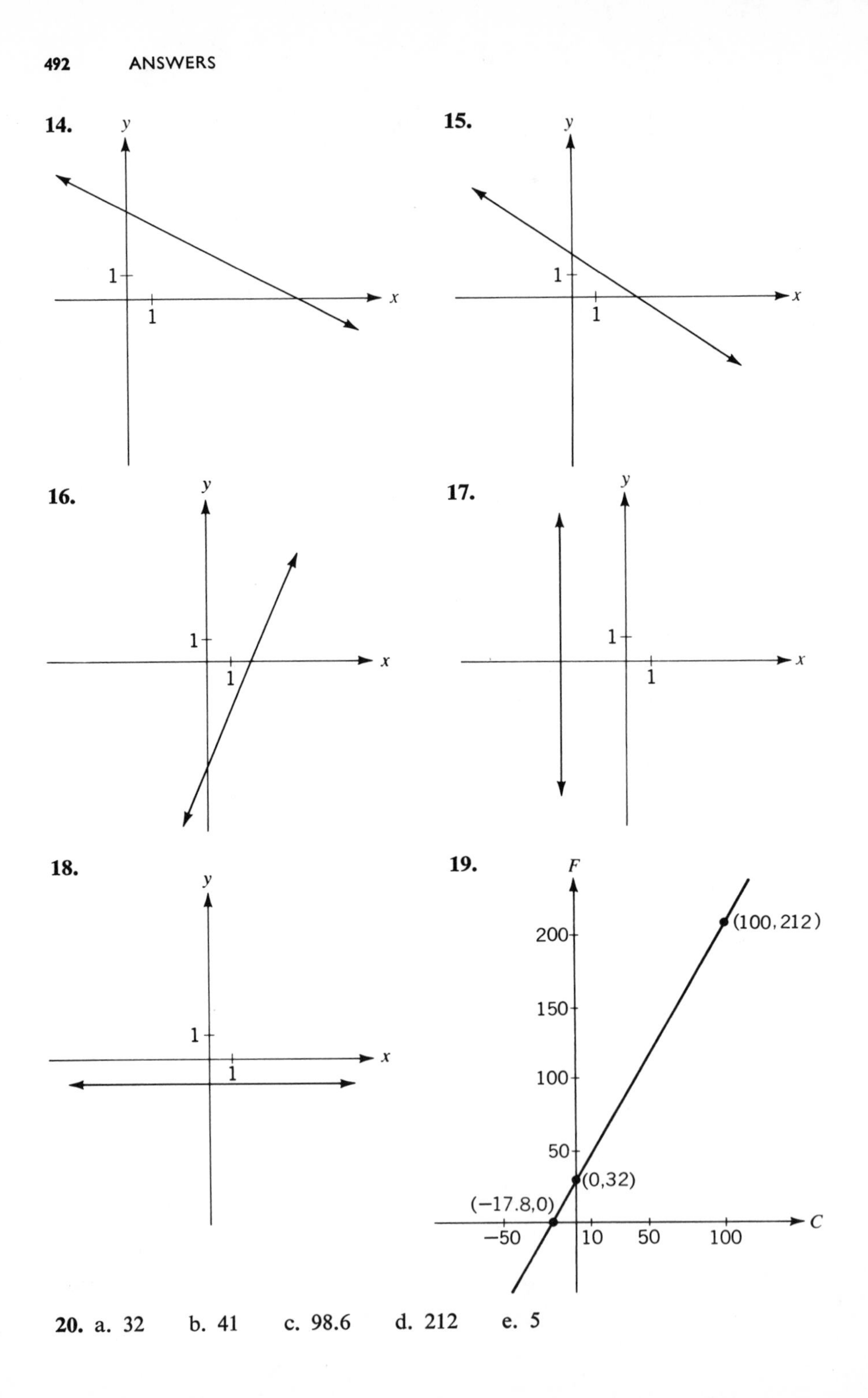

20. a. 32 b. 41 c. 98.6 d. 212 e. 5

21. a. -17.8 b. 37.8 c. 15 d. 60 e. -30

22. a.

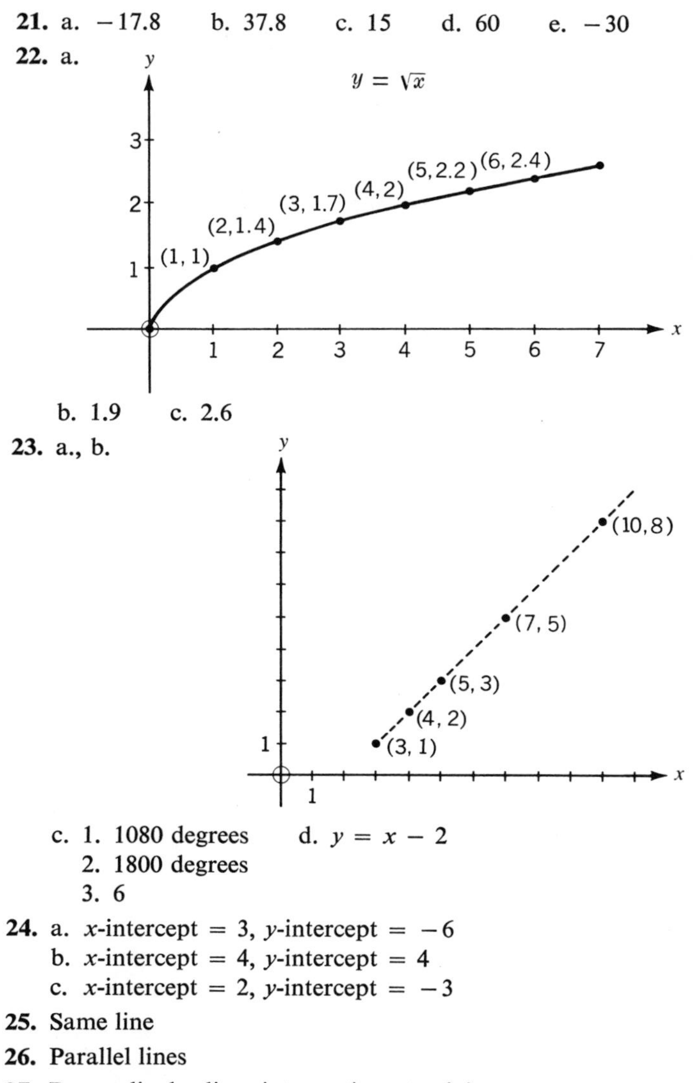

b. 1.9 c. 2.6

23. a., b.

c. 1. 1080 degrees
2. 1800 degrees
3. 6

d. $y = x - 2$

24. a. x-intercept $= 3$, y-intercept $= -6$
b. x-intercept $= 4$, y-intercept $= 4$
c. x-intercept $= 2$, y-intercept $= -3$

25. Same line

26. Parallel lines

27. Perpendicular lines intersecting at origin

28. Same line

29. Parallel lines

30. Same line

Exercises 5.3 A, page 311

1. -2

2. -1

3. $\frac{3}{5}$

4. $\frac{1}{2}$

5. $\frac{1}{3}$

6. $m = -2, b = 5$

7. $m = 2, b = -4$

8. $m = 2, b = -8$

9. $m = -\frac{3}{2}, b = 2$

10. $m = \frac{5}{2}, b = -5$

11. $m = \frac{3}{2}, b = \frac{3}{2}$

12. $m = 1, b = \frac{3}{5}$

13. Slope is undefined, no y-intercept

14. $m = 0, b = 5$

15. $m = 1, b = 0$

16. (a) and (c)

17. (b) and (d)

18.

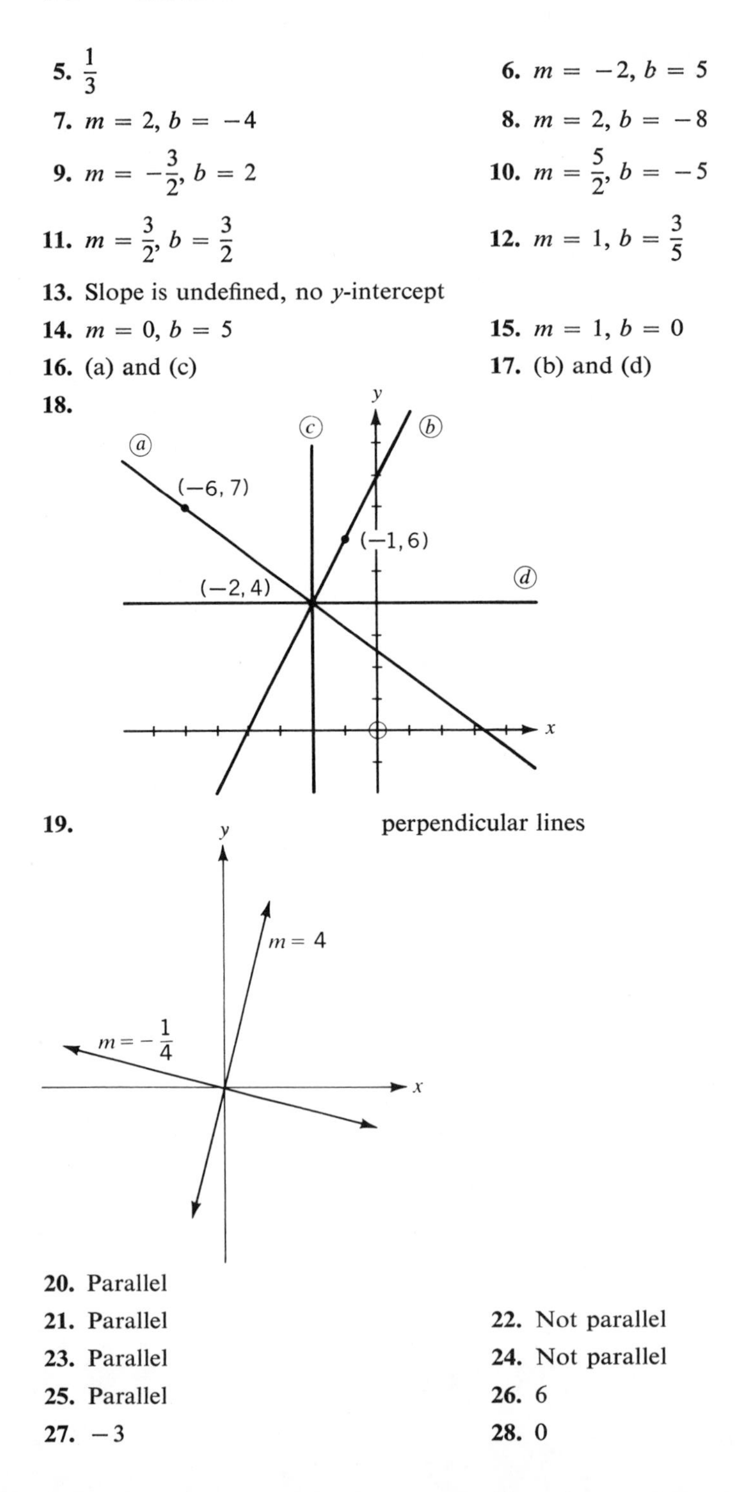

19. perpendicular lines

20. Parallel

21. Parallel

22. Not parallel

23. Parallel

24. Not parallel

25. Parallel

26. 6

27. −3

28. 0

29. -3

30. $\frac{3}{2}$

31.

32.

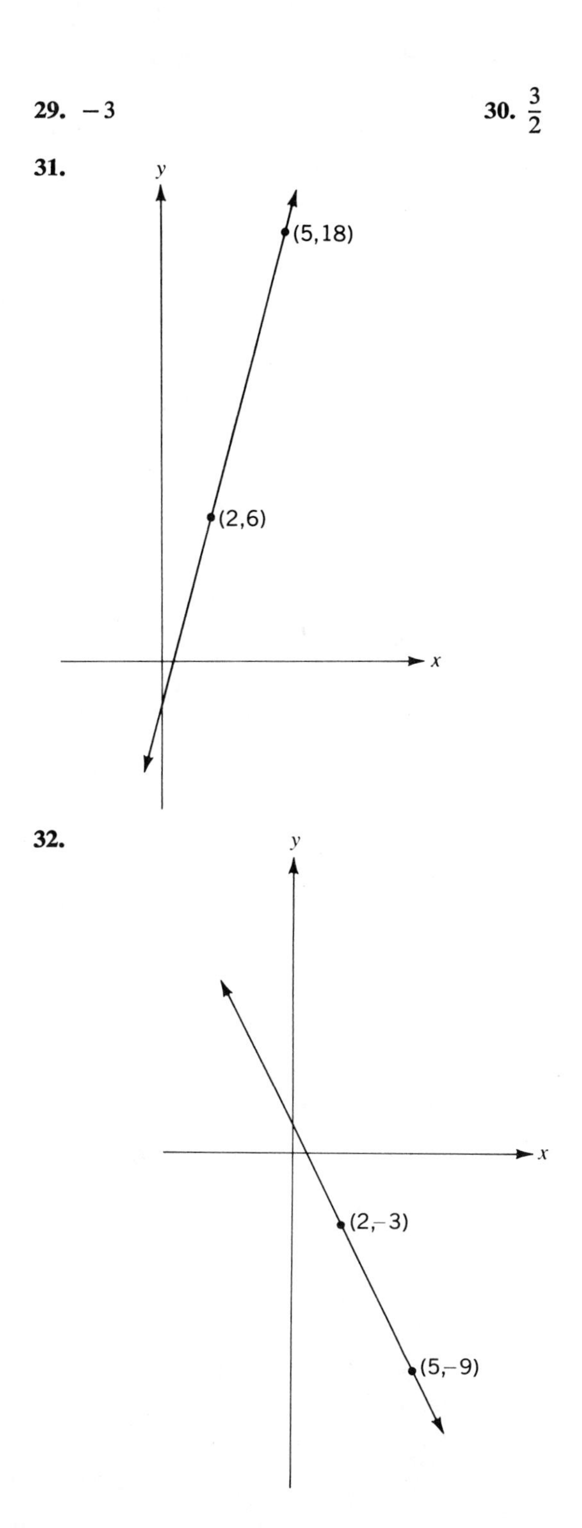

33.

34.

35.

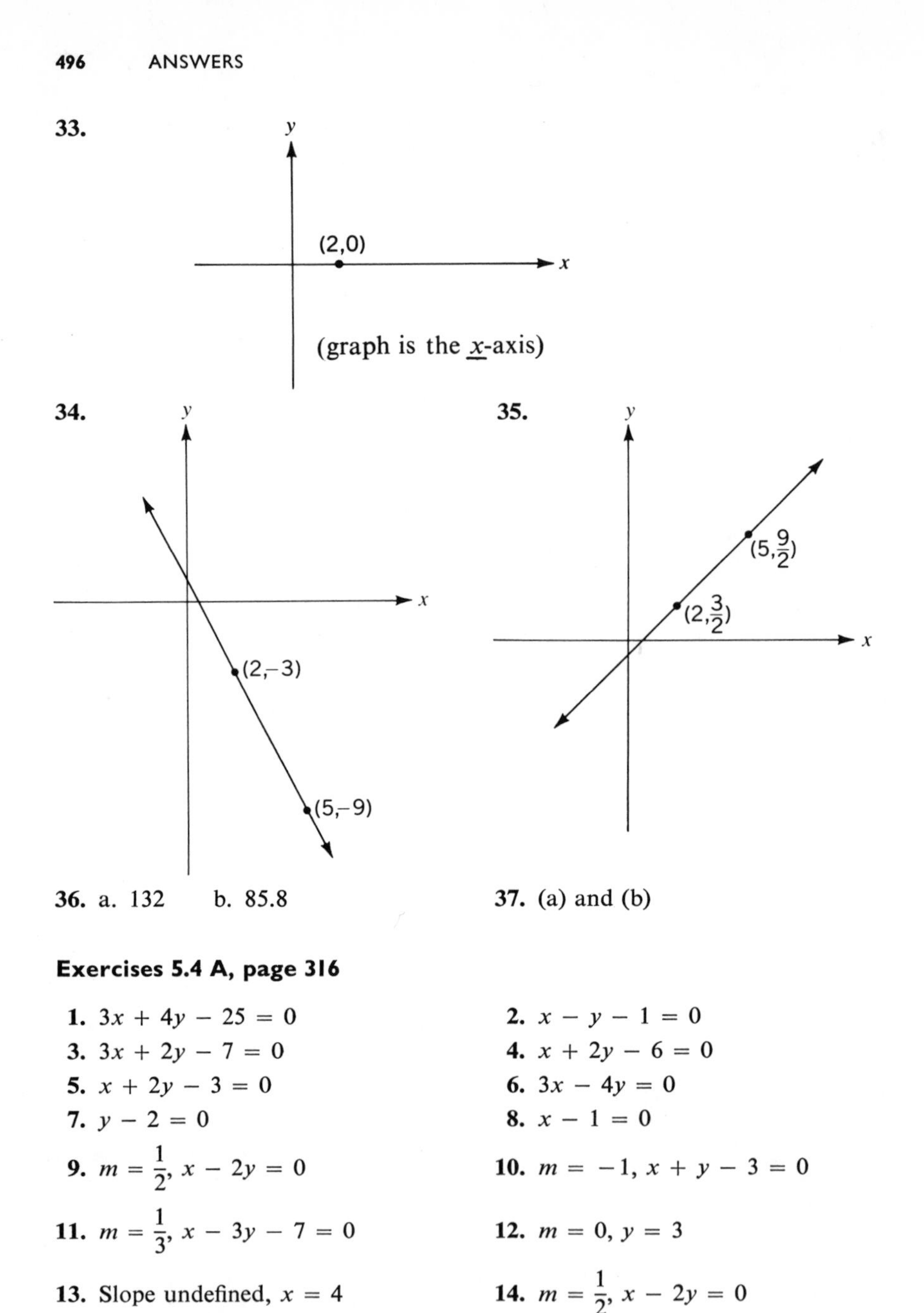

36. a. 132 b. 85.8

37. (a) and (b)

Exercises 5.4 A, page 316

1. $3x + 4y - 25 = 0$

2. $x - y - 1 = 0$

3. $3x + 2y - 7 = 0$

4. $x + 2y - 6 = 0$

5. $x + 2y - 3 = 0$

6. $3x - 4y = 0$

7. $y - 2 = 0$

8. $x - 1 = 0$

9. $m = \frac{1}{2}$, $x - 2y = 0$

10. $m = -1$, $x + y - 3 = 0$

11. $m = \frac{1}{3}$, $x - 3y - 7 = 0$

12. $m = 0$, $y = 3$

13. Slope undefined, $x = 4$

14. $m = \frac{1}{2}$, $x - 2y = 0$

15. $y = 2x + 3$; $(3, 9)$, $(5, 13)$

16. $y = 2x - 3$; $(4, 5)$, $(6, 9)$

17. $x = 5$ or $x = -5$

18. $y = 2$ or $y = -2$

19. $y = 2x$

20. $x = -2$

21. $y = 0$

22. $3x - 4y = 0$

23. $y = -2x + 6$

24. $3x + 4y = 24$ or $4x + 3y = 24$

25. Yes, slope of line AB = slope of line BC

26. No, slope of line $PQ \neq$ slope of line QR

27. 12

28. a., b.

c. $F = \frac{1}{2}L - \frac{5}{2}$ d. $\frac{1}{2}$

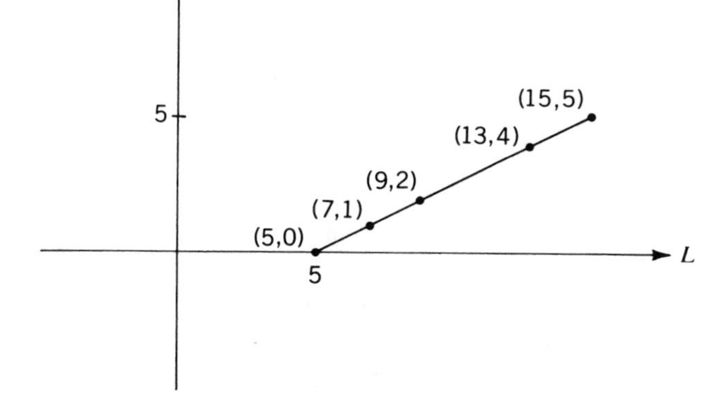

29. a., b.

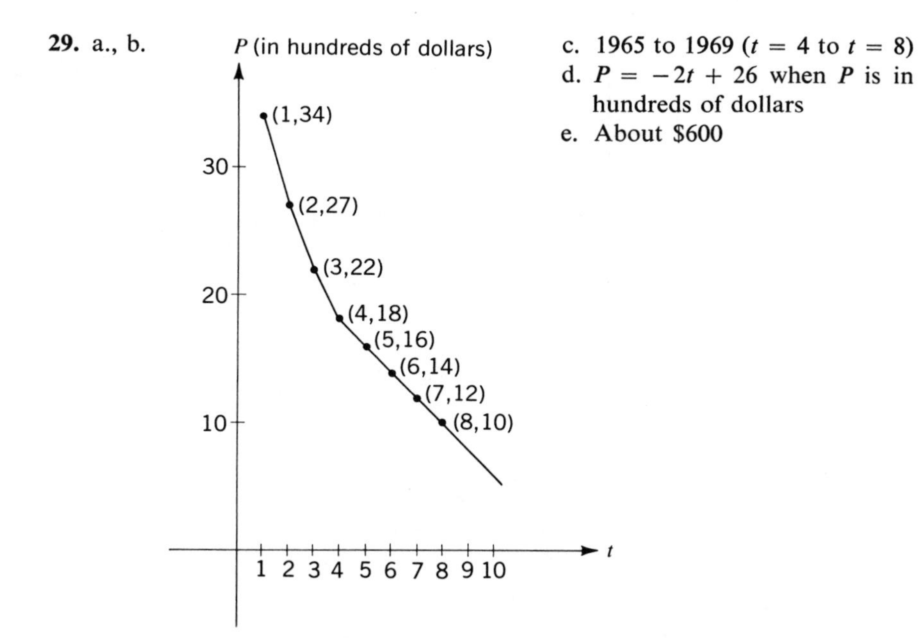

c. 1965 to 1969 ($t = 4$ to $t = 8$)

d. $P = -2t + 26$ when P is in hundreds of dollars

e. About $600

Exercises 5.5 A, page 327

1. $\{4, 8\}$

2. $\{0\}$

3. $\varnothing$

4. $\{(4, 8)\}$

5. $\{(0, 0)\}$

6. $\{1, 2, 3, 4, 6, 12\}$

7. $\{(6, 4)\}$

8. $\{(2, 8)\}$

9. $\{(7, 3)\}$

10. $\varnothing$

11. a. $m_1 = -1, m_2 = +1$ b. $b_1 = \frac{4}{3}, b_2 = -\frac{4}{3}$
c. Intersect d. Exactly one solution

12. a. $m_1 = -1, m_2 = -1$ b. $b_1 = \frac{4}{3}, b_2 = \frac{5}{2}$
c. Parallel d. No solution

13. a. $m_1 = -1, m_2 = -1$ b. $b_1 = \frac{4}{3}, b_2 = \frac{4}{3}$
c. Coincide d. Line of solutions

14. a. m_1 undefined, $m_2 = 0$ b. No y-intercept, $b_2 = 4$
c. Intersect d. Exactly one solution

15. a. m_1 undefined, $m_2 = -3$ b. No y-intercept, $b_2 = 0$
c. Intersect d. Exactly one solution

16. $(4, 3)$

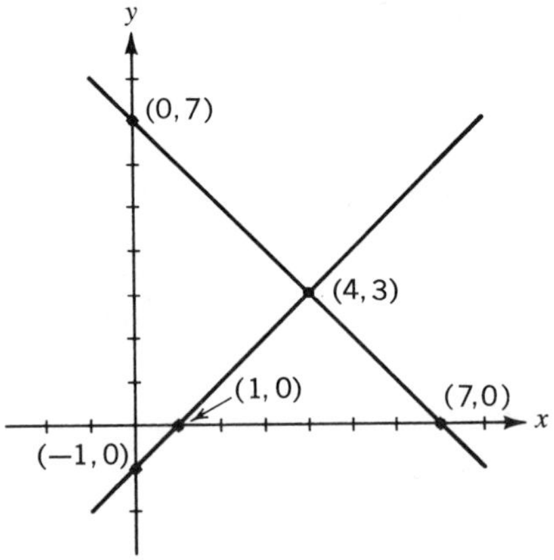

17. $(-3, -4)$

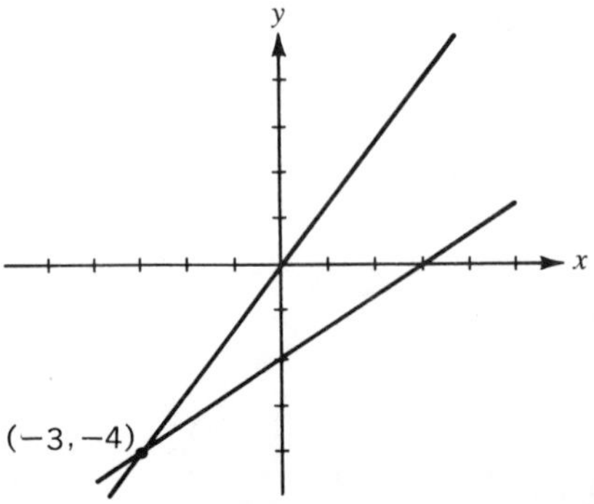

18. $\left(\frac{18}{5}, -\frac{1}{5}\right)$

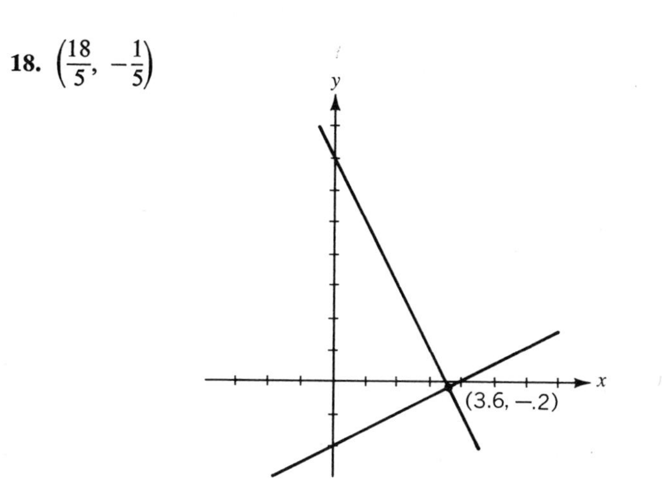

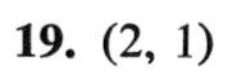

19. (2, 1)

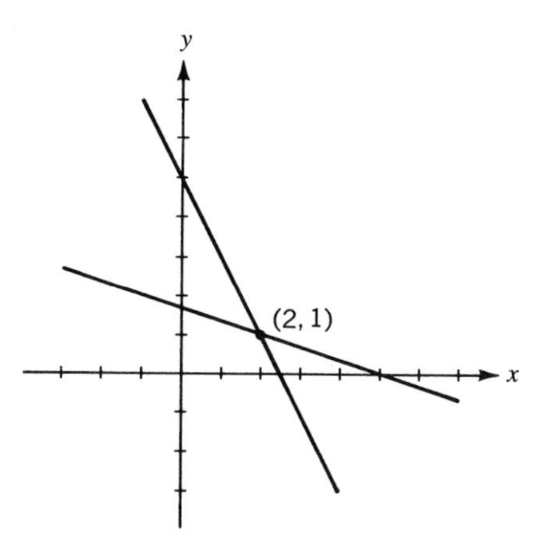

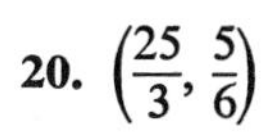

20. $\left(\frac{25}{3}, \frac{5}{6}\right)$

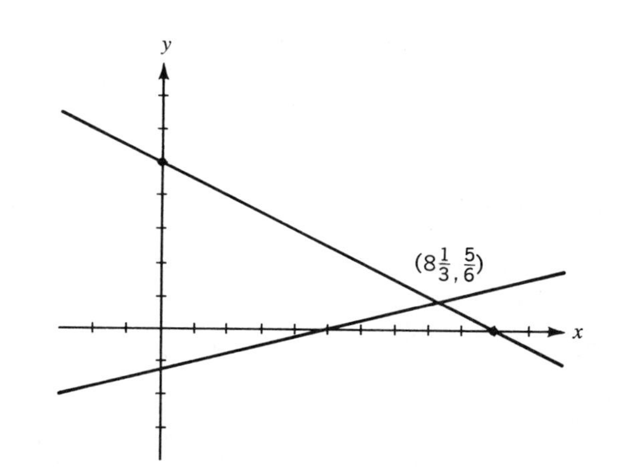

21. 6th year

22. a.

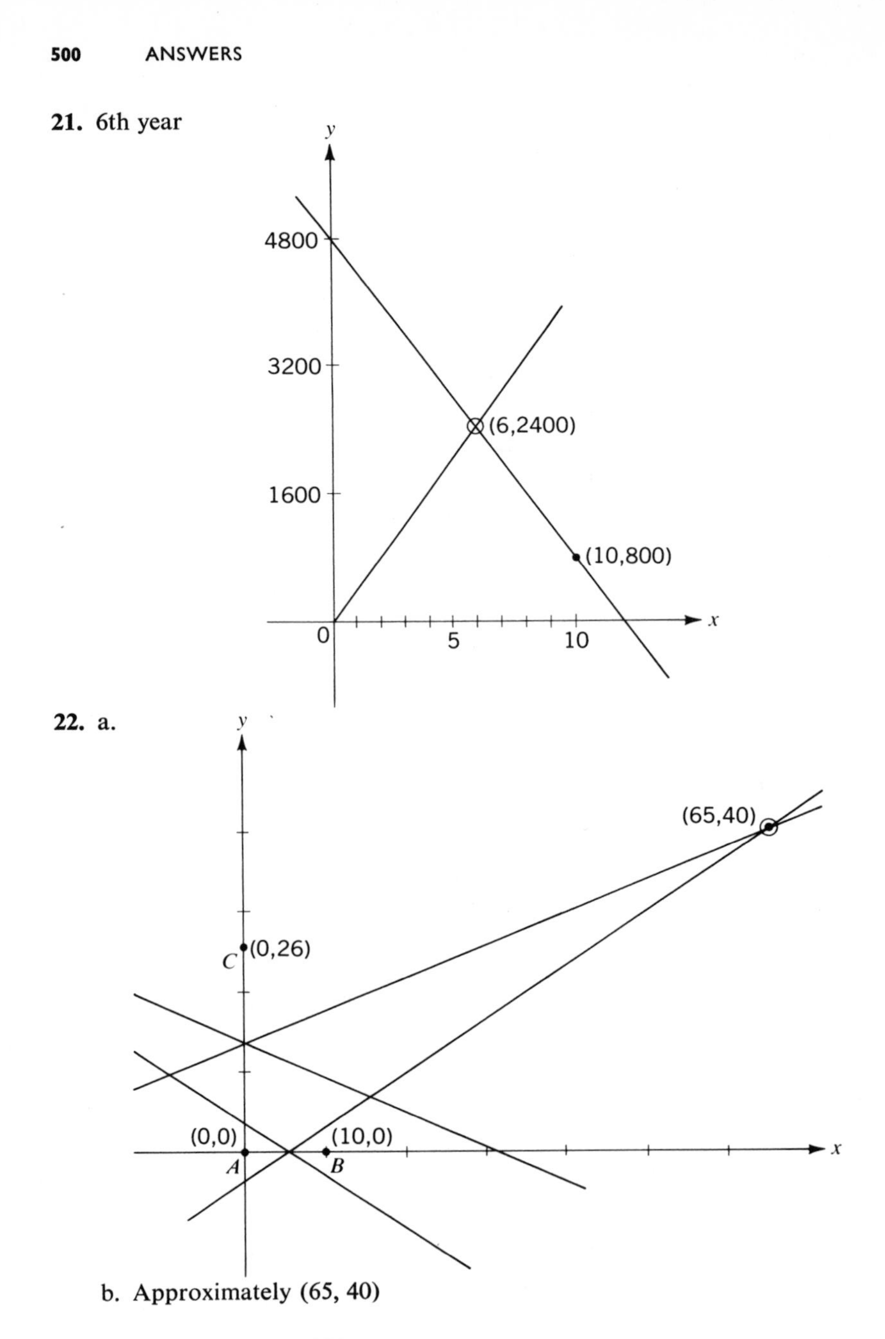

b. Approximately (65, 40)

Exercises 5.6 A, page 334

1. $(5, -2)$

2. $(2, 1)$

3. $(-3, -7)$

4. $(-6, 4)$

5. $\left(\frac{1}{2}, -3\right)$

6. $(7, -4)$

7. $(-1, 2)$
8. $\left(\frac{7}{3}, \frac{5}{2}\right)$
9. $\left(0, -\frac{36}{5}\right)$
10. $\left(-\frac{35}{2}, -\frac{75}{2}\right)$
11. $(6, -3)$
12. $\left(\frac{1}{2}, \frac{1}{2}\right)$
13. All ordered pairs (x, y) such that $4x + 10y - 3 = 0$
14. $\varnothing$
15. $\left(\frac{23}{5}, -\frac{1}{5}\right)$

Exercises 5.7 A, page 337

1. $(2, 4)$
2. $(3, 5)$
3. $(-3, -3)$
4. $(5, 3)$
5. $(-2, -11)$
6. $\left(4, -\frac{1}{2}\right)$
7. $(7, -1)$
8. $\left(4, -\frac{1}{2}\right)$
9. $\varnothing$
10. $\{(x, y) \mid 3y = x - 1\}$ (Infinitely many solutions)
11. $\left(-\frac{2}{5}, \frac{1}{5}\right)$
12. $\left(\frac{11}{5}, -\frac{4}{5}\right)$
13. $(2, 0)$
14. $\left(\frac{13}{5}, 6\right)$
15. $(-1, 2)$

Exercises 5.8 A, page 340

1. $\frac{23}{2}$ and $\frac{17}{2}$
2. 9 quarters, 13 dimes
3. 50 pounds at \$1.20, 100 pounds at 90 cents
4. 12 miles
5. \$6500 at 5 percent, \$3500 at $6\frac{1}{2}$ percent
6. Plane 175 mph, wind 25 mph
7. 17, 37
8. Current $1\frac{1}{2}$ mph, crew 6 mph
9. Bread, 42 cents, butter 77 cents
10. Fixed fee, 89 cents
    Additional fee, 8 cents

Chapter 5 Review Exercises, page 343

1. Refer to text.
2. (b) and (e)
3. a. $m = -\frac{3}{2}, b = 6$ b. $m = -\frac{5}{2}, b = 5$
   c. $m = 2, b = -10$ d. $m = \frac{1}{3}, b = -\frac{2}{3}$

4. a. $y = -\frac{3}{2}x + 6$ b. $y = -\frac{5}{2}x + 5$

c. $y = 2x - 10$ d. $y = \frac{1}{3}x - \frac{2}{3}$

5. a. $2x - 9y + 6 = 0$ b. $7x + 3y + 3 = 0$
c. $3x - 2y + a = 0$ d. $y = 5$

6. a. $x - 2y + 6 = 0$ b. $3x + y - 8 = 0$
c. $y + 1 = 0$ d. $x + 3y + 6 = 0$

7. $2x - 3y + 14 = 0$

8. (b) and (c)

9. a. $\{(\frac{1}{2}, 0)\}$ b. $\{(-3, 2)\}$ c. $\{(5, 2)\}$ d. $\varnothing$

10. a. $3x - y - 11 = 0$ b. $y = 2$ c. $3x + 5y = 0$ d. $x = 5$

11. a.

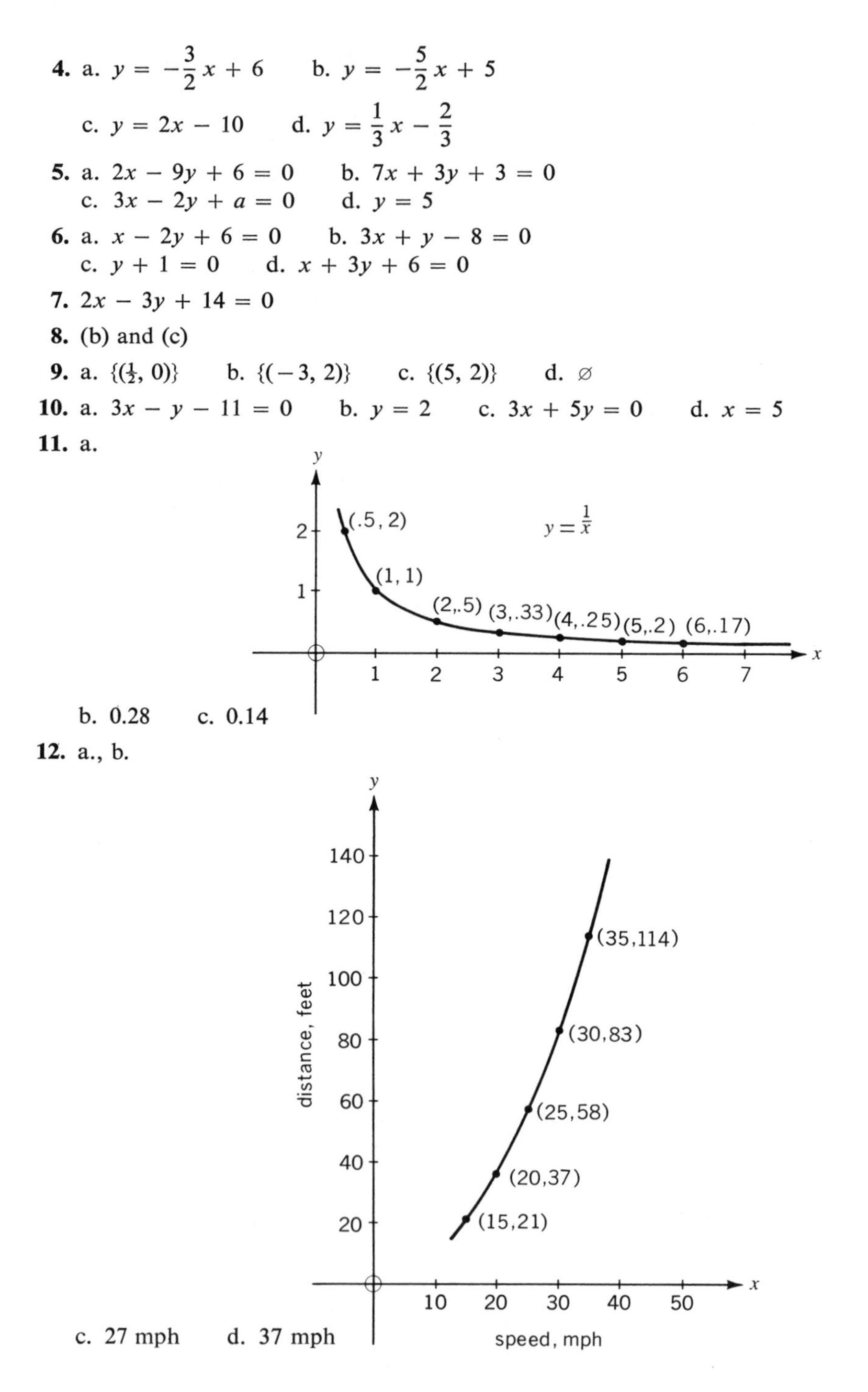

b. 0.28 c. 0.14

12. a., b.

c. 27 mph d. 37 mph

13. a. $(-4, 3)$ b. $(3.5, -3)$

14. a. $(-4, -3)$ b. $(3, 0)$

15. a. 3 b. 5 c. 1 d. 2 e. 4

16. \$9, \$4

17. 26 mph

18. 20 hours

19. a. and **20.** a.

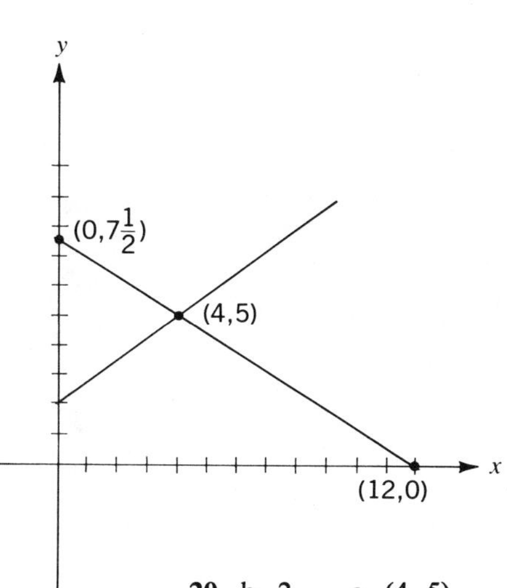

19. b. $\frac{15}{2}$ c. 12

20. b. 2 c. $(4, 5)$

Exercises 6.1 A, page 352

1. x^8, theorem 2

2. x^7, theorem 1

3. x^5, theorem 3

4. $-8y^3$, theorem 4

5. $\frac{81}{y^4}$, theorem 5

6. $\frac{1}{x^4}$, theorem 3

7. $10^4 = 10{,}000$, theorem 4

8. $-5(2^6) = -320$, theorem 2

9. $2^6 = 64$, theorem 5

10. $(-a)^{10} = a^{10}$, theorem 1

11. $-125x^6$

12. $-25x^6$

13. a^3b^8

14. $\frac{27}{x^3}$

15. $3y^8$

16. 64

17. $-x^4$

18. 1

19. $125x^3y^6$

20. x^7y^6

21. $\frac{36a^2c^2}{25b^4}$

22. $-32a^{10}b^{25}$

23. -1

24. 100,000,000

25. 1000

26. $\frac{8}{25}$

27. 40,000
28. 1
29. 32
30. x^{n-1}
31. y^{4n}
32. $\frac{x^{2n}}{5^{2n}}$
33. 1
34. x^n
35. x^{2n+1}

Exercises 6.2 A, page 358

1. $\frac{1}{2}$
2. $\frac{1}{125}$
3. 16
4. 100,000
5. 9
6. 625
7. $\frac{1}{1000} = 0.001$
8. $\frac{5}{3}$
9. $\frac{81}{10,000} = 0.0081$
10. 27
11. $\frac{1}{16}$
12. $\frac{1}{100} = 0.01$
13. $\frac{1}{100,000} = 0.000\ 01$
14. 1
15. 1,000,000
16. $\frac{1}{100,000} = 0.000\ 01$
17. 1
18. $\frac{1}{1000} = 0.001$
19. $\frac{1}{64}$
20. $\frac{81}{25}$
21. -4
22. $\frac{25}{4}$
23. $\frac{243}{2}$
24. 1600
25. 1
26. 6
27. -15
28. 64
29. 10
30. $\frac{10}{7}$
31. $\frac{1}{1000} = 0.001$
32. $\frac{7}{1000} = 0.007$
33. $\frac{1}{8x^3}$
34. $\frac{2}{x^3}$
35. $\frac{16}{x^2}$
36. $\frac{x^3}{125}$
37. $4x^2$
38. $\frac{4x^2}{x^2 + 4x + 4}$

39. y^2

40. $\frac{x^2}{2}$

41. 1

42. $\frac{x^2}{y}$

43. 1

44. $\frac{xy}{x + y}$

45. 2

46. $\frac{xy^9}{4}$

47. $-\frac{1}{x}$

48. $\frac{a^5}{b}$

49. $\frac{y^6}{9}$

50. $\frac{3}{y^5}$

51. x

52. $\frac{x - 1}{x}$

53. $\frac{x^n}{y^n}$

54. $\frac{1}{x^{10n}}$

55. $\frac{y^{3n}}{x^{3n}}$

Exercises 6.3 A, page 367

1. 4

2. -6

3. 3

4. $\frac{2}{5}$

5. 20

6. 3

7. 1

8. 13

9. 17

10. $\frac{5}{7}$

11. 6.71

12. 3.16

13. 80.62

14. 0.22

15. 4.47

16. 3.27

17. 4.06

18. 1.72

19. -0.26

20. 3.46

21. 13

22. 12

23. $\sqrt{2}$

24. 8

25. $\sqrt{8} \approx 2.83$

26. $\sqrt{2} \approx 1.41$

27. $\sqrt{32} \approx 5.66$

28. $t^2 - 1$

29. $4u^2 + v^2$

30. $x^2 + 3$

31. $\{5\}$

32. $\{5, -5\}$

33. $\{6, -6\}$

34. $\{-6\}$

35. $\varnothing$

Exercises 6.4 A, page 372

1. 4
2. -3
3. 2
4. -8
5. 7
6. -1
7. 10
8. 15
9. 5
10. $\frac{1}{2}$
11. $\frac{3}{5}$
12. 10
13. 100
14. 0.1
15. $25x^2$
16. 4.48
17. 3.14
18. 3.85
19. 2.29
20. 3.42
21. 0.79
22. 0.56
23. $\frac{3}{4}$
24. 3
25. 2.52
26. $\{4\}$
27. $\{-4\}$
28. $\{4\}$
29. $\{-4\}$
30. $\{2\}$
31. $\{2, -2\}$
32. $\varnothing$

Exercises 6.5 A, page 377

1. $8 = 8$, theorem 1
2. $2 = 2$, theorem 2
3. $4 = 4$, theorem 3
4. $10 = 10$, theorem 4
5. $5 = 5$, theorem 5
6. 2
7. 4
8. 2
9. 3
10. 2
11. $\frac{3}{5}$
12. $\frac{3}{4}$
13. 16
14. 512
15. $\frac{1}{4}$
16. $\frac{1}{125}$
17. $\frac{1}{1000} = 0.001$
18. 9
19. 25
20. $\frac{36}{49}$
21. $\frac{1}{8}$
22. 8
23. 4
24. -4
25. $\frac{1}{4}$
26. $-\frac{1}{4}$

27. 15,625

28. $\frac{1}{8}$

29. $\sqrt{10}$

30. $\sqrt[3]{4}$

31. $\sqrt[4]{14}$

32. 4

33. $\sqrt[3]{5}$

34. 100,000

35. $\sqrt{10}$

36. $\frac{1}{10} = 0.1$

37. $10\sqrt[4]{10}$

38. $\sqrt{5}$

39. 2

40. $\sqrt{10}$

41. 4

42. $y\sqrt{x}$

43. $\frac{1}{64x^2}$

44. $\frac{4}{x^2}$

45. $\frac{3y}{x^2}$

46. $\frac{1}{1{,}000{,}000xy^5}$

47. $x + y + 2\sqrt{xy}$

48. $x - 3$

49. $2x + 5$

50. 5

Exercises 6.6 A, page 382

1. 10

2. 10

3. $6x$

4. y^2

5. $2\sqrt{3}$

6. $5\sqrt{5}$

7. $9\sqrt{5}$

8. $3\sqrt{x}$

9. $2x\sqrt{6}$

10. $2y\sqrt{10y}$

11. x^2

12. $y^3\sqrt{y}$

13. $6x\sqrt{x}$

14. $2x^2\sqrt{10x}$

15. $4\sqrt{2x}$

16. $22\sqrt{3}$

17. $7x\sqrt{7x}$

18. $30\sqrt{2}$

19. $6x^3\sqrt{6}$

20. $200x^2y\sqrt{2y}$

21. 3

22. 13

23. 2

24. -1

25. 6

26. $3\sqrt[3]{2}$

27. $-3\sqrt[3]{2}$

28. $2\sqrt[3]{9}$

29. $2\sqrt[3]{2}$

30. 2

31. 9

32. $10x$

33. $2\sqrt[3]{4x}$

34. $3\sqrt[3]{4y^2}$

35. $2y\sqrt[3]{5y}$

36. x

37. $2y^2\sqrt[3]{2}$

38. $6y\sqrt[3]{2y^2}$

39. $6x$

40. $6\sqrt[3]{3}$

Exercises 6.7 A, page 386

1. $\frac{\sqrt{2}}{2}$
2. $\frac{\sqrt{6}}{6}$
3. $\frac{\sqrt{15}}{6}$
4. $\frac{\sqrt{5}}{5}$
5. $\frac{\sqrt{5}}{10}$
6. $\frac{\sqrt{2}}{5}$
7. $\frac{\sqrt{6}}{10}$
8. $\frac{4\sqrt{3}}{3}$
9. $2\sqrt{7}$
10. $\frac{\sqrt{2}}{2x}$
11. $\frac{\sqrt{2x}}{x}$
12. $\frac{2\sqrt{2y}}{y}$
13. $\frac{\sqrt{2x}}{2x^2}$
14. $3\sqrt{5}$
15. $\frac{\sqrt{2x}}{2x}$
16. $\sqrt{2} + 1$
17. $\sqrt{7} - \sqrt{3}$
18. $-(2\sqrt{5} + 5)$
19. $\frac{\sqrt{11} - 1}{2}$
20. $\frac{2(\sqrt{x} + 1)}{x - 1}$
21. $\frac{\sqrt{2}}{4}$
22. $\frac{5}{7}$
23. 1
24. $\sqrt{x} + \sqrt{y}$
25. $\sqrt{x} - \sqrt{2y}$

Exercises 6.8 A, page 389

1. $8\sqrt{5}$
2. $3\sqrt{2}$
3. $6\sqrt{6}$
4. $\sqrt{10}$
5. $3\sqrt{5} + 5\sqrt{3}$
6. $3\sqrt{3}$
7. $2\sqrt{14x}$
8. $2\sqrt{2x} + 3\sqrt{3x}$
9. $4\sqrt{3}$
10. $\sqrt{10}$
11. $12\sqrt{6}$
12. $11\sqrt{2x}$
13. $\sqrt{13x}$
14. $5x\sqrt{7}$
15. $6x$
16. $6y$
17. $\frac{3\sqrt{10}}{10}$
18. $\frac{17\sqrt{2}}{6}$
19. $5\sqrt{5} + 2\sqrt{3}$
20. $4\sqrt{x} - 5\sqrt{y}$

21. $\frac{9\sqrt{42}}{20}$

22. $3\sqrt{3x} - 3\sqrt{2x}$

23. $4\sqrt{14} - \frac{\sqrt{2}}{2}$

24. $5 + 2\sqrt{6}$

25. $30 - 12\sqrt{6}$

26. $9x + 6x\sqrt{2}$

27. $6 + 9\sqrt{2}$

28. $-5\sqrt{3} - 5\sqrt{2}$

29. 0

30. $4 - \sqrt{5}$

Exercises 6.9 A, page 393

1. 9.29×10^7
2. 6.6×10^{21}
3. 1.14×10^7
4. 9.5×10^{-8}
5. 2.4×10^{-9}
6. 2.21×10^9
7. 6.1×10^{-4}
8. 1.2×10^5
9. 2.205×10^{-3}
10. 6.67×10^{-8}
11. 2300
12. -460
13. 0.000 0180
14. 864,000
15. 0.000 000 0303
16. 0.01745
17. 0.16667
18. 0.000 000 000 480
19. 1,870,000,000
20. 6,300,000,000,000,000,000
21. 1080
22. 0.01
23. 0.03
24. 300,000
25. 0.048
26. 1.047
27. 42
28. About 8.9 hours
29. 2.88×10^{-10}
30. 1.33×10^{-9}

Chapter 6 Review Exercises, page 397

1. a. $2\sqrt{3}$ b. 10
2. a. 8.54 b. 0.02 c. 2.92 d. 0.03
3. a. $-3\sqrt[3]{3}$ b. 7 c. $\frac{2}{3}$ d. 0.0013
4. a. $8\sqrt{3}$ b. $6\sqrt{5}$ c. $2\sqrt[3]{18}$ d. $12\sqrt{15}$
5. a. $\frac{3\sqrt{5}}{5}$ b. $\frac{\sqrt{15}}{5}$ c. $6\sqrt{3}$ d. $2\sqrt{2} - 5$
   e. $-(\sqrt{3} + \sqrt{6})$
6. a. $5\sqrt{6}$ b. $5\sqrt{2} - 20$ c. $8\sqrt{7} - 3\sqrt{2}$
   d. 288 e. $\frac{-\sqrt{6}}{2}$ f. $\frac{31\sqrt{30}}{30} + 3\sqrt{3}$
7. a. a^8b^{13} b. $\frac{-18a^8b^7}{25m^8n^3}$ c. $\frac{a^8b^4}{192}$ d. $-y^4$

8. a. $\frac{1}{m^4}$ b. p^2 c. $\frac{x^2z}{y}$ d. $\frac{a+b}{ab}$

9. a. $\sqrt[4]{x}$ b. $2\sqrt{a}$ c. $\sqrt[4]{x^3}$ d. $\frac{1}{\sqrt[3]{x^2y^2}} = \frac{\sqrt[3]{xy}}{xy}$

10. a. $y^{1/3}$ b. $x^{1/2}y^{5/6}$ c. $xy^{1/2}$ d. $(xy)^{2/3}$

11. a. 1 b. 5 c. $\frac{1}{8}$ d. $\frac{29}{6}$

12. a. 0 b. 0 c. 18

13. a. 7 b. $3\sqrt{5}$ c. $4\sqrt{6}$

14. a. 64 b. $\frac{399}{4}$

15. a. 0.09 b. 80,000

Exercises 7.1 A, page 406

1. $2i$
2. $9i$
3. xi
4. $\frac{1}{2}i$
5. $i\sqrt{2}$
6. $2i\sqrt{2}$
7. $40i$
8. $-40i$
9. $6i\sqrt{2}$
10. $-6i\sqrt{2}$
11. $4 + 5i$
12. $6 - 7i$
13. $7 + 6i$
14. $8 - 2i\sqrt{3}$
15. $5 + 0i$
16. $0 + 0i$
17. $0 + 2i$
18. $0 + 13i$
19. $x = 4, y = 2$
20. $x = 4, y = -5$
21. $x = 0, y = -5$
22. $x = 4, y = 0$
23. $x = 5, y = -1$
24. $x = 1, y = 2$
25. $7 + 2i$
26. $5 + 7i$
27. $-2 - 3i$
28. $1 - 5i$
29. $5 + 3i$
30. $-3 + 5i$
31. $9 - i\sqrt{3}$
32. $-4\sqrt{2} - i\sqrt{6}$
33. $-1 - 4i$
34. $0 + 0i$
35. $0 + 9i$
36. $x \geq 4$
37. $x \leq 9$
38. $x \leq -4$ or $x \geq 4$
39. $-3 \leq x \leq 3$
40. $x < -5$ or $x > 5$
41. $b = 0$

Exercises 7.2 A, page 410

1. -10
2. $-6\sqrt{2}$
3. $6i$
4. $i\sqrt{6}$

5. $-6i$
6. $-6 + 8i$
7. $6 + 15i$
8. $6 + 6i$
9. $12 - 5i$
10. $22 + 7i$
11. 20
12. 1
13. $5 + 5i$
14. $15 - 90i$
15. $5 + 12i$
16. $-4 + 78i$
17. $17 + i$
18. $22 - 7i$
19. $3 + i$
20. $5 - 15i$
21. $5 - 15i$
22. $9 - 40i$
23. 0
24. 0
25. 1
26. $(3 + i)^2 - 6(3 + i) + 10 = (8 + 6i) - 18 - 6i + 10 = 0$
$(3 - i)^2 - 6(3 - i) + 10 = (8 - 6i) - 18 + 6i + 10 = 0$
27. $4i^2 + 2i^2 + 6 = -4 - 2 + 6 = 0$
28. $x = 0, y = 0$
29. $x = 3$ or $x = -3$
30. $x = -1$
31. i
32. -1
33. $-i$
34. 1
35. $-i$
36. -1
37. 0
38. $-i$

Exercises 7.3 A, page 413

1. $5 - 7i$
2. $6 + 4i$
3. $2 - i$
4. $4 + 3i$
5. $-i$
6. i
7. 3
8. $-1 - i$
9. $\frac{3}{13} + \frac{-2}{13}i$
10. $\frac{-5}{34} + \frac{-3}{34}i$
11. i
12. $2 - \frac{3}{2}i$
13. $\frac{3}{5} + \frac{3}{10}i$
14. $-\frac{1}{2} - \frac{\sqrt{3}}{2}i$
15. $-2 - \sqrt{3}$
16. $\frac{4\sqrt{2}}{3} + \frac{1}{3}i$
17. $\frac{a^2 - b^2}{a^2 + b^2} + \frac{2ab}{a^2 + b^2}i$
18. $\frac{3}{2} - \frac{1}{2}i$
19. $\frac{8}{5} + 0i$
20. $\frac{3}{10} + \frac{9}{10}i$ or $0.3 + 0.9i$

Exercises 7.4 A, page 418

1.–10.

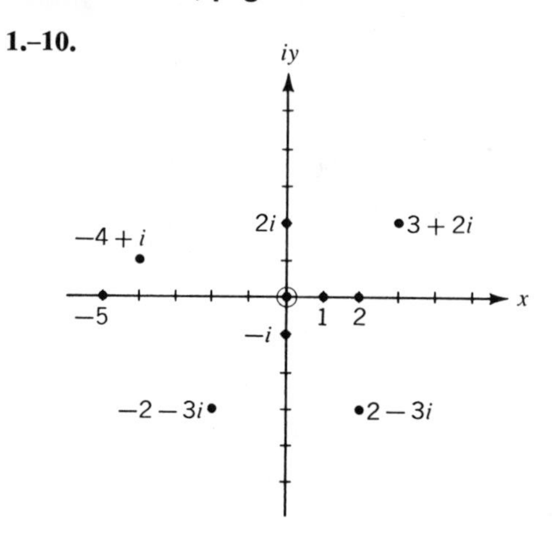

11. a. $8 + 6i$ b. 10 mph c. 3 miles

12. $-8 + 16i$; $8\sqrt{5}$; -2

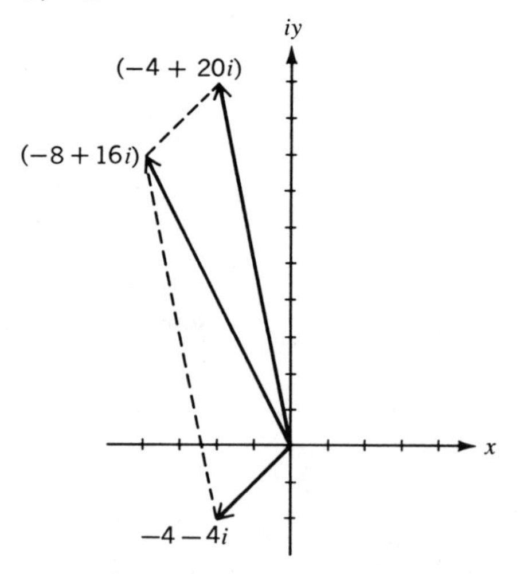

Exercises 7.5 A, page 424

1. $-5, 1$

2. $0, 4$

3. $-c, 2c$

4. $-3, -2$

5. $\frac{5}{2}, -\frac{5}{2}$

6. $0, -3$

7. $-2, -1$

8. $\frac{2}{3}, -1$

9. $2i, -2i$

10. $-3, 2$

11. $-5, 4$

12. $0, 4$

13. -5

14. $0, k$

15. $\sqrt{5}, -\sqrt{5}$

16. $\frac{a}{2}, -\frac{a}{2}$

17. $3, 1$

18. $2 + n, 2 - n$

19. $a + n, a - n$

20. $\frac{-b + 5}{a}, \frac{-b - 5}{a}$

21. $\frac{d}{c}$

22. $8a, -2a$

23. $0, 1, 4$

24. $-2b$

25. $5, -5$

26. $4, -4$

27. $\frac{1}{2}, -1$

28. c, d

29. 2

30. -5

31. $t = \pm\sqrt{\frac{2s}{g}}$

32. $r = \pm\sqrt{\frac{\pi R^2 - A}{\pi}}$

33. $n_1 = \pm\sqrt{\frac{kn_2^2}{1 - k}} = \pm n_2\sqrt{\frac{k}{1 - k}}$

34. $c = 0$ or $c = d$

35. $d = L$ or $d = \frac{L}{4}$

Exercises 7.6 A, page 429

1. $5 + \sqrt{2}, 5 - \sqrt{2}$

2. $-3 + 2i, -3 - 2i$

3. $2 + \sqrt{3}, 2 - \sqrt{3}$

4. $1 + i\sqrt{2}, 1 - i\sqrt{2}$

5. $3, -1$

6. $5, 3$

7. $4 + i, 4 - i$

8. $1 + 2\sqrt{5}, 1 - 2\sqrt{5}$

9. $\frac{1 + \sqrt{5}}{2}, \frac{1 - \sqrt{5}}{2}$

10. $1, -\frac{2}{3}$

11. $1 + i, 1 - i$

12. $3, 1$

13. $0, -4$

14. $\frac{3 + \sqrt{2}}{2}, \frac{3 - \sqrt{2}}{2}$

15. $7 + 2i, 7 - 2i$

16. $\frac{4}{3}, -\frac{1}{2}$

17. $-3 + \sqrt{7}, -3 - \sqrt{7}$

18. $-3 + 4i, -3 - 4i$

19. $5 + i\sqrt{70}, 5 - i\sqrt{70}$

20. $4, -1$

21. $5 + i, 5 - i$

22. $2 + \sqrt{2}, 2 - \sqrt{2}$

23. $5y, y$

24. $2\sqrt{y^2 + 1}, -2\sqrt{y^2 + 1}$

25. $\frac{2}{5}\sqrt{25 - x^2}, \frac{-2}{5}\sqrt{25 - x^2}$

26. $\frac{-x}{2} + \frac{ix\sqrt{3}}{2}, \frac{-x}{2} - \frac{ix\sqrt{3}}{2}$

27. $\frac{v + \sqrt{v^2 - 64s}}{32}, \frac{v - \sqrt{v^2 - 64s}}{32}$

28. $\dfrac{-E + \sqrt{E^2 + 4RP}}{2R}, \dfrac{-E - \sqrt{E^2 + 4RP}}{2R}$

29. $b + \sqrt{b^2 - c}, b - \sqrt{b^2 - c}$

30. $\dfrac{-b + \sqrt{b^2 - 4ac}}{2a}, \dfrac{-b - \sqrt{b^2 - 4ac}}{2a}$

Exercises 7.7 A, page 436

1. a. $2x^2 - 3x - 1 = 0; a = 2, b = -3, c = 1$
   b. 17, real and unequal
2. a. $4x^2 - 12x + 9 = 0; a = 4, b = -12, c = 9$
   b. 0, real and equal
3. a. $x^2 - 5 = 0; a = 1, b = 0, c = -5$
   b. 20, real and unequal
4. a. $x^2 + 5x = 0; a = 1, b = 5, c = 0$
   b. 25, real and unequal
5. a. $x^2 - 4x + 8 = 0; a = 1, b = -4, c = 8$
   b. -16, imaginary
6. $-\dfrac{2}{3}, -1$
7. $2, -\dfrac{1}{3}$
8. $\dfrac{3 + \sqrt{5}}{2}, \dfrac{3 - \sqrt{5}}{2}$
9. $\dfrac{-3 + \sqrt{13}}{2}, \dfrac{-3 - \sqrt{13}}{2}$
10. $\dfrac{3 + i\sqrt{3}}{2}, \dfrac{3 - i\sqrt{3}}{2}$
11. $\dfrac{-1 + i\sqrt{3}}{2}, \dfrac{-1 - i\sqrt{3}}{2}$
12. $\dfrac{2 + \sqrt{7}}{2}, \dfrac{2 - \sqrt{7}}{2}$
13. $\dfrac{1 + i\sqrt{5}}{3}, \dfrac{1 - i\sqrt{5}}{3}$
14. $\dfrac{-1 + \sqrt{7}}{3}, \dfrac{-1 - \sqrt{7}}{3}$
15. $4i, -4i$
16. $0, \dfrac{5}{3}$
17. $\dfrac{1}{18}$
18. $\dfrac{5 + \sqrt{5}}{2}, \dfrac{5 - \sqrt{5}}{2}$
19. $6, \dfrac{3}{2}$
20. $5 + 5\sqrt{5}, 5 - 5\sqrt{5}$
21. $-p \pm \sqrt{p^2 - q}$
22. $\dfrac{(1 + \sqrt{5})n}{2}$
23. $\dfrac{y}{3}, -3y$
24. $2 + \sqrt{10 - 2x}, 2 - \sqrt{10 - 2x}$
25. $\dfrac{-v \pm \sqrt{v^2 + 2sg}}{g}$
26. $\dfrac{2k + 1 + \sqrt{4k + 1}}{2}, \dfrac{2k + 1 - \sqrt{4k + 1}}{2}$
27. $\dfrac{a \pm \sqrt{a^2 - 4ab}}{2}$
28. $\dfrac{-\pi s + \sqrt{\pi^2 s^2 + 4A\pi}}{2\pi}$

Exercises 7.8 A, page 444

1. $\frac{3\sqrt{2}}{2}$
2. Width = 4 inches, length = 13 inches
3. 4 or $-\frac{1}{4}$
4. 6 hours
5. 12 inches by 24 inches
6. (3 and 4) or (28 and -21)
7. 13 feet, 5 inches
8. $4\frac{1}{2}$ mph
9. 20 feet
10. About 2.07 inches
11. 5.2×10^{-5} (*Note:* x must be positive.)
12. 30 mph
13. Faster, 4 days at \$50 = \$200
Slower, 12 days at \$20 = \$240
Together, \$210
Answer: faster man alone
14. 60 inches = 5 feet
15. 2 seconds, 544 feet per second (=) 371 mph
16. 2500 amperes or 250 amperes
(In reality, only 2500 amperes is an acceptable answer.)

Chapter 7 Review Exercises, page 449

1. a. $2 + 4i$ b. $1 + 8i$ c. $2 + i\sqrt{3}$ d. $18 + i$ e. $-48 - 14i$
f. $\frac{-1}{13} + \frac{-5}{13}i$
2. a. $x = 3, y = 2$ b. $x = \frac{2}{3}, y = -\frac{4}{3}$
3. Resultant: $12 - 5i$; magnitude: 13; direction: $\frac{-5}{12}$
4. $\left(\frac{3\sqrt{3}}{2} + \frac{3i}{2}\right)^2 = \frac{27}{4} + \frac{18\sqrt{3}i}{4} + \frac{9i^2}{4} = \frac{18}{4} + \frac{18\sqrt{3}i}{4} = \frac{9}{2} + \frac{9\sqrt{3}i}{2}$
5. a. $\{10, -4\}$ b. $\left\{-\frac{1}{6}, \frac{-5}{2}\right\}$ c. $\left\{0, \frac{4}{3}\right\}$ d. $\{3i, -3i\}$
6. a. $\{-5, -3\}$ b. $\{1 + i\sqrt{2}, 1 - i\sqrt{2}\}$
c. $\{2 + \sqrt{5}, 2 - \sqrt{5}\}$ d. $\{-5 + 2\sqrt{2}, -5 - 2\sqrt{2}\}$
7. a. $\left\{\frac{-5 + \sqrt{33}}{4}, \frac{-5 - \sqrt{33}}{4}\right\}$ b. $\left\{\frac{1 + \sqrt{17}}{4}, \frac{1 - \sqrt{17}}{4}\right\}$
c. $\left\{\frac{5}{2}, -\frac{4}{3}\right\}$ d. $\left\{\frac{8 + 2\sqrt{10}}{3}, \frac{8 - 2\sqrt{10}}{3}\right\}$

8. a. $\left\{\frac{\sqrt{42}}{6}, \frac{-\sqrt{42}}{6}\right\}$ b. $\{-2 + 2i, -2 - 2i\}$

c. $\left\{-2, \frac{5}{2}\right\}$ d. $\left\{\frac{-5 + \sqrt{13}}{6}, \frac{-5 - \sqrt{13}}{6}\right\}$

9. a. 9 b. ± 2 c. $\frac{5}{4}$ d. ± 4

10. a. $y = 5x$ or $y = -2x$ b. $y = (-1 \pm \sqrt{2})x^2$

11. 30 minutes

12. i or $-i$

13. 5 and 6

14. 3 feet by 15 feet

15. 180 mph

16. 27 inches, 36 inches

17. $\frac{cv + \sqrt{c^2v^2 - 2gH}}{g}$

18. $\frac{-\pi h + \sqrt{\pi^2 h^2 + 2\pi T}}{2\pi}$

INDEX

TABLES OF POWERS AND ROOTS

No.	*Squares*	*Cubes*	*Square Roots*	*Cube Roots*	*No.*	*Squares*	*Cubes*	*Square Roots*	*Cube Roots*
1	1	1	1.000	1.000	**51**	2 601	132,651	7.141	3.708
2	4	8	1.414	1.259	**52**	2 704	140,608	7.211	3.732
3	9	27	1.732	1.442	**53**	2 809	148,877	7.280	3.756
4	16	64	2.000	1.587	**54**	2 916	157,464	7.348	3.779
5	25	125	2.236	1.709	**55**	3 025	166,375	7.416	3.802
6	36	216	2.449	1.817	**56**	3 136	175,616	7.483	3.825
7	49	343	2.645	1.912	**57**	3 249	185,193	7.549	3.848
8	64	512	2.828	2.000	**58**	3 364	195,112	7.615	3.870
9	81	729	3.000	2.080	**59**	3 481	205,379	7.681	3.892
10	100	1 000	3.162	2.154	**60**	3 600	216,000	7.745	3.914
11	121	1 331	3.316	2.223	**61**	3 721	226,981	7.810	3.936
12	144	1 728	3.464	2.289	**62**	3 844	238,328	7.874	3.957
13	169	2 197	3.605	2.351	**63**	3 969	250,047	7.937	3.979
14	196	2 744	3.741	2.410	**64**	4 096	262,144	8.000	4.000
15	225	3 375	3.872	2.466	**65**	4 225	274,625	8.062	4.020
16	256	4 096	4.000	2.519	**66**	4 356	287,496	8.124	4.041
17	289	4 913	4.123	2.571	**67**	4 489	300,763	8.185	4.061
18	324	5 832	4.242	2.620	**68**	4 624	314,432	8.246	4.081
19	361	6 859	4.358	2.668	**69**	4 761	328,509	8.306	4.101
20	400	8 000	4.472	2.714	**70**	4 900	343,000	8.366	4.121
21	441	9 261	4.582	2.758	**71**	5 041	357,911	8.426	4.140
22	484	10,648	4.690	2.802	**72**	5 184	373,248	8.485	4.160
23	529	12,167	4.795	2.843	**73**	5 329	389,017	8.544	4.179
24	576	13,824	4.898	2.884	**74**	5 476	405,224	8.602	4.198
25	625	15,625	5.000	2.924	**75**	5 625	421,875	8.660	4.217
26	676	17,576	5.099	2.962	**76**	5 776	438,976	8.717	4.235
27	729	19,683	5.196	3.000	**77**	5 929	456,533	8.774	4.254
28	784	21,952	5.291	3.036	**78**	6 084	474,552	8.831	4.272
29	841	24,389	5.385	3.072	**79**	6 241	493,039	8.888	4.290
30	900	27,000	5.477	3.107	**80**	6 400	512,000	8.944	4.308
31	961	29,791	5.567	3.141	**81**	6 561	531,441	9.000	4.326
32	1 024	32,768	5.656	3.174	**82**	6 724	551,368	9.055	4.344
33	1 089	35,937	5.744	3.207	**83**	6 889	571,787	9.110	4.362
34	1 156	39,304	5.830	3.239	**84**	7 056	592,704	9.165	4.379
35	1 225	42,875	5.916	3.271	**85**	7 225	614,125	9.219	4.396
36	1 296	46,656	6.000	3.301	**86**	7 396	636,056	9.273	4.414
37	1 369	50,653	6.082	3.332	**87**	7 569	658,503	9.327	4.431
38	1 444	54,872	6.164	3.361	**88**	7 744	681,472	9.380	4.447
39	1 521	59,319	6.244	3.391	**89**	7 921	704,969	9.433	4.464
40	1 600	64,000	6.324	3.419	**90**	8 100	729,000	9.486	4.481
41	1 681	68,921	6.403	3.448	**91**	8 281	753,571	9.539	4.497
42	1 764	74,088	6.480	3.476	**92**	8 464	778,688	9.591	4.514
43	1 849	79,507	6.557	3.503	**93**	8 649	804,357	9.643	4.530
44	1 936	85,184	6.633	3.530	**94**	8 836	830,584	9.695	4.546
45	2 025	91,125	6.708	3.556	**95**	9 025	857,375	9.746	4.562
46	2 116	97,336	6.782	3.583	**96**	9 216	884,736	9.797	4.578
47	2 209	103,823	6.855	3.608	**97**	9 409	912,673	9.848	4.594
48	2 304	110,592	6.928	3.634	**98**	9 604	941,192	9.899	4.610
49	2 401	117,649	7.000	3.659	**99**	9 801	970,299	9.949	4.626
50	2 500	125,000	7.071	3.684	**100**	10,000	1,000,000	10.000	4.641

SECOND EDITION

Vivian Shaw Groza

Susanne M. Shelley

Sacramento City College

Rinehart Press / Holt, Rinehart and Winston

SAN FRANCISCO

Library of Congress Cataloging in Publication Data

Groza, Vivian Shaw.
Modern elementary algebra for college students.

1. Algebra. I. Shelley, Susanne, joint author.
II. Title.
QA152.2.G76 1974 512.9′042 73-10200
ISBN 0-03-007456-8

5643 Paradise Drive
Corte Madera, Calif. 94925

A division of Holt, Rinehart and Winston, Inc.

PRINTED IN THE UNITED STATES OF AMERICA

6 7 038 9 8 7 6 5 4